LRFD
Steel Design

Second Edition

THE PWS SERIES IN CIVIL ENGINEERING

LRFD
Steel Design
Second Edition

William T. Segui
The University of Memphis

PWS Publishing

An Imprint of Brooks/Cole Publishing Company
I(T)P® An International Thomson Publishing Company

Pacific Grove • Albany • Belmont • Bonn • Boston • Cincinnati • Detroit • Johannesburg • London • Madrid • Melbourne • Mexico City • New York • Paris • Singapore • Tokyo • Toronto • Washington

Sponsoring Editor: *Suzanne Jeans*
Marketing Team: *Nathan Wilbur, Michele Mootz*
Editorial Associate: *Tricia Kelly, Major Mugrage*
Production Editor: *Marlene Thom*
Production: *Publishers' Design and Production Services, Inc.*
Manuscript Editor: *Jerrold A. Moore*
Manufacturing Buyer: *Vena Dyer*

Interior Design and Illustration: *Publishers' Design and
 Production Services, Inc.*
Cover Design: *Laurie Albrech*
Cover Photo: *Craig Aurness/Westlight*
Typesetting: *Publishers' Design and Production Services, Inc.*
Printing and Binding: *R.R. Donnelley & Sons, Crawfordsville*

COPYRIGHT© 1999, 1994 by Brooks/Cole Publishing Company
A division of International Thomson Publishing Inc.
I(T)P   The ITP logo is a registered trademark used herein under license.

For more information, contact PWS Publishing at Brooks/Cole Publishing Company:

BROOKS/COLE PUBLISHING COMPANY
511 Forest Lodge Road
Pacific Grove, CA 93950
USA

International Thomson Editores
Seneca 53
Col. Polanco
11560 México, D.F., México

International Thomson Publishing Europe
Berkshire House 168–173
High Holborn
London WCIV 7AA
England

International Thomson Publishing GmbH
Königswinterer Strasse 418
53227 Bonn
Germany

Nelson ITP
102 Dodds Street
South Melbourne, 3205
Victoria, Australia

International Thomson Publishing Asia
221 Henderson Road
#05–10 Henderson Building
Singapore 0315

Nelson Canada
1120 Birchmount Road
Scarborough, Ontario
Canada M1K 5G4

International Thomson Publishing Japan
Hirakawacho Kyowa Building, 3F
2–2–1 Hirakawacho
Chiyoda-ku, Tokyo 102
Japan

Printed in the United States of America

10   9   8   7   6   5   4   3   2   1

Library of Congress Cataloging-in-Publication Data

Segui, William T.
    LRFD steel design / William T. Segui.—2nd ed.
        p.   cm. — (The PWS series in engineering)
    Includes bibliographical references and index.
    ISBN 0-534-95155-4
    1. Building, iron and steel.   2. Load factor design.   3. Steel.
Structural.   I. Title.   II. Series.
TA684.5424   1998
624.1′821—dc21   98-29403
                              CIP

Contents

Preface

LRFD Steel Design is a basic textbook in structural steel design for undergraduate civil engineering students at the junior or senior level. Its primary function is that of a textbook, although practicing civil engineers who need a review of current practice and the current AISC Specification will find it useful as a reference. Students using this book should have a background in mechanics of materials and analysis of statically determinate structures.

Several changes were made to the first edition of *LRFD Steel Design* to make it consistent with the 1993 AISC Specification and the second edition of the LRFD Manual of Steel Construction. These changes include incorporation of the column base plate design procedure used in the *Manual*, use of the AISC equation for the modulus of elasticity of concrete, and the *Manual's* treatment of construction loads for composite beam design. To make the book as current as possible, the provisions of the latest high-strength bolt specifications are used, even though the AISC Specification is based on an earlier bolt specification. These provisions include the equations for combined shear and tension in bolts and bearing strength.

Other changes include the addition of problems at the end of Chapters 1 and 2, mostly new problems for the other chapters, and a numbering scheme for the problems that helps identify the corresponding section in the textbook. Some material has been rewritten for clarity, and some new examples and other material have been added, including a discussion of tabulated constants for noncompact beams in Chapter 5, Beams. Chapter 5 also includes the Steel Joist Institute procedure for using LRFD procedures with the joist load tables and a new section summarizing the bending strength of I- and H-shapes bent about the strong axis. In Chapters 7 and 8, slip-critical bolt strength is now based on factored loads.

Depending on the level of competence of the student, *LRFD Steel Design* can be used for one or two courses of three semester hours each. A suggested two-course sequence is as follows: a first course covering Chapters 1 through 7 and a second course covering Chapters 8 through 10, supplemented by comprehensive design assignments. This division of topics has been used successfully for several years at The University of Memphis.

The emphasis of this book is on the design of building components in accordance with the provisions of the AISC LRFD Specification and the LRFD Manual of Steel Construction. Although the AASHTO and Area Specifications are referred to occasionally, no examples or assigned problems are based on these documents.

Prior to the introduction of the Load and Resistance Factor Design Specification by AISC in 1986, the dominant design approach for structure steel was allowable stress design. The trend today is toward load and resistance factor design, but because allowable stress design is still in use, students should have some familiarity with it. To that end, Appendix B provides a brief introduction to that topic.

It is absolutely essential that students have a copy of the Manual of Steel Construction. In order to promote familiarity with it, material from the *Manual* is not reproduced in this book so that the reader will be required to refer to the *Manual*. All notation in *LRFD Steel Design* is consistent with that in the *Manual*, and AISC equation numbers are used in tandem with sequential numbering of other equations according to the textbook chapter.

U.S. customary units are used throughout, with no introduction of SI units. Metrication is inevitable and will be the basis of future AISC Specifications and Manuals, but the change has not yet taken place in the steel construction industry. Although construction projects for some government agencies now require the use of SI units on all contract documents, their use is still not wide-spread enough to be considered standard.

As far as design procedures are concerned, the application of fundamental principles is encouraged. Although this book is oriented toward practical design, sufficient theory is included to avoid a "cookbook" approach. Direct design methods are used where feasible, but no complicated design formulas have been developed. Instead, trial and error, with "educated guesses," is the rule. Tables, curves, and other design aids from the *Manual* are used, but they have a role that is subordinate to the use of basic equations. Assigned problems provide practice with both approaches, and where appropriate, the required approach is specified in the statement of the problem.

In keeping with the objective of providing a basic textbook, a large number of assigned problems are given at the end of each chapter. Answers to selected problems are given at the back of the book, and an instructor's manual with solutions is available.

A fairly comprehensive treatment of roof trusses is given in Chapter 3, since components of trusses are dealt with in subsequent chapters. Column base plates are considered in Chapter 5, Beams, rather than in Chapter 4, Columns, because base plate design requires a consideration of bending strength, and coverage of the topic is deferred until bending has been discussed.

I would like to express my appreciation to the following people who reviewed the manuscript for this edition and provided helpful comments: Robert E. Abendroth, Iowa State University; Gary Consolazio, Rutgers University–Busch Campus; Souhail Elhouar, Bradley University; Duane S. Ellifritt, University of Florida; Hany J. Farran, California State University–Pomona; Richard Fellar, University of Pittsburgh–Johnstown; Gary T. Fry, Texas A & M University; Richard M. Gutkowski, Colorado State University; Rola L. Idriss, New Mexico State University; Roger A. LaBoube, University of Missouri–Rolla; Kincho H. Law, Stanford University; Eric Lui, Syracuse University; B. K. Rao, Idaho State University; Hadid Sadid, Idaho State University; Wallace W. Sanders, Iowa State University; Edwin R. Schmeckpeper, University of Idaho; and George Tsiatas, University of Rhode Island. In addition, Abraham J. Rokach and Lewis B. Burgett of the American Institute of Steel Construction were helpful in providing updates on AISC Specification revisions as well as other assistance. Suzanne Jeans, Pamela Rockwell, and Marlene Thom of PWS and Brooks/Cole provided invaluable assistance in the final production of the book. Thanks also to Yupeng Wang and Timothy Mays for their assistance in the preparation of the solutions manual and to Jamie Evans for checking the examples.

Finally, I want to thank my wife, Angela, for her support and for her valuable suggestions and help in proofreading the manuscript for this book.

William T. Segui

1 Introduction

STRUCTURAL DESIGN

The structural design of buildings, whether of structural steel or reinforced concrete, requires the determination of the overall proportions and dimensions of the supporting framework and the selection of the cross sections of individual members. In most cases the functional design, including the establishment of the number of stories and the floor plan, will have been done by an architect, and the structural engineer must work within the constraints imposed by this design. Ideally, the engineer and architect will collaborate throughout the design process to complete the project in an efficient manner. In effect, however, the design can be summed up as follows: The architect decides how the building should look; the engineer must make sure that it doesn't fall down. Although this distinction is an oversimplification, it affirms the first priority of the structural engineer: safety. Other important considerations include serviceability (how well the structure performs in terms of appearance and deflection) and economy. An economical structure requires an efficient use of materials and construction labor. Although this objective can usually be accomplished by a design that requires a minimum amount of material, savings can often be realized by using more material if it results in a simpler, more easily constructed project.

A good design requires the evaluation of several framing plans — that is, different arrangements of members and their connections. In other words, several alternative designs should be prepared and their costs compared. For each framing plan investigated, the individual components must be designed. To do so requires the structural analysis of the frame(s) of the building and the computation of forces and bending moments in the individual members. Armed with this information, the structural designer can then select the appropriate cross section. Before any analysis, however, a decision must be made on the primary building material to be used; it will usually be reinforced concrete, structural steel, or both. Ideally, alternative designs should be prepared with each.

The emphasis in this book will be on the design of individual structural steel members and their connections. The structural engineer must select and evaluate the overall structural system in order to produce an efficient and economical design but cannot do so without a thorough understanding of the design of the components (the "building blocks") of the structure. Thus component design is the focus of this book.

Before discussing structural steel, we need to examine various types of structural members. Figure 1.1 shows a truss with vertical concentrated forces applied at the joints along the top chord. In keeping with the usual assumptions of truss analysis — pinned connections and loads applied only at the joints — each component of the truss will be a two-force member, subject to either axial compression or tension. For simply supported

■ **FIGURE 1.1**

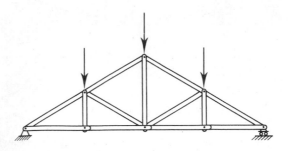

trusses loaded as shown — a typical loading condition — each of the top chord members will be in compression, and the bottom chord members will be in tension. The web members will either be in tension or compression, depending on their location and orientation and on the location of the loads.

Other types of members can be illustrated with the rigid frame of Figure 1.2a. The members of this frame are rigidly connected by welding and can be assumed to form a continuous structure. At the supports, the members are welded to a rectangular plate that is bolted to a concrete footing. Placing several of these frames in parallel and connecting them with additional members that are then covered with roofing material and walls produces a typical building system. Many important details have not been mentioned, but many small commercial buildings are constructed essentially in this manner. The design and analysis of each frame in the system begins with the idealization of the frame as a two-dimensional structure as shown in Figure 1.2b. Because the frame has a plane of symmetry parallel to the page, we are able to treat the frame as two-dimensional and represent the frame members by their centerlines. (Although it is not shown in Figure 1.1, this same idealization is made with trusses, and the members are usually represented by their centerlines.) Note that the supports are represented as hinges (pins), not as fixed supports. If there is a possibility that the footing will be able to undergo a slight rotation, or if the connection is flexible enough to allow a slight rotation, the support must be considered to be pinned. One assumption made in the usual methods of structural analysis is that deformations are very small, which means that only a slight rotation of the support is needed to qualify it as a pinned connection.

Once the geometry and support conditions of the idealized frame have been established, the loading must be determined. This determination usually involves apportioning a share of the total load to each frame. If the hypothetical structure under consideration is subjected to a uniformly distributed roof load, the portion carried by one frame will be a uniformly distributed line load measured in force per unit length as shown in Figure 1.2b. Typical units would be kips per foot.

For the loading shown in Figure 1.2b, the frame will deform as indicated by the dashed line (drawn to a greatly exaggerated scale). The individual members of the frame can be classified according to the type of behavior represented by this deformed shape. The horizontal members *AB* and *BC* are subjected primarily to bending, or flexure, and

■ **FIGURE 1.2**

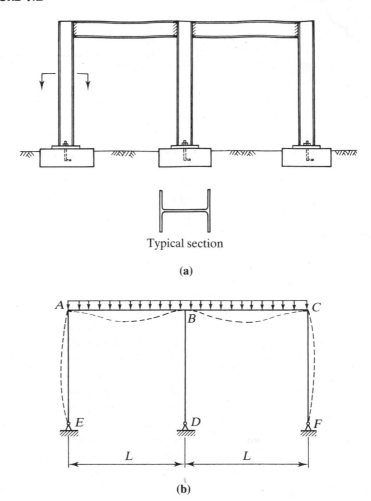

Typical section

(a)

(b)

are called *beams*. The vertical member *BD* is subjected to couples transferred from each beam, but for the symmetrical frame shown, they are equal and opposite, thereby canceling each other. Thus member *BD* is subjected only to axial compression arising from the vertical loads. In buildings, vertical compression members such as these are referred to as *columns*. The other two vertical members, *AE* and *CF,* must resist not only axial compression from the vertical loads, but also a significant amount of bending. Such members are called *beam-columns*. In reality, all members, even those classified as beams or columns, will be subjected to both bending and axial load, but in many cases, one of the effects is minor and can be neglected.

In addition to the members described, this book covers the design of connections and the following special members: composite beams, composite columns, and plate girders.

1.2 LOADS

The forces that act on a structure are called *loads*. They belong to one of two broad categories: *dead load* and *live load*. Dead loads are those that are permanent, including the weight of the structure itself, which is sometimes called the *self-weight*. In addition to the weight of the structure, dead loads in a building include the weight of nonstructural components such as floor coverings, partitions, and suspended ceilings (with light fixtures, mechanical equipment, and plumbing). All of the loads mentioned thus far are forces resulting from gravity and are referred to as *gravity loads*. Live loads, which can also be gravity loads, are those that are not as permanent as dead loads. They may or may not be acting on the structure at any given time, and the location may not be fixed. Examples of live loads include furniture, equipment, and occupants of buildings. In general, the magnitude of a live load is not as well defined as that of a dead load, and it usually must be estimated. In many cases, a structural member must be investigated for various positions of a live load so that a potential failure condition is not overlooked.

If a live load is applied slowly and is not removed and reapplied an excessive number of times, the structure can be analyzed as if the load were static. If the load is applied suddenly, as would be the case when the structure supports a moving crane, the effects of impact must be accounted for. If the load is applied and removed many times over the life of the structure, fatigue stress becomes a problem, and its effects must be accounted for. Impact loading occurs in relatively few buildings, notably industrial buildings, and fatigue loading is rare, with thousands of load cycles over the life of the structure required before fatigue becomes a problem. For these reasons, all loading conditions in this book will be treated as static, and fatigue will not be considered.

Wind exerts a pressure or suction on the exterior surfaces of a building; because of its transient nature, it properly belongs in the category of live loads. Because of the relative complexity of determining wind loads, however, wind is usually considered a separate category of loading. Because lateral loads are most detrimental to tall structures, wind loads are usually not as important for low buildings, but uplift on light roof systems can be critical. Although wind is present most of the time, wind loads of the magnitude considered in design are infrequent and are not considered to be fatigue loads.

Earthquake loads are another special category and need to be considered only in those geographic locations where there is a reasonable probability of occurrence. A structural analysis of the effects of an earthquake requires an analysis of the structure's response to the ground motion produced by the earthquake. Simpler methods are sometimes used in which the effects of the earthquake are simulated by a system of horizontal loads, similar to those resulting from wind pressure, acting at each floor level of the building.

Snow is another live load that is treated as a separate category. Adding to the uncertainty of this load is the complication of drift, which can cause much of the load to accumulate over a relatively small area.

Other types of live load are often treated as separate categories, such as hydrostatic pressure and soil pressure, but the cases we have enumerated are the ones ordinarily encountered in the design of structural steel building frames and their members.

1.3   **BUILDING CODES**

Buildings must be designed and constructed according to the provisions of a building code, which is a legal document containing requirements related to such things as structural safety, fire safety, plumbing, ventilation, and accessibility to the physically disabled. A building code has the force of law and is administered by a governmental entity such as a city, a county, or, for some large metropolitan areas, a consolidated government. Building codes do not give design procedures, but they do specify the design requirements and constraints that must be satisfied. Of particular importance to the structural engineer is the prescription of minimum live loads for buildings. Although the engineer is encouraged to investigate the actual loading conditions and attempt to determine realistic values, the structure must be able to support these specified minimum loads.

Although some large cities have their own building codes, many municipalities will modify a "model" building code to suit their particular needs and adopt it as modified. Model codes are written by various nonprofit organizations in a form that can be easily adopted by a governmental unit. Currently, there are three national model codes: the *Uniform Building Code* (ICBO, 1997), the *Standard Building Code* (SBCC, 1997), and the *BOCA National Building Code* (BOCA, 1996). The *Uniform Building Code* is the most widely used code in the United States, and it is essentially the only one used west of the Mississippi. The *Standard Building Code* is commonly used in the southeastern states, and the *BOCA National Building Code* is used in the northeastern part of the country. (BOCA is an acronym for Building Officials Conference of America.) A related document, similar in form to a building code, is ASCE 7-95, *Minimum Design Loads for Buildings and Other Structures* (ASCE, 1996). This standard provides load requirements in a format suitable for adoption as part of a building code.

At present, the various code-writing bodies are involved in a joint effort to produce a unified building code. The present codes are very similar in content, and with each revision become even more so. A single building code would simplify the work of engineers who design structures in various regions of the United States, as well as in other countries. The new code, to be called the *International Building Code,* is scheduled for completion by the year 2000.

1.4   **DESIGN SPECIFICATIONS**

In contrast to building codes, design specifications give more specific guidance for the design of structural members and their connections. They present the guidelines and criteria that enable a structural engineer to achieve the objectives mandated by a building code. Design specifications represent what is considered to be good engineering practice based on the latest research. They are periodically revised and updated by the issuance of supplements or completely new editions. As with model building codes, design specifications are written in a legal format by nonprofit organizations. They have no legal standing on their own, but by presenting design criteria and limits in the form of legal mandates and prohibitions, they can easily by adopted, by reference, as part of a building code.

The specifications of most interest to the structural steel designer are those published by the following organizations.

1. **American Institute of Steel Construction (AISC):** This specification provides for the design of structural steel buildings and their connections. It is the one of primary concern in this book and we discuss it in detail (AISC, 1993).

2. **American Association of State Highway and Transportation Officials (AASHTO):** This specification covers the design of highway bridges and related structures. It provides for all structural materials normally used in bridges, including steel, reinforced concrete, and timber (AASHTO, 1992, 1994).

3. **American Railway Engineering Association (AREA):** This document covers the design of railway bridges and related structures (AREA, 1992).

4. **American Iron and Steel Institute (AISI):** This specification deals with cold-formed steel, which we discuss in Section 1.6 of this book (AISI, 1996).

1.5 STRUCTURAL STEEL

The earliest use of iron, the chief component of steel, was for small tools in approximately 4000 B.C. (Murphy, 1957). This material was in the form of wrought iron, produced by heating ore in a charcoal fire. In the latter part of the eighteenth century and in the early nineteenth century, cast iron and wrought iron were used in various types of bridges. Steel, an alloy of primarily iron and carbon, with fewer impurities and less carbon than cast iron, was first used in heavy construction in the nineteenth century. With the advent of the Bessemer converter in 1855, steel began to displace wrought iron and cast iron in construction. In the United States, the first structural steel railroad bridge was the Eads bridge, constructed in 1874 in St. Louis, Missouri (Tall, 1964). In 1884 the first building with a steel frame was completed in Chicago.

The characteristics of steel that are of the most interest to structural engineers can be examined by plotting the results of a tensile test. If a test specimen is subjected to an axial load P, as shown in Figure 1.3a, the stress and strain can be computed as follows:

$$f = \frac{P}{A} \text{ and } \varepsilon = \frac{\Delta L}{L}$$

where
f = axial tensile stress
A = cross-sectional area
ε = axial strain
L = length of specimen
ΔL = change in length

If the load is increased in increments from zero to the point of fracture, and stress and strain are computed at each step, a stress–strain curve such as the one shown in Figure 1.3b can be plotted. This curve is typical of a class of steel known as *ductile,* or *mild, steel.* The relationship between stress and strain is linear up to the proportional limit; the material is said to follow *Hooke's law.* A peak value, the upper yield point, is quickly reached after that, followed by a leveling off at the lower yield point. The stress then remains constant, even though the strain continues to increase. At this stage of loading, the test specimen continues to elongate as long as the load is not removed, even though the load cannot be increased. This constant stress region is called the *yield plateau,* or *plastic range.* At a strain of approximately 12 times the strain at yield, strain hardening begins, and additional load (and stress) is required to cause additional elongation (and strain). A maximum value of stress is reached, after which the specimen begins to "neck down" as the stress decreases with increasing strain, and fracture occurs.

■ **FIGURE 1.3**

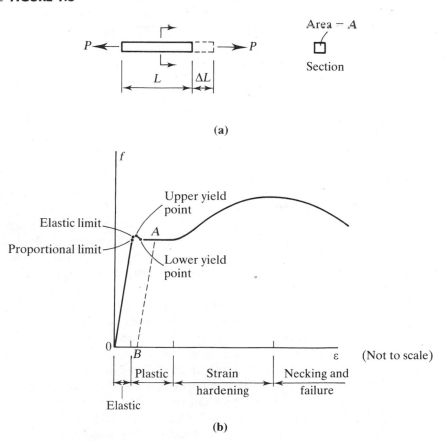

(a)

(b)

Although the cross section is reduced during loading (the Poisson effect), the original cross-sectional area is used to compute all stresses. Stress computed in this way is known as *engineering stress*. If the original length is used to compute the strain, it is called *engineering strain*.

Steel exhibiting the behavior shown in Figure 1.3b is called *ductile* because of its ability to undergo large deformations before fracturing. Ductility can be measured by the elongation, defined as

$$e = \frac{L_f - L_0}{L_0} \times 100 \qquad (1.1)$$

where

e = elongation (expressed as a percent)

L_f = length of the specimen at fracture

L_0 = original length

The elastic limit of the material is a stress that lies between the proportional limit and the upper yield point. Up to this stress, the specimen can be unloaded without permanent deformation; the unloading will be along the linear portion of the diagram, the same path followed during loading. This part of the stress–strain diagram is called the *elastic range*. Beyond the elastic limit, unloading will be along a straight line parallel to the initial linear part of the loading path, and there will be a permanent strain. For example, if the load is removed at point A in Figure 1.3b, the unloading will be along line AB, resulting in the permanent strain OB.

Figure 1.4 shows an idealized version of this stress–strain curve. The proportional limit, elastic limit, and the upper and lower yield points are all very close to one another and are treated as a single point called the *yield point,* defined by the stress F_y. The other point of interest to the structural engineer is the maximum value of stress that can be attained, called the *ultimate tensile strength F_u*. The shape of this curve is typical of all mild structural steels, which are different from one another primarily in the values of F_y and F_u. The ratio of stress to strain within the elastic range, denoted E and called *Young's modulus,* or *modulus of elasticity,* is the same for all structural steels and has a value of 29,000,000 psi (pounds per square inch) or 29,000 ksi (kips per square inch).

Figure 1.5 shows a typical stress–strain curve for high-strength steels, which are less ductile than the mild steels discussed thus far. Although there is a linear elastic portion and a distinct tensile strength, there is no well-defined yield point or yield plateau. To use these higher strength steels in a manner consistent with the use of ductile steels, some value of stress must be chosen as a value for F_y so that the same procedures and formulas can be used with all structural steels. Although there is no yield point, one needs to be defined. As previously shown, when a steel is stressed beyond its elastic limit and then unloaded, the path followed to zero stress will not be the original path from zero stress; it will be along a line having the slope of the linear portion of the path followed during loading — that is, a slope equal to E, the modulus of elasticity. Thus there will be a residual strain, or permanent set, after unloading. The yield stress for steel with a stress–strain curve of the type shown in Figure 1.5 is called the *yield strength* and is

■ **FIGURE 1.4**

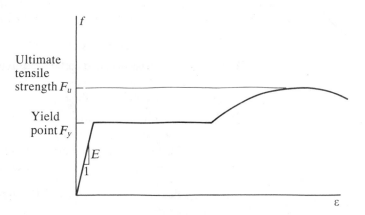

detined as the stress at the point of unloading that corresponds to a permanent strain of some arbitrarily defined amount. A strain of 0.002 is usually selected, and this method of determining the yield strength is called the *0.2% offset method.* As previously mentioned the two properties usually needed in structural steel design are F_u and F_y, regardless of the shape of the stress–strain curve and regardless of how F_y was obtained. For this reason, the generic term *yield stress* is used, and it can mean either yield point or yield strength.

The various properties of structural steel, including strength and ductility, are determined by its chemical composition. Steel is an alloy, its principal component being iron. Another component of all structural steels, although in much smaller amounts, is

■ **FIGURE 1.5**

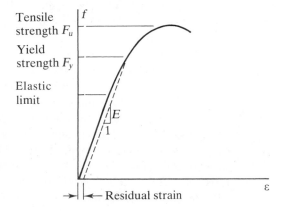

carbon, which contributes to strength but reduces ductility. Other components of some grades of steel include copper, manganese, nickel, chromium, molybdenum, and silicon. Structural steels can be grouped according to their composition as follows.

1. **Plain carbon steels:** mostly iron and carbon, with less than 1% carbon
2. **Low-alloy steels:** iron and carbon plus other components (usually less than 5%). The additional components are primarily for increasing strength, which is accomplished at the expense of a reduction in ductility.
3. **High-alloy or specialty steels:** similar in composition to the low-alloy steels but with a higher percentage of the components added to iron and carbon. These steels are higher in strength than the plain carbon steels and also have some special quality, such as resistance to corrosion.

Different grades of structural steel are identified by the designation assigned them by the American Society for Testing and Materials (ASTM). This organization develops standards for defining materials in terms of their composition, properties, and performance, and it prescribes specific tests for measuring these attributes (ASTM, 1996a). The most commonly used structural steel today is a mild steel designated as ASTM A36, or A36 for short. It has a stress–strain curve of the type shown in Figures 1.3b and 1.4 and has the following tensile properties.

Yield stress: $F_y = 36{,}000$ psi (36 ksi)

Tensile strength: $F_u = 58{,}000$ psi to 80,000 psi (58 ksi to 80 ksi)

A36 steel is classified as a plain carbon steel, and it has the following components (other than iron).

Carbon: 0.26% (maximum)
Phosphorous: 0.04% (maximum)
Sulfur: 0.05% (maximum)

These percentages are approximate, the exact values depending on the form of the finished steel product. A36 is a ductile steel, with an elongation as defined by Equation 1.1 of 20% based on an undeformed original length of 8 inches.

Steel producers who provide A36 steel must certify that it meets the ASTM standard. The values for yield stress and tensile strength shown are minimum requirements; they may be exceeded and usually are to a certain extent. The tensile strength is given as a range of values because this property cannot be achieved to the same degree of precision as the yield stress.

A steel with a yield stress of more than 36 ksi is usually considered to be a high-strength steel. The most frequently used high-strength steels are those with a yield stress of 50 ksi and a tensile strength of either 65 ksi or 70 ksi, although steel with a yield stress of 100 ksi is available. As an example, ASTM A242 is a low-alloy, corrosion-resistant steel available in yield stresses of 42, 46, and 50 ksi with corresponding tensile strengths of 63, 67, and 70 ksi. Its composition is as follows:

Carbon:	0.15% (maximum)
Manganese:	1.00% (maximum)
Phosphorus:	0.15% (maximum)
Sulfur:	0.05% (maximum)
Copper:	0.20% (minimum)

A242 steel is not as ductile as A36; the elongation, based on an original length of 8 inches, is 18%, compared with 20% for A36.

1.6   STANDARD CROSS-SECTIONAL SHAPES

In the design process outlined earlier, one of the objectives — and the primary emphasis of this book — is the selection of the appropriate cross sections for the individual members of the structure being designed. Most often, this selection will entail choosing a standard cross-sectional shape that is widely available rather than requiring the fabrication of a shape with unique dimensions and properties. The selection of an "off-the-shelf" item will almost always be the most economical choice, even if it means using slightly more material. The largest category of standard shapes includes those produced by *hot-rolling*. In this manufacturing process, which takes place in a mill, molten steel is taken from the furnace and poured into a *continuous casting* system where the steel solidifies but is never allowed to cool completely. The hot steel passes through a series of rollers that squeeze the material into the desired cross-sectional shape. Rolling the steel while it is still hot allows it to be deformed with no resulting loss in ductility, as would be the case with cold-working. During the rolling process, the member increases in length and is cut to standard lengths, usually a maximum of 65 to 75 feet, which are subsequently cut (in a fabricating shop) to the lengths required for a particular structure.

Cross sections of some of the more commonly used hot-rolled shapes are shown in Figure 1.6. The dimensions and designations of the standard available shapes are defined in the ASTM standards (ASTM, 1996b). The *W-shape,* also called a *wide-flange shape,* consists of two parallel flanges separated by a single web. The orientation of these elements is such that the cross section has two axes of symmetry. A typical designation would be "W18 × 50," where W indicates the type of shape, 18 is the nominal depth parallel to the web, and 50 is the weight in pounds per foot of length. The nominal depth is the approximate depth expressed in whole inches. For some of the lighter shapes, it is equal to the depth to the nearest inch, but this is not a general rule for the W-shapes. All of the W-shapes of a given nominal size can be grouped into families that have the same depth from inside-of-flange to inside-of-flange but with different flange thicknesses.

The *American Standard,* or *S-shape,* is similar to the W-shape in having two parallel flanges, a single web, and two axes of symmetry. The difference is in the proportions: The flanges of the W are wider in relation to the web than are the flanges of the S. In addition, the outside and inside faces of the flanges of the W-shape are parallel,

■ **FIGURE 1.6**

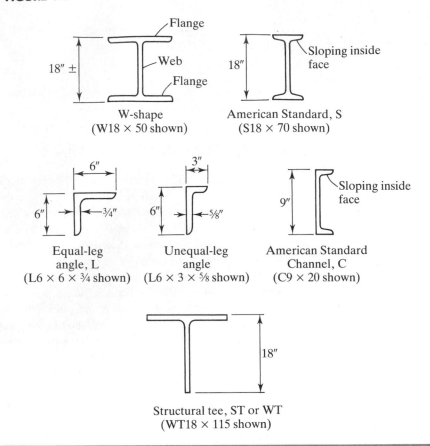

W-shape
(W18 × 50 shown)

American Standard, S
(S18 × 70 shown)

Equal-leg
angle, L
(L6 × 6 × ¾ shown)

Unequal-leg
angle
(L6 × 3 × ⅝ shown)

American Standard
Channel, C
(C9 × 20 shown)

Structural tee, ST or WT
(WT18 × 115 shown)

whereas the inside faces of the flanges of the S-shape slope with respect to the outside faces. An example of the designation of an S-shape is "S18 × 70," with the S indicating the type of shape, and the two numbers giving the depth in inches and the weight in pounds per foot. This shape was formerly called an *I-beam*.

The angle shapes are available in either equal-leg or unequal-leg versions. A typical designation would be "L6 × 6 × ¾" or "L6 × 3 × ⅝." The three numbers are the lengths of each of the two legs as measured from the corner, or heel, to the toe at the other end of the leg, and the thickness, which is the same for both legs. In the case of the unequal-leg angle, the longer leg dimension is always given first. Although this designation provides all of the dimensions, it does not provide the weight per foot.

The *American Standard Channel,* or *C-shape,* has two flanges and a web, with only one axis of symmetry; it carries a designation such as "C9 × 20." This notation is similar to that for W- and S-shapes, with the first number giving the total depth in inches parallel to the web and the second number the weight in pounds per linear foot. For the channel, however, the depth is exact rather than nominal. The inside faces of the flanges are sloping, just as with the American Standard shape. Miscellaneous Channels — for example, the MC10 × 25 — are similar to American Standard Channels.

The *Structural Tee* is produced by splitting a W-, M-, or S-shape at middepth. This shape is sometimes referred to as a *split-tee*. The prefix of the designation is either WT, MT, or ST, depending on which shape is the "parent." For example, a WT18 × 115 has a nominal depth of 18 inches and a weight of 115 pounds per foot, and is cut from a W36 × 230. Similarly, an ST10 × 32.7 is cut from an S20 × 65.4, and an MT3 × 10 is cut from an M6 × 20.

Not shown in Figure 1.6 are two hot-rolled shapes similar to the W-shape: the HP- and M-shapes. The HP shape, used for piles, has parallel flange surfaces, approximately the same width and depth, and equal flange and web thicknesses. The "M" stands for miscellaneous, and it is a shape that does not fit exactly into any of the W, HP, or S categories. Both the M- and HP-shapes are designated in the same manner as the W-shape: for example, M14 × 18 and HP14 × 117.

Other frequently used cross-sectional shapes are shown in Figure 1.7. *Bars* can have circular, square, or rectangular cross sections. If the width of a rectangular shape is 8 inches or less, it is classified as a bar and is usually designated with the width given before the thickness; for example, a bar 8 × ⅜. If the width is more than 8 inches, the shape is classified as a *plate* and is usually designated with the thickness first, as in a plate ½ × 10. Bars and plates are formed by hot-rolling.

Also shown in Figure 1.7 are hollow shapes, which can be produced by either bending plate material into the desired shape and welding the seam or by hot-working to produce a seamless shape. These shapes are designated as HSS, for hollow steel sections.

■ **FIGURE 1.7**

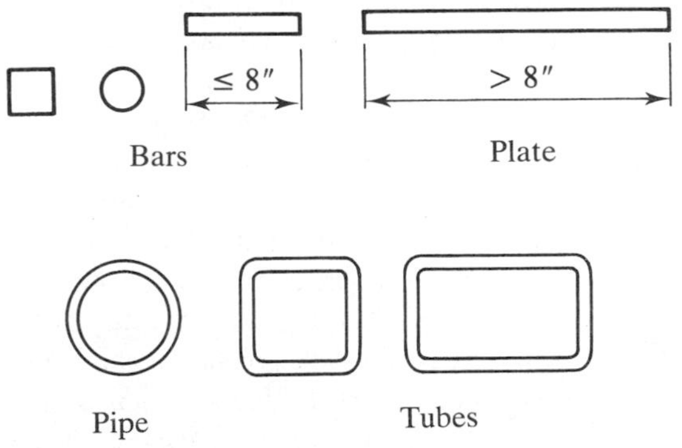

Bars   Plate

Pipe   Tubes

■ **FIGURE 1.8**

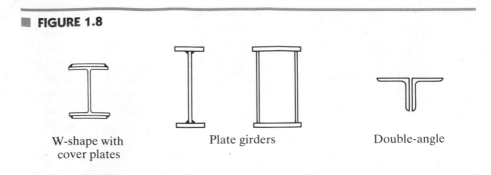

W-shape with   Plate girders   Double-angle
cover plates

Most hollow steel sections available in the United States today are produced by cold-forming and welding (Sherman, 1997). Hollow shapes include circular sections (called *pipe*) and tubes, both square and rectangular.

Other shapes are available, but those just described are the ones most frequently used. In most cases, one of these standard shapes will satisfy design requirements. If the requirements are especially severe, then a built-up section, such as one of those shown in Figure 1.8, may be needed. Sometimes a standard shape is augmented by additional cross-sectional elements, as when a cover plate is welded to one or both flanges of a W-shape. Building up sections is an effective way of strengthening an existing structure that is being rehabilitated or modified for some use other than the one for which it was designed. Sometimes a built-up shape must be used because none of the standard rolled shapes are large enough; that is, the cross section does not have enough area or moment of inertia. In such cases, plate girders can be used. These can be I-shaped sections, with two flanges and a web, or box sections, with two flanges and two webs. The components can be welded together, and can be designed to have exactly the properties needed. Built-up shapes can also be created by attaching two or more standard rolled shapes together. A widely used combination is a pair of angles placed back-to-back and connected at intervals along their length. This is called a *double-angle shape*. There are many other possibilities, some of which we illustrate throughout this book.

Another category of steel products for structural applications is cold-formed steel. Structural shapes of this type are created by bending thin material such as sheet steel or plate into the desired shape without heating. Typical cross sections are shown in Figure 1.9. Only relatively thin material can be used, and the resulting shapes are suitable only for light applications. An advantage of this product is its versatility, since almost any conceivable cross-sectional shape can easily be formed. In addition, cold-working will increase the yield point of the steel, and under certain conditions it may be accounted for in design (AISI, 1996). This increase comes at the expense of reduced ductility, however. Because of the thinness of the cross-sectional elements, the problem of instability (discussed in Chapters 4 and 5) is a particularly important factor in the design of cold-formed steel structures.

■ **FIGURE 1.9**

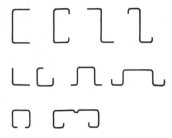

■ **PROBLEMS**

NOTE   The following problems illustrate the concepts of stress and strain covered in Section 1.5. The materials cited in these problems are not necessarily steel.

1.5-1   A tensile test was performed on a metal specimen with a circular cross section. The diameter was measured to be 0.550 inch. Two marks were made along the length of the specimen and were measured to be 2.030 inches apart. This distance is defined as the *gage length,* and all length measurements are made between the two marks. The specimen was loaded to failure. Fracture occurred at a load of 28,500 pounds. The specimen was then reassembled, and the diameter and gage length were measured to be 0.430 inch and 2.300 inch. Determine the

   a. ultimate tensile stress in ksi.

   b. elongation as a percentage.

   c. reduction in cross-sectional area as a percentage.

1.5-2   A tensile test was performed on a metal specimen having a circular cross section with a diameter of ½ inch. The *gage length* (the length over which the elongation is measured) is 2 inches. For a load of 13.5 kips, the elongation was 4.66×10^{-3} inches. If the load is assumed to be within the linear elastic range of the material, determine the modulus of elasticity.

1.5-3   A tensile test was performed on a metal specimen having a circular cross section with a diameter of 0.510 inches. For each increment of load applied, the strain was directly determined by means of a *strain gage* attached to the specimen. The results are shown in Table 1.5.1.

   a. Prepare a table of stress and strain.

   b. Plot these data to obtain a stress–strain curve. Do not connect the data points; draw a *best-fit* straight line through them.

   c. Determine the modulus of elasticity as the slope of the best-fit line.

TABLE 1.5.1

Load (lb)	Strain $\times 10^6$ (in./in.)
0	0
250	37.1
500	70.3
1000	129.1
1500	230.1
2000	259.4
2500	272.4
3000	457.7
3500	586.5

1.5-4   A tensile test was performed on a metal specimen with a diameter of ½ inch and a gage length (the length over which the elongation is measured) of 4 inches. The data were plotted on a load-displacement graph, P vs. ΔL. A best-fit line was drawn through the points, and the slope of the straight-line portion was calculated to be $P/\Delta L = 1392$ kips/in. What is the modulus of elasticity?

1.5-5   The results of a tensile test are shown in Table 1.5.2. The test was performed on a metal specimen with a circular cross section. The diameter was ⅜ inch and the gage length (the length over which the elongation is measured) was 2 inch.

    a. Use the data in Table 1.5.2 to produce a table of stress and strain values.

    b. Plot the stress–strain data and draw a best-fit curve.

    c. Compute the modulus of elasticity from the initial slope of the curve.

    d. Estimate the yield stress.

TABLE 1.5.2

Load (lb)	Elongation $\times 10^6$ (in.)
0	0
550	350
1100	700
1700	900
2200	1350
2800	1760
3300	2200
3900	2460
4400	2860
4900	3800
4970	5300
5025	7800

1.5-6 The data in Table 1.5.3 were obtained from a tensile test of a metal specimen with a rectangular cross section of 0.2011 in.^{2} in area and a gage length (the length over which the elongation is measured) of 2.000 inch. The specimen was not loaded to failure.

    a. Generate a table of stress and strain values.

    b. Plot these values and draw a best-fit line to obtain a stress–strain curve.

    c. Determine the modulus of elasticity from the slope of the linear portion of the curve.

    d. Estimate the value of the proportional limit.

    e. Use the 0.2% offset method to determine the yield stress.

TABLE 1.5.3

Load (kips)	Elongation $\times 10^3$ (in.)
0	0
1	0.160
2	0.352
3	0.706
4	1.012
5	1.434
6	1.712
7	1.986
8	2.286
9	2.612
10	2.938
11	3.274
12	3.632
13	3.976
14	4.386
15	4.640
16	4.988
17	5.432
18	5.862
19	6.362
20	7.304
21	8.072
22	9.044
23	11.310
24	14.120
25	20.044
26	29.106

2 Concepts in Structural Steel Design

2.1 DESIGN PHILOSOPHIES

As discussed earlier, the design of a structural member entails the selection of a cross section that will safely and economically resist the applied loads. Economy usually means minimum weight — that is, the minimum amount of steel. This amount corresponds to the cross section with the smallest weight per foot, which is the one with the smallest cross-sectional area. Although other considerations, such as ease of construction, may ultimately affect the choice of member size, the process begins with the selection of the lightest cross-sectional shape that will do the job. Having established this objective, the engineer must decide how to do it safely, which is where different approaches to design come into play. There are essentially three different philosophies; we discuss them here in a general way, saving specifics for later chapters.

In *allowable stress design,* a member is selected that has cross-sectional properties such as area and moment of inertia large enough to prevent the maximum stress from exceeding an allowable, or permissible, stress. This allowable stress will be in the elastic range of the material and will be less than the yield stress F_y (see Figure 1.4). A typical value might be $0.60F_y$. The allowable stress is obtained by dividing either the yield stress F_y or the ultimate tensile strength F_u by a factor of safety. This approach to design is also called *elastic design* or *working stress design*. Working stresses are those resulting from the working loads, which are the applied loads. Working loads are also known as *service* loads. A properly designed member will be stressed to no more than the allowable stress when subjected to working loads.

Plastic design is based on a consideration of failure conditions rather than working load conditions. A member is selected by using the criterion that the structure will fail at a load substantially higher than the working load. Failure in this context means either collapse or extremely large deformations. The term *plastic* is used because, at failure, parts of the member will be subjected to very large strains — large enough to put the member into the plastic range (see Figure 1.3b). When the entire cross section becomes plastic at enough locations, "plastic hinges" will form at those locations, creating a *collapse mechanism*. As the actual loads will be less than the failure loads by a factor of safety known as the *load factor,* members designed this way are not unsafe, despite being designed based on what happens at failure. The design procedure is roughly as follows.

1. Multiply the working loads (service loads) by the load factor to obtain the failure loads.

18

2. Determine the cross-sectional properties needed to resist failure under these loads. (A member with these properties is said to have sufficient strength and would be at the verge of failure when subjected to the factored loads.)

3. Select the lightest cross-sectional shape that has these properties.

Members designed by plastic theory would reach the point of failure under the factored loads but are safe under actual working loads.

Load and resistance factor design (LRFD) is similar to plastic design in that strength, or the failure condition, is considered. Load factors are applied to the service loads, and a member is selected that will have enough strength to resist the factored loads. In addition, the theoretical strength of the member is reduced by the application of a resistance factor. The criterion that must be satisfied in the selection of a member is

$$\text{Factored load} \leq \text{factored strength} \tag{2.1}$$

In this expression, the factored load is actually the sum of all service loads to be resisted by the member, each multiplied by its own load factor. For example, dead loads will have load factors that are different from those for live loads. The factored strength is the theoretical strength multiplied by a resistance factor. Equation 2.1 can therefore be written as

$$\sum (\text{Loads} \times \text{load factors}) \leq \text{resistance} \times \text{resistance factor} \tag{2.2}$$

The factored load is a failure load greater than the total actual service load, so the load factors are usually greater than unity. However, the factored strength is a reduced, usable strength, and the resistance factor is usually less than unity. The factored loads are the loads that bring the structure or member to its limit. In terms of safety, this *limit state* can be fracture, yielding, or buckling, and the factored resistance is the useful strength of the member, reduced from the theoretical value by the resistance factor. The limit state can also be one of serviceability, such as a maximum acceptable deflection.

2.2 AMERICAN INSTITUTE OF STEEL CONSTRUCTION SPECIFICATION

Because the emphasis of this book is on the design of structural steel building members and their connections, the specification of the American Institute of Steel Construction is the design specification of most importance here. It is written and kept current by an AISC committee comprising structural engineering practitioners, educators, steel producers, and fabricators. New editions are published periodically, and supplements are issued when interim revisions are needed. Allowable stress design has been the primary method used for structural steel buildings since the first AISC Specification was issued in 1923, although recent editions have contained provisions for plastic design. In 1986, AISC issued the first specification for load and resistance factor design of structural steel buildings and a companion *Manual of Steel Construction*. The purpose of these two documents was to provide an alternative to allowable stress design, much as plastic design is an alternative. The second edition of the *Manual* (AISC, 1994), includes the 1993 AISC Specification. The provisions of the LRFD Specification are based on research reported in eight papers published in 1978 in the structural journal of the American Society of Civil Engineers (Ravindra and Galambos; Yura, Galambos, and Ravindra; Bjorhovde, Galambos, and Ravindra; Cooper, Galambos, and Ravindra; Hansell, et al.; Fisher, et al.; Ravindra, Cornell, and Galambos; Galambos and Ravindra, 1978). Unless otherwise stated, references to the AISC Specification and the *Manual of Steel Construction* will be to the LRFD versions.

Load and resistance factor design is not a recent concept; since 1974 it has been used in Canada, where it is known as *limit states design*. It is also the basis of most European building codes. In the United States, LRFD has been an accepted method of design for reinforced concrete for years and is the primary method authorized in the American Concrete Institute's Building Code, where it is known as *strength design* (ACI, 1995). Highway bridge design standards provide for both allowable stress design (AASHTO, 1992) and load and resistance-factor design (AASHTO, 1994).

The AISC Specification is published as a stand-alone document, but it is also part of the *Manual of Steel Construction,* which we discuss in the next section. Except for specialized steel products such as cold-formed steel, which is covered by a different specification (AISI, 1996), the AISC Specification is the standard by which virtually all structural steel buildings in this country are designed and constructed. Hence the student of structural steel design must have ready access to this document. The details of the Specification will be covered in the chapters that follow, but we discuss the overall organization here.

The Specification consists of four parts: the main body, the appendixes, the Numerical Values section, and the Commentary. The body is alphabetically organized into Chapters A through M. Within each chapter, major headings are labeled with the chapter designation followed by a number. Further subdivisions are numerically labeled. For example, the types of structural steel authorized are listed in Chapter A, "General Provisions," under Section A3. Material, and, under it, Section 1. Structural Steel. The main body of the Specification is followed by appendixes to selected chapters. The appendixes are labeled B, E, F, G, H, J, and K to correspond to the chapters that are referenced. This section is followed by the Numerical Values section, which contains tables of numerical values for some of the Specification requirements. The Numerical Values section is followed by the Commentary, which gives background and elaboration on many of the provisions of the Specification. Its organizational scheme is the same as that of the Specification, so material applicable to a particular section can be easily located. The appendixes, the Numerical Values section, and the Commentary are considered to be official parts of the Specification and carry the same authority as material in the main body.

2.3 LOAD AND RESISTANCE FACTORS USED IN THE AISC SPECIFICATION

Equation 2.2 can be written more precisely as

$$\Sigma\gamma_i Q_i \leq \phi R_n \tag{2.3}$$

where

Q_i = a load effect (a force or a moment)

γ_i = a load factor

R_n = the nominal resistance, or strength, of the component under consideration

ϕ = resistance factor

The factored resistance ϕR_n is called the *design strength*. The summation on the left side of Equation 2.3 is over the total number of load effects (including, but not limited to, dead load and live load), where each load effect can be associated with a different load factor. Not only can each load effect have a different load factor, but also the value of the load factor for a particular load effect will depend on the combination of loads under consideration. Load combinations to be considered are given in Chapter A, "General Provisions," of the Specification as

$$1.4D \qquad (A4\text{-}1)$$

$$1.2D + 1.6L + 0.5(L_r \text{ or } S \text{ or } R) \qquad (A4\text{-}2)$$

$$1.2D + 1.6(L_r \text{ or } S \text{ or } R) + (0.5L \text{ or } 0.8W) \qquad (A4\text{-}3)$$

$$1.2D + 1.3W + 0.5L + 0.5(L_r \text{ or } S \text{ or } R) \qquad (A4\text{-}4)$$

$$1.2D \pm 1.0E + 0.5L + 0.2S \qquad (A4\text{-}5)$$

$$0.9D \pm (1.3W \text{ or } 1.0E) \qquad (A4\text{-}6)$$

where

D = dead load

L = live load due to equipment and occupancy

L_r = roof live load

S = snow load

R = rain or ice load*

W = wind load

E = earthquake load

The identifier to the right of each of these load combinations is the numbering scheme used by AISC for equations and expressions, with the letter representing the chapter, the first number the section, and the second number the sequence within that section.

As previously mentioned, the load factor for a particular load effect is not the same in all load combinations. For example, in A4-2, γ for the live load L is 1.6, whereas in A4-3, it is 0.5. The reason is that the live load is being taken as the dominant effect in A4-2, and one of the three effects, L_r, S, or R, will be dominant in A4-3. In each combination, one of the effects is considered to be at its "lifetime maximum" value and the others at their "arbitrary point in time" values. These load factors and load combinations are the ones recommended in *Minimum Design Loads for Buildings and Other Structures* (ASCE, 1996) and are based on extensive statistical studies.

---

*This load does not include *ponding,* a phenomenon that we discuss in Chapter 5.

The resistance factor ϕ for each type of resistance is given by AISC in the Specification chapter dealing with that resistance. These factors range in value from 0.75 to 1.0.

Before considering the theoretical basis of the load and resistance factors, we illustrate their application with an axially loaded compression member.

■ EXAMPLE 2.1

A column (compression member) in the upper story of a building is subject to the following loads.

Dead load: 109 kips compression

Floor live load: 46 kips compression

Roof live load: 19 kips compression

Snow: 20 kips compression

a. Determine the controlling AISC load combination and the corresponding factored load.

b. If the resistance factor, ϕ, is 0.85, what is the required *nominal* strength?

SOLUTION

Even though a load may not be acting directly on a member, it can still cause a load effect in the member. This is true of both snow and roof live load in this example. Although this building is subjected to wind, the resulting forces on the structure are resisted by members other than this particular column.

a. The controlling load combination is the one that produces the largest factored load. We evaluate each expression that involves dead load, D, live load resulting from equipment and occupancy, L, roof live load, L_r, and snow, S.

(A4-1): $1.4D = 1.4(109) = 152.6$ kips

(A4-2): $1.2D + 1.6L + 0.5(L_r$ or S or $R)$. Because S is larger than L_r and $R = 0$, we need to evaluate this combination only once, using S.

$1.2D + 1.6L + 0.5S = 1.2(109) + 1.6(46) + 0.5(20) = 214.4$ kips

(A4-3): $1.2D + 1.6(L_r$ or S or $R) + (0.5L$ or $0.8W)$. In this combination, we use S instead of L_r, and both R and W are zero.

$1.2D + 1.6S + 0.5L = 1.2(109) + 1.6(20) + 0.5(46) = 185.8$ kips

(A4-4): $1.2D + 1.3W + 0.5L + 0.5(L_r$ or S or $R)$. This expression reduces to $1.2D + 0.5L + 0.5S$, and by inspection, we can see that it produces a smaller result than combination A4-3.

(A4-5): $1.2D \pm 1.0E + 0.5L + 0.2S$: As $E = 0$, this expression reduces to $1.2D + 0.5L + 0.2S$, which produces a smaller result than combination A4-4.

(A4-6): $0.9D \pm (1.3W$ or $1.0E)$: This expression reduces to $0.9D$, which is smaller than any of the other combinations.

ANSWER   The controlling combination is A4-2, and the factored load is 214 kips.

        b. If the factored load obtained in part (a) is substituted into the fundamental
           LRFD relationship, Equation 2.3, we obtain

$$\Sigma \lambda_i Q_i \leq \phi R_n$$
$$214.4 \leq 0.85 R_n$$
$$R_n \geq 252.2 \text{ kips}$$

ANSWER   The required nominal strength is 252 kips

2.4   PROBABILISTIC BASIS OF LOAD AND RESISTANCE FACTORS

Both the load and the resistance factors specified by AISC are based on probabilistic
concepts. The resistance factors account for uncertainties in material properties, design
theory, and fabrication and construction practices. Although a complete treatment of
probability theory is beyond the scope of this book, we present a brief summary of the
basic concepts here.

    Experimental data can be represented in the form of a histogram, or bar graph, as
shown in Figure 2.1, with the abscissa representing sample values, or events, and the
ordinate representing either the number of samples having a certain value or the frequency of occurrence of a certain value. Each bar can represent a single sample value or
a range of values. If the ordinate is the percentage of values rather than the actual number of values, the graph is referred to as a *relative* frequency distribution. In such a case
the sum of the ordinates will be 100%. If the abscissa values are random events, and
enough samples are used, each ordinate can be interpreted as the probability, expressed

---

■ **FIGURE 2.1**

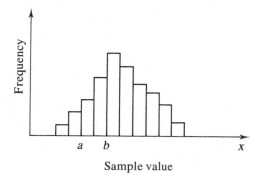

Sample value

as a percentage, of that sample value or event occurring. The relative frequency can also be expressed in decimal form, with values between 0 and 1.0. Thus the sum of the ordinates will be unity, and if each bar has a unit width, the total area of the diagram will also be unity. This result implies a probability of 1.0 that an event will fall within the boundaries of the diagram. Furthermore, the probability that a certain value or something smaller will occur is equal to the area of the diagram to the left of that value. The probability of an event having a value falling between a and b in Figure 2.1 equals the area of the diagram between a and b.

Before proceeding, some definitions are in order. The *mean* $\bar{x}$ of a set of sample values, or *population,* is the arithmetic average, or

$$\bar{x} = \frac{1}{n} \sum_{i=1}^{n} x_i$$

where x_i is a sample value and n is the number of values. The *median* is the middle value of x, and the *mode* is the most frequently occurring value. The *variance, v,* is a measure of the overall variation of the data from the mean and is defined as

$$v = \frac{1}{n} \sum_{i-1}^{n} (x_i - \bar{x})^2$$

The *standard deviation s* is the square root of the variance, or

$$s = \sqrt{\frac{1}{n} \sum_{i=1}^{n} (x_i - \bar{x})^2}$$

Like the variance, the standard deviation is a measure of the overall variation, but it has the same units and the same order of magnitude as the data. The *coefficient of variation, V,* is the standard deviation divided by the mean, or

$$V = \frac{s}{\bar{x}}$$

If the actual frequency distribution is replaced by a theoretical continuous function that closely approximates the data, it is called a *probability density function.* Such a function is illustrated in Figure 2.2. Probability functions are designed so that the total area under the curve is unity. That is, for a function $f(x)$,

$$\int_{-\infty}^{+\infty} f(x)dx = 1.0$$

which means that the probability that one of the sample values or events will occur is 1.0. The probability of one of the events between a and b in Figure 2.2 equals the area under the curve between a and b, or

$$\int_{a}^{b} f(x)dx$$

■ **FIGURE 2.2**

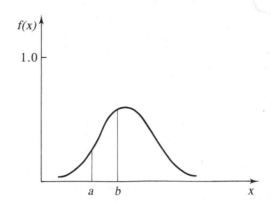

When a theoretical probability density function is used, the following notation is conventional:

μ = mean

σ = standard deviation

The probabilistic basis of the load and resistance factors used by AISC is presented in the ASCE structural journal and is summarized here (Ravindra and Galambos, 1978). Load effects Q and resistances R are random variables and depend on many factors. Loads can be estimated or obtained from measurements and inventories of actual structures, and resistances can be computed or determined experimentally. Discrete values of Q and R from observations can be plotted as frequency distribution histograms or represented by theoretical probability density functions. We use this latter representation in the material that follows.

If the probability density functions for load effects Q and resistances R are plotted on the same graph, as in Figure 2.3, the region corresponding to $Q > R$ represents failure, and $Q < R$ represents survival. If the distributions of Q and R are combined into one function, $R - Q$, positive values of $R - Q$ correspond to survival. Equivalently, if a probability density function of R/Q, the factor of safety, is used, survival is represented by values of R/Q greater than 1.0. The corresponding probability of failure is the probability that R/Q is less than 1; that is,

$$P_F = P\left[\left(\frac{R}{Q}\right) < 1\right]$$

Taking the natural logarithm of both sides of the inequality, we have

$$P_F = P\left[\ln\left(\frac{R}{Q}\right) < \ln 1\right] = P\left[\ln\left(\frac{R}{Q}\right) < 0\right]$$

■ **FIGURE 2.3**

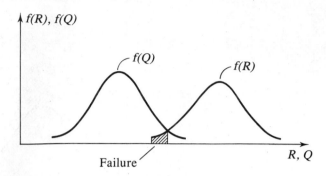

The frequency distribution curve of ln (R/Q) is shown in Figure 2.4. The *standardized* form of the variable ln (R/Q) can be defined as

$$U = \frac{\ln\left(\dfrac{R}{Q}\right) - \left[\ln\left(\dfrac{R}{Q}\right)\right]_m}{\sigma_{\ln(R/Q)}}$$

where

$$\left[\ln\left(\frac{R}{Q}\right)\right]_m = \text{the mean value of } \ln\left(\frac{R}{Q}\right)$$

$$\sigma_{\ln(R/Q)} = \text{standard deviation of } \ln\left(\frac{R}{Q}\right)$$

■ **FIGURE 2.4**

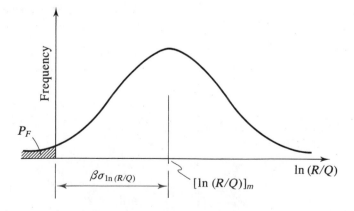

This transformation converts the abscissa U to multiples of standard deviations and places the mean of U at $U = 0$. The probability of failure can then be written as

$$P_F = P\left[\ln\left(\frac{R}{Q}\right) < 0\right] = P\left(\left\{U\sigma_{\ln(R/Q)} + \left[\ln\left(\frac{R}{Q}\right)\right]_m\right\} < 0\right)$$

$$= P\left\{U < -\frac{\left[\ln\left(\frac{R}{Q}\right)\right]_m}{\sigma_{\ln(R/Q)}}\right\} = F_u\left\{-\frac{\left[\ln\left(\frac{R}{Q}\right)\right]_m}{\sigma_{\ln(R/Q)}}\right\}$$

where F_u is the *cumulative distribution function* of U, or the probability that U will not exceed the argument of the function. If we let

$$\beta = \frac{\left[\ln\left(\frac{R}{Q}\right)\right]_m}{\sigma_{\ln(R/Q)}} \tag{2.4}$$

then

$$\left[\ln\left(\frac{R}{Q}\right)\right]_m = \beta\sigma_{\ln(R/Q)}$$

The variable β can be interpreted as the number of standard deviations from the origin that the mean value of $\ln(R/Q)$ is. For safety, the mean value *must* be more than zero, and as a consequence, β is called the *safety index* or *reliability index*. The larger this value, the larger will be the margin of safety. This means that the probability of failure, represented by the shaded area in Figure 2.4 labeled P_F, will be smaller. The reliability index is a function of both the load effect Q and the resistance R. Use of the same reliability index for all types of members subjected to the same type of loading gives the members relatively uniform strength. The "target" values of β shown in Table 2.1, selected and used in computing both load and resistance factors for the AISC Specification, were based on the recommendations of Ravindra and Galambos (1978), who also showed that

$$\phi = \frac{R_m}{R_n} e^{-0.55\beta V_R} \tag{2.5}$$

where

R_m = mean value of the resistance R

R_n = nominal or theoretical resistance

V_R = coefficient of variation of R

■ **TABLE 2.1** Target Values of β

	Loading Condition		
Type of Component	$D + (L$ or $S)$	$D + L + W$	$D + L + E$
Members	3.0	2.5	1.75
Connections	4.5	4.5	4.5

Equation 2.5 is the expression for the resistance factor ϕ as given in the Commentary to the Specification.

2.5 MANUAL OF STEEL CONSTRUCTION

Anyone engaged in structural steel design in the United States must have access to AISC's *Manual of Steel Construction* (AISC, 1994). This publication contains the AISC Specification and numerous design aids in the form of tables and graphs as well as a "catalog" of the most widely available structural shapes. This textbook was written under the assumption that you would have access to the *Manual* at all times. To encourage use of the *Manual,* we did not reproduce its tables and graphs in this book.

The *Manual* consists of two volumes. Volume I, subtitled "Structural Members, Specifications and Codes," contains Parts 1–7 and deals primarily with member design. Volume II, subtitled "Connections," contains Parts 8–12 and is devoted to connection design. The primary focus of this textbook is on Volume I, which is divided into seven parts.

Part 1. Dimensions and Properties: This part contains details on standard rolled shapes, pipe, and structural tubing, including all necessary cross-sectional dimensions and properties such as area and moment of inertia. Information is also given on their availability in various strengths. The steels covered are some of those authorized by the AISC Specification for use in building construction and include the following.

ASTM A36: Carbon structural steel

ASTM A529: High-strength, carbon–manganese structural steel

ASTM A572: High-strength, low-alloy structural steel

ASTM A242: Corrosion-resistant, high-strength, low-alloy structural steel

ASTM A588: Corrosion-resistant, high-strength, low-alloy structural steel

ASTM A852: Quenched and tempered low-alloy structural steel plate

ASTM A514: High-strength, quenched and tempered alloy structural steel plate

Part 2. Essentials of LRFD: This part is a condensed introduction to the basics of load and resistance factor design of steel structures. Numerical examples are included.

Part 3. Column Design: This part contains numerous tables to facilitate the design of both axially loaded compression members and beam-columns. Most of these tables are specialized for steels with yield stresses of 36 ksi and 50 ksi. In addition, design examples illustrating the use of the tables are given.

Part 4. Beam and Girder Design: This part, as with Part 3, contains many design aids, including both graphs and tables. Many of these deal with the requirements of the AISC Specification, but some — such as the Beam Diagrams and Formulas — pertain to structural analysis. This part also contains discussion of beam and girder design procedures and design examples.

Part 5. Composite Design: This part covers composite members, usually beams or columns, that are structural components composed of two materials: structural steel and reinforced concrete. Ordinarily, composite beams are used when a system of parallel beams supports a reinforced concrete floor slab. In this application, elements welded to the top flange are embedded in the concrete, forming the connection between the two materials. Composite columns consist of either structural steel shapes encased in reinforced concrete or hollow shapes filled with concrete. This part contains background information, design aids, and examples.

Part 6. Specifications and Codes: This part contains the AISC Specification and Commentary, a specification for high-strength bolts (RCSC, 1994), and other documents.

Part 7. Miscellaneous Data and Mathematical Tables: This part deals with wire, sheet steel, and various properties of steel and other building materials. Also included are mathematical formulas and conversion factors for different systems of units.

Volume II, comprising Parts 8–13, contains tables to assist in the design of bolted and welded connections along with tables that provide details on "standard" connections. (Part 13 is a list of construction industry organizations.)

The AISC Specification is only a small part of the *Manual*. Many of the terms and constants used in other parts of the *Manual* are presented to facilitate the design process and are not necessarily part of the Specification. In some instances, the recommendations are only "rules of thumb" based on common practice, not requirements of the Specification. Although such information is not in conflict with the Specification, it is important to recognize what is a *requirement* (when adopted by a building code) and what is not.

2.6   DESIGN COMPUTATIONS AND PRECISION

The computations required in engineering design and analysis are done with either an electronic calculator or a digital computer, which may be a microcomputer, minicomputer, or mainframe computer. When doing manual computations with the aid of an electronic calculator, an engineer must make a decision regarding the degree of precision needed. The problem of how many significant figures to use in engineering computations has no simple solution. Recording too many significant digits is misleading and can imply an unrealistic degree of precision. Conversely, recording too few figures can lead to chaos and meaningless results. The question of precision was mostly academic before the early 1970s, when the chief calculating tool was the slide rule. The guiding principle at that time was to read and record numbers as accurately as possible, which meant three or four significant figures.

There are many inherent inaccuracies and uncertainties in structural design, including variations in material properties and loads; load estimates sometimes border on educated guesses. It hardly makes sense to perform computations with 12 significant figures and record the answer to that degree of precision when the yield stress is given to the nearest 1000 psi, which is two significant figures. To avoid results that are even less precise, however, it is reasonable to assume that the given parameters of a problem, such as the yield stress, are exact and then decide on the degree of precision required in subsequent calculations.

A further complication arises when electronic calculators are used. If all of the computations for a problem are done in one continuous series of operations on a calculator, the number of significant figures used is whatever the calculator uses, perhaps 10 or 12. But if intermediate values are rounded, recorded, and used in subsequent computations, then a consistent number of significant figures will not have been used. Furthermore, the manner in which the computations are grouped will influence the final result. In general, the result will be no more accurate than the least accurate number used in the computation — and sometimes less because of round-off error. For example, consider a number calculated on a 12-digit calculator and recorded to four significant figures. If this number is multiplied by a similarly obtained and rounded number, the product will be precise to four significant figures at most, regardless of the number of digits displayed on the calculator. It is not reasonable to record this number to more than four significant figures.

It is also unreasonable to record the results of every calculator multiplication or division to a predetermined number of significant figures in order to have a consistent degree of precision throughout. A reasonable approach is to perform operations on the calculator in any convenient manner and record intermediate values to whatever degree of precision is deemed adequate (without clearing the intermediate value from the calculator if it can be used in the next computation). The final results should then be expressed to a precision consistent with this procedure. It is difficult to determine what this precision should be for the typical structural steel design problem. More precision than three or four significant figures is probably unrealistic in most cases, and results based on less than three may be too approximate to be of any value. In this book we record intermediate values to three or four digits (usually four), depending on the circumstances, and record final results to three digits.

Although three digits is not a high degree of precision, it should not be made worse by sloppy numerical practices. When numbers are rounded, they should be rounded properly and not truncated. Thus, 1.666666 rounded to three significant figures is 1.67, not 1.66. Care should also be taken to differentiate between number of significant figures and number of decimal places; the two are not always the same. For example,

0.00323 is expressed to three significant figures, whereas 0.003 is expressed to three decimal places but only one significant figure.

■ PROBLEMS

NOTE   All given loads are service loads

Load and Resistance Factors Used in the AISC Specification

2.3-1   A column in a building is subjected to the following load effects:

9 kips compression from dead load

5 kips compression from roof live load

6 kips compression from snow

7 kips compression from 3 inches of rain accumulated on the roof

8 kips compression from wind

a. Determine the factored load to be used in the design of the column. Which AISC load combination controls?

b. What is the required *design* strength of the column?

c. If the resistance factor ϕ is 0.85, what is the required *nominal* strength of the column?

2.3-2   Repeat Problem 2.3-1 without the possibility of rain accumulation on the roof.

2.3-3   A beam is part of the framing system for the floor of an office building. The floor is subjected to both dead loads and live loads. The maximum moment caused by the service dead load is 45 ft-kips, and the maximum moment for the service live load is 63 ft-kips (these moments occur at the same location on the beam and can therefore be combined).

a. Determine the maximum factored bending moment. What is the controlling AISC load combination?

b. If the resistance factor ϕ is 0.9, what is the required nominal moment strength in ft-kips?

2.3-4   A tension member must be designed for a service dead load of 18 kips and a service live load of 2 kips. Determine the maximum factored load and the controlling AISC load combination.

2.3-5   A flat roof is subject to the following uniformly distributed loads: a dead load of 21 psf (pounds per square foot of roof surface), a roof live load of 12 psf, a snow load of 13.5 psf, and a wind load of 22 psf *upward*. (Although the wind itself is in a horizontal direction, the force that it exerts on this roof is upward. It will be upward regardless of wind direction. The dead, live, and snow loads are *gravity loads* and act *downward*.) Compute the factored load in pounds per square foot. Which AISC load combination controls?

3 Tension Members

INTRODUCTION

Tension members are defined as structural elements that are subjected to axial tensile forces. They are used in various types of structures and include truss members, cables in suspension and cable-stayed bridges, bracing for buildings and bridges, and cables in suspended roof systems. Any cross-sectional configuration may be used, because for any material, the only determinant of strength is the cross-sectional area. Circular rods and rolled angle shapes are commonly used. Shapes built up from rolled shapes or a combination of rolled shapes and plates are sometimes used when large loads must be resisted. The most common built-up configuration is probably the double-angle section, shown in Figure 3.1 along with other typical cross sections. Because the use of this section is so widespread, tables of properties of various combinations of angles are included in the AISC *Manual of Steel Construction*.

The stress in an axially loaded tension member is given by

$$f = \frac{P}{A}$$

where P is the magnitude of the load, and A is the cross-sectional area normal to the load. The stress as given by this equation is exact, provided the cross section under consideration is not adjacent to the point of application of the load, where the distribution of stress is not uniform.

■ **FIGURE 3.1**

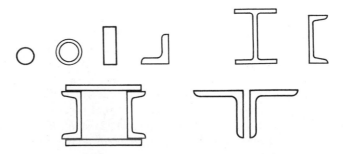

■ **FIGURE 3.2**

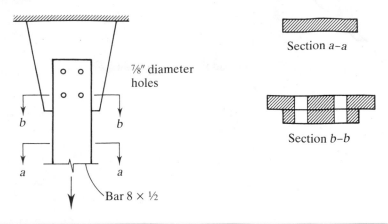

If the cross-sectional area of a tension member varies along its length, the stress is a function of the particular section under consideration. The presence of holes in a member also influences the stress at a cross section through the hole or holes. At these locations, the cross-sectional area will be reduced by an amount equal to the area removed by the holes. Tension members are frequently connected at their ends with bolts, as shown in Figure 3.2. The tension member, an $8 \times \frac{1}{2}$ bar, is connected to a *gusset plate,* a connection element whose purpose is to transfer the load from the member. The area of the bar at section *a–a* is $(\frac{1}{2})(8) = 4$ in.^{2}, but the area at section *b–b* is only $4 - (2)(\frac{1}{2})(\frac{7}{8}) = 3.12$ in.^{2} and will be more highly stressed. This reduced area is referred to as the *net area,* or *net section,* and the unreduced area is the *gross area.*

The typical design problem is to select a member with sufficient cross-sectional area so that the factored load does not exceed the design strength (the nominal strength multiplied by the resistance factor). A closely related problem is that of analysis, or review, of a given member, wherein the design strength is computed and compared with the factored load. In general, analysis is a direct procedure, but design is an iterative process and may require some trial and error.

Tension members are covered in Chapter D of the Specification. Requirements that are common with other types of members are covered in Chapter B, "Design Requirements."

3.2   DESIGN STRENGTH

A tension member can fail by reaching one of two limit states: excessive deformation or fracture. To prevent excessive deformation, initiated by yielding, the load on the gross section must be small enough that the stress on the gross section is less than the yield stress F_y. To prevent fracture, the stress on the net section must be less than the tensile strength F_u. In each case, the stress P/A must be less than a limiting stress F or

$$\frac{P}{A} < F$$

Thus the load P must be less than FA, or

$$P < FA$$

The left side of this expression is the applied factored load, and the right side is the strength. The *nominal* strength in yielding is

$$P_n = F_y A_g$$

and the nominal strength in fracture is

$$P_n = F_u A_e$$

where A_e is the *effective* net area, which may be equal to either the net area or, in some cases, a smaller area. We discuss effective net area in Section 3.3.

Although yielding will first occur on the net cross section, the deformation within the length of the connection will generally be smaller than the deformation in the remainder of the tension member. The reason is that the net section exists over a relatively small length of the member, and the total elongation is a product of the length and the strain (a function of the stress). Most of the member will have an unreduced cross section, so attainment of the yield stress on the gross area will result in larger total elongation. It is this larger deformation, not the first yield, that is the limit state.

The resistance factor $\phi = \phi_t$ is smaller for fracture than for yielding, reflecting the more serious nature of reaching the limit state of fracture.

For yielding, $\phi_t = 0.90$

For fracture, $\phi_t = 0.75$

Equation 2.3,

$$\sum \gamma_i Q_i \leq \phi R_n$$

can be written for tension members as

$$\sum \gamma_i Q_i \leq \phi_t P_n$$

or

$$P_u \leq \phi_t P_n$$

where P_u is the governing combination of factored loads. Because there are two limit states, both of the following conditions must be satisfied:

$$P_u \leq 0.90 F_y A_g$$
$$P_u \leq 0.75 F_u A_e$$

The smaller of these is the design strength of the member.

You can find values for F_y, F_u, and $0.75F_u$ in Table 2 in the Numerical Values section of the AISC Specification. However, some of the high-strength steels are grouped in such a way that you cannot always associate a given material with a particular yield stress and tensile strength by inspection. You can obtain this information from Tables 1-1 and 1-2 in Part 1 of the *Manual*. For example, to determine the properties of a W33 × 221 of ASTM A242 steel, first refer to Table 1-2 and determine the shape group to which this section belongs — in this case, Group 3. In Table 1-1, all the steels available for Group 3 shapes are represented by shaded areas under the Group 3 heading. For A242 steel, the only alloy available for Group 3 shapes has a yield stress F_y of 46 ksi and a tensile strength F_u of 67 ksi. The yield stress and tensile strength of steel pipe and structural tubing are given in Table 1-4.

The exact amount of area to be deducted from the gross area to account for the presence of bolt holes depends on the fabrication procedure. The usual practice is to drill or punch standard holes (i.e., not oversized) with a diameter $\frac{1}{16}$ inch larger than the fastener diameter. To account for possible roughness around the edges of the hole, Section B2 of the AISC Specification (in the remainder of this book, references to the Specification will be in the form AISC B2) requires the addition of $\frac{1}{16}$ inch to the actual hole diameter. This amounts to using an effective hole diameter $\frac{1}{8}$ inch larger than the fastener diameter. In the case of slotted holes, $\frac{1}{16}$ inch should be added to the actual *width* of the hole. You can find details related to standard, oversized, and slotted holes in AISC J3.2, "Size and Use of Holes" (in Chapter J, "Connections, Joints, and Fasteners").

■ EXAMPLE 3.1

A bar $5 \times \frac{1}{2}$ of A36 steel is used as a tension member. It is connected to a gusset plate with four $\frac{5}{8}$-inch-diameter bolts as shown in Figure 3.3. Assume that the effective net area A_e equals the actual net area and compute the design strength.

SOLUTION For yielding of the gross section,

$$A_g = 5(\tfrac{1}{2}) = 2.5 \text{ in.}^2$$

The *nominal* strength is

$$P_n = F_y A_g = 36(2.5) = 90 \text{ kips}$$

and the *design* strength is

$$\phi_t P_n = 0.90(90) = 81 \text{ kips}$$

For fracture of the net section,

$$
\begin{aligned}
A_n &= A_g - A_{\text{holes}} \\
&= 2.5 - (\tfrac{1}{2})(\tfrac{3}{4}) \times 2 \text{ holes} \\
&= 2.5 - 0.75 = 1.75 \text{ in.}^2 \\
A_e &= A_n = 1.75 \text{ in.}^2 \quad \text{(for this example, } A_e \text{ does not always equal } A_n)
\end{aligned}
$$

■ **FIGURE 3.3**

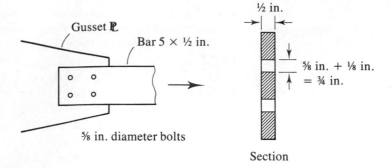

The *nominal* strength is

$$P_n = F_u A_e = 58(1.75) = 101.5 \text{ kips}$$

and the *design* strength is

$$\phi_t P_n = 0.75(101.5) = 76.1 \text{ kips}$$

The smaller value controls.

ANSWER   Design strength = 76.1 kips   ■

The effects of stress concentrations at holes appear to have been overlooked. In reality, stresses at holes can be as high as three times the average stress on the net section, and at fillets of rolled shapes they can be more than twice the average (McGuire, 1968). Because of the ductile nature of structural steel, the usual design practice is to neglect such localized overstress. After yielding begins at a point of stress concentration, additional stress is transferred to adjacent areas of the cross section. This stress redistribution is responsible for the "forgiving" nature of structural steel. Its ductility permits the initially yielded zone to deform without fracture as the stress on the remainder of the cross section continues to increase. Under certain conditions, however, steel may lose its ductility and stress concentrations can precipitate brittle fracture. These situations include fatigue loading and extremely low temperature.

■ **EXAMPLE 3.2**

A single angle tension member, an L3½ × 3½ × ⅜, is connected to a gusset plate with ⅞-inch-diameter bolts, as shown in Figure 3.4. A36 steel is used. The service loads are 35 kips dead load and 15 kips live load. Investigate this member for compliance with the AISC Specification. Assume that the effective net area is 85% of the computed net area (we cover computation of effective net area in Section 3.3).

■ **FIGURE 3.4**

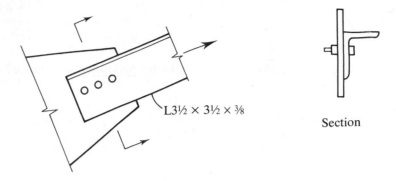

L3½ × 3½ × ⅜

Section

SOLUTION   The load combinations are

(A4-1): $1.4D = 1.4(35) = 49$ kips

(A4-2): $1.2D + 1.6L = 1.2(35) + 1.6(15) = 66$ kips

The second combination controls; $P_u = 66$ kips.
The design strengths are

Gross section: $A_g = 2.48$ in.^{2} (from Part 1 of the *Manual*)

$$\phi_t P_n = \phi_t F_y A_g = 0.90(36)(2.48) = 80.4 \text{ kips}$$

Net section: $A_n = 2.48 - \frac{3}{8}(\frac{7}{8} + \frac{1}{8}) = 2.105$ in.^{2}

$A_e = 0.85(2.105) = 1.789$ in.^{2} (in this example)

$\phi_t P_n = \phi_t F_u A_e = 0.75(58)(1.789) = 77.8$ kips   (controls)

ANSWER   Since $P_u < \phi_t P_n$ (66 kips < 77.8 kips), the member is satisfactory.   ■

In Example 3.2, load combination A4-2 controls. When only dead load and live loads are present, this combination will always control when the dead load is less than eight times the live load. In future examples, we do not check 1.4D (combination A4-1) when it obviously does not control.

3.3   **EFFECTIVE NET AREA**

Of the several factors influencing the performance of a tension member, the manner in which it is connected is the most important. A connection almost always weakens the member, and the measure of its influence is called the *joint efficiency*. This factor is a function of the ductility of the material, fastener spacing, stress concentrations at holes, fabrication procedure, and a phenomenon known as *shear lag*. All contribute to reducing the effectiveness of the member, but shear lag is the most important.

Shear lag occurs when some elements of the cross section are not connected, as when only one leg of an angle is bolted to a gusset plate, as shown in Figure 3.5. The consequence of this partial connection is that the connected element becomes overloaded and the unconnected part is not fully stressed. Lengthening the connected region will reduce this effect. Research reported by Munse and Chesson (1963) suggests that shear lag be accounted for by using a reduced, or effective, net area. Because shear lag affects both bolted and welded connections, the effective net area concept applies to both types of connections.

For bolted connections, the effective net area is

$$A_e = UA_n$$

and for welded connections, it is

$$A_e = UA_g$$

where the reduction factor U is given by

$$U = 1 - \frac{\bar{x}}{L} \leq 0.9 \qquad \text{(AISC Equation B3-2)}$$

In this expression, $\bar{x}$ is the distance from the centroid of the connected area to the plane of the connection, and L is the length of the connection. If a member has two symmetrically located planes of connection, $\bar{x}$ is measured from the centroid of the nearest one-half of the area. Figure 3.6 illustrates $\bar{x}$ for various cross sections.

In addition to this definition of $\bar{x}$, formulated by Munse and Chesson (1963), the Commentary to the AISC Specification shows additional approaches to computing $\bar{x}$, as suggested by Easterling and Giroux (1993).

The length L in AISC Equation B3-2 is the length of the connection in the direction of the load, as shown in Figure 3.7. For bolted connections, it is measured from the center of the bolt at one end of the connection to the center of the bolt at the other

■ **FIGURE 3.5**

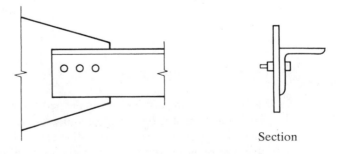

Section

■ **FIGURE 3.6**

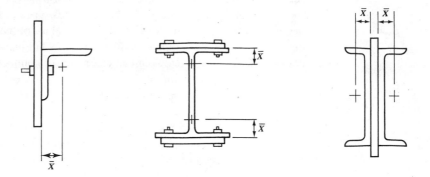

■ **FIGURE 3.7**

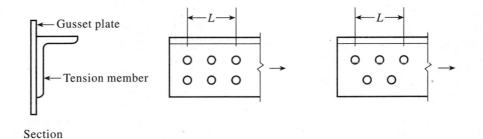

(a) Bolted

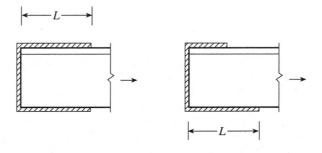

(b) Welded

end. For welds, it is measured from one end of the connection to the other. If there are segments of different lengths in the direction of the load, the longest segment is used.

Based on average values of $\bar{x}/L$ for various types of bolted tension member connections, the Commentary to AISC B3 gives values of the reduction factor U that may be used in lieu of the computed value of $1 - \bar{x}/L$. These average values of U for bolted connections are based on two broad categories of connections: those with two fasteners per line in the direction of the applied load and those with three or more per line. Only three different values are given; they correspond to the following conditions.

1. For W, M, and S shapes that have a width-to-depth ratio of at least ⅔ (and tee shapes cut from them) and are connected through the flanges with at least three fasteners per line in the direction of applied load,

    $$U = 0.90$$

2. For all other shapes (including built-up shapes) with at least three fasteners per line,

    $$U = 0.85$$

3. For all members with only two fasteners per line,

    $$U = 0.75$$

Application of these rules is illustrated in Figure 3.8.

Average U values can also be used for welded connections. Although not explicitly stated in the Commentary, that is the intent (AISC, 1989b). The rules are the same, except that the provision corresponding to two fasteners per line does not apply. The average values for welded connections are as follows.

1. For W, M, or S shapes with a width-to-depth ratio of at least ⅔ (and tee shapes cut from them) and connected at the flanges,

    $$U = 0.90$$

2. For all other shapes,

    $$U = 0.85$$

Special Cases for Welded Connections

Only when some elements of the cross section are not connected will A_e be less than A_n. For tension members such as single plates and bars (as in Example 3.1), the effective net area is taken as the full computed net area. There is an exception to this rule, however: For *plates or bars* connected by longitudinal welds at their ends, as shown in Figure 3.9,

$$A_e = UA_g$$

■ **FIGURE 3.8**

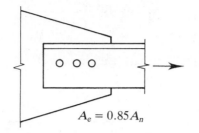

$A_e = 0.85A_n$

(Single or double angle)

(a)

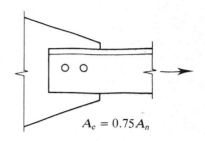

$A_e = 0.75A_n$

(Single or double angle)

(b)

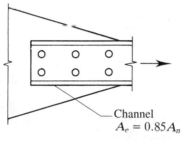

Channel
$A_e = 0.85A_n$

(c)

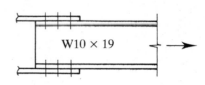

W10 × 19

$\dfrac{b_f}{d} = 0.393 < \,^2/_3$

$A_c = 0.85A_n$

(d)

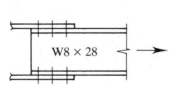

W8 × 28

$\dfrac{b_f}{d} = 0.811 > \,^2/_3$

$A_e = 0.90A_n$

(e)

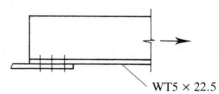

WT5 × 22.5

$\dfrac{b_f}{d} = 0.794 > \,^2/_3$ (for parent W-shape)

$A_e = 0.90A_n$

(f)

■ **FIGURE 3.9**

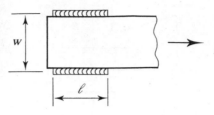

where
$$U = 1.0, \quad \text{for } \ell \geq 2w$$
$$= 0.87, \quad \text{for } 1.5w \leq \ell < 2w$$
$$= 0.75, \quad \text{for } w \leq \ell < 1.5w$$

ℓ = length of the pair of welds $\geq w$

w = distance between the welds (which may be taken as the width of the plate or bar)

AISC B3 also gives one other special case for welded connections. For any member connected by *transverse welds alone,*

A_e = area of the connected element of the cross section

Figure 3.10 illustrates the difference between transverse and longitudinal welds. Connections by transverse welds alone are uncommon.

■ **FIGURE 3.10**

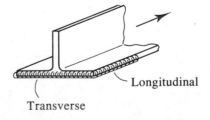

Longitudinal

Transverse

■ **EXAMPLE 3.3**

Determine the effective net area for the tension member shown in Figure 3.11.

SOLUTION

$$A_n = A_g - A_{holes}$$
$$= 5.75 - \tfrac{1}{2}(\tfrac{5}{8} + \tfrac{1}{8})(2) = 5.00 \text{ in.}^2$$

Only one element (one leg) of the cross section is connected, so the net area must be reduced. From the properties tables in Part 1 of the *Manual,* the distance from the centroid to the outside face of the leg of an $L6 \times 6 \times \tfrac{1}{2}$ is

$$\bar{x} = 1.68 \text{ in.}$$

The length of the connection is

$$L = 3 + 3 = 6 \text{ in.}$$

$$\therefore U = 1 - \left(\frac{\bar{x}}{L}\right) = 1 - \left(\frac{1.68}{6}\right) = 0.720 < 0.9$$

$$A_e = UA_n = 0.720(5.00) = 3.60 \text{ in.}^2$$

The average value of U from the Commentary could also be used. Because this shape is not a W, M, S, or tee and has more than two bolts in the direction of the load, the reduction factor U can be taken as 0.85, and

$$A_e = UA_n = 0.85(5.00) = 4.25 \text{ in.}^2$$

Either U value is acceptable, but the value obtained from AISC Equation B3-2 is more accurate. However, the average values of U are useful during preliminary design, when actual section properties and connection details are not known.

■ **FIGURE 3.11**

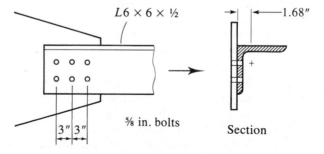

$L6 \times 6 \times \tfrac{1}{2}$

1.68"

⅝ in. bolts

3" 3"

Section

■ **EXAMPLE 3.4**

If the tension member of Example 3.3 is welded as shown in Figure 3.12, determine the effective net area.

SOLUTION

As in Example 3.3, only part of the cross section is connected, and a reduced effective net area must be used. The connection is made with a combination of longitudinal and transverse welds, so it is not one of the special cases for welded members.

$$U = 1 - \left(\frac{\bar{x}}{L}\right) = 1 - \left(\frac{1.68}{5.5}\right) = 0.695 < 0.9$$

ANSWER

$$A_e = UA_g = 0.695(5.75) = 4.00 \text{ in.}^2 \qquad ■$$

■ **FIGURE 3.12**

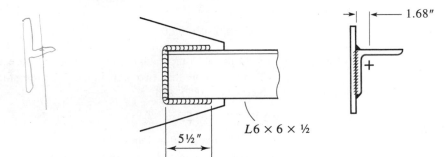

$5\frac{1}{2}''$

$L6 \times 6 \times \frac{1}{2}$

$1.68''$

3.4	**STAGGERED FASTENERS**

If a tension member connection is made with bolts, the net area will be maximized if the fasteners are placed in a single line. Sometimes space limitations, such as a limit on dimension a in Figure 3.13a, necessitate using more than one line. If so, the reduction in cross-sectional area is minimized if the fasteners are arranged in a staggered pattern, as shown. Sometimes staggered fasteners are required by the geometry of a connection such as the one shown in Figure 3.13b. In either case, any cross section passing through holes will pass through fewer holes than if the fasteners are not staggered.

If the amount of stagger is small enough, the influence of an offset hole may be felt by a nearby cross section, and fracture along an inclined path such as *abcd* in Figure 3.13c is possible. In such a case, the relationship $f = P/A$ does not apply, and stresses on the inclined portion *b–c* are a combination of tensile and shearing stresses. Several approximate methods have been proposed to account for the effects of staggered holes. Cochran

■ **FIGURE 3.13**

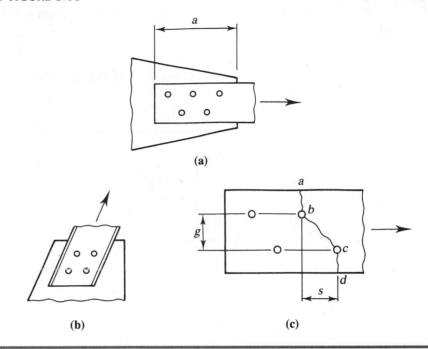

(a)

(b)   (c)

(1922) suggested the use of a net area equal to the product of the plate thickness and a net width. The area is computed as follows: Decide on a possible failure line to investigate and let the net width equal the gross width minus the quantity

$$d' = d - \frac{s^2}{4g} \tag{3.1}$$

for each staggered hole in the chain, minus d for each unstaggered hole. In this expression, d is the hole diameter, s (pitch) is the spacing of two adjacent holes in the direction parallel to the load, and g (gage) is the transverse spacing. The AISC Specification uses this same approximation, but in a slightly different form. Section B2 requires that the net width be computed by subtracting the sum of the hole diameters from the gross width and adding, for each inclined line in the chain, the quantity $s^2/4g$. That is,

$$w_n = w_g - \sum d + \sum \frac{s^2}{4g}$$

When more than one failure pattern is conceivable, all possibilities should be investigated, and the one corresponding to the smallest load capacity should be used. Note that this method will not accommodate failure patterns with lines parallel to the applied load.

■ **EXAMPLE 3.5**

Compute the smallest net area for the plate shown in Figure 3.14. The holes are for 1-inch-diameter bolts.

■ **FIGURE 3.14**

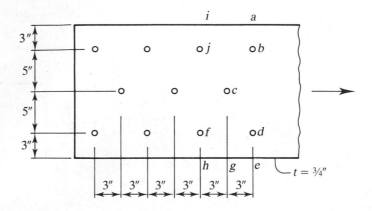

SOLUTION

The effective hole diameter is $1 + \frac{1}{8} = 1\frac{1}{8}$ in. For line *abde*,

$$w_n = 16 - 2(1.125) = 13.75 \text{ in.}$$

For line *abcde*,

$$w_n = 16 - 3(1.125) + \frac{2(3)^2}{4(5)} = 13.52 \text{ in.}$$

The second condition governs:

ANSWER $A_n = tw_n = 0.75(13.52) = 10.14 \text{ in.}^2$ ■

As each fastener resists an equal share of the load (an assumption used in the design of simple connections; see Chapter 7), different potential failure lines may be subjected to different loads. For example, line *abcde* in Figure 3.14 must resist the full load, whereas *ijfh* will be subjected to $\frac{8}{11}$ of the applied load. The reason is that $\frac{3}{11}$ of the load will have been transferred from the member before *ijfh* receives any load.

When lines of fasteners are present in both legs of an angle, and the fasteners in these lines are staggered with respect to one another, the net area is found by first "unfolding" the angle to obtain an equivalent plate. This plate is then analyzed like any

■ **FIGURE 3.15**

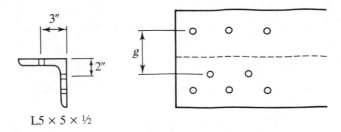

L5 × 5 × ½

other. The unfolding is done along the middle surface, giving a gross width of plate that equals the sum of the leg lengths minus the angle thickness. AISC B2 specifies that any gage line crossing the heel of the angle be reduced by an amount that equals the angle thickness. Thus the distance g in Figure 3.15, to be used in the $s^2/4g$ term, would be $3 + 2 - \frac{1}{2} = 4\frac{1}{2}$ inches.

■ **EXAMPLE 3.6**

Find the design tensile strength of the angle shown in Figure 3.16. A36 steel is used, and holes are for $\frac{7}{8}$-inch-diameter bolts.

SOLUTION   Compute the net width:

$$w_g = 8 + 6 - \frac{1}{2} = 13.5 \text{ in.}$$

■ **FIGURE 3.16**

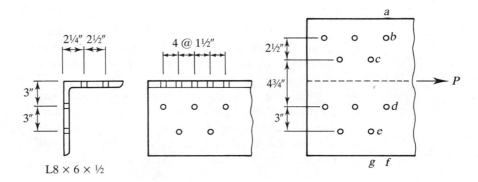

L8 × 6 × ½

Effective hole diameter $= \frac{7}{8} + \frac{1}{8} = 1$ in.

For line *abdf*,

$$w_n = 13.5 - 2(1) = 11.5 \text{ in.}$$

For line *abceg*,

$$w_n = 13.5 - 3(1) + \frac{(1.5)^2}{4(2.5)} = 10.73 \text{ in.}$$

Because $\frac{1}{10}$ of the load has been transferred from the member by the fastener at *d*, this potential failure line must resist only $\frac{9}{10}$ of the load. Therefore the net width of 10.73 inch should be multiplied by $\frac{10}{9}$ to obtain a net width that can be compared with those lines that resist the full load. Use $w_n = 10.73(\frac{10}{9}) = 11.92$ inch. For line *abcdeg*,

$$g_{cd} = 3 + 2.25 - 0.5 = 4.75 \text{ in.}$$

$$w_n = 13.5 - 4(1) + \frac{(1.5)^2}{4(2.5)} + \frac{(1.5)^2}{4(4.75)} + \frac{(1.5)^2}{4(3)} = 10.03 \text{ in.}$$

The last case controls:

$$A_n = t(w_n) = 0.5(10.03) = 5.015 \text{ in.}^2$$

Both legs of the angle are connected, so

$$A_e = A_n = 5.015 \text{ in.}^2$$

The design strength based on fracture is

$$\phi_t P_n = 0.75 F_u A_e = 0.75(58)(5.015) = 218 \text{ kips}$$

The design strength based on yielding is

$$\phi_t P_n = 0.90 F_y A_g = 0.90(36)(6.75) = 219 \text{ kips}$$

ANSWER   Fracture controls; design strength $= 218$ kips. ■

Note that the product of the calculated gross width and angle thickness is the exact gross area of the angle, just as if the legs of the angle were rectangular. The reason is that the rectangular shape can be obtained by removing area at the heel fillet and adding it at the toes.

The AISC Specification furnishes no guidance when staggered rows of fasteners are placed in rolled shapes other than angles. The equivalent-plate concept is complicated by the presence of elements of different thicknesses in shapes such as channels and wide flanges. In such cases, the use of areas (rather than widths) and reduced hole diameters, given by Equation 3.1, is recommended.

In Example 3.7, all of the holes are in one element of the cross section.

■ **EXAMPLE 3.7**

Determine the smallest net area for the American Standard Channel shown in Figure 3.17. The holes are for ⅝-inch-diameter bolts.

■ **FIGURE 3.17**

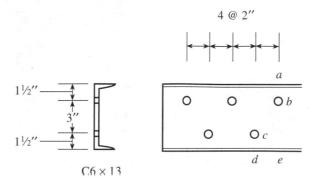

1½″

3″

1½″

C6 × 13

4 @ 2″

a
b
c
d   e

SOLUTION

$$A_n = A_g - \sum t_w \times (d \text{ or } d')$$

$$d = \text{bolt diameter} + \frac{1}{8} = \frac{5}{8} + \frac{1}{8} = \frac{3}{4} \text{ in.}$$

Line *abe:*

$$A_n = A_g - t_w d = 3.83 - 0.437\left(\frac{3}{4}\right) = 3.50 \text{ in.}^2$$

Line *abcd:*

$$A_n = A_g - t_w(d \text{ for hole at } b) - t_w(d' \text{ for hole at } c)$$

$$= 3.83 - 0.437\left(\frac{3}{4}\right) - 0.437\left[\frac{3}{4} - \frac{(2)^2}{4(3)}\right] = 3.32 \text{ in.}^2$$

ANSWER Smallest net area = 3.32 in.²

When holes are present in more than one element of the cross section, a slightly different strategy is called for. Although shapes other than angles cannot be "unfolded" the way an angle can, the need to visualize the shape as an unfolded plate remains. A method for doing so is illustrated in Figure 3.18 and in Example 3.8.

■ **FIGURE 3.18**

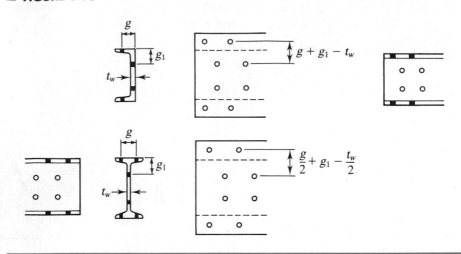

■ **EXAMPLE 3.8**

Find the design strength of the S-shape shown in Figure 3.19. The holes are for ¾-inch-diameter bolts. Use A36 steel.

SOLUTION   Compute the net area:

$$A_n = A_g - \sum (t \times \text{hole diameter})$$

Effective hole diameter $= \dfrac{3}{4} + \dfrac{1}{8} = \dfrac{7}{8}$

■ **FIGURE 3.19**

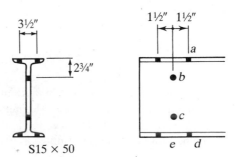

For line *ad*,

$$A_n = 14.7 - 4\left(\frac{7}{8}\right)(0.622) = 12.52 \text{ in.}^2$$

For line *abcd*, the gage distance for use in the $s^2/4g$ term is

$$\frac{g}{2} + g_1 - \frac{t_w}{2} = \frac{3.5}{2} + 2.75 - \frac{0.550}{2} = 4.225 \text{ in.}$$

Treating the holes at *b* and *c* as the staggered holes gives

$$A_n = 14.7 - 4(0.622)\frac{7}{8} - 2(0.550)\left[\frac{7}{8} - \frac{(1.5)^2}{4(4.225)}\right] = 11.71 \text{ in.}^2$$

Line *abcd* controls. As all elements of the cross section are connected,

$$A_e = A_n = 11.71 \text{ in.}^2$$

For the net section,

$$\phi_t P_n = 0.75 F_u A_e = 0.75(58)(11.71) = 509 \text{ kips}$$

For the gross section,

$$\phi_t P_n = 0.90 F_y A_g = 0.90(36)(14.7) = 476 \text{ kips}$$

Yielding of the gross section controls.

ANSWER   Design strength = 476 kips.

3.5   **BLOCK SHEAR**

For certain connection configurations, a segment or "block" of material at the end of the member can tear out. For example, the connection of the single angle tension member shown in Figure 3.20 is susceptible to this phenomenon, called *block shear*. For the case illustrated, the shaded block would tend to fail by shear along the longitudinal section *ab* and tension on the transverse section *bc*. This topic is not covered explicitly in AISC Chapter D ("Tension Members"), but the introductory paragraph directs you to Chapter J ("Connections, Joints, and Fasteners"), Section J4.3 ("Block Shear Rupture Strength").

The procedure is based on the assumption that one of the two failure surfaces fractures and the other yields. That is, fracture on the shear surface is accompanied by yielding on the tension surface, or fracture on the tension surface accompanies yielding on the shear surface. Both surfaces contribute to the total strength, and the resistance to block shear will be the sum of the strengths of the two surfaces.

The nominal strength in tension is $F_u A_{nt}$ for fracture and $F_y A_{gt}$ for yielding, where A_{nt} and A_{gt} are the net and gross areas along the tension surface (*bc* in Figure 3.20). Taking

■ **FIGURE 3.20**

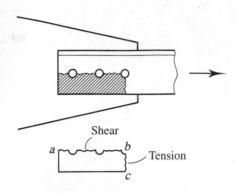

the shear yield stress and ultimate stress as 60% of the values for tension, the nominal strength for shear fracture is $0.6F_uA_{nv}$, and the strength for shear yielding is $0.6F_yA_{gv}$, where A_{nv} and A_{gv} are the net and gross areas along the shear surface (*ab* in Figure 3.20).

There are two possible failure modes. For shear yield and tension fracture, the design strength is

$$\phi R_n = \phi[0.6F_yA_{gv} + F_uA_{nt}]$$ (AISC Equation J4-3a)

For shear fracture and tension yield,

$$\phi R_n = \phi[0.6F_uA_{nv} + F_yA_{gt}]$$ (AISC Equation J4-3b)

In both cases, $\phi = 0.75$. Because the limit state is fracture, the controlling equation will be the one that has the larger fracture term.

■ **EXAMPLE 3.9**

Check the block-shear design strength of the tension member shown in Figure 3.21. The holes are for ⅞-inch-diameter bolts, and A36 steel is used.

SOLUTION   The shear areas are

$$A_{gv} = \frac{3}{8}(7.5) = 2.812 \text{ in.}^2$$

and, since there are 2.5 hole diameters,

$$A_{nv} = \frac{3}{8}[7.5 - 2.5(1.0)] = 1.875 \text{ in.}^2$$

■ **FIGURE 3.21**

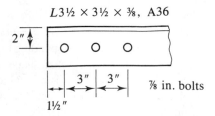

$L3\frac{1}{2} \times 3\frac{1}{2} \times \frac{3}{8}$, A36

7⁄8 in. bolts

The tension areas are

$$A_{gt} = \frac{3}{8}(1.5) = 0.5625 \text{ in.}^2$$

$$A_{nt} = \frac{3}{8}[1.5 - 0.5(1.0)] = 0.375 \text{ in.}^2$$

Using AISC Equation J4-3a yields

$$\phi R_n = \phi[0.6F_y A_{gv} + F_u A_{nt}]$$
$$= 0.75[0.6(36)(2.812) + 58(0.375)]$$
$$= 0.75[60.74 + 21.75] = 61.9 \text{ kips}$$

Using AISC Equation J4-3b yields

$$\phi R_n = \phi[0.6F_u A_{nv} + F_y A_{gt}]$$
$$= 0.75[0.6(58)(1.875) + 36(0.5625)]$$
$$= 0.75[65.25 + 20.25] = 64.1 \text{ kips}$$

The second equation has the larger fracture term (the one involving F_u); therefore the second equation governs.

ANSWER Design strength for block shear = 64.1 kips.

■

3.6 DESIGN OF TENSION MEMBERS

The design of a tension member involves finding a member with adequate gross and net areas. If the member has a bolted connection, the selection of a suitable cross section requires an accounting for the area lost because of holes. For a member with a rectangular cross section, the calculations are relatively straightforward. If a rolled shape is to be used, however, the area to be deduced cannot be predicted in advance because the member's thickness at the location of the holes is not known.

A secondary consideration in the design of tension members is slenderness. If a structural member has a small cross section in relation to its length, it is said to be slender. A more precise measure is the slenderness ratio L/r, where L is the member length and r is the minimum radius of gyration of the cross-sectional area. The minimum radius of gyration is the one corresponding to the minor principal axis of the cross section. This value is tabulated for all rolled shapes in the properties tables in Part 1 of the *Manual*.

Although slenderness is critical to the strength of a compression member, it is inconsequential for a tension member. In many situations, however, it is good practice to limit the slenderness of tension members. If the axial load in a slender tension member is removed and small transverse loads are applied, undesirable vibrations or deflections might occur. These conditions could occur, for example, in a slack bracing rod subjected to wind loads. For this reason, AISC B7 suggests a maximum slenderness ratio of 300. It is only a recommended value because slenderness has no structural significance for tension members, and the limit may be exceeded when special circumstances warrant it. This limit does not apply to cables, and the Specification explicitly excludes rods.

The central problem of all member design, including tension member design, is to find a cross section for which the sum of the factored loads does not exceed the member strength; that is,

$$\sum \gamma Q \leq \phi R_n$$

For tension members, this expression takes the form

$$P_u \leq \phi_t P_n \quad \text{or} \quad \phi_t P_n \geq P_u$$

where P_u is the sum of the factored loads. To prevent yielding,

$$0.90 F_y A_g \geq P_u \quad \text{or} \quad A_g \geq \frac{P_u}{0.90 F_y}$$

To avoid fracture,

$$0.75 F_u A_e \geq P_u \quad \text{or} \quad A_e \geq \frac{P_u}{0.75 F_u}$$

The slenderness ratio limitation will be satisfied if

$$r \geq \frac{L}{300}$$

where r is the minimum radius of gyration of the cross section and L is the member length.

■ EXAMPLE 3.10

A tension member with a length of 5 feet 9 inches must resist a service dead load of 18 kips and a service live load of 52 kips. Select a member with a rectangular cross section. Use A36 steel and assume a connection with one line of ⅞-inch-diameter bolts.

SOLUTION

$$P_u = 1.2D + 1.$$

$$\text{Required } A_g = \frac{P_u}{0.90F_y} = \frac{}{0.90(}$$

$$\text{Required } A_e = \frac{P_u}{0.75F_u} = \frac{104.8}{0.75(58)} = \quad 104.8 \text{ kips}$$

Because $A_e = A_n$ for this member, the gross area corres
is
quired net area

$$A_g = A_n + A_{hole}$$

$$= 2.409 + \left(\frac{7}{8} + \frac{1}{8}\right)t = 2.409 + t$$

Try $t = 1$ in.

$$A_g = 2.409 + 1(1) = 3.409 \text{ in.}^2$$

Because $3.409 > 3.235$, the required gross area is 3.409 in.^{2}, and

$$w_g = \frac{A_g}{t} = \frac{3.409}{1} = 3.409 \text{ in.}$$

Round to the nearest $\frac{1}{8}$ inch and try a $1 \times 3\frac{1}{2}$ cross section. Check the slenderness ratio:

$$I_{min} = \frac{3.5(1)^3}{12} = 0.2917 \text{ in.}^4$$

$$A = 1(3.5) = 3.5 \text{ in.}^2$$

From $I = Ar^2$, we obtain

$$r_{min} = \sqrt{\frac{I_{min}}{A}} = \sqrt{\frac{0.2917}{3.5}} = 0.2887 \text{ in.}$$

$$\text{Maximum } \frac{L}{r} = \frac{5.75(12)}{0.2887} = 239 < 300 \text{ (OK)}$$

ANSWER Use a $3\frac{1}{2} \times 1$ bar.

The member in Example 3.10 is less than 8 inches wide and thus is classified as bar rather than a plate. Bars should be specified to the nearest $\frac{1}{4}$ inch in width and the nearest $\frac{1}{8}$ inch in thickness (the precise classification system is given in Part the *Manual* under the heading "Bars and Plates").

■ **FIGURE 3.23**

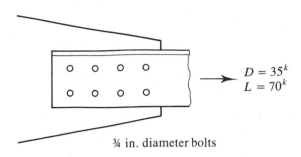

¾ in. diameter bolts

The radius of gyration should be at least

$$\frac{L}{300} = \frac{15(12)}{300} = 0.600 \text{ in.}$$

To find the lightest shape that satisfies these criteria, we search for the unequal-leg angle that has the smallest acceptable gross area and then check the effective net area. The radius of gyration can be checked by inspection. There are two lines of bolts, so the connected leg must be at least 5 inches long (see the table of Usual Gages for Angles, Figure 3.22). We start at the end of the table of properties for single angles in Part 1 of the *Manual* and list several of the lightest shapes in order of increasing area (not the same order in which they appear in the table). The following shapes are possibilities.

L6 × 4 × ½: A_g = 4.75 in.² and r_{min} = 0.870 in.

L5 × 3½ × ⅝: A_g = 4.92 in.² and r_{min} = 0.751 in.

L8 × 4 × ⁷⁄₁₆: A_g = 5.06 in.² and r_{min} = 0.869 in.

L7 × 4 × ½: A_g = 5.25 in.² and r_{min} = 0.872 in.

(Note that for an angle the *x*- and *y*-axes shown in the table are not the principal axes and that $r_{min} = r_z$. For a double-angle shape, however, the *x*- and *y*-axes *will* be the principal axes.)

Try L6 × 4 × ½. This shape has the exact gross area needed (to three significant figures, the precision to which the area is given in the table).

$$A_n = A_g - A_{holes} = 4.75 - 2\left(\frac{7}{8}\right)\left(\frac{1}{2}\right) = 3.875 \text{ in.}^2$$

Because the length of the connection is not known, AISC Eq. B3-2 cannot be used to compute the shear lag factor *U*. Therefore we use *U* = 0.85, the average value from the Commentary. (Example 7.8 in Chapter 7, Simple Connections, shows how known connection details and an average *U* value can be used to obtain a trial shape, after which the value of *U* from AISC Equation B3-2 can be computed.)

$$A_e = UA_n = 0.85(3.875) = 3.29 \text{ in.}^2 < 3.540 \text{ in.}^2 \quad \text{(N.G.)*}$$

Try L5 × 3½ × ⅝

$$A_n = 4.92 - 2\left(\frac{7}{8}\right)\left(\frac{5}{8}\right) = 3.826 \text{ in.}^2$$

$$A_e = 0.85(3.826) = 3.25 \text{ in.}^2 < 3.540 \text{ in.}^2 \quad \text{(N.G.)}$$

Although this shape has more gross area than the previous one, there is no improvement in the net area. The reason is that more area is deducted for the holes because of the thicker leg. Try L8 × 4 × ⁷⁄₁₆.

$$A_n = 5.06 - 2\left(\frac{7}{8}\right)\left(\frac{7}{16}\right) = 4.924 \text{ in.}^2$$

$$A_e = UA_n = 0.85(4.924) = 3.65 \text{ in.}^2 > 3.54 \text{ in.}^2 \quad \text{(OK)}$$

ANSWER   This shape satisfies all requirements, so use an L8 × 4 × ⁷⁄₁₆, connected through the 8-inch leg. ■

When structural shapes or plates are connected to form a built-up shape, they must be connected not only at the ends of the member, but also at intervals along its length. A continuous connection is not required. This type of connection is called *stitching,* and the fasteners used are termed *stitch bolts.* The usual practice is to locate the points of stitching so that L/r for any component part does not exceed L/r for the built-up member. AISC D2 recommends that built-up shapes whose component parts are separated by intermittent fillers be connected at the fillers at intervals such that the maximum L/r for any component does not exceed 300. Built-up shapes consisting of plates or a combination of plates and shapes are addressed in AISC Section J3.5 of Chapter J ("Connections, Joints, and Fasteners"). In general, the spacing of fasteners or welds should not exceed 24 times the thickness of the thinner plate, or 12 inches. If the member is of "weathering" steel subject to atmospheric corrosion, the maximum spacing is 14 times the thickness, or 7 inches.

3.7   THREADED RODS AND CABLES

When slenderness is not a consideration, rods with circular cross sections and cables are often used as tension members. The distinction between the two is that rods are solid and cables are made from individual strands wound together in ropelike fashion. Rods and cables are frequently used in suspended roof systems and as hangers or sus-

---

*The notation N.G. means "No Good."

pension members in bridges. Rods are also used in bracing systems; in some cases, they are pretensioned to prevent them from going slack when external loads are removed. Figure 3.24 illustrates typical rod and cable connection methods.

When the end of a rod is to be threaded, an upset end is sometimes used. This is an enlargement of the end in which the threads are to be cut. Threads reduce the cross-sectional area, and upsetting the end produces a larger gross area to start with. Standard upset ends with threads will actually have more net area in the threaded portion than in the unthreaded part. Upset ends are relatively expensive, however, and in most cases unnecessary.

The effective cross-sectional area in the threaded portion of a rod is called the *stress area* and is a function of the unthreaded diameter and the number of threads per inch. The ratio of stress area to nominal area varies but has a lower bound of approximately 0.75. The nominal tensile strength of the threaded rod can therefore be written as

$$P_n = A_s F_u$$
$$= 0.75 A_b F_u$$

where

A_s = stress area

A_b = nominal (unthreaded) area

This expression gives the nominal strength presented in Table J3.2 in Section J3.6 of the AISC Specification. The resistance factor for this case is $\phi_t = 0.75$.

■ FIGURE 3.24

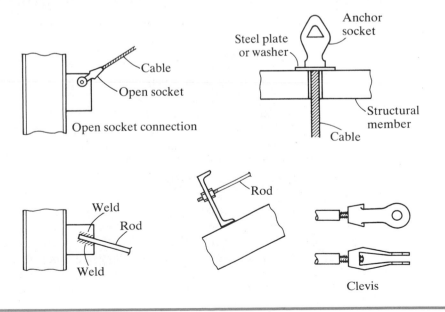

Cable

Open socket

Open socket connection

Anchor socket

Steel plate or washer

Structural member

Cable

Weld

Rod

Weld

Rod

Clevis

If upset ends are used, the tensile capacity at the major thread diameter must be greater than F_y times the unthreaded body area (AISC Table J3.2, footnote c).

■ **EXAMPLE 3.12**

A threaded rod is to be used as a bracing member that must resist a service tensile load of 2 kips dead load and 6 kips live load. What size rod is required if A36 steel is used?

SOLUTION

The factored load is

$$P_u = 1.2(2) + 1.6(6) = 12 \text{ kips}$$

Because $\phi_t P_n \geq P_u$,

$$\phi_t(0.75F_u)A_g \geq P_u$$

$$\text{Required } A_g = \frac{P_u}{\phi_t(0.75)F_u} = \frac{12}{0.75(0.75)(58)} = 0.3678 \text{ in.}^2$$

From $A_g = \dfrac{\pi d^2}{4}$,

$$\text{Required } d = \sqrt{\frac{4(0.3678)}{\pi}} = 0.684 \text{ in.}$$

ANSWER

Use a ¾-inch-diameter threaded rod ($A_g = 0.442$ in.²) ■

To prevent damage during construction, rods should not be too slender. Although there is no specification requirement, a common practice is to use a minimum diameter of ⅝ inch.

Flexible cables, in the form of strands or wire rope, are used in applications where high strength is required and rigidity is unimportant. In addition to their use in bridges and cable roof systems, they are also used in hoists and derricks, as guy lines for towers, and as longitudinal bracing in metal building systems. The difference between strand and wire rope is illustrated in Figure 3.25. A strand consists of individual wires wound helically around a central core, and a wire rope is made of several strands laid helically around a core.

Selection of the correct cable for a given loading is usually based on both strength and deformation considerations. In addition to ordinary elastic elongation, an initial stretching is caused by seating or shifting of the individual wires, which results in a permanent stretch. For this reason, cables are often prestretched. Wire rope and strand are made from steels of much higher strength than structural steels and are not covered by the AISC Specification. The breaking strengths of various cables, as well as details of

■ **FIGURE 3.25**

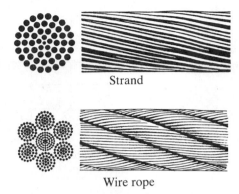

Strand

Wire rope

available fixtures for connections, can be obtained from manufacturers' literature. *Cable Roof Structures* contains useful data of this type (Bethlehem Steel, 1968).

3.8 TENSION MEMBERS IN ROOF TRUSSES

Many of the tension members that structural engineers design are components of trusses. For this reason, some general discussion of roof trusses is in order. A more comprehensive treatment of the subject is given by Lothars (1972).

When trusses are used in buildings, they usually function as the main supporting elements of roof systems where long spans are required. They are used when the cost and weight of a beam would be prohibitive. (A truss may be thought of as a deep beam with much of the web removed.) Roof trusses are often used in industrial or mill buildings, although construction of this type has largely given way to rigid frames. Typical roof construction with trusses supported by load-bearing walls is illustrated in Figure 3.26. In this type of construction, one end of the connection of the truss to the walls usually can be considered as pinned and the other as roller-supported. Thus the truss can be analyzed as an externally statically determinate structure. The supporting walls can be reinforced concrete, concrete block, brick, or a combination of these materials.

Roof trusses normally are spaced uniformly along the length of the building and are tied together by longitudinal beams called *purlins* and by x-bracing. The primary function of the purlins is to transfer loads to the top chord of the truss, but they can also act as part of the bracing system. Bracing is usually provided in the planes of both the top and bottom chords, but it is not required in every bay because lateral forces can be transferred from one braced bay to the other through the purlins.

Ideally, purlins are located at the truss joints so that the truss can be treated as a pin-connected structure loaded only at the joints. Sometimes, however, the roof deck cannot span the distance between joints, and intermediate purlins may be needed. In such

■ **FIGURE 3.26**

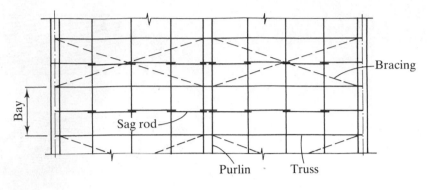

Plan

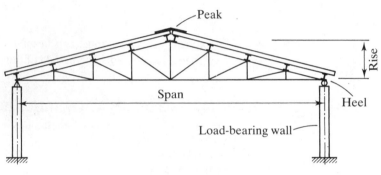

Elevation

cases the top chord will be subjected to significant bending as well as axial compression and must be designed as a beam-column (Chapter 6).

Sag rods are tension members used to provide lateral support for the purlins. Most of the loads applied to the purlins are vertical, so there will be a component parallel to a sloping roof, which will cause the purlin to bend (sag) in that direction (Figure 3.27).

Sag rods can be located at the midpoint, the third points, or at more frequent intervals along the purlins, depending on the amount of support needed. The interval is a function of the truss spacing, the slope of the top chord, the resistance of the purlin to this type of bending (most shapes used for purlins are very weak in this respect), and the amount of support furnished by the roofing. If a metal deck is used, it will usually be rigidly attached to the purlins, and sag rods may not be needed. Sometimes, however, the weight of the purlin itself is enough to cause problems, and sag rods may be needed to provide support during construction before the deck is in place.

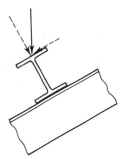

If sag rods are used, they are designed to support the component of roof loads parallel to the roof. Each segment between purlins is assumed to support everything below it; thus the top rod is designed for the load on the roof area tributary to the rod, from the heel of the truss to the peak, as shown in Figure 3.28. Although the force will be different in each segment of rod, the usual practice is to use one size throughout. The extra amount of material in question is insignificant, and the use of the same size for each segment eliminates the possibility of a mix-up during construction.

A possible treatment at the peak or ridge is shown in Figure 3.29a. The tie rod between ridge purlins must resist the load from all of the sag rods on either side. The tensile force in this horizontal member has as one of its components the force in the upper sag-rod segment. A free-body diagram of one ridge purlin illustrates this effect, as shown in Figure 3.29b.

■ **FIGURE 3.28**

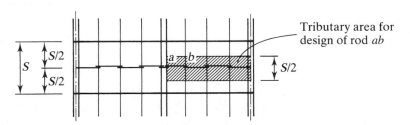

Tributary area for design of rod *ab*

■ **FIGURE 3.29**

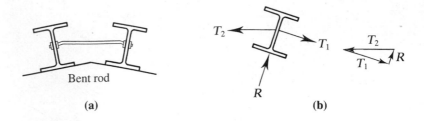

(a)   (b)

■ **EXAMPLE 3.13**

Fink trusses spaced at 20 feet on centers support W6 × 12 purlins, as shown in Figure 3.30a. The purlins are supported at their midpoints by sag rods. Use A36 steel and design the sag rods and the tie rod at the ridge for the following service loads.

■ **FIGURE 3.30**

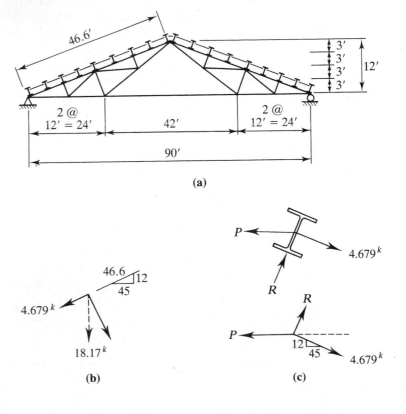

(a)

(b)   (c)

Metal deck: 2 psf

Built-up roof: 5 psf

Snow: 18 psf of horizontal projection of the roof surface

Purlin weight: 12 pounds per foot (lb/ft) of length

SOLUTION   Calculate loads.

Tributary width for each sag rod = 20/2 = 10 ft

Tributary area for deck and built-up roof = 10(46.6) = 466 ft²

Dead load (deck and roof) = (2 + 5)(466) = 3262 lb

Total purlin weight = 12(10)(9) = 1080 lb

Total dead load = 3262 + 1080 = 4342 lb

Tributary area for snow load = 10(45) = 450 ft²

Total snow load = 18(450) = 8100 lb

Check load combinations.

(A4-2): $1.2D + 0.5S = 1.2(4342) + 0.5(8100) = 9260$ lb

(A4-3): $1.2D + 1.6S = 1.2(4342) + 1.6(8100) = 18{,}170$ lb

Combination A4-3 controls. (By inspection, A4-1, A4-4, and A4-5 will not govern.) For the component parallel to the roof (Figure 3.30b),

$$T = (18.17)\frac{12}{46.6} = 4.679 \text{ kips}$$

$$\text{Required } A_g = \frac{T}{\phi_t(0.75F_u)} = \frac{4.679}{0.75(0.75)(58)} = 0.1434 \text{ in.}^2$$

ANSWER   Use a ⅝-inch-diameter threaded rod ($A_g = 0.3068$ in.²).
Tie rod at the ridge (Figure 3.30c):

$$P = (4.679)\frac{46.6}{45} = 4.845 \text{ kips}$$

$$\text{Required } A_g = \frac{4.845}{0.75(0.75)(58)} = 0.1485 \text{ in.}^2$$

ANSWER   Use a ⅝-inch-diameter threaded rod ($A_g = 0.3068$ in.²). ■

For the usual truss geometry and loading, the bottom chord will be in tension, and the top chord will be in compression. Some web members will be in tension and others

■ **FIGURE 3.31**

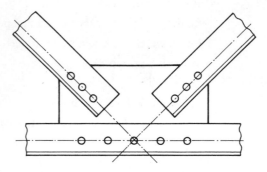

will be in compression. When wind effects are included and consideration is given to different wind directions, the force in some web members may alternate between tension and compression. In this case, the affected member must be designed to function as both a tension member and a compression member.

In bolted trusses, double-angle sections are frequently used for both chord and web members. This design facilitates the connection of members meeting at a joint by permitting the use of a single gusset plate, as illustrated in Figure 3.31. When structural tee-shapes are used as chord members in welded trusses, the web angles can usually be welded to the stem of the tee. If the force in a web member is small, single angles can be used, although doing so eliminates the plane of symmetry from the truss and causes the web member to be eccentrically loaded. Chord members are usually fabricated as continuous pieces and spliced if necessary.

The fact that chord members are continuous and joints are bolted or welded would seem to invalidate the usual assumption that the truss is pin-connected. Joint rigidity does introduce some bending moment into the members, but it is usually small and considered to be a secondary effect. The usual practice is to ignore it. Bending caused by loads directly applied to members between the joints, however, must be taken into account. We consider this condition in Chapter 6, "Beam-Columns."

The *working lines* of the members in a properly detailed truss intersect at the *working point* at each joint. For a bolted truss, the bolt lines are the working lines, and in welded trusses the centroidal axes of the welds are the working lines. An assumption in the usual truss-analysis procedure is that member lengths are measured from working point to working point.

■ **EXAMPLE 3.14**

Select a structural tee for the bottom chord of the Warren roof truss shown in Figure 3.32a. The trusses are welded and spaced at 20 feet. Assume that the bottom chord connection is made with 9-inch-long longitudinal fillet welds at the flange. Use A36 steel and the following load data (wind is not considered in this example).

■ **FIGURE 3.32**

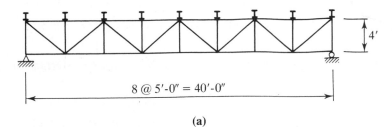

(a)

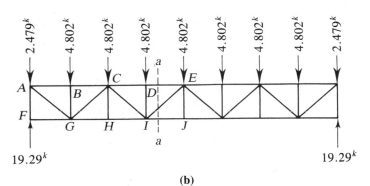

(b)

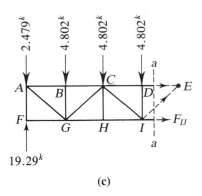

(c)

Purlins: M8 × 6.5

Snow: 20 psf of horizontal projection

Metal deck: 2 psf

Roofing: 4 psf

Insulation: 3 psf

SOLUTION   Calculate loads:

Snow = 20(40)(20) = 16,000 lb

Dead load (exclusive of purlins) =	Deck,	2 psf
	Roof,	4
	Insulation,	3
	Total,	9 psf

Total dead load = 9(40)(20) = 7200 lb

Total purlin weight = 6.5(20)(9) = 1170 lb

Estimate the truss weight as 10% of the other loads:

0.10(16,000 + 7200 + 1170) = 2437 lb

Loads at an interior joint are

$$D = \frac{7200}{8} + \frac{2437}{8} + 6.5(20) = 1335 \text{ lb}$$

$$S = \frac{16,000}{8} = 2000 \text{ lb}$$

At an exterior joint, the tributary roof area is half of that at an interior joint. The corresponding loads are

$$D = \frac{7200}{2(8)} + \frac{2437}{2(8)} + 6.5(20) = 732.3 \text{ lb}$$

$$S = \frac{16,000}{2(8)} = 1000 \text{ lb}$$

Load combination A4-3 will control:

$$P_u = 1.2D + 1.6S$$

At an interior joint,

$$P_u = 1.2(1.335) + 1.6(2.0) = 4.802 \text{ kips}$$

At an exterior joint,

$$P_u = 1.2(0.7323) + 1.6(1.0) = 2.479 \text{ kips}$$

The loaded truss is shown in Figure 3.32b.

The bottom chord is designed by determining the force in each member of the bottom chord and selecting a cross section to resist the largest force. In this example, the force in member *IJ* will control. For the free body left of section *a–a* shown in Figure 3.32c,

$$\sum M_E = 19.29(20) - 2.479(20) - 4.802(15 + 10 + 5) - 4F_{IJ} = 0$$

$$F_{IJ} = 48.02 \text{ kips}$$

For the gross section,

$$\text{Required } A_g = \frac{F_{IJ}}{0.90F_y} = \frac{48.02}{0.90(36)} = 1.48 \text{ in.}^2$$

For the net section,

$$\text{Required } A_e = \frac{F_{IJ}}{0.75F_u} = \frac{48.02}{0.75(58)} = 1.10 \text{ in.}^2$$

Try a WT5 × 6:

$$A_g = 1.77 \text{ in.}^2 > 1.48 \text{ in.}^2 \qquad \text{(OK)}$$

The connection is made with longitudinal welds and the member is not a plate or bar, so this connection is not one of the special cases for welded members subject to shear lag.

$$U = 1 - \left(\frac{\bar{x}}{L}\right) = 1 - \left(\frac{1.36}{9}\right) = 0.849 < 0.9$$

$$A_e = UA_g = 0.849(1.77) = 1.50 \text{ in.}^2 > 1.10 \text{ in.}^2 \qquad \text{(OK)}$$

If we assume that the bottom chord is braced at the panel points,

$$\frac{L}{r} = \frac{5(12)}{0.785} = 76.43 < 300 \qquad \text{(OK)}$$

ANSWER Use a WT5 × 6. ■

3.9   PIN-CONNECTED MEMBERS

When a member is to be pin-connected, a hole is made in both the member and the parts to which it is connected, and a pin is placed through the holes. This provides a connection that is as moment-free as can be fabricated. Tension members connected in this manner are subject to several types of failure, which are covered in AISC D3 and discussed in the following paragraphs.

The eyebar is a special type of pin-connected member in which the end containing the pin hole is enlarged, as shown in Figure 3.33. The design strength is based on yielding of the gross section. Detailed rules for proportioning eyebars are given in

■ **FIGURE 3.33**

AISC D3 and are not repeated here. These requirements are based on experience and test programs for forged eyebars, but they are conservative when applied to eyebars thermally cut from plates (the present fabrication method). Eyebars were widely used in the past as single tension members in bridge trusses or were linked in chainlike fashion in suspension bridges. They are rarely used today.

Pin-connected members should be designed for the following limit states (see Figure 3.34).

1. **Tension** on the net effective area (Figure 3.34a):

$$\phi_t = 0.75, \qquad P_n = 2tb_{eff}F_u \qquad \text{(AISC Equation D3-1)}$$

2. **Shear** on the effective area (Figure 3.34b):

$$\phi_{sf} = 0.75, \qquad P_n = 0.6A_{sf}F_u \qquad \text{(AISC Equation D3-2)}$$

3. **Bearing.** This requirement is given in Chapter J ("Connections, Joints, and Fasteners") (Figure 3.34c):

$$\phi = 0.75, \qquad P_n = 1.8F_yA_{pb} \qquad \text{(AISC Equation J8-1)}$$

4. **Tension** on the gross section:

$$\phi = 0.90, \qquad P_n = F_yA_g \qquad \text{(AISC Equation D1-1)}$$

where

t = thickness of connected part

$b_{eff} = 2t + 0.63 \leq b$

b = distance from edge of pin hole to edge of member, perpendicular to direction of force

$A_{sf} = 2t(a + d/2)$

a = distance from edge of pin hole to edge of member, parallel to direction of force

d = pin diameter

A_{pb} = projected bearing area = dt

Additional requirements for the relative proportions of the pin and the member are covered in AISC D3.

■ **FIGURE 3.34**

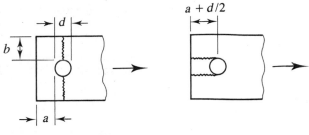

(a) Fracture of net section   (b) Longitudinal shear

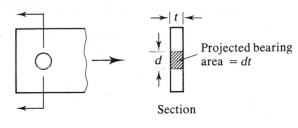

(c) Bearing

■ **PROBLEMS**

Design Strength

3.2-1   A bar 7 × ⅜ tension member is connected with three 1-inch-diameter bolts, as shown in Figure P3.2-1. The steel is A36. Assume that $A_e = A_n$ and compute the design strength.

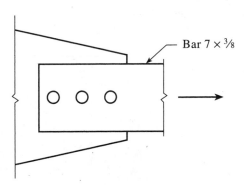

FIGURE P3.2-1

3.2-2   A bar $6 \times \frac{3}{8}$ tension member is welded to a gusset plate, as shown in Figure P3.2-2. The steel has a yield stress $F_y = 50$ ksi and an ultimate tensile stress $F_u = 65$ ksi. Assume that $A_e = A_g$ and compute the design strength.

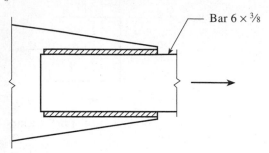

Bar $6 \times \frac{3}{8}$

FIGURE P3.2-2

3.2-3   A bar $8 \times \frac{1}{2}$ tension member is connected with six 1-inch-diameter bolts, as shown in Figure P3.2-3. The steel is A242 Grade 42. Assume that $A_e = A_n$ and compute the design strength.

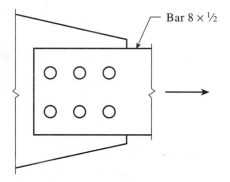

Bar $8 \times \frac{1}{2}$

FIGURE P3.2-3

3.2-4   The tension member shown in Figure P3.2-4 must resist a service dead load of 25 kips and a service live load of 45 kips. Does the member have enough strength? The steel is A588 and the bolts are $1\frac{1}{8}$ inches in diameter. Assume that $A_e = A_n$.

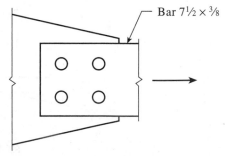

Bar $7\frac{1}{2} \times \frac{3}{8}$

FIGURE P3.2-4

Effective Net Area

3.3-1 Compute the effective net area A_e for each case shown in Figure P3.3-1.

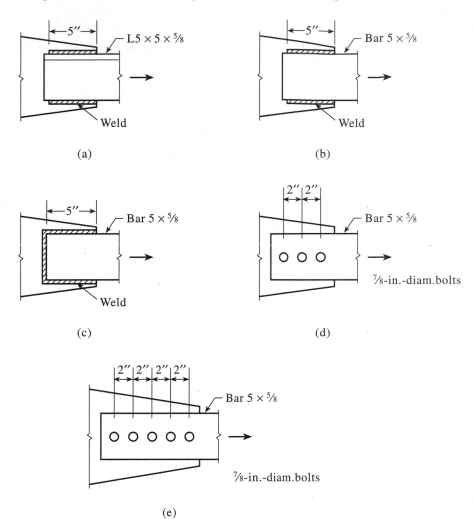

(a)

(b)

(c)

(d)

(e)

FIGURE P3.3-1

3.3-2 A single-angle tension member is connected to a gusset plate, as shown in Figure P3.3-2. The yield stress is $F_y = 50$ ksi and the ultimate tensile stress is $F_u = 70$ ksi. The bolts are ⅞ inch in diameter.

    a. Determine the design strength. Use AISC Equation B3-2 for U.

    b. Repeat part (a), but use the average value of U from the Commentary.

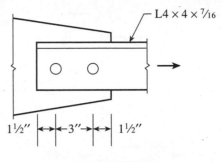

FIGURE P3.3-2

3.3-3 An L4 × 3 × ⅜ tension member is welded to a gusset plate, as shown in Figure P3.3-3. A36 steel is used.

    a. Determine the design strength. Use AISC Equation B3-2 for U.

    b. Repeat part (a), but use the average value of U from the Commentary.

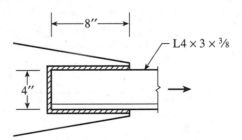

FIGURE P3.3-3

3.3-4 An L5 × 5 × ½ tension member of A242 steel is connected to a gusset plate with six ¾-inch-diameter bolts, as shown in Figure P3.3-4.

    a. Use the average value of U from the Commentary and compute the design strength.

    b. If the member is subjected to dead load and live load only, what is the maximum total service load that can be applied if the ratio of live load to dead load is 2.0?

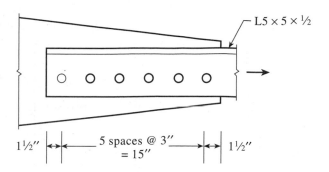

FIGURE P3.3-4

3.3-5 An L6 × 4 × ⅝ tension member of A36 steel is connected to a gusset plate with 1-inch-diameter bolts, as shown in Figure P3.3-5. It is subjected to the following service loads: dead load = 50 kips, live load = 100 kips, and wind load = 45 kips. Use AISC Equation B3-2 for U and determine whether this member is adequate.

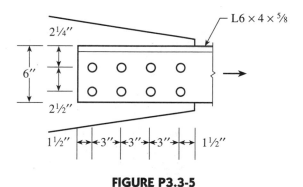

FIGURE P3.3-5

3.3-6 A bar 5 × ¼ is used as a tension member and is connected by a pair of longitudinal welds along its edges. The welds are each 7 inches long. A36 steel is used. What is the design strength?

3.3-7 A W12 × 35 of A36 steel is connected through its flanges with ⅞-inch-diameter bolts, as shown in Figure P3.3-7. Use the average value of U from the Commentary and compute the design tensile strength.

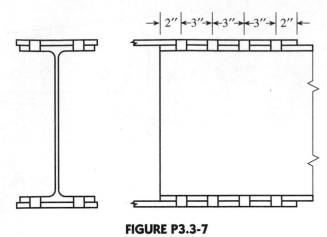

FIGURE P3.3-7

3.3-8   A WT6 × 17.5 is welded to a plate as shown in Figure P3.3-8. F_y = 50 ksi and F_u = 70 ksi.

a. Use AISC Equation B3-2 for U and compute the design tensile strength.

b. Determine whether the member can resist the following service loads: D = 75 kips, L_r = 40 kips, S = 50 kips, and W = 70 kips.

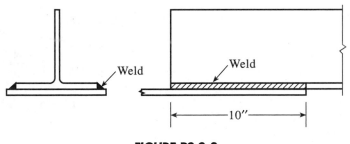

FIGURE P3.3-8

Staggered Fasteners

3.4-1   The tension member shown in Figure P3.4-1 is a ½ × 10 plate of A36 steel. The connection is with ⅞-inch-diameter bolts. Compute the design strength.

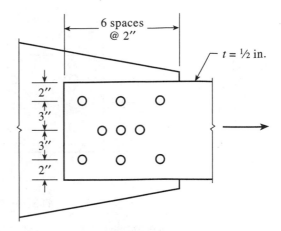

FIGURE P3.4-1

3.4-2 A tension member is composed of two $\frac{1}{2} \times 10$ plates. They are connected to a gusset plate with the gusset plate between the two tension member plates, as shown in Figure P3.4-2. A36 steel and $\frac{3}{4}$-inch-diameter bolts are used. Determine the design strength.

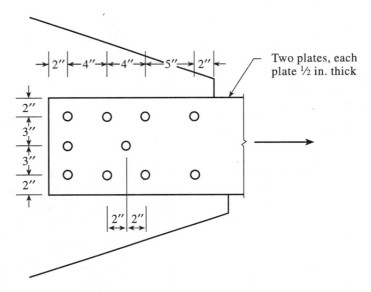

FIGURE P3.4-2

3.4-3 Compute the design strength of the tension member shown in Figure P3.4-3. The bolts are $\frac{1}{2}$ inch in diameter. A36 steel is used.

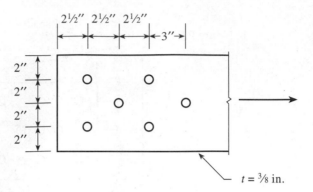

FIGURE P3.4-3

3.4-4 A C9 × 20 tension member is connected with $1\frac{1}{8}$-inch-diameter bolts, as shown in Figure P3.4-4. $F_y = 50$ ksi and $F_u = 70$ ksi. The member is subjected to the following service loads: dead load = 36 kips and live load = 110 kips. Use an average value of U from the Commentary and determine whether the member has enough strength.

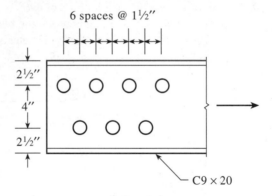

FIGURE P3.4-4

3.4-5 A double-angle shape, 2L7 × 4 × ⅜, is used as a tension member. The two angles are connected to a gusset plate with ⅞-inch-diameter bolts through the 7-inch legs, as shown in Figure P3.4-5. A572 Grade 50 steel is used. Use an average value of U from the Commentary and compute the design strength.

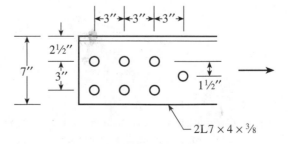

FIGURE P3.4-5

3.4-6 An L4 × 4 × $\frac{7}{16}$ tension member is connected with ¾-inch-diameter bolts, as shown in Figure P3.4-6. Both legs of the angle are connected. If A36 steel is used, what is the design strength?

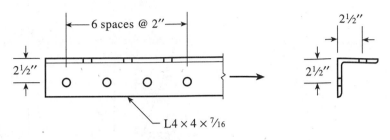

FIGURE P3.4-6

Block Shear

3.5-1 Compute the block shear strength of the tension member shown in Figure P3.5-1. ASTM A572 Grade 50 steel is used. The bolts are $\frac{7}{8}$ inch in diameter.

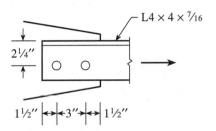

FIGURE P3.5-1

3.5-2 Determine the block shear strength of the tension member shown in Figure P3.5-2. The bolts are 1 inch in diameter, and A36 steel is used.

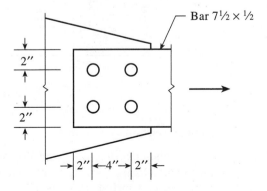

FIGURE P3.5-2

3.5-3   Determine the block shear strength (consider both the tension member and the gusset plate) of the connection shown in Figure P3.5-3. The bolts are ¾ inch in diameter, and A36 steel is used for all components.

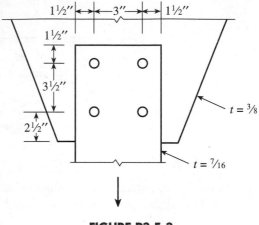

FIGURE P3.5-3

3.5-4   Compute the maximum factored load that can be applied to the connection shown in Figure P3.5-4. Consider all limit states. Use an average value of U from the Commentary. ASTM A572 Grade 50 steel is used for the tension member, the gusset plate is A36 steel, and the holes are for ¾-inch bolts.

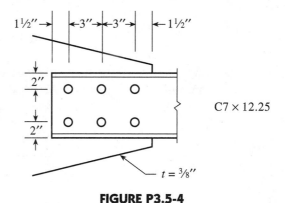

FIGURE P3.5-4

Design of Tension Members

3.6-1   Select a single-angle tension member of A36 steel to resist a dead load of 28 kips and a live load of 84 kips. The length of the member is 18 feet, and it will be connected with a single line of 1-inch-diameter bolts, as shown in Figure P3.6-1. There will be more than two bolts in this line.

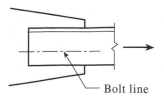

FIGURE P3.6-1

3.6-2   Select the lightest American Standard Channel that will safely support a *factored* tensile load of 200 kips. The member is 20 feet long, and there is one line of three 1-inch-diameter bolts in each flange at the connection. Use A36 steel.

3.6-3   Select a double-angle tension member to resist a *factored* load of 180 kips. The member will be connected with two lines of ⅞-inch-diameter bolts placed at the usual gage distances (see Figure 3.22), as shown in Figure P3.6-3. There will be more than two bolts in each line. The member is 25 feet long and will be connected to a ⅜-inch-thick gusset plate. Use A572 Grade 50 steel.

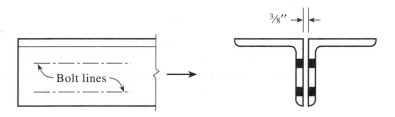

FIGURE P3.6-3

3.6-4   Select an American Standard Channel for the following tensile loads: dead load = 54 kips, live load = 80 kips, and wind load = 75 kips. The connection will be with longitudinal welds. The length is 17.5 feet. Use F_y = 50 ksi and F_u = 65 ksi.

3.6-5   Select an American Standard Channel to resist a *factored* tensile load of 180 kips. The length is 15 feet, and there will be two lines of ⅞-inch-diameter bolts in the web, as shown in Figure P3.6-5. There will be more than two bolts in each line. Use A36 steel.

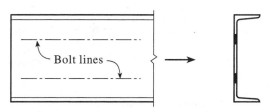

FIGURE P3.6-5

3.6-6 Select a W shape with a nominal depth of 10 inches (a W10) to resist a dead load of 175 kips and a live load of 175 kips. The connection will be through the flanges with two lines of 1¼-inch-diameter bolts in each flange, as shown in Figure P3.6-6. Each line contains more than two bolts. The length of the member is 30 feet. Use A242 steel.

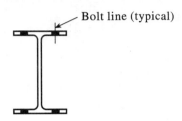

FIGURE P3.6-6

Threaded Rods and Cables

3.7-1 Select a threaded rod to resist a service dead load of 45 kips and a service live load of 5 kips. Use A36 steel.

3.7-2 A W14 × 48 is supported by two tension rods *AB* and *CD*, as shown in Figure P3.7-2. The 20-kip load is a service live load. Use A36 steel and select threaded rods for the following load cases.

    a. The 20-kip load cannot move from the location shown.

    b. The 20-kip load can be located anywhere between the two rods.

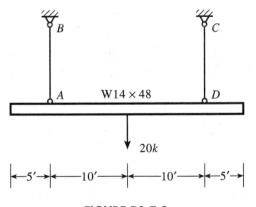

FIGURE P3.7-2

3.7-3 As shown in Figure P3.7-3, member *AB* is used to brace the pin-connected structure against the horizontal load. Select a threaded rod of A36 steel. The 10-kip load is factored.

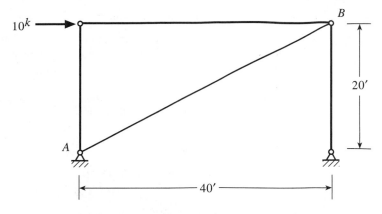

FIGURE P3.7-3

3.7-4 What size threaded rod is required for member *AB,* as shown in Figure P3.7-4? The load is a service live load. (Neglect the weight of member *CB.*) Use A36 steel.

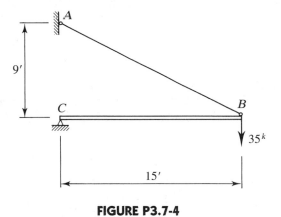

FIGURE P3.7-4

3.7-5 A pipe is supported at 10-foot intervals by a bent, threaded rod, as shown in Figure P3.7-5. If a 10-inch-diameter standard weight steel pipe full of water is used, what size rod is required? Use A36 steel.

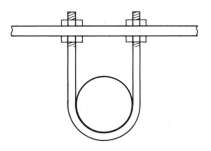

FIGURE P3.7-5

Tension Members in Roof Trusses

3.8-1 Use A36 steel and select a structural tee for the top chord of the welded roof truss shown in Figure P3.8-1. All connections are made with longitudinal plus transverse welds. The spacing of trusses in the roof system is 12 feet 6 inches. Design for the following loads.

Snow: 20 psf of horizontal projection

Roofing: 12 psf

MC8 × 8.5 purlins

Truss weight: 1000 lb (estimated)

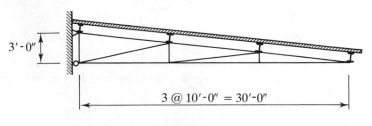

FIGURE P3.8-1

3.8-2 Select single-angle shapes for the web tension members of the truss loaded as shown in Figure P3.8-2. The loads are *factored loads*. All connections are with longitudinal welds. Use A572 Grade 50 steel.

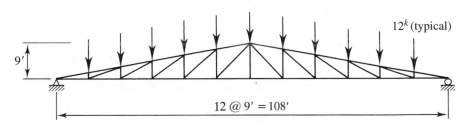

FIGURE P3.8-2

3.8-3 Compute the factored joint loads for the truss of Problem 3.8-2 for the following conditions.

Trusses spaced at 15 ft

Weight of roofing = 12 psf

Snow load = 18 psf of horizontal projection

W10 × 33 purlins located only at the joints

Total estimated truss weight = 5000 lb

3.8-4   Design the tension members of the roof truss shown in Figure P3.8-4. Use double-angle shapes throughout and assume ⅜-inch-thick gusset plates and welded connections (either longitudinal or longitudinal plus transverse). The trusses are spaced at 25 feet. Use A572 Grade 50 steel and design for the following loads.

Metal deck: 4 psf of roof surface

Built-up roof: 12 psf of roof surface

Purlins: 6 psf of roof surface (estimated)

Snow: 18 psf of horizontal projection

Truss weight: 5 psf of horizontal projection (estimated)

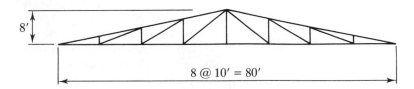

FIGURE P3.8-4

3.8-5   Design sag rods for the truss of Problem 3.8-4. Assume that, once attached, the metal deck will provide lateral support for the purlins; therefore the sag rods need to be designed for the purlin weight only. Use A36 steel.

4 Compression Members

4.1 DEFINITION

Compression members are structural elements that are subjected only to axial compressive forces; that is, the loads are applied along a longitudinal axis through the centroid of the member cross section, and the stress can be taken as $f_a = P/A$, where f_a is considered to be uniform over the entire cross section. This ideal state is never achieved in reality, however, and some eccentricity of the load is inevitable. Bending will result, but it can usually be regarded as secondary and can be neglected if the theoretical loading condition is closely approximated. Bending cannot be neglected if there is a *computed* bending moment. We consider situations of this type in Chapter 6, "Beam-Columns."

The most common type of compression member occurring in buildings and bridges is the *column,* a vertical member whose primary function is to support vertical loads. In many instances these members are also called upon to resist bending, and in these cases the member is a *beam-column*. Compression members are also used in trusses and as components of bracing systems. Smaller compression members not classified as columns are sometimes referred to as *struts*.

4.2 COLUMN THEORY

Consider the long, slender compression member shown in Figure 4.1a. If the axial load P is slowly applied, it will ultimately become large enough to cause the member to become unstable and assume the shape indicated by the dashed line. The member is said to have buckled, and the corresponding load is called the *critical buckling load*. If the member is stockier, as shown in Figure 4.1b, a larger load will be required to bring the member to the point of instability. For extremely stocky members, failure may occur by compressive yielding rather than buckling. Prior to failure, the compressive stress P/A will be uniform over the cross section at any point along the length, whether the failure is by yielding or by buckling. The load at which buckling occurs is a function of slenderness, and for very slender members this load could be quite small.

If the member is so slender (we give a precise definition of slenderness shortly) that the stress just before buckling is below the proportional limit — that is, the member is still elastic — the critical buckling load is given by

$$P_{cr} = \frac{\pi^2 EI}{L^2} \tag{4.1}$$

86

■ FIGURE 4.1

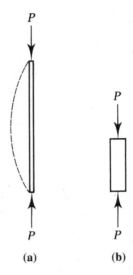

where E is the modulus of elasticity of the material, I is the moment of inertia of the cross-sectional area with respect to the minor principal axis, and L is the length of the member between points of support. For Equation 4.1 to be valid, the member must be elastic, and its ends must be free to rotate but not translate laterally. This end condition is satisfied by hinges or pins, as shown in Figure 4.2. This remarkable relationship was first formulated by Swiss mathematician Leonhard Euler and published in 1759. The critical

■ FIGURE 4.2

load is sometimes referred to as the *Euler load* or the *Euler buckling load*. The validity of Equation 4.1 has been demonstrated convincingly by numerous tests. Its derivation has become somewhat of a standard exercise in textbooks on differential equations and is given here to illustrate the importance of the end conditions.

For convenience, the member will be oriented with its longitudinal axis along the *x*-axis of the coordinate system given in Figure 4.3. The roller support is to be interpreted as restraining the member from translating either up or down. An axial compressive load is applied and gradually increased. If a temporary transverse load is applied so as to deflect the member into the shape indicated by the dashed line, the member will return to its original position when this temporary load is removed if the axial load is less than the critical buckling load. The critical buckling load, P_{cr}, is defined as the load that is just large enough to maintain the deflected shape when the temporary transverse load is removed.

The differential equation giving the deflected shape of an elastic member subjected to bending is

$$\frac{d^2y}{dx^2} = -\frac{M}{EI} \tag{4.2}$$

where *x* locates a point along the longitudinal axis of the member, *y* is the deflection of the axis at that point, and *M* is the bending moment at the point. *E* and *I* were previously defined, but here the moment of inertia *I* is with respect to the axis of bending (buckling). This equation was derived by Jacob Bernoulli and independently by Euler, who specialized it for the column buckling problem (Timoshenko, 1953). If we begin at the point of buckling, then from Figure 4.3 the bending moment is $P_{cr}y$. Equation 4.2 can then be written as

$$y'' + \frac{P_{cr}}{EI}\,y = 0$$

■ **FIGURE 4.3**

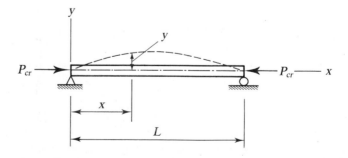

where the prime denotes differentiation with respect to x. This is a second-order, linear, ordinary differential equation with constant coefficients and has the solution

$$y = A \cos(cx) + B \sin(cx)$$

where

$$c = \sqrt{\frac{P_{cr}}{EI}}$$

and A and B are constants. These constants are evaluated by applying the following boundary conditions:

At $x = 0$, $y = 0$: $\quad 0 = A \cos(0) + B \sin(0) \quad\quad A = 0$

At $x = L$, $y = 0$: $\quad 0 = B \sin(cL)$

This last condition requires that $\sin(cL)$ be zero if B is not to be zero (the trivial solution, corresponding to $P = 0$). For $\sin(cL) = 0$,

$$cL = 0, \pi, 2\pi, 3\pi, \ldots = n\pi, \quad n = 0, 1, 2, 3, \ldots$$

From

$$c = \sqrt{\frac{P_{cr}}{EI}}$$

we obtain

$$cL = \left(\sqrt{\frac{P_{cr}}{EI}}\right) L = n\pi, \quad \frac{P_{cr}}{EI} L^2 = n^2 \pi^2 \quad \text{and} \quad P_{cr} = \frac{n^2 \pi^2 EI}{L^2}$$

The various values of n correspond to different buckling modes; $n = 1$ represents the first mode, $n = 2$ the second, and so on. A value of zero gives the trivial case of no load. These buckling modes are illustrated in Figure 4.4. Values of n larger than 1 are not possible unless the compression member is physically restrained from deflecting at the points where the reversal of curvature would occur.

The solution to the differential equation is therefore

$$y = B \sin\left(\frac{n \pi x}{L}\right)$$

and the coefficient B is indeterminate. This result is a consequence of approximations made in formulating the differential equation; a linear representation of a nonlinear phenomenon was used.

For the usual case of a compression member with no supports between its ends, $n = 1$ and the Euler equation is written as

$$P_{cr} = \frac{\pi^2 EI}{L^2} \tag{4.3}$$

■ **FIGURE 4.4**

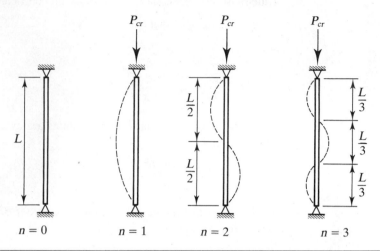

It is convenient to rewrite Equation 4.3 as

$$P_{cr} = \frac{\pi^2 EI}{L^2} = \frac{\pi^2 EAr^2}{L^2} = \frac{\pi^2 EA}{(L/r)^2}$$

where A is the cross-sectional area and r is the radius of gyration with respect to the axis of buckling. The ratio L/r is the slenderness ratio and is the measure of a member's slenderness, with large values corresponding to slender members.

If the critical load is divided by the cross-sectional area, the critical buckling stress is obtained:

$$F_{cr} = \frac{P_{cr}}{A} = \frac{\pi^2 E}{(L/r)^2} \tag{4.4}$$

At this compressive stress, buckling will occur about the axis corresponding to r. Buckling will take place as soon as the load reaches the value given by Equation 4.3, and the column will become unstable about the principal axis corresponding to the largest slenderness ratio. It usually is the axis with the smaller moment of inertia (we examine exceptions to this condition later). Thus the minimum moment of inertia and radius of gyration of the cross section should be used in Equations 4.3 and 4.4.

■ **EXAMPLE 4.1**

A W12 × 50 is used as a column to support an axial compressive load of 145 kips. The length is 20 feet, and the ends are pinned. Without regard to load or resistance factors, investigate this member for stability. (The grade of steel need not be known: The critical buckling load is a function of the modulus of elasticity, not the yield stress or ultimate tensile strength.)

SOLUTION For a W12 × 50,

$$\text{Minimum } r = r_y = 1.96 \text{ in.}$$

$$\text{Maximum } \frac{L}{r} = \frac{20(12)}{1.96} = 122.4$$

$$P_{cr} = \frac{\pi^2 EA}{(L/r)^2} = \frac{\pi^2 (29,000)(14.7)}{(122.4)^2} = 280.8 \text{ kips}$$

ANSWER Because the applied load of 145 kips is less than P_{cr}, the column remains stable and has an overall factor of safety against buckling of 280.8/145 = 1.94. ■

Early researchers soon found that Euler's equation did not give reliable results for stocky, or less slender, compression members. The reason is that the small slenderness ratio for members of this type causes a large buckling stress (from Equation 4.4). If the stress at which buckling occurs is greater than the proportional limit of the material, the relation between stress and strain is not linear, and the modulus of elasticity E can no longer be used. (In Example 4.1, the stress at buckling is $P_{cr}/A = 280.8/14.7 = 19.10$ ksi, which is well below the proportional limit for any grade of structural steel.) This difficulty was initially resolved by Friedrich Engesser, who proposed in 1889 the use of a variable tangent modulus, E_t, in Equation 4.3. For a material with a stress–strain curve like the one shown in Figure 4.5, E is not a constant for stresses greater than the proportional limit F_{pl}. The tangent modulus E_t is defined as the slope of the tangent to the stress–strain curve for values of f between F_{pl} and F_y. If the compressive stress at buckling, P_{cr}/A, is in this region, it can be shown that

$$P_{cr} = \frac{\pi^2 E_t I}{L^2} \tag{4.5}$$

Equation 4.5 is identical to the Euler equation, except that E_t is substituted for E.

The stress–strain curve shown in Figure 4.5 is different from those shown earlier for ductile steel (in Figures 1.3 and 1.4) because it has a pronounced region of nonlinearity. This curve is typical of a compression test of a short length of W-shape called a *stub column,* rather than the result of testing a tensile specimen. The nonlinearity is primarily the result of the presence of residual stresses in the W-shape. When a hot-rolled shape cools after rolling, all elements of the cross section do not cool at the same rate. The tips of the flanges, for example, cool faster than the junction of the flange and the web. This uneven cooling induces stresses that remain permanently. Other factors, such as welding and cold-bending to create curvature in a beam, can contribute to the residual stress, but the cooling process is its chief source.

Note that E_t is smaller than E, and for the same L/r corresponds to a smaller critical load, P_{cr}. Because of the variability of E_t, computation of P_{cr} in the inelastic range by the use of Equation 4.5 is difficult. In general, a trial-and-error approach must be used, and a compressive stress–strain curve such as the one shown in Figure 4.5 must be used to determine E_t for trial values of P_{cr}. For this reason, most design specifications, including the AISC Specification, contain empirical formulas for inelastic columns.

Engesser's tangent modulus theory had its detractors, who pointed out several inconsistencies. Engesser was convinced by their arguments, and in 1895 he refined his theory to incorporate a reduced modulus, which has a value between E and E_t. Test results, however, always agreed more closely with the tangent modulus theory. Shanley (1947) resolved the apparent inconsistencies in the original theory, and today the tangent modulus formula, Equation 4.5, is accepted as the correct one for inelastic buckling. Although the load predicted by this equation is actually a lower bound on the true value of the critical load, the difference is slight (Bleich, 1952).

For any material, the critical buckling stress can be plotted as a function of slenderness, as shown in Figure 4.6. The tangent modulus curve is tangent to the Euler curve at the point corresponding to the proportional limit of the material. The composite curve, called a *column strength curve,* completely describes the stability of any column of a given material. Other than F_y, E, and E_t, which are properties of the material, the strength is a function only of the slenderness ratio.

■ **FIGURE 4.5**

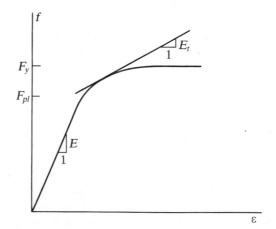

■ **FIGURE 4.6**

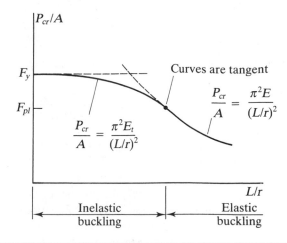

Effective Length

Both the Euler and tangent modulus equations are based on the following assumptions.

1. The column is perfectly straight, with no initial crookedness.
2. The load is axial, with no eccentricity.
3. The column is pinned at both ends.

The first two conditions mean that there is no bending moment in the member before buckling. As mentioned previously, some accidental moment will be present, but in most cases it can be neglected. The requirement for pinned ends, however, is a serious limitation, and provisions must be made for other support conditions. The pinned-end condition requires that the member be restrained from lateral translation, but not rotation, at the ends. Constructing a frictionless pin connection is virtually impossible, so even this support condition can only be closely approximated at best. Obviously, all columns must be free to deform axially.

Other end conditions can be accounted for in the derivation of Equation 4.3. In general, the bending moment will be a function of x, resulting in a nonhomogeneous differential equation. The boundary conditions will be different from those in the original derivation, but the overall procedure will be the same. The form of the resulting equation for P_{cr} will also be the same. For example, consider a compression member pinned at one end and fixed against rotation and translation at the other, as shown in

Figure 4.7. The Euler equation for this case, derived in the same manner as Equation 4.3, is

$$P_{cr} = \frac{2.05\pi^2 EI}{L^2}$$

or

$$P_{cr} = \frac{2.05\pi^2 EA}{(L/r)^2} = \frac{\pi^2 EA}{(0.70L/r)^2}$$

Thus this compression member has the same load capacity as a column that is pinned at both ends and is only 70% as long as the given column. Similar expressions can be found for columns with other end conditions.

The column-buckling problem can also be formulated in terms of a fourth-order differential equation instead of Equation 4.2. This proves to be convenient when dealing with boundary conditions other than pinned ends.

For convenience, the equations for critical buckling load will be written as

$$P_{cr} = \frac{\pi^2 EA}{(KL/r)^2} \quad \text{or} \quad P_{cr} = \frac{\pi^2 E_t A}{(KL/r)^2} \qquad (4.6a/4.6b)$$

■ **FIGURE 4.7**

where KL is the *effective length,* and K is the *effective length factor.* The effective length factor for the fixed-pinned compression member is 0.70. For the most favorable condition of both ends fixed against rotation and translation, $K = 0.5$. Values of K for these and other cases can be determined with the aid of Table C-C2.1 in the Commentary to the AISC Specification. The three conditions mentioned thus far are included, as well as some for which end translation is possible. Two values of K are given: a theoretical value and a recommended design value to be used when the ideal end condition is approximated. Hence, unless a "fixed" end is perfectly fixed, the more conservative design values are to be used. Only under the most extraordinary circumstances would the use of the theoretical values be justified. Note, however, that the theoretical and recommended design values are the same for conditions (d) and (f) in Commentary Table C-C2.1. The reason is that any deviation from a perfectly frictionless hinge or pin introduces rotational restraint and tends to reduce K. Therefore use of the theoretical values in these two cases is conservative.

The use of the effective length KL in place of the actual length L in no way alters any of the relationships discussed so far. The column strength curve shown in Figure 4.6 is unchanged except for renaming the abscissa KL. The critical buckling stress corresponding to a given length, actual or effective, remains the same.

4.3   AISC REQUIREMENTS

The basic requirements for compression members are covered in Chapter E of the AISC Specification. The relationship between loads and strength (Equation 2.3) takes the form

$$P_u \leq \phi_c P_n$$

where

P_u = sum of factored loads

P_n = nominal compressive strength = $A_g F_{cr}$

F_{cr} = critical buckling stress

ϕ_c = resistance factor for compression members = 0.85

Instead of expressing the critical buckling stress F_{cr} as a function of the slenderness ratio KL/r, the Specification uses the slenderness parameter

$$\lambda_c = \frac{KL}{r\pi} \sqrt{\frac{F_y}{E}} \qquad\qquad \text{(AISC Equation E2-4)}$$

which incorporates the material properties but is nondimensional. For elastic columns, Equation 4.4 can be written as

$$F_{cr} = \frac{\pi^2 E}{(KL/r)^2} = \frac{1}{\lambda_c^2} F_y$$

To account for the effects of initial crookedness, this value is reduced as follows:

$$F_{cr} = \frac{0.877}{\lambda_c^2} F_y$$

For inelastic columns, the tangent modulus equation, Equation 4.6b, is replaced by

$$F_{cr} = (0.658^{\lambda_c^2})F_y$$

which also accounts for initial crookedness. Thus a direct solution can be obtained, avoiding the trial-and-error approach inherent in the use of the tangent modulus equation. If the boundary between elastic and inelastic columns is taken as $\lambda_c = 1.5$, the AISC equations for critical buckling stress can be summarized as follows.

For $\lambda_c \leq 1.5$,

$$F_{cr} = (0.658^{\lambda_c^2})F_y \qquad \text{(AISC Equation E2-2)}$$

For $\lambda_c > 1.5$,

$$F_{cr} = \frac{0.877}{\lambda_c^2} F_y \qquad \text{(AISC Equation E2-3)}$$

These requirements are represented graphically in Figure 4.8.

AISC Equations E2-2 and E2-3 are a condensed version of five equations that cover five ranges of λ_c (Galambos, 1988). These equations are based on experimental and theoretical studies that account for the effects of residual stresses and an initial out-of-straightness of $L/1500$, where L is the member length.

■ **FIGURE 4.8**

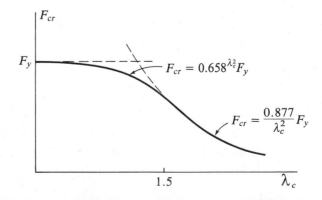

AISC B7 recommends a maximum slenderness ratio KL/r of 200 for compression members. Though only a suggested limit, it is a practical upper limit because columns that are any more slender will have little strength and not be economical.

■ **EXAMPLE 4.2**

Compute the design compressive strength of a W14 × 74 with a length of 20 feet and pinned ends. A36 steel is used.

SOLUTION   Slenderness ratio:

$$\text{Maximum } \frac{KL}{r} = \frac{KL}{r_y} = \frac{1.0(20 \times 12)}{2.48} = 96.77 < 200 \qquad (\text{OK})$$

$$\lambda_c = \frac{KL}{r\pi} \sqrt{\frac{F_y}{E}} = \frac{96.77}{\pi} \sqrt{\frac{36}{29,000}} = 1.085$$

$$\lambda_c < 1.5 \qquad \therefore F_{cr} = (0.658)^{\lambda_c^2} F_y$$

$$= (0.658)^{(1.085)^2} (36) = 21.99 \text{ ksi}$$

$$P_n = A_g F_{cr} = 21.8(21.99) = 479.5 \text{ kips}$$

$$\phi_c P_n = 0.85(479.5) = 408 \text{ kips}$$

ANSWER   Design compressive strength = 408 kips.   ■

In Example 4.2, $r_y < r_x$, and there is excess strength in the x-direction. Square structural tubes are efficient shapes for compression members because $r_y = r_x$ and the strength is the same for both axes. Hollow circular shapes are sometimes used as compression members for the same reason.

The mode of failure considered so far is referred to as *flexural* buckling, as the member is subjected to flexure, or bending, when it becomes unstable. For some cross-sectional configurations, the member will fail by twisting (torsional buckling) or by a combination of twisting and bending (flexural-torsional buckling). We consider these infrequent cases in Section 4.6.

Local Stability

The strength corresponding to any buckling mode cannot be developed, however, if the elements of the cross section are so thin that *local* buckling occurs. This type of instability is a localized buckling or wrinkling at an isolated location. If it occurs, the cross section is no longer fully effective, and the member has failed. I- and H-shaped cross sections with thin flanges or webs are susceptible to this phenomenon, and their use should be avoided whenever possible. Otherwise, the compressive strength given by AISC Equations E2-2 and E2-3 must be reduced. The measure of this suscepti- bility is the width–thickness ratio of each cross-sectional element. Two types of el- ements must be considered: unstiffened elements, which are unsupported along one edge parallel to the direction of load, and stiffened elements, which are supported along both edges.

Limiting values of width–thickness ratios are given in AISC B5, "Local Buckling," where cross-sectional shapes are classified as *compact, noncompact,* or *slender,* accord- ing to the values of the ratios. For uniformly compressed elements, as in an axially loaded compression member, the strength must be reduced if the shape has any slender elements. The width–thickness ratio is given the generic name of λ. Depending on the particular cross-sectional element, λ is either the ratio b/t or h/t_w, both of which are defined presently. If λ is greater than the specified limit, denoted λ_r, the shape is slender, and the potential for local buckling must be accounted for. (We postpone a discussion of the compact and noncompact categories until Chapter 5, "Beams.") For I- and H-shapes, the projecting flange is considered to be an unstiffened element, and its width can be taken as half the full nominal width. Using AISC notation gives

$$\lambda = \frac{b}{t} = \frac{b_f/2}{t_f} = \frac{b_f}{2t_f}$$

where b_f and t_f are the width and thickness of the flange. The upper limit is

$$\lambda_r = \frac{95}{\sqrt{F_y}}$$

The webs of I- and H-shapes are stiffened elements, and the stiffened width is the dis- tance between the roots of the flanges. The width–thickness parameter is

$$\lambda = \frac{h}{t_w}$$

where h is the distance between the roots of the flanges, and t_w is the web thickness. The upper limit is

$$\lambda_r = \frac{253}{\sqrt{F_y}}$$

Values of the ratios $b_f/2t_f$ and h/t_w are tabulated in the dimensions and properties ta- bles in Part 1 of the *Manual*.

■ **FIGURE 4.9**

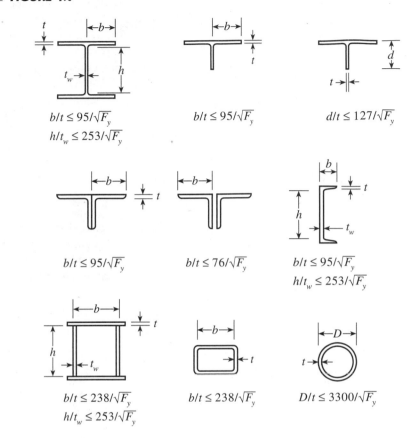

Stiffened and unstiffened elements of various cross-sectional shapes are illustrated in Figure 4.9. The appropriate compression member limit, λ_r, from AISC B5 is given for each case.

■ **EXAMPLE 4.3**

Investigate the column of Example 4.2 for local stability.

SOLUTION   For a W14 × 74, $b_f = 10.07$ in., $t_f = 0.785$ in., and

$$\frac{b_f}{2t_f} = \frac{10.07}{2(0.785)} = 6.4$$

Values of $b_f/2t_f$ are also tabulated in the properties tables.

$$\frac{95}{\sqrt{F_y}} = \frac{95}{\sqrt{36}} = 15.8 > 6.4 \qquad \text{(OK)}$$

$$\frac{h}{t_w} = 25.3 \qquad \text{(from properties tables)}$$

$$\frac{253}{\sqrt{F_y}} = \frac{253}{\sqrt{36}} = 42.2 > 25.3 \qquad \text{(OK)}$$

ANSWER   Local instability is not a problem.   ■

It is permissible to use a cross-sectional shape that does not satisfy the width–thickness ratio requirements, but such a member may not be permitted to carry as large a load as one that does satisfy the requirements. In other words, the design strength could be reduced because of local buckling. The general procedure for making this investigation is as follows.

- If the width–thickness ratio λ is greater than λ_r, refer to Appendix B of the Specification and compute a reduction factor Q.

- Compute λ_c as usual: $\lambda_c = \dfrac{KL}{r\pi} \sqrt{\dfrac{F_y}{E}}$

- If $\lambda_c Q \le 1.5$, $F_{cr} = Q(0.658^{Q\lambda_c^2})F_y$ (AISC Eq. A-B5-15)

- If $\lambda_c Q > 1.5$, $F_{cr} = \left[\dfrac{0.877}{\lambda_c^2}\right]F_y$ (AISC Eq. A-B5-16)

- The design strength is $\phi_c P_n = 0.85A_g F_{cr}$.

In most cases, a rolled shape that satisfies the width–thickness ratio requirements can be found, and this procedure will not be necessary. In this book, we consider only compression members with $\lambda \le \lambda_r$.

Tables for Compression Members

The *Manual* contains many useful tables for analysis and design. For compression members whose strength is governed by flexural buckling (the type considered so far), Tables 3-36, 3-50, and 4 in the Numerical Values section of the Specification and the column load tables in Part 3 of the *Manual,* "Column Design," are the most useful. Table 3-36 gives values of $\phi_c F_{cr}$ as a function of KL/r for $F_y = 36$ ksi. Table 3-50 does the same for $F_y = 50$ ksi, and Table 4 gives $\phi_c F_{cr}/F_y$ as a function of λ_c. (All *Manual* tables for $F_y = 50$ ksi, such as Table 3-50, are distinguished from tables for $F_y = 36$ ksi by gray shading.) The column load tables give the design strength of selected shapes for various values of the effective length. Tables 3-36 and 3-50 stop at the recommended upper limit of $KL/r = 200$, and the column load tables include values of KL up to those corresponding to $KL/r = 200$. The use of each of these tables is illustrated in the following example.

■ EXAMPLE 4.4

Compute the design strength of the compression member of Example 4.2 with the aid of (a) Table 3-36, (b) Table 4, and (c) the column load tables.

SOLUTION

a. From Example 4.2, $KL/r = 96.77$. For $F_y = 36$ ksi, Table 3-36 is used. Values of $\phi_c F_{cr}$ are given only for integer values of KL/r; for decimal values, KL/r may be rounded *up*, or linear interpolation may be used. For uniformity, we use interpolation in this book for all tables unless otherwise indicated. For $KL/r = 96.77$,

$$\phi_c F_{cr} = 18.69 \text{ ksi}$$
$$\phi_c P_n = \phi_c A_g F_{cr} = A_g(\phi_c F_{cr}) = 21.8(18.69) = 407 \text{ kips}$$

b. From Example 4.2, $\lambda_c = 1.085$ and from Table 4, for $\lambda_c = 1.085$,

$$\phi_c \frac{F_{cr}}{F_y} = 0.519$$

$$\phi_c P_n = A_g \left(\phi_c \frac{F_{cr}}{F_y} \right) F_y = 21.8(0.519)(36) = 407 \text{ kips}$$

c. The column load tables in Part 3 of the *Manual* give the design strength $\phi_c P_n$ for selected W, HP, pipe, tube, double-angle, WT, and single-angle shapes. The tabular values for the symmetrical shapes (W, HP, pipe, and tube) were calculated by using the minimum radius of gyration for each shape. From Example 4.2, $K = 1.0$, so

$$KL = 1.0(20) = 20 \text{ ft}$$

For a W14 × 74, A36 steel, and $KL = 20$ ft,

$$\phi_c P_n = 407 \text{ kips}$$

■

The values from Tables 3-36, 3-50, and 4 are based on flexural buckling and AISC Equations E2-2 and E2-3. Thus local stability is assumed, and width–thickness ratio limits must not be exceeded. The design strengths given in the column load tables account for the reduction required when the width–thickness limits are exceeded.

From a practical standpoint, if a compression member to be analyzed can be found in the column load tables, then these tables should be used. Otherwise, Tables 3-36 and 3-50 offer a computational advantage over Table 4. If the yield strength is other than 36 ksi or 50 ksi, neither Table 3-36 nor 3-50 can be used.

4.4   DESIGN

The selection of an economical rolled shape to resist a given compressive load is simple with the aid of the column load tables. Enter the table with the effective length and move horizontally until you find the desired design strength (or something slightly larger). In some cases, you must continue the search to be certain that you have found the lightest shape. Usually the category of shape (W, WT, etc.) will have been decided upon in advance. Often the overall nominal dimensions will also be known because of architectural or other requirements. As pointed out earlier, all tabulated values correspond to a slenderness ratio of 200 or less. The tabulated unsymmetrical shapes — the structural tees and the single and double-angles — require special consideration and are covered in Section 4.6.

■ EXAMPLE 4.5

A compression member is subjected to service loads of 165 kips dead load and 535 kips live load. The member is 26 feet long and pinned at each end. Use A36 steel and select a W14 shape.

SOLUTION   Calculate the factored load:

$$P_u = 1.2D + 1.6L = 1.2(165) + 1.6(535) = 1054 \text{ kips}$$

$$\therefore \quad \text{Required design strength } \phi_c P_n = 1054 \text{ kips}$$

From the column load tables for $KL = 26$ ft, a W14 × 176 has a design strength of 1150 kips.

ANSWER   Use a W14 × 176.   ■

■ EXAMPLE 4.6

Select the lightest W-shape that can resist a factored compressive load P_u of 190 kips. The effective length is 24 feet. Use ASTM A572 Grade 50 steel.

SOLUTION   The appropriate strategy here is to find the lightest shape for each nominal size and then choose the lightest overall. The choices are as follows.

W4, W5, and W6: None of the tabulated shapes will work.

W8:   W8 × 58, $\phi_c P_n = 194$ kips

W10:   W10 × 49, $\phi_c P_n = 239$ kips

W12:   W12 × 53, $\phi_c P_n = 247$ kips

W14:   W14 × 61, $\phi_c P_n = 276$ kips

Note that the load capacity is not proportional to the weight (or cross-sectional area). Although the W8 × 58 has the smallest design strength of the four choices, it is the second heaviest.

ANSWER   Use a W10 × 49.   ■

For shapes not in the column load tables, a trial-and-error approach must be used. The general procedure is to assume a shape and then compute its design strength. If the strength is too small (unsafe) or too large (uneconomical), another trial must be made. A systematic approach to making the trial selection is as follows.

1. Assume a value for the critical buckling stress F_{cr}. Examination of AISC Equations E2-2 and E2-3 shows that the theoretically maximum value of F_{cr} is the yield stress F_y.

2. From the requirement that $\phi_c P_n \geq P_u$, let

$$\phi_c A_g F_{cr} \geq P_u \ \text{ and } \ A_g \geq \frac{P_u}{\phi_c F_{cr}}$$

3. Select a shape that satisfies this area requirement.

4. Compute F_{cr} and $\phi_c P_n$ for the trial shape.

5. Revise if necessary. If the design strength is very close to the required value, the next tabulated size can be tried. Otherwise, repeat the entire procedure, using the value of F_{cr} found for the current trial shape as a value for Step 1.

6. Check local stability (check width–thickness ratios). Revise if necessary.

■ EXAMPLE 4.7

Select a W18 shape of A36 steel that can resist a factored load of 1054 kips. The effective length KL is 26 feet.

SOLUTION   Try $F_{cr} = 24$ ksi (two-thirds of F_y):

$$\text{Required } A_g = \frac{P_u}{\phi_c F_{cr}} = \frac{1054}{0.85(24)} = 51.7 \text{ in.}^2$$

Try W18 × 192:

$$A_g = 56.4 \text{ in.}^2 > 51.7 \text{ in.}^2$$

$$\frac{KL}{r_{min}} = \frac{26(12)}{2.79} = 111.8 < 200 \quad \text{(OK)}$$

$$\lambda_c = \frac{KL}{r\pi}\sqrt{\frac{F_y}{E}} = \frac{111.8}{\pi}\sqrt{\frac{36}{29,000}} = 1.254 < 1.5$$

∴ use AISC Equation E2-2

$$F_{cr} = (0.658^{\lambda_c^2})F_y = (0.658)^{(1.254)^2}(36) = 18.64 \text{ ksi}$$

$$\phi_c P_n = 0.85 A_g F_{cr} = 0.85(56.4)(18.64) = 894 \text{ kips} < 1054 \text{ kips} \quad \text{(N.G.)}$$

Try $F_{cr} = 18.64$ ksi (the value just computed for the W18 × 192):

$$\text{Required } A_g = \frac{P_u}{\phi_c F_{cr}} = \frac{1054}{0.85(18.64)} = 66.5 \text{ in.}^2$$

Try W18 × 234:

$$A_g = 68.8 \text{ in.}^2 > 66.5 \text{ in.}^2$$

$$\frac{KL}{r_{min}} = \frac{26(12)}{2.85} = 109.5 < 200 \quad \text{(OK)}$$

$$\lambda_c = \frac{KL}{r\pi}\sqrt{\frac{F_y}{E}} = \frac{109.5}{\pi}\sqrt{\frac{36}{29,000}} = 1.228 < 1.5$$

∴ use AISC Equation E2-2

$$F_{cr} = (0.658^{\lambda_c^2})F_y = (0.658)^{(1.228)^2}(36) = 19.15 \text{ ksi}$$

$$\phi_c P_n = 0.85 A_g F_{cr} = 0.85(68.8)(19.15) = 1120 \text{ kips} > 1054 \text{ kips} \quad \text{(OK)}$$

This shape is not listed in the column load tables, so the width–thickness ratios must be checked:

$$\frac{b_f}{2t_f} = 2.8 < \frac{95}{\sqrt{36}} = 15.8 \quad \text{(OK)}$$

$$\frac{h}{t_w} = 13.8 < \frac{253}{\sqrt{36}} = 42.2 \quad \text{(OK)}$$

ANSWER   Use a W18 × 234.

If Table 3-36 or Table 3-50 is used, trial values of $\phi_c F_{cr}$ can be conveniently used in the equation

$$\text{Required } A_g = \frac{P_u}{\phi_c F_{cr}}$$

4.5   MORE ON EFFECTIVE LENGTH

We introduced the concept of effective length in Section 4.2, "Column Theory." All compression members are treated as pin-ended regardless of the actual end conditions, but with an effective length KL that may differ from the actual length. With this modification, the load capacity of compression members is a function of only the slenderness parameter, λ_c. For a given material, it is a function of the slenderness ratio, KL/r.

If a compression member is supported differently with respect to each of its principal axes, the effective length will be different for the two directions. In Figure 4.10,

■ **FIGURE 4.10**

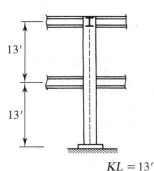

$KL = 13'$

(a) Minor Axis Buckling

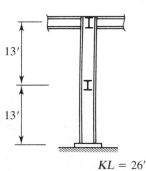

$KL = 26'$

(b) Major Axis Buckling

a W-shape is used as a column and is braced by horizontal members in two perpendicular directions at the top. These members prevent translation of the column in all directions, but the connections, the details of which are not shown, permit small rotations to take place. Under these conditions, the member can be treated as pin-connected at the top. For the same reasons, the connection to the support at the bottom may also be treated as a pin connection. Generally speaking, a rigid, or fixed, condition is very difficult to achieve, and unless some special provisions are made, ordinary connections will usually closely approximate a hinge or pin connection. At midheight, the column is braced, but only in one direction. Again, the connection prevents translation, but no restraint against rotation is furnished. This brace prevents translation perpendicular to the weak axis of the cross section but provides no restraint perpendicular to the strong axis. As shown schematically in Figure 4.10, if the member were to buckle about the major axis, the effective length would be 26 feet, whereas buckling about the minor axis would have to be in the second buckling mode, corresponding to an effective length of 13 feet. Because its strength is inversely proportional to the square of the slenderness ratio, a column will buckle in the direction corresponding to the largest slenderness ratio, so $K_x L/r_x$ must be compared with $K_y L/r_y$. In Figure 4.10, the ratio $26(12)/r_x$ must be compared with $13(12)/r_y$ (where r_x and r_y are in inches), and the larger ratio would be used for the determination of the nominal axial compressive strength P_n.

■ EXAMPLE 4.8

A W12 × 65, 24 feet long, is pinned at both ends and braced in the weak direction at the third points, as shown in Figure 4.11. A36 steel is used. Determine the design compressive strength.

■ **FIGURE 4.11**

x-direction

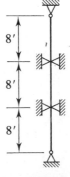

y-direction

COMPARE SLENDERNESS
RATIOS

SOLUTION

$$\frac{K_x L}{r_x} = \frac{24(12)}{5.28} = 54.55$$

$$\frac{K_y L}{r_y} = \frac{8(12)}{3.02} = 31.79$$

$K_x L/r_x$, the larger value, controls. From Table 3-36 with $KL/r = 54.55$

$$\phi_c F_{cr} = 26.17 \text{ ksi}$$

$$\phi_c P_n = A_g(\phi_c F_{cr}) = 19.1(26.17) = 500 \text{ kips}$$

ANSWER   Design strength $= 500$ kips.   ■

The design strengths given in the column load tables are based on the effective length with respect to the *y*-axis. A procedure for using the tables with $K_x L$, however, can be developed by examining how the tabular values were obtained. Starting with a value of *KL,* the value of $\phi_c P_n$ was obtained by a procedure similar to the following.

- *KL* was divided by r_y to obtain KL/r_y.

- The slenderness parameter, $\lambda_c = \dfrac{KL}{r_y \pi} \sqrt{\dfrac{F_y}{E}}$, was computed.

- F_{cr} was computed.

- The design strength, $\phi_c P_n = 0.85 A_g F_{cr}$, was computed.

Thus the tabulated strengths are based on the values of *KL* being equal to $K_y L$. If the capacity with respect to *x*-axis buckling is desired, the table can be entered with

$$KL = \frac{K_x L}{r_x / r_y}$$

and the tabulated load will be based on

$$\frac{KL}{r_y} = \frac{K_x L/(r_x/r_y)}{r_y} = \frac{K_x L}{r_x}$$

The ratio r_x/r_y is given in the column load tables for each shape listed.

■ EXAMPLE 4.9

The compression member shown in Figure 4.12 is pinned at both ends and supported in the weak direction at midheight. A service load of 400 kips, with equal parts of dead and live load, must be supported. Use A36 steel and select the lightest W-shape.

■ **FIGURE 4.12**

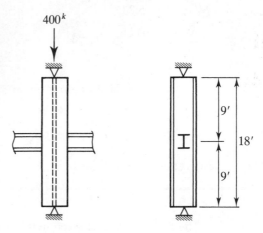

SOLUTION   Factored load $= P_u = 1.2(200) + 1.6(200) = 560$ kips

Assume that the weak direction controls and enter the column load tables with $KL = 9$ feet. Beginning with the smallest shapes, the first one found that will work is a W10 × 77 with a design strength of 632 kips.

Check the strong axis:

$$\frac{K_x L}{r_x / r_y} = \frac{18}{1.73} = 10.40 \text{ ft} > 9 \text{ ft} \qquad \therefore \; K_x L \text{ controls for this shape.}$$

Enter the tables with $KL = 10.4$ feet. A W10 × 77 is still the lightest W10, with a design strength of 612 kips (interpolated).

Continue the search and investigate a W12 × 72:

$$\frac{K_x L}{r_x / r_y} = \frac{18}{1.75} = 10.3 \text{ ft} > 9 \text{ ft}$$

$K_x L$ again controls, and the design strength is 592 kips.

Determine the lightest W14. The lightest one with a possibility of working is a W14 × 74. It is heavier than the lightest one found so far, so it will not be considered.

ANSWER   Use a W12 × 72.   ■

Whenever possible, the designer should provide extra support for the weak direction of a column. Otherwise, the member is inefficient: It has an excess of strength in one direction. When K_xL and K_yL are different, K_yL will control unless r_x/r_y is smaller than K_xL/K_yL. When the two ratios are equal, the column has equal strength in both directions. For the W-shapes in the column load tables, r_x/r_y ranges between 1.6 and 1.8 except for some of the lighter shapes.

■ **EXAMPLE 4.10**

The column shown in Figure 4.13 is subjected to a factored axial load of 840 kips. Use A36 steel and select a W-shape.

SOLUTION $K_xL = 20$ ft and maximum $K_yL = 8$ ft

The effective length K_xL will control when

$$\frac{K_xL}{r_x/r_y} > K_yL$$

or when

$$K_xL > \left(\frac{r_x}{r_y}\right)(K_yL)$$

■ **FIGURE 4.13**

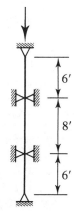

Support in
strong direction

Support in
weak direction

20′

6′

8′

6′

In this case,

$$\frac{K_x L}{K_y L} = \frac{20}{8} = 2.5 \text{ or } K_x L = 2.5 K_y L$$

Because $K_x L$ is so much larger than $K_y L$, $K_x L$ probably controls. The reason is that most of the tabulated values of r_x/r_y are less than 2.5, so a $K_x L$ of $2.5 K_y L$ is likely to be larger than $(r_x/r_y)K_y L$. Try $r_x/r_y = 1.7$:

$$\frac{K_x L}{r_x/r_y} = \frac{20}{1.7} = 11.76 > K_y L$$

Rounding this result to $KL = 12$ ft and entering the column load tables, try a W10 × 112 ($\phi_c P_n = 865$ kips):

$$\text{Actual } \frac{K_x L}{r_x/r_y} = \frac{20}{1.74} = 11.5 \text{ ft} < 12 \text{ ft} \qquad \therefore \phi_c P_n > \text{ required } 840 \text{ kips}$$

(By interpolation, $\phi_c P_n = 876$ kips.) Check a W12 × 106:

$$\frac{K_x L}{r_x/r_y} = \frac{20}{1.76} = 11.4 \text{ ft}$$

For $KL = 12$ ft,

$$\phi_c P_n = 853 \text{ kips} > 840 \text{ kips} \qquad \text{(OK)}$$

Investigate W14 shapes. For $r_x/r_y = 1.7$ (the approximate ratio for all likely possibilities),

$$\frac{K_x L}{r_x/r_y} = \frac{20}{1.7} = 11.76 \text{ ft} > K_y L = 8 \text{ ft}$$

For $KL = 12$ ft, a W14 × 109, with a capacity of 905 kips, is the lightest W14. As 12 feet is a conservative approximation of the actual effective length, this shape is satisfactory.

ANSWER   Use a W12 × 106 (lightest of the three possibilities).   ■

For isolated columns that are not part of a continuous frame, Table C-C2.1 in the Commentary to the Specification will usually suffice. Consider, however, the rigid frame in Figure 4.14. The columns in this frame are not independent members but part of a continuous structure. Except for those in the lower story, the columns are restrained at both ends by their connection to beams and other columns. This frame is also unbraced,

■ **FIGURE 4.14**

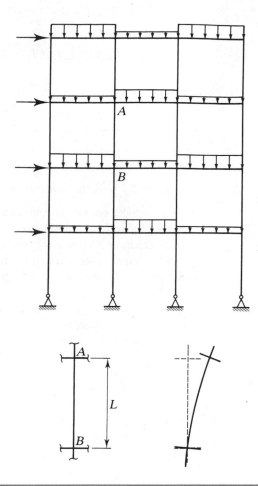

meaning that horizontal displacements of the frame are possible, and all columns are subject to sidesway. If Table C-C2.1 is used for this frame, the lower story columns are best approximated by condition (f), and a value of $K = 2$ might be used. For a column such as AB, a value of $K = 1.2$, corresponding to condition (c), could be selected. A more rational procedure, however, will account for the degree of restraint provided by connecting members.

The rotational restraint provided by the beams, or girders, at the end of a column is a function of the rotational stiffnesses of the members intersecting at the joint. The rotational stiffness of a member is proportional to EI/L, where I is the moment of inertia of the cross section with respect to the axis of bending. Gaylord, Gaylord, and

Stallmeyer (1992) show that the effective length factor K depends on the ratio of column stiffness to girder stiffness at each end of the member, which can be expressed as

$$G = \frac{\sum E_c I_c \, / \, L_c}{\sum E_g I_g \, / \, L_g} = \frac{\sum I_c \, / \, L_c}{\sum I_g \, / \, L_g} \tag{4.7}$$

where

$\sum E_c I_c / L_c$ = sum of the stiffnesses of all columns at the end of the column under consideration

$\sum E_g I_g / L_g$ = sum of the stiffnesses of all girders at the end of the column under consideration

$E_c = E_g = E$, the modulus of elasticity of structural steel.

If a very slender column is connected to girders having large cross sections, the girders will effectively prevent rotation of the column. The ends of the column are approximately fixed, and K is relatively small. This condition corresponds to small values of G given by Equation 4.7. However, the ends of stiff columns connected to flexible beams can more freely rotate and approach the pinned condition, giving relatively large values of G and K.

The relationship between G and K has been quantified in the Jackson–Mooreland Alignment Charts (Johnston, 1976), which are reproduced in Figure C-C2.2 in the Commentary. To obtain a value of K from one of the nomograms, first calculate the value of G at each end of the column, letting one value be G_A and the other be G_B. Connect G_A and G_B with a straight line, and read the value of K on the middle scale. The effective length factor obtained in this manner is with respect to the axis of bending, which is the axis perpendicular to the plane of the frame. A separate analysis must be made for buckling about the other axis. Normally the beam-to-column connections in this direction will not transmit moment, sidesway is prevented by bracing, and K can be taken as 1.0.

■ EXAMPLE 4.11

The rigid frame shown in Figure 4.15 is unbraced. Each member is oriented so that its web is in the plane of the frame. Determine the effective length factor K_x for columns AB and BC.

SOLUTION   Column AB:

For joint A,

$$G = \frac{\sum I_c / L_c}{\sum I_g / L_g} = \frac{833/12 + 1070/12}{1350/20 + 1830/18} = \frac{158.6}{169.2} = 0.94$$

■ **FIGURE 4.15**

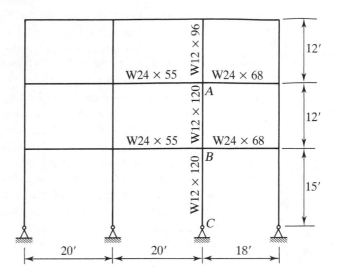

For joint B,

$$G = \frac{\Sigma\, I_c/L_c}{\Sigma\, I_g/L_g} = \frac{1070/12 + 1070/15}{169.2} = \frac{160.5}{169.2} = 0.95$$

ANSWER   From the alignment chart for sidesway uninhibited, with $G_A = 0.94$ and $G_B = 0.95$, $K_x = 1.3$ for column AB.

Column *BC:*

For joint *B,* as before,

$G = 0.95.$

For joint *C,* at a pin connection the situation is analogous to that of a very stiff column attached to infinitely flexible girders — that is, girders of zero stiffness. The ratio of column stiffness to girder stiffness would therefore be infinite for a perfectly frictionless hinge. This end condition can only be approximated in practice, so the discussion accompanying the alignment chart recommends that G be taken as 10.0.

ANSWER   From the alignment chart with $G_A = 0.95$ and $G_B = 10.0$, $K_x = 1.85$ for column BC.

■

As pointed out in Example 4.11, for a pinned support, G should be taken as 10.0; for a fixed support, G should be taken as 1.0. The latter support condition corresponds to an infinitely stiff girder and a flexible column, corresponding to a theoretical value of $G = 0$. The discussion accompanying the alignment chart in the Commentary recommends a value of $G = 1.0$ because true fixity will rarely be achieved.

Unbraced frames are able to support lateral loads because of their moment-resisting joints. Often the frame is augmented by a bracing system of some sort; such frames are called *braced frames*. The additional resistance to lateral loads can take the form of diagonal bracing, as illustrated in Figure 4.16, or rigid shear walls. In either case, the tendency for columns to sway is blocked within a given panel, or bay, for the full height of the frame. This support forms a cantilever structure that is resistant to horizontal displacements and also provides horizontal support for the other bays. Depending on the size of the structure, more than one bay may require bracing. Columns that are members of braced frames are prevented from sidesway and have some degree of rotational restraint at their ends. Thus they are in a category that lies somewhere between cases (a) and (d) in Table C-C2.1 of the Commentary, and K is between 0.5 and 1.0. A value of

■ **FIGURE 4.16**

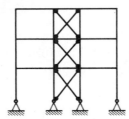 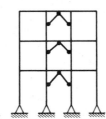

(a) **Diagonal bracing**

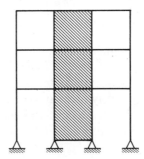

(b) **Shear Walls**
(masonry, reinforced concrete,
or steel plate)

1.0 is therefore always conservative for members of braced frames and is the value prescribed by AISC C2.1 unless an analysis is made. Such an analysis can be made with the alignment chart for braced frames. Use of this nomogram would result in an effective length factor somewhat less than 1.0, and some savings could be realized.*

As with any design aid, the alignment charts should be used only under the conditions for which they were derived. These conditions are discussed in Section C2 of the Commentary to the Specification and are not enumerated here. Most of the conditions will usually be approximately satisfied; if they are not, the deviation will be on the conservative side. One condition that usually is not satisfied is the requirement that all behavior be elastic. If the slenderness parameter λ_c is less than 1.5, the column will buckle inelastically, and the effective length factor obtained from the alignment chart will be overly conservative. A large number of columns are in this category. A convenient procedure for determining K for inelastic columns allows the alignment charts to be used (Yura, 1971 and Disque, 1973). To demonstrate the procedure, we begin with the critical buckling load for an inelastic column given by Equation 4.6b. Dividing it by the cross-sectional area gives the buckling stress:

$$F_{cr} = \frac{\pi^2 E_t}{(KL/r)^2}$$

The rotational stiffness of a column in this state would be proportional to $E_t I_c/L_c$, and the appropriate value of G for use in the alignment chart is

$$G_{\text{inelastic}} = \frac{\sum E_t I_c/L_c}{\sum EI_g/L_g} = \frac{E_t}{E} G_{\text{elastic}}$$

Because E_t is less than E, $G_{\text{inelastic}}$ is less than G_{elastic}, and the effective length factor K will be reduced, resulting in a more economical design. To evaluate E_t/E, called the *stiffness reduction factor* (SRF), consider the following relationship for a column with pinned ends:

$$\frac{F_{cr(\text{inelastic})}}{F_{cr(\text{elastic})}} = \frac{\pi^2 E_t/(L/r)^2}{\pi^2 E/(L/r)^2} = \frac{E_t}{E} \tag{4.8}$$

AISC uses an approximation for the inelastic portion of the column strength curve, so Equation 4.8 is an approximation when AISC Equations E2-2 and E2-3 are used for F_{cr}. If we let

$$F_{cr} = \frac{P_{cr}}{A} \approx \frac{P_u/\phi_c}{A}$$

---

*If a frame is braced against sidesway, the beam-to-column connections need not be moment-resisting, and the bracing system could be designed to resist all sidesway tendency. If the connections are not moment-resisting, however, there will be no continuity between columns and girders, and the alignment chart cannot be used. For this type of braced frame, K_x should be taken as 1.0.

then $F_{cr(\text{inelastic})}/F_{cr(\text{elastic})}$ will be a function of $P_u/(\phi_c A)$. For example, for $P_u/(\phi_c A) = 26$ ksi and $F_y = 36$ ksi,

$$F_{cr(\text{inelastic})} \approx 26 = 0.658^{\lambda_c^2} F_y = 0.658^{\lambda_c^2} (36)$$

$$\lambda_c^2 = 0.7776$$

$$F_{cr(\text{elastic})} = \frac{0.877}{\lambda_c^2} F_y = \frac{0.877}{0.7776} (36) = 40.7$$

The stiffness reduction factor is therefore

$$\text{SRF} = \frac{F_{cr(\text{inelastic})}}{F_{cr(\text{elastic})}} = \frac{26}{40.7} = 0.639$$

Because ϕ_c is a constant, the stiffness reduction factor can also be expressed as a function of P_u/A. Values of the stiffness reduction factor SRF as a function of P_u/A are given in Table 3-1 in Part 3 of the *Manual*.

■ **EXAMPLE 4.12**

A rigid unbraced frame is shown in Figure 4.17. All members are oriented so that bending is about the strong axis. Lateral support is provided at each joint by simply connected bracing in the direction perpendicular to the frame. Determine the effective length factors with respect to each axis for member *AB*. The factored axial load on this member is 180 kips, and A36 steel is used.

SOLUTION   Compute elastic *G* factors:

For joint *A*,

$$G_A \approx \frac{\Sigma(I_c/L_c)}{\Sigma(I_g/L_g)} = \frac{170/12}{88.6/20 + 88.6/18} = \frac{14.17}{9.35} = 1.52$$

For joint *B*,

$$G_B \approx \frac{\Sigma(I_c/L_c)}{\Sigma(I_g/L_g)} = \frac{2(170/12)}{199/20 + 199/18} = \frac{28.3}{21.0} = 1.35$$

From the alignment chart for unbraced frames, $K_x = 1.43$, based on elastic behavior, so

$$\lambda_c = \frac{K_x L}{r_x \pi} \sqrt{\frac{F_y}{E}} = \frac{1.43(12)(12)}{4.19\pi} \sqrt{\frac{36}{29,000}} = 0.5512$$

■ **FIGURE 4.17**

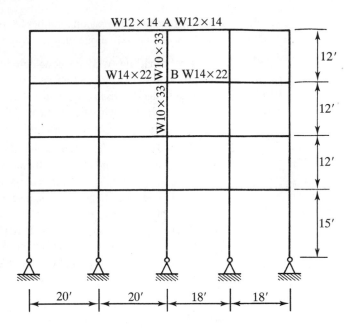

Because λ_c is less than 1.5, the inelastic K factor should be used, giving

$$\frac{P_u}{A} = \frac{180}{9.71} = 18.5 \text{ ksi}$$

OF *MODIFIER* *G AT PT. A*
1

From Table 3-1 in Part 3 of the *Manual,* the stiffness reduction factor SRF = 0.83.

For joint *A,*

$$G_{\text{inelastic}} = \text{SRF} \times G_{\text{elastic}} = 0.83(1.52) = 1.26$$

For joint *B,* *MODIFIER*

$$G_{\text{inelastic}} = 0.83(1.35) = 1.12$$

ANSWER From the alignment chart, $K_x = 1.37$. Because of the support conditions normal to the frame, K_y can be taken as 1.0. ■

If the end of a column is fixed ($G = 1.0$) or pinned ($G = 10.0$), the value of G should *not* be multiplied by the stiffness reduction factor.

4.6	**TORSIONAL AND FLEXURAL-TORSIONAL BUCKLING**

When an axially loaded compression member becomes unstable overall (that is, not locally unstable), it can buckle in one of three ways, as shown in Figure 4.18).

1. **Flexural buckling.** We have considered this type of buckling up to now. It is a deflection caused by bending, or flexure, about the axis corresponding to the largest slenderness ratio (Figure 4.18a). This is usually the minor principal axis — the one with the smallest radius of gyration. Compression members with any type of cross-sectional configuration can fail in this way.

2. **Torsional buckling.** This type of failure is caused by twisting about the longitudinal axis of the member. It can occur only with doubly symmetrical cross sections with very slender cross-sectional elements (Figure 4.18b). Standard hot-rolled shapes are not susceptible to torsional buckling, but members built up from thin plate elements may be and should be investigated. The cruciform shape shown is particularly vulnerable to this type of buckling. This shape can be fabricated from plates as shown in the figure, or built up from four angles placed back to back.

■ **FIGURE 4.18**

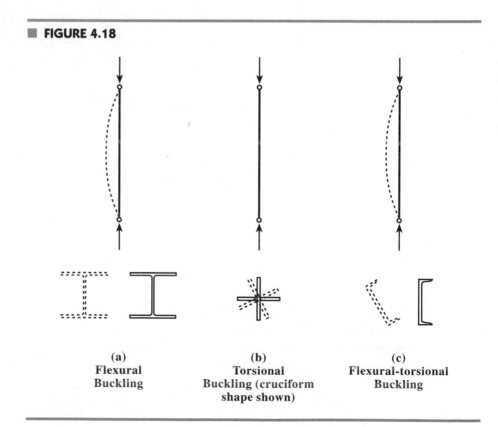

(a)	(b)	(c)
Flexural Buckling	**Torsional Buckling (cruciform shape shown)**	**Flexural-torsional Buckling**

3. **Flexural-torsional buckling.** This type of failure is caused by a combination of flexural buckling and torsional buckling. The member bends and twists simultaneously (Figure 4.18c). This type of failure can occur only with unsymmetrical cross sections, both those with one axis of symmetry — such as channels, structural tees, double-angle shapes, and equal-leg single angles — and those with no axis of symmetry, such as unequal-leg single angles.

The AISC Specification requires an analysis of torsional or flexural-torsional buckling when appropriate. Section E3 of the Specification covers double-angle and tee-shaped members, and Appendix E3 provides a more general approach that can be used for any unsymmetrical shape.

We discuss the procedure prescribed in Appendix E3 first. It is based on the use of a slenderness parameter λ_e instead of λ_c. We derive λ_e as follows. From the Euler buckling stress,

$$F_e = \frac{\pi^2 E}{(KL/r)^2}$$

the slenderness ratio can be written as

$$\frac{KL}{r} = \sqrt{\frac{\pi^2 E}{F_e}}$$

If F_e is defined as the elastic buckling stress corresponding to the controlling mode of failure, whether flexural, torsional, or flexural-torsional, then the corresponding slenderness ratio is

$$\left(\frac{KL}{r}\right)_e = \sqrt{\frac{\pi^2 E}{F_e}}$$

and the corresponding slenderness parameter is

$$\lambda_e = \frac{(KL/r)_e}{\pi}\sqrt{\frac{F_y}{E}} = \sqrt{\frac{F_y(KL/r)_e^2}{\pi^2 E}} = \sqrt{\frac{F_y}{F_e}}$$

An outline of the overall procedure is as follows.

1. Determine F_e for torsional elastic buckling or flexural-torsional elastic buckling from the equations given in Appendix E3.
2. Compute the effective slenderness parameter, λ_e.
3. Compute the critical stress F_{cr} from the usual equations (AISC Equations E2-2 and E2-3), but use λ_e instead of λ_c. The design strength is then

$$\phi_c P_n = \phi_c A_g F_{cr}$$

where ϕ_c is 0.85, the same as for flexural buckling.

The equations for F_e given in AISC Appendix E3 are based on the well-established theory given in *Theory of Elastic Stability* (Timoshenko and Gere, 1961). Except for some changes in notation, they are the same equations as those given in that work, with no simplifications. For doubly symmetrical shapes (torsional buckling),

$$F_e = \left[\frac{\pi^2 E C_w}{(K_z L)^2} + GJ \right] \frac{1}{I_x + I_y} \qquad \text{(AISC Equation A-E3-5)}$$

For singly symmetrical shapes (flexural-torsional buckling),

$$F_e = \frac{F_{ey} + F_{ez}}{2H} \left(1 - \sqrt{1 - \frac{4 F_{ey} F_{ez} H}{(F_{ey} + F_{ez})^2}} \right) \qquad \text{(AISC Equation A-E3-6)}$$

For shapes with *no* axis of symmetry (flexural-torsional buckling),

$$(F_e - F_{ex})(F_e - F_{ey})(F_e - F_{ez}) - F_e^2 (F_e - F_{ey})(x_0 / \bar{r}_0)^2$$

$$-F_e^2 (F_e - F_{ex})(y_0 / \bar{r}_0)^2 = 0 \qquad \text{(AISC Equation A-E3-7)}$$

This last equation is a cubic; F_e is the smallest root. Fortunately, there will be little need for solving this equation because completely unsymmetrical shapes are rarely used as compression members. The previously undefined terms used in these three equations are defined as:

C_w = warping constant (in.)

K_z = effective length factor for *torsional* buckling, which is based on the amount of end restraint against twisting about the longitudinal axis.

G = shear modulus (ksi)

J = torsional constant (equal to the polar moment of inertia only for circular cross sections) (in.^{4})

$$F_{ex} = \frac{\pi^2 E}{(K_x L / r_x)^2} \qquad \text{(AISC Equation A-E3-10)}$$

$$F_{ey} = \frac{\pi^2 E}{(K_y L / r_y)^2} \qquad \text{(AISC Equation A-E3-11)}$$

where y is the axis of symmetry for singly symmetric shapes.

$$F_{ez} = \left[\frac{\pi^2 E C_w}{(K_z L)^2} + GJ \right] \frac{1}{A \bar{r}_0^2} \qquad \text{(AISC Equation A-E3-12)}$$

$$H = 1 - \left(\frac{x_0^2 + y_0^2}{\bar{r}_0^2} \right) \qquad \text{(AISC Equation A-E3-9)}$$

where x_0 and y_0 are the coordinates of the shear center of the cross section with respect to the centroid (in inches). The shear center is the point on the cross section through which a transverse load on a beam must pass if the member is to bend without twisting. We cover the shear center in more detail in Chapter 5, "Beams."

$$\bar{r}_0^2 = x_0^2 + y_0^2 + \frac{I_x + I_y}{A} \qquad \text{(AISC Equation A-E3-8)}$$

Values of the constants used in the three equations for F_e can be found in the tables of torsion properties and flexural-torsional properties in Part 1 of the *Manual*. For W, M, S, and HP shapes, J and C_w are given. The values of J, C_w, $\bar{r}_0$, and H are given for channels, single angles, and structural tees. The tables for double angles give values of $\bar{r}_0$ and H (J and C_w are double the values given for single angles).

As previously pointed out, the need for a torsional buckling analysis of a doubly symmetrical shape will be rare. Similarly, shapes with no axis of symmetry are rarely used for compression members, and flexural-torsional buckling analysis of these types of members will seldom, if ever, need to be done. For these reasons, we limit further consideration to flexural-torsional buckling of shapes with one axis of symmetry. Furthermore, the most commonly used of these shapes, the double angle, is a built-up shape, and we postpone consideration of it until Section 4.7.

For singly symmetrical shapes, the flexural-torsional buckling stress F_e is found from AISC Equation A-E3-6. In this equation, y is defined as the axis of symmetry (regardless of the orientation of the member), and flexural-torsional buckling will take place only about this axis (flexural buckling about this axis will not occur). The x-axis is subject only to flexural buckling. Therefore, for singly symmetrical shapes, there are two possibilities for the strength: either flexural-torsional buckling about the y-axis (the axis of symmetry) or flexural buckling about the x-axis. To determine which one controls, compute the strength corresponding to each axis and use the smaller value.

■ EXAMPLE 4.13

Compute the design compressive strength of a WT13.5 × 80.5. The effective length with respect to the x-axis is 25 feet 6 inches, the effective length with respect to the y-axis is 20 feet, and the effective length with respect to the z-axis is 20 feet. A36 steel is used.

SOLUTION Compute the flexural buckling strength for the x-axis:

$$\frac{K_x L}{r_x} = \frac{25.5(12)}{3.96} = 77.27$$

$$\lambda_c = \frac{KL}{r\pi}\sqrt{\frac{F_y}{E}} = \frac{77.27}{\pi}\sqrt{\frac{36}{29,000}} = 0.8666 < 1.5$$

∴ use AISC Equation E2-2

$$F_{cr} = (0.658)^{\lambda_c^2} F_y = (0.658)^{(0.8666)^2}(36) = 26.29 \text{ ksi}$$

$$\phi_c P_n = \phi_c A_g F_{cr} = 0.85(23.7)(26.29) = 530 \text{ kips}$$

Compute the flexural-torsional buckling strength about the y-axis:

$$F_{ey} = \frac{\pi^2 E}{(K_y L/r_y)^2} = \frac{\pi^2 (29,000)}{(74.07)^2} = 52.17 \text{ ksi}$$

$$F_{ez} = \left[\frac{\pi^2 E C_w}{(K_z L)^2} + GJ \right] \frac{1}{A\bar{r}_0^2}$$

$$= \left[\frac{\pi^2 (29,000)(42.7)}{(20 \times 12)^2} + 11,200(7.31) \right] \frac{1}{23.7(5.67)^2} = 107.7 \text{ ksi}$$

$$F_{ey} + F_{ez} = 52.17 + 107.7 = 159.9 \text{ ksi}$$

$$F_e = \frac{F_{ey} + F_{ez}}{2H} \left[1 - \sqrt{1 - \frac{4 F_{ey} F_{ez} H}{(F_{ey} + F_{ez})^2}} \right]$$

$$= \frac{159.9}{2(0.813)} \left[1 - \sqrt{1 - \frac{4(52.17)(107.7)(0.813)}{(159.9)^2}} \right] = 45.81 \text{ ksi}$$

$$\lambda_e = \sqrt{\frac{F_y}{F_e}} = \sqrt{\frac{36}{45.81}} = 0.8865$$

Because this value is less than 1.5, use AISC Equation E2-2 with λ_e instead of λ_c:

$$F_{cr} = (0.658^{\lambda_e^2}) F_y = (0.658)^{(0.8865)^2} (36) = 25.91 \text{ ksi}$$

$$\phi_c P_n = \phi_c A_g F_{cr} = 0.85(23.7)(25.91) = 522 \text{ kips} \qquad \text{(controls)}$$

ANSWER Design strength = 522 kips.

Note that, once F_{cr} and F_e have been computed, the computations for flexural buckling about the x-axis and flexural-torsional buckling about the y-axis are identical. Thus after F_{cr} and F_e have been computed, both λ_c and λ_e can be computed, and the smaller value can be used to compute the strength. This eliminates the need to compute the strength for both axes.

The procedure for flexural-torsional buckling analysis of double-angles and tees given in AISC Section E3 is a modification of the procedure given in AISC Appendix E3. There is also some notational change: F_e becomes F_{crft}, F_{ey} becomes F_{cry}, and F_{ez} becomes F_{crz}. The stress F_{cry} is found from AISC E2 and is based on flexural buckling about the y-axis.

To obtain F_{crz}, we can drop the first term of AISC Equation A-E3-12 to give

$$F_{crz} = \frac{GJ}{A\bar{r}_0^2}$$

This deletion is permissible, because for double-angles and tees, the first term is negligible compared to the second term.

The flexural buckling stress F_{cry} is computed with the usual equations of AISC Chapter E, using KL/r corresponding to the y-axis (the axis of symmetry).

The nominal strength can then be computed as

$$P_n = A_g F_{crft},$$

where

$$F_{crft} = \left(\frac{F_{cry} + F_{crz}}{2H} \right) \left[1 - \sqrt{1 - \frac{4F_{cry}F_{crz}H}{(F_{cry} + F_{crz})^2}} \right] \qquad \text{(AISC Equation E3-1)}$$

All other terms from Appendix E3 remain unchanged. This procedure, to be used with double-angles and tees only, is more accurate than the procedure given in Appendix E3.

■ EXAMPLE 4.14

Compute the design strength of the shape in Example 4.13 by using the equations of AISC E3.

SOLUTION

From Example 4.13, the flexural buckling strength for the x-axis is 530 kips, and $K_y L/r_y = 74.07$. From AISC Equation E2-4, the slenderness parameter is

$$\lambda_c = \frac{KL}{r\pi} \sqrt{\frac{F_y}{E}} = \frac{74.07}{\pi} \sqrt{\frac{36}{29,000}} = 0.8307 < 1.5$$

From AISC Equation E2-2,

$$F_{cr} = F_{cry} = (0.658^{\lambda_c^2})F_y = (0.658)^{(0.8307)^2}(36) = 26.97 \text{ ksi}$$

From AISC E3,

$$F_{crz} = \frac{GJ}{A\bar{r}_0^2} = \frac{11,200(7.31)}{23.7(5.67)^2} = 107.5 \text{ ksi}$$

$$F_{cry} + F_{crz} = 26.97 + 107.5 = 134.5 \text{ ksi}$$

$$F_{crft} = \left(\frac{F_{cry} + F_{crz}}{2H} \right) \left[1 - \sqrt{1 - \frac{4F_{cry}F_{crz}H}{(F_{cry} + F_{crz})^2}} \right]$$

$$= \frac{134.5}{2(0.813)} \left[1 - \sqrt{1 - \frac{4(26.97)(107.5)(0.813)}{(134.5)^2}} \right] = 25.48 \text{ ksi}$$

$$\phi_c P_n = \phi_c A_g F_{crft} = 0.85(23.7)(25.48) = 513 \text{ kips} \qquad \text{(controls)}$$

ANSWER   Design strength = 513 kips.

The results of Examples 4.13 and 4.14 show that the error in using Appendix E3 for this shape is on the unconservative side. The procedure used in Example 4.14, which is based on AISC Specification E3, should always be used for double angles and tees. In practice, however, the strength of most double angles and tees can be found in the column load tables. These tables are based on the recommended procedure of AISC E3 and can be used to verify the result of Example 4.14. The tables give two values of the design strength, one based on flexural buckling about the *x*-axis and one based on flexural-torsional buckling about the *y*-axis.

Tables are also provided for single-angle compression members. The design strengths given are not based on flexural-torsional buckling theory, but on an approximation given in the separate specification for single-angle members in Part 6 of the *Manual,* Specifications and Codes.

4.7 BUILT-UP MEMBERS

If the cross-sectional properties of a built-up compression member are known, its analysis is the same as for any other compression member, provided the component parts of the cross section are properly connected. AISC E4 contains many details related to this connection, with separate requirements for members composed of two or more rolled shapes and for members composed of plates or a combination of plates and shapes. Before considering the connection problem, we will review the computation of cross-sectional properties of built-up shapes.

The design strength of a built-up compression member is a function of the slenderness parameter λ_c. Hence the principal axes and the corresponding radii of gyration about these axes must be determined. For homogeneous cross sections, the principal axes coincide with the centroidal axes. The procedure is illustrated in Example 4.15. The components of the cross section are assumed to be properly connected.

■ EXAMPLE 4.15

The column shown in Figure 4.19 is fabricated by welding a ⅜-inch by 4-inch cover plate to the flange of a W18 × 35. A36 steel is used for both components. The effective length is 15 feet with respect to both axes. Assume that the components are con-

■ FIGURE 4.19

W18 × 35

TABLE 4.1

Component	A	y	Ay
Plate	1.500	0.1875	0.2812
W	10.3	9.225	95.02
Σ	11.8		95.30

nected in such a way that the member is fully effective and compute the design strength based on flexural buckling.

SOLUTION With the addition of the cover plate, the shape is slightly unsymmetrical, but the flexural-torsional effects will be negligible.

The vertical axis of symmetry is one of the principal axes, and its location need not be computed. The horizontal principal axis will be found by application of the *principle of moments:* The sum of moments of component areas about any axis (in this example, a horizontal axis along the top of the plate will be used) must equal the moment of the total area. We use Table 4.1 to keep track of the computations.

$$\bar{y} = \frac{\Sigma\, Ay}{\Sigma\, A} = \frac{95.30}{11.8} = 8.076 \text{ in.}$$

With the location of the horizontal centroidal axis known, the moment of inertia with respect to this axis can be found by using the *parallel-axis theorem:*

$$I = \bar{I} + Ad^2$$

where
$\bar{I}$ = moment of inertia about the centroidal axis of a component area

A = area of the component

I = moment of inertia about an axis parallel to the centroidal axis of the component area

d = perpendicular distance between the two axes

The contributions from each component area are computed and summed to obtain the moment of inertia of the composite area. Table 4.2, an expanded version of Table 4.1, includes these computations. For the vertical axis,

$$I_y = \frac{1}{12}(3/8)(4)^3 + 15.3 = 17.30 \text{ in.}^4 \qquad \text{(controls)}$$

$$r_y = \sqrt{\frac{I_y}{A}} = \sqrt{\frac{17.30}{11.8}} = 1.211 \text{ in.}$$

TABLE 4.2

Component	A	y	Ay	$\bar{I}$	d	$\bar{I} + Ad^2$
Plate	1.500	0.1875	0.2812	0.01758	7.889	93.37
W	10.3	9.225	95.02	510	1.149	523.6
Σ	11.8		95.30			$617.0 = I_x$

$$\lambda_c = \frac{KL}{r\pi}\sqrt{\frac{F_y}{E}} = \frac{15(12)}{1.211\pi}\sqrt{\frac{36}{29,000}} = 1.667 > 1.5$$

$$F_{cr} = \left[\frac{0.877}{\lambda_c^2}\right]F_y = \left[\frac{0.877}{(1.667)^2}\right](36) = 11.36 \text{ ksi}$$

$$\phi_c P_n = \phi_c A_g F_{cr} = 0.85(11.8)(11.36) = 114 \text{ kips}$$

ANSWER   Design strength = 114 kips.   ■

Connection Requirements for Built-Up Members Composed of Rolled Shapes

The most common built-up shape is one that is composed of rolled shapes, namely, the double-angle shape. This type of member will be used to illustrate the requirements for this category of built-up members. Figure 4.20 shows a truss compression member connected to gusset plates at each end. To maintain the back-to-back separation of the angles along the length, fillers, or spacers, of the same thickness as the gusset plate are placed between the angles at equal intervals. The intervals must be small enough that the member functions as a unit. If the member buckles about the x-axis (flexural buckling) the connectors are not subjected to any calculated load, and the connection problem is simply one of maintaining the relative positions of the two components. To ensure that the built-up member acts as a unit, AISC requires that the slenderness of an individual component be no greater than three-fourths of the slenderness of the built-up member; that is,

$$\frac{a}{r_i} \leq \frac{3}{4}\frac{KL}{r}$$

where

a = spacing of the connectors

r_i = minimum radius of gyration of the component

KL/r = maximum slenderness ratio of the built-up member.

■ **FIGURE 4.20**

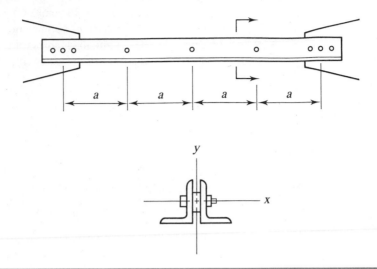

If the member buckles about the axis of symmetry — that is, it is subjected to flexural-torsional buckling about the y-axis — the connectors are subjected to shearing forces. This condition can be visualized by considering two planks used as a beam, as shown in Figure 4.21. If the planks are unconnected, they will slip along the surface of contact when loaded and will function as two separate beams. When connected by bolts (or any other fasteners, such as nails), the two planks will behave as a unit, and the resistance to slip will be provided by shear in the bolts. This behavior takes place in the double-angle shape when bending about its y-axis. If the plank beam is oriented so that bending takes place about its other axis (the b-axis), then both planks bend in exactly the same manner, and there is no slippage and hence no shear. This behavior is analogous

■ **FIGURE 4.21**

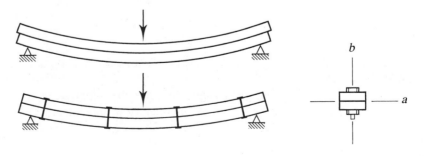

to bending about the x-axis of the double-angle shape. When the fasteners are subjected to shear, a modified slenderness ratio larger than the actual value may be required.

AISC E4 considers two categories of connectors: (1) snug-tight bolts and (2) welds or fully tightened bolts. We cover these connection methods in detail in Chapter 7, "Simple Connections." The column load tables for double-angles are based on the use of welds or fully tightened bolts. For this category,

$$\left(\frac{KL}{r}\right)_m = \sqrt{\left(\frac{KL}{r}\right)_0^2 + 0.82\frac{\alpha^2}{(1+\alpha^2)}\left(\frac{a}{r_{ib}}\right)^2} \qquad \text{(AISC Equation E4-2)}$$

where

$\quad (KL/r)_0 =$ original unmodified slenderness ratio

$\quad (KL/r)_m =$ modified slenderness ratio

$\qquad r_{ib} =$ radius of gyration of component about axis parallel to member axis of buckling

$\qquad \alpha =$ separation ratio $= \dfrac{h}{2r_{ib}}$

$\qquad h =$ distance between component centroids (perpendicular to member axis of buckling)

When the connectors are snug-tight bolts,

$$\left(\frac{KL}{r}\right)_m = \sqrt{\left(\frac{KL}{r}\right)_0^2 + \left(\frac{a}{r_i}\right)^2} \qquad \text{(AISC Equation E4-1)}$$

The column load tables for double-angle shapes show the number of intermediate connectors required for the given y-axis flexural-torsional buckling strength. The number of connectors needed for the x-axis flexural buckling strength must be determined from the requirement that the slenderness of one angle between connectors must not exceed three-fourths of the overall slenderness of the double-angle shape; that is,

$$\frac{a}{r_i} \le \frac{3}{4}\frac{KL}{r}$$

■ EXAMPLE 4.16

Compute the design strength of the compression member shown in Figure 4.22. Two angles, 5 × 3 × ½, are oriented with the long legs back-to-back and separated by ⅜ inch. The effective length KL is 16 feet, and there are three fully tightened intermediate connectors. A36 steel is used.

■ **FIGURE 4.22**

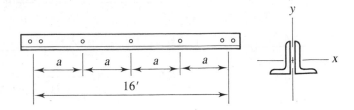

SOLUTION   Compute the flexural buckling strength for the x-axis:

$$\frac{KL}{r_x} = \frac{16(12)}{1.59} = 120.8$$

$$\lambda_c = \frac{KL}{r\pi}\sqrt{\frac{F_y}{E}} = \frac{120.8}{\pi}\sqrt{\frac{36}{29,000}} = 1.355 < 1.5$$

∴ use AISC Equation E2-2

$$F_{cr} = (0.658)^{\lambda_c^2} F_y = (0.658)^{(1.355)^2}(36) = 16.69 \text{ ksi}$$

$$\phi_c P_n = \phi_c A_g F_{cr} = 0.85(7.5)(16.69) = 106 \text{ kips}$$

For the y-axis,

$$\frac{KL}{r_y} = \frac{16(12)}{1.25} = 153.6$$

To determine the flexural-torsional buckling strength for the y-axis, use the modified slenderness ratio, based on the spacing of the connectors. The spacing of the connectors is

$$a = \frac{16(12)}{4 \text{ spaces}} = 48 \text{ in.}$$

Then,

$$\frac{a}{r_i} = \frac{a}{r_z} = \frac{48}{0.648} = 74.07 < 0.75(153.6) = 115.2 \qquad (OK)$$

$$r_{ib} = r_y = 0.829 \text{ in.}$$

$$h = 2(0.75) + \frac{3}{8} = 1.875 \text{ in.}$$

$$\alpha = \frac{h}{2r_{ib}} = \frac{1.875}{2 \times 0.829} = 1.131$$

From AISC Equation E4-2, the modified slenderness ratio is

$$\left(\frac{KL}{r}\right)_m = \sqrt{\left(\frac{KL}{r}\right)_0^2 + 0.82 \frac{\alpha^2}{(1 + \alpha^2)} \left(\frac{a}{r_{ib}}\right)^2}$$

$$= \sqrt{(153.6)^2 + 0.82 \frac{(1.131)^2}{\left[1 + (1.131)^2\right]} \left(\frac{48}{0.829}\right)^2} = 158.5$$

This value should be used in place of KL/r_y for the computation of F_{cry}:

$$\lambda_c = \frac{KL}{r\pi} \sqrt{\frac{F_y}{E}} = \frac{158.5}{\pi} \sqrt{\frac{36}{29,000}} = 1.778 > 1.5$$

$\therefore$ use AISC Equation E2-3

$$F_{cry} = \left[\frac{0.877}{\lambda_c^2}\right] F_y = \left[\frac{0.877}{(1.778)^2}\right](36) = 9.987 \text{ ksi}$$

$$F_{crz} = \frac{GJ}{A\bar{r}_0^2} = \frac{11,200(2 \times 0.3220)}{7.50(2.52)^2} = 151.4 \text{ ksi}$$

$$F_{cry} + F_{crz} = 9.987 + 151.4 = 161.4 \text{ ksi}$$

$$F_{crft} = \left(\frac{F_{cry} + F_{crz}}{2H}\right)\left[1 - \sqrt{1 - \frac{4F_{cry}F_{crz}H}{(F_{cry} + F_{crz})^2}}\right]$$

$$= \frac{161.4}{2(0.645)}\left[1 - \sqrt{1 - \frac{4(9.987)(151.4)(0.645)}{(161.4)^2}}\right] = 9.748 \text{ ksi}$$

$$\phi_c P_n = \phi_c A_g F_{crft} = 0.85(7.50)(9.748) = 62.1 \text{ kips} \quad \text{(controls)}$$

Note that the results of this example compare favorably with the values given in the column load tables.

ANSWER   The design strength is 62 kips.   ■

■ EXAMPLE 4.17

Design a 14-foot-long compression member to resist a factored load of 50 kips. Use a double-angle shape with the short legs back-to-back, separated by ⅜ inch. The member will be braced at midlength against buckling about the x-axis (the axis parallel to the long legs). Specify the number of intermediate connectors needed (the midlength brace will provide one such connector). Use A36 steel.

SOLUTION   From the column load tables, select 2L $3\frac{1}{2} \times 3 \times \frac{1}{4}$, weighing 10.8 lb/ft. The capacity of this shape is 51 kips, based on buckling about the y-axis with an effective length of 14 feet. (The strength corresponding to flexural buckling about the x-axis is 60 kips, based on an effective length of $\frac{14}{2}$, or 7 feet.)

Bending about the y-axis subjects the fasteners to shear, so a sufficient number of fasteners must be provided to account for this action. The table reveals that 3 intermediate connectors are required.

ANSWER   Use 2L $3\frac{1}{2} \times 3 \times \frac{1}{4}$ with three intermediate connectors within the 14-foot length. ■

Connection Requirements for Built-Up Members Composed of Plates or Both Plates and Shapes

When a built-up member consists of two or more rolled shapes separated by a substantial distance, plates must be used to connect the shapes. AISC E4 contains many details regarding the connection requirements and the proportioning of the plates. Additional connection requirements are given for other built-up compression members composed of plates or plates and shapes.

■ **PROBLEMS**

AISC Requirements

4.3-1   Determine the design strength of the compression member shown in Figure P4.3-1, in each of the following ways.

a. Use AISC Equation E2-2 or E2-3.

b. Use Table 3-36 or 3-50 in the Numerical Values section of the Specification.

c. Use Table 4 in the Numerical Values section of the Specification.

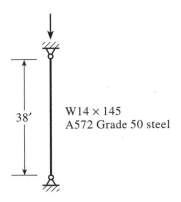

38′   W14 × 145
A572 Grade 50 steel

FIGURE P4.3-1

4.3-2 Determine the design strength of the compression member shown in Figure P4.3-2, in each of the following ways.

  a. Use AISC Equation E2-2 or E2-3.

  b. Use Table 4 in the Numerical Values section of the Specification. (Neither Table 3-36 nor 3-50 can be used. Why not?)

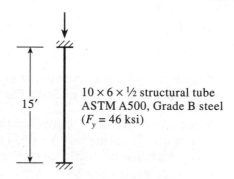

15′ $10 \times 6 \times \frac{1}{2}$ structural tube
ASTM A500, Grade B steel
($F_y = 46$ ksi)

FIGURE P4.3-2

4.3-3 Compute the axial compressive design strength of the member shown in Figure P4.3-3. A36 steel is used.

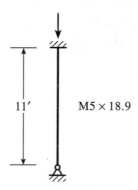

11′   M5 × 18.9

FIGURE P4.3-3

4.3-4 Compute the axial compressive design strength of the member shown in Figure P4.3-4.

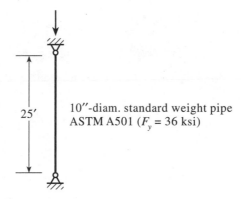

FIGURE P4.3-4

4.3-5   Compute the compressive design strength of the member shown in Figure P4.3-5.

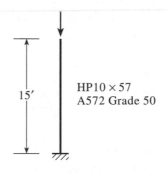

FIGURE P4.3-5

4.3-6   Compute the compressive design strength of a tube 12 × 10 × ½ of A500 Grade B steel (F_y = 46 ksi). The effective length with respect to both axes is 20 feet. Use the column load tables to verify your answer.

4.3-7   Compute the compressive design strength of a W14 × 43 of A572 Grade 50 steel. The effective length with respect to both axes is 16 feet. Use the column load tables to verify your answer.

Design

4.4-1   Choose a W-shape for a column with an effective length of 13 feet to resist a factored load of 570 kips. Use A572 Grade 50 steel and the column load tables in Part 3 of the *Manual.*

4.4-2   Select a W-shape of A36 steel for the member shown in Figure P4.4-2. Use the column load tables. The load consists of a service dead load of 260 kips and a service live load of 520 kips.

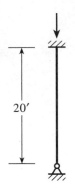

FIGURE P4.4-2

4.4-3 A column with an effective length of 20 feet must resist a factored axial compressive load of 205 kips.

    a. Select the lightest *square* structural tube. Use F_y = 46 ksi and the column load tables.

    b. Select the lightest *rectangular* structural tube. Use F_y = 46 ksi and the column load tables.

4.4-4 A pipe column with an effective length of 15 feet will be used to resist a service dead load of 113 kips and a service live load of 112 kips. Use A36 steel and the column load tables to

    a. select a standard steel pipe.

    b. select an extra-strong steel pipe.

    c. select a double-extra strong steel pipe.

If material weight is the sole determinant of cost, which of the three pipes is the most economical?

4.4-5 Select the lightest HP shape from the column load tables. The load shown in Figure P4.4-5 is factored. Use ASTM A572 Grade 50 steel.

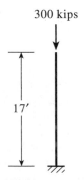

FIGURE P4.4-5

4.4-6 Select a W16 shape for the conditions of Problem 4.4-1.

4.4-7 Select a W21 shape for the conditions of Problem 4.4-5.

4.4-8 Choose a square structural tube $2\frac{1}{2} \times 2\frac{1}{2}$ *or smaller* for the conditions shown in Figure P4.4-8. Use $F_y = 46$ ksi.

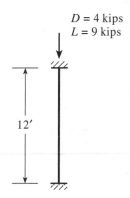

$D = 4$ kips
$L = 9$ kips

12'

FIGURE P4.4-8

Effective Length

4.5-1 A W14 × 38 of A36 steel is used as a compression member. It is 30 feet long and pinned at each end. In addition, it is braced against buckling in the weak direction at 10-foot intervals. What is the maximum factored compressive load that this member can support?

4.5-2 A structural tube 14 × 6 × ⅝ is used as a compression member. It is 20 feet long and is pinned at each end. In addition, it is braced against buckling in the weak direction at midheight. The steel is ASTM A500 Grade B ($F_y = 46$ ksi). Compute the compressive design strength.

4.5-3 A compression member must be designed to resist a service dead load of 180 kips and a service live load of 320 kips. The member will be 28 feet long and pinned at each end. In addition, it will be supported in the weak direction at a point 12 feet from the top. Use A572 Grade 50 steel and select a W-shape.

4.5-4 Select a structural tube of A500 Grade B steel ($F_y = 46$ ksi) to support a service dead load of 14 kips and a service live load of 30 kips. The member will be 30 feet long and pinned at each end. In addition, it will be braced against buckling in the weak direction at the one-third points.

4.5-5 The frame shown in Figure P4.5-5 is unbraced against sidesway. The columns are W10 × 33 and the girder is a W12 × 26. ASTM A572 grade 50 steel is used for all members. The members are oriented so that bending is about the *x*-axis. Assume that $K_y = 1.0$.

    a. Use the alignment chart to determine K_x for the columns. Use the stiffness reduction factor if applicable (P_u = 50 kips for each column).

    b. Compute the compressive design strength of the columns.

    c. Estimate K_x from Table C-C2.1 in the Commentary and compare your estimate with the results of part (a).

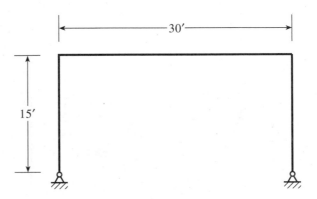

FIGURE P4.5-5

4.5-6   The frame shown in Figure P4.5-6 is unbraced against sidesway. Relative moments of inertia of the members have been assumed for preliminary design purposes. Use the alignment chart and determine K_x for members *AB, BC, DE,* and *EF.*

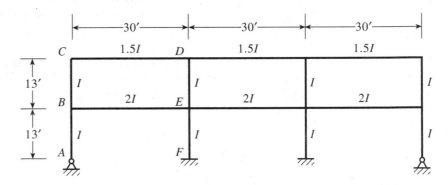

FIGURE P4.5-6

4.5-7   The frame shown in Figure P4.5-7 is unbraced against sidesway. Assume that all columns are W14 × 61 and that all girders are W18 × 76. ASTM A572 grade 50 steel is used for all members. The members are oriented so that bending is about the *x*-axis. Assume that K_y = 1.0.

    a. Use the alignment chart to determine K_x for member *GF*. Use the stiffness reduction factor if applicable (P_u = 350 kips for member *GF*).

    b. Compute the compressive design strength of member *GF*.

c. Estimate K_x from Table C-C2.1 in the Commentary and compare your estimate with the results of part (a).

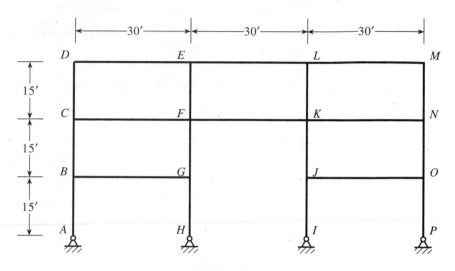

FIGURE P4.5-7

4.5-8 The frame shown in Figure P4.5-8 is braced against sidesway by x-bracing, which is not part of the frame. The columns are W12 × 40 and the girder is a W16 × 40. A36 steel is used for all members. The members are oriented so that bending is about the x-axis. Use $K_y = 0.8$.

a. Use the alignment chart to determine K_x for the columns. Use the stiffness reduction factor if applicable ($P_u = 200$ kips for each column).

b. Compute the compressive design strength of the columns.

c. Estimate K_x from Table C-C2.1 in the Commentary and compare your estimate with the results of part (a).

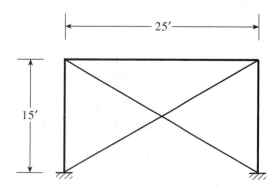

FIGURE P4.5-8

4.5-9   The frame shown in Figure P4.5-9 is unbraced against sidesway. The columns are square structural tubes $6 \times 6 \times \frac{5}{8}$, and the beams are W12 × 22. ASTM A500 grade B steel ($F_y = 46$ ksi) is used for the columns, and $F_y = 50$ ksi for the beams. The beams are oriented so that bending is about the x-axis. Assume that $K_y = 1.0$.

    a. Use the alignment chart to determine K_x for column AB. Use the stiffness reduction factor if applicable ($P_u = 100$ kips for column AB).

    b. Compute the compressive design strength of column AB.

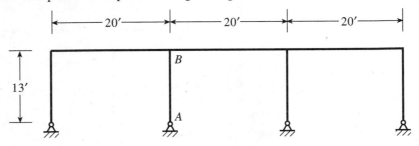

FIGURE P4.5-9

4.5-10   The rigid frame shown in Figure P4.5-10 is unbraced in the plane of the frame. In the direction perpendicular to the frame, the frame is braced at the joints. The connections at these points of bracing are simple (moment-free) connections. Roof girders are W14 × 30, and floor girders are W16 × 36. Member BC is a W10 × 45. Use A36 steel and select a W-shape for AB. Assume that the controlling load combination causes no moment in AB and that the factored axial load is 150 kips.

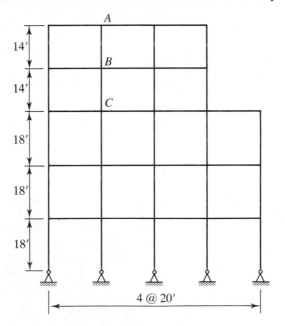

FIGURE P4.5-10

Torsional and Flexural-Torsional Buckling

4.6-1 Compute the compressive design strength for a WT12 × 65.5 with an effective length of 18 feet with respect to each axis. Use A36 steel and the procedure of AISC Section E3 (*not* the appendix).

4.6-2 Use A572 Grade 50 steel and compute the design strength of the column shown in Figure P4.6-2. The member ends are fixed in all directions (*x, y,* and *z*).

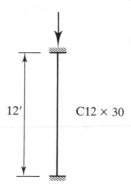

FIGURE P4.6-2

4.6-3 Select an American Standard Channel for the compression member shown in Figure P4.6-3. Use A36 steel. The member ends are fixed in all directions (*x, y,* and *z*).

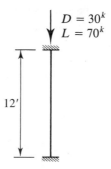

FIGURE P4.6-3

Built-Up Members

4.7-1 Verify the values of r_x and r_y given in the properties tables in Part 1 of the *Manual* for a pair of angles, 5 × 3½ × ½, with the long legs back-to-back. The angles will be connected to a ⅜-inch thick gusset plate.

4.7-2   Verify the values of r_x and r_y given in the properties tables in Part 1 of the *Manual* for the combination section consisting of a W12 × 26 and a C10 × 15.3.

4.7-3   Verify the values of r_x and r_y given in the properties tables in Part 1 of the *Manual* for the combination section consisting of a C12 × 20.7 and an L3½ × 3 × ¼ with the long leg turned out.

4.7-4   A column for a multistory building is fabricated from ASTM A588 plates as shown in Figure P4.7-4. Compute the axial compressive design strength based on flexural buckling (do not consider torsional buckling). Assume that the components of the cross section are connected in such a way that the section is fully effective.

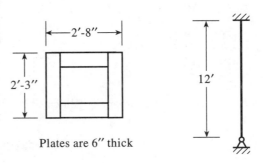

Plates are 6″ thick

FIGURE P4.7-4

4.7-5   Compute the axial compressive design strength based on flexural buckling for the built-up shape shown in Figure P4.7-5. (do not consider torsional buckling). Assume that the components of the cross section are connected in such a way that the section is fully effective. ASTM A242 steel is used.

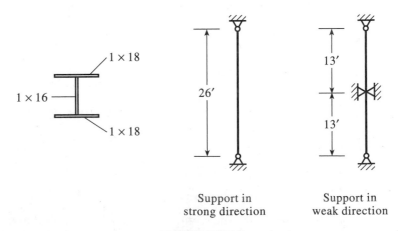

Support in strong direction

Support in weak direction

FIGURE P4.7-5

4.7-6 Two plates $\%_{16} \times 10$ are welded to a W10 × 49 to form a built-up shape, as shown in Figure P4.7-6. Assume that the components are connected so that the cross section is fully effective. A36 steel is used, and $K_xL = K_yL = 25$ ft.

    a. Compute the axial compressive design strength based on flexural buckling (do not consider torsional buckling).

    b. What is the percentage increase in strength from the unreinforced W10 × 49?

FIGURE P4.7-6

4.7-7 A structural tee shape is fabricated by splitting an HP14 × 117, as shown in Figure P4.7-7. Compute the axial compressive design strength based on flexural buckling (do not consider flexural-torsional buckling). Account for the area of the fillets at the web-to-flange junction. A36 steel is used. The effective length is 10 feet with respect to both axes.

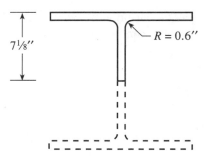

FIGURE P4.7-7

4.7-8 A column cross section is built up from four L5 × 5 × ¾, as shown in Figure P4.7-8. The angles are held in position by *lacing* bars, whose primary function is to hold the angles in position. The lacing is not considered to contribute to the cross-sectional area, which is why it is shown by a dashed line. AISC Section E4 covers the design of lacing. Compute the compressive design strength for A572 Grade 50 steel and an effective length of 30 feet with respect to both axes. Investigate flexural buckling only (no torsional buckling).

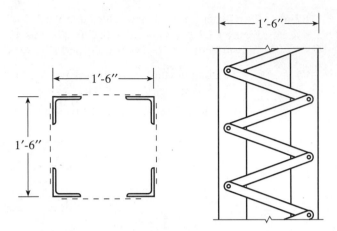

FIGURE P4.7-8

4.7-9   Compute the design strength for the following double-angle shape: 2L6 × 4 × ⅝, long legs ⅜-inch back-to-back, F_y = 50 ksi. The effective length KL is 18 feet for all axes, and there are two intermediate fully tightened bolts. Use the procedure for AISC Section E3 (*not* the appendix). Compare the flexural and the flexural-torsional buckling strengths.

4.7-10   For the conditions shown in Figure P4.7-10, select a double-angle section (⅜-inch gusset plate connection). Use A36 steel. Specify the number of intermediate connectors.

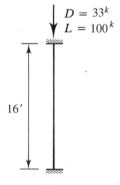

$D = 33^k$
$L = 100^k$

16′

FIGURE P4.7-10

4.7-11   Select a WT section for the compression member shown in Figure P4.7-11. The load is the total service load, with a live-to-dead load ratio of 2.5:1. Use F_y = 50 ksi.

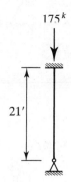

175^k

$21'$

FIGURE P4.7-11

4.7-12 Select a double-angle shape for the top chord of the truss of Problem 3.8-2 (see AISC C2.1). Assume ⅜-inch gusset plates, and use A36 stccl.

5 Beams

5.1 **INTRODUCTION**

Beams are structural members that support transverse loads and are therefore subjected primarily to flexure, or bending. If a substantial amount of axial load is also present, the member is referred to as a *beam-column* (beam-columns are considered in Chapter 6). Although some degree of axial load will be present in any structural member, in many practical situations this effect is negligible and the member can be treated as a beam. Beams are usually thought of as being oriented horizontally and subjected to vertical loads, but that is not necessarily the case. A structural member is considered to be a beam if it is loaded so as to cause bending.

Commonly used cross-sectional shapes include the W-, S-, and M-shapes. Channel shapes are sometimes used, as are beams built up from plates, in the form of I-, H-, or box shapes. For reasons to be discussed later, doubly symmetric shapes such as the standard rolled W-, M-, and S-shapes are the most efficient.

Shapes that are built up from plate elements are usually thought of as plate girders, but the AISC Specification distinguishes beams from plate girders on the basis of the width–thickness ratio of the web. Figure 5.1 shows both a hot-rolled shape and a built-up shape along with the dimensions to be used for the width–thickness ratios. If

$$\frac{h}{t_w} \leq \frac{970}{\sqrt{F_y}}$$

the member is to be treated as a beam, regardless of whether it is a rolled shape or is built-up. This category is considered in Chapter F of the Specification, "Beams and Other Flexural Members," and is the subject of the present chapter of this textbook. If

$$\frac{h}{t_w} > \frac{970}{\sqrt{F_y}}$$

■ **FIGURE 5.1**

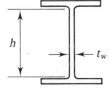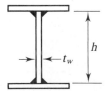

the member is considered to be a plate girder and is covered in Chapter G of the Specification, "Plate Girders." In this textbook, we cover plate girders in Chapter 10. Because of the slenderness of the web, plate girders require special consideration over and above that needed for beams.

All of the standard hot-rolled shapes found in the *Manual* are in the first category (beams). Most built-up shapes are classified as plate girders, but some are beams by the AISC definition.

For beams, the basic relationship between load effects and strength is

$$M_u \leq \phi_b M_n$$

where

M_u = controlling combination of factored load moments

ϕ_b = resistance factor for beams = 0.90

M_n = nominal moment strength

The design strength, $\phi_b M_n$, is sometimes called the *design moment*.

5.2   BENDING STRESS AND THE PLASTIC MOMENT

To be able to determine the nominal moment strength M_n, we must first examine the behavior of beams throughout the full range of loading, from very small loads to the point of collapse. Consider the beam shown in Figure 5.2a, which is oriented so that bending is about the major principal axis (for an I- and H-shape, it will be the *x*–*x* axis). For a linear elastic material and small deformations, the distribution of bending stress will be as shown in Figure 5.2b, with the stress assumed to be uniform across the width of the beam. (Shear is considered separately in Section 5.7.) From elementary mechanics of materials, the stress at any point can be found from the flexure formula:

$$f_b = \frac{My}{I_x} \tag{5.1}$$

where M is the bending moment at the cross section under consideration, y is the perpendicular distance from the neutral plane to the point of interest, and I_x is the moment of inertia of the area of the cross section with respect to the neutral axis. For a homogeneous material, the neutral axis coincides with the centroidal axis. Equation 5.1 is based on the assumption of a linear distribution of strains from top to bottom, which in turn is based on the assumption that cross sections that are plane before bending remain plane after bending. In addition, the beam cross section must have a vertical axis of symmetry, and the loads must be in the longitudinal plane containing this axis. Beams that do not satisfy these criteria are considered in Section 5.13. The maximum stress will occur at the extreme fiber, where y is maximum. Thus there are two maxima: maximum compressive stress in the top fiber and maximum tensile stress in the bottom

■ **FIGURE 5.2**

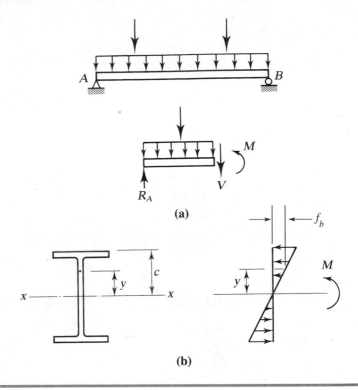

fiber. If the neutral axis is an axis of symmetry, these two stresses will be equal in magnitude. For maximum stress, Equation 5.1 takes the form:

$$f_{max} = \frac{Mc}{I_x} = \frac{M}{I_x/c} = \frac{M}{S_x} \tag{5.2}$$

where c is the perpendicular distance from the neutral axis to the extreme fiber, and S_x is the elastic section modulus of the cross section. For any cross-sectional shape, the section modulus will be a constant. For an unsymmetrical cross section, S_x will have two values: one for the top extreme fiber and one for the bottom. Values of S_x for standard rolled shapes are tabulated in the dimensions and properties tables in the *Manual*.

Equations 5.1 and 5.2 are valid as long as the loads are small enough that the material remains within its linear elastic range. For structural steel, this means that the stress f_{max} must not exceed F_y and that the bending moment must not exceed

$$M_y = F_y S_x$$

where M_y is the bending moment that brings the beam to the point of yielding.

In Figure 5.3, a simply supported beam with a concentrated load at midspan is shown at successive stages of loading. Once yielding begins, the distribution of stress on the cross section will no longer be linear, and yielding will progress from the extreme fiber toward the neutral axis. At the same time, the yielded region will extend longitudinally from the center of the beam as the bending moment reaches M_y at more locations. These yielded regions are indicated by the dark areas in Figure 5.3c and d.

■ **FIGURE 5.3**

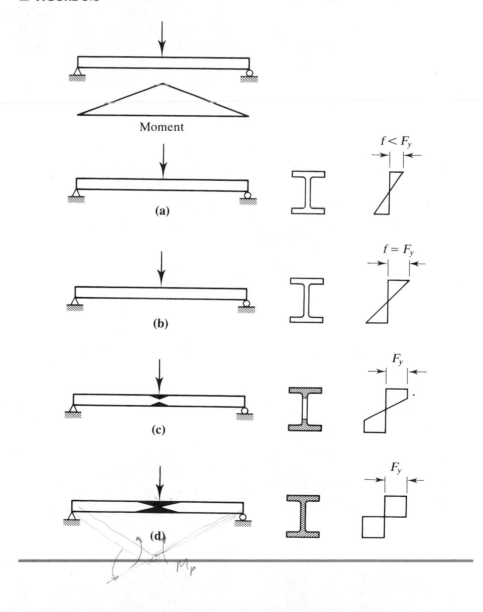

In Figure 5.3b, yielding has just begun. In Figure 5.3c, the yielding has progressed into the web, and in Figure 5.3d the entire cross section has yielded. The additional moment required to bring the beam from stage b to stage d is, on the average, approximately 12% of the yield moment, M_y, for W-shapes. When stage d has been reached, any further increase in the load will cause collapse, since all elements of the cross section have reached the yield plateau of the stress–strain curve and unrestricted plastic flow will occur. A *plastic hinge* is said to have formed at the center of the beam, and this hinge along with the actual hinges at the ends of the beam constitute an unstable mechanism. During plastic collapse, the mechanism motion will be as shown in Figure 5.4. Structural analysis based on a consideration of collapse mechanisms is called *plastic analysis*. An introduction to plastic analysis and design is presented in Appendix A of this book.

The plastic moment capacity, which is the moment required to form the plastic hinge, can easily be computed from a consideration of the corresponding stress distribution. In Figure 5.5, the compressive and tensile stress resultants are shown, where A_c is the cross-sectional area subjected to compression, and A_t is the area in tension. These are the areas above and below the plastic neutral axis, which is not necessarily the same as the elastic neutral axis. From equilibrium of forces.

$$C = T$$
$$A_c F_y = A_t F_y$$
$$A_c = A_t$$

Thus the plastic neutral axis divides the cross section into two equal areas. For shapes that are symmetrical about the axis of bending, the elastic and plastic neutral axes are the same. The plastic moment M_p is the resisting couple formed by the two equal and opposite forces, or

$$M_p = F_y(A_c)a = F_y(A_t)a = F_y\left(\frac{A}{2}\right)a = F_y Z$$

where

A = total cross-sectional area

a = distance between the centroids of the two half-areas

$Z = \left(\dfrac{A}{2}\right)a$ = plastic section modulus

■ **FIGURE 5.4**

Plastic hinge

■ **FIGURE 5.5**

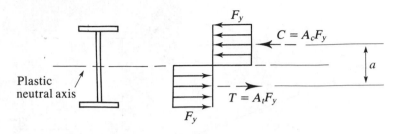

■ **EXAMPLE 5.1**

For the built-up shape shown in Figure 5.6, determine (a) the elastic section modulus S and the yield moment M_y, and (b) the plastic section modulus Z and the plastic moment M_p. Bending is about the x-axis, and the steel is A572 Grade 50.

SOLUTION

a. Because of symmetry, the elastic neutral axis (the x-axis) is located at mid-depth of the cross section (the location of the centroid). The moment of inertia of the cross section can be found by using the parallel axis theorem, and the results of the calculations are summarized in Table 5.1.

TABLE 5.1

Component	$\bar{I}$	A	d	$\bar{I} + Ad^2$
Flange	0.6667	8	6.5	338.7
Flange	0.6667	8	6.5	338.7
Web	72	—	—	72.0
Sum				749.4

The elastic section modulus is

$$S = \frac{I}{c} = \frac{749.4}{1 + (12/2)} = \frac{749.4}{7} = 107 \text{ in.}^3$$

and the yield moment is

$$M_y = F_y S = 50(107) = 5350 \text{ in.-kips} = 446 \text{ ft-kips}$$

ANSWER $S = 107 \text{ in.}^3$ and $M_y = 446 \text{ ft-kips}$

■ **FIGURE 5.6**

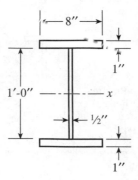

b. Because this shape is symmetrical about the x-axis, this axis divides the cross section into equal areas and is therefore the plastic neutral axis. The centroid of the top half-area can be found by the principle of moments. Taking moments about the neutral axis of the entire cross section (Figure 5.6) and tabulating the computations in Table 5.2, we get

$$\bar{y} = \frac{\Sigma \, Ay}{\Sigma \, A} = \frac{61}{11} = 5.545 \text{ in.}$$

TABLE 5.2

Component	A	y	Ay
Flange	8	6.5	52
Web	3	3	9
Sum	11		61

Figure 5.7 shows that the moment arm of the internal resisting couple is

$$a = 2\bar{y} = 2(5.545) = 11.09 \text{ in.}$$

and that the plastic section modulus is

$$\left(\frac{A}{2}\right)a = 11(11.09) = 122 \text{ in.}^3$$

The plastic moment is

$$M_p = F_y Z = 50(122) = 6100 \text{ in.-kips} = 508 \text{ ft-kips}$$

■ **FIGURE 5.7**

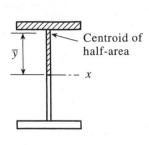 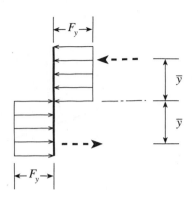

ANSWER $Z = 122$ in.^{3} and $M_p = 508$ ft-kips   ■

■ **EXAMPLE 5.2**

Compute the plastic moment, M_p, for a W10 × 60 of A36 steel.

SOLUTION   From the dimensions and properties tables in Part 1 of the *Manual,*

$$A = 17.6 \text{ in.}^2$$

$$\frac{A}{2} = \frac{17.6}{2} = 8.8 \text{ in.}^2$$

The centroid of the half-area can be found in the tables for WT-shapes, which are cut from W-shapes. The relevant shape here is the WT5 × 30, and the distance from the outside face of the flange to the centroid is 0.884 inch, as shown in Figure 5.8.

$$a = d - 2(0.884) = 10.22 - 2(0.884) = 8.452 \text{ in.}$$

$$Z = \left(\frac{A}{2}\right)a = 8.8(8.452) = 74.38 \text{ in.}^3$$

This result compares favorably with the value of 74.6 given in the dimensions and properties tables (the difference results from rounding of the tabular values).

■ **FIGURE 5.8**

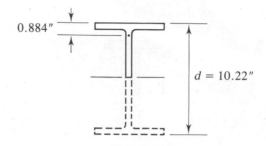

ANSWER $M_p = F_y Z = 36(74.38) = 2678 \text{ in.-kips} = 223 \text{ ft-kips.}$

| 5.3 | **STABILITY** |

If a beam can be counted on to remain stable up to the fully plastic condition, the nominal moment strength can be taken as the plastic moment capacity; that is,

$$M_n = M_p$$

Otherwise, M_n will be less than M_p.

As with a compression member, instability can be in an overall sense or it can be local. Overall buckling is illustrated in Figure 5.9a. When a beam bends, the compression region (above the neutral axis) is analogous to a column, and in a manner similar to a column, will buckle if the member is slender enough. Unlike a column, however, the compression portion of the cross section is restrained by the tension portion, and the outward deflection (flexural buckling) is accompanied by twisting (torsion). This form of instability is called *lateral-torsional buckling* (LTB). Lateral-torsional buckling can be prevented by lateral bracing of the compression zone, preferably the compression flange, at sufficiently close intervals. This bracing is shown schematically in Figure 5.9b. As we shall see, the moment strength depends in part on the unbraced length, which is the distance between points of lateral support.

Whether the beam can sustain a moment large enough to bring it to the fully plastic condition also depends on whether the cross-sectional integrity is maintained. This integrity will be lost if one of the compression elements of the cross section buckles. This type of buckling can be either compression flange buckling, called *flange local buckling* (FLB), or buckling of the compression part of the web, called *web local buckling* (WLB). As discussed in Chapter 4, "Compression Members," whether either type of local buckling occurs will depend on the width–thickness ratios of the compression elements of the cross section.

■ **FIGURE 5.9**

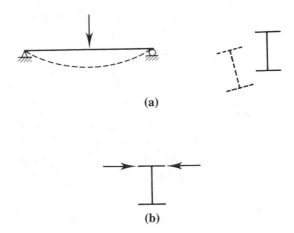

(a)

(b)

Figure 5.10 further illustrates the effects of local and lateral-torsional buckling. Five separate beams are represented on this graph of load versus central deflection. Curve 1 is the load-deflection curve of a beam that becomes unstable (in any way) and loses its load-carrying capacity before first yield (see Figure 5.3b) is attained. Curves 2 and 3 correspond to beams that can be loaded past first yield but not far enough for the formation of a plastic hinge and the resulting plastic collapse. If plastic collapse can be reached, the load-deflection curve will have the appearance of either curve 4 or curve 5. Curve 4 is for the case of uniform moment over the full length of the beam,

■ **FIGURE 5.10**

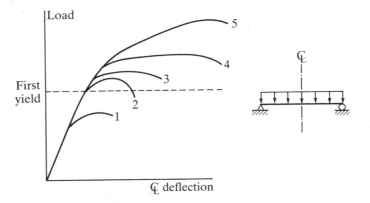

and curve 5 is for a beam with a variable bending moment (moment gradient). Safe designs can be achieved with beams corresponding to any of these curves, but curves 1 and 2 represent inefficient use of material.

CLASSIFICATION OF SHAPES

AISC classifies cross-sectional shapes as compact, noncompact, or slender, depending on the values of the width–thickness ratios. For I- and H-shapes, the ratio for the projecting flange (an *unstiffened* element) is $b_f/2t_f$, and the ratio for the web (a *stiffened* element), is h/t_w. The classification of shapes is found in Section B5 of the Specification, "Local Buckling," in Table B5.1. It can be summarized as follows. Let

λ = width–thickness ratio

λ_p = upper limit for compact category

λ_r = upper limit for noncompact category

Then

if $\lambda \leq \lambda_p$ and the flange is continuously connected to the web, the shape is compact;

if $\lambda_p < \lambda \leq \lambda_r$, the shape is noncompact; and

if $\lambda > \lambda_r$, the shape is slender.

The category is based on the worst width–thickness ratio of the cross section. For example, if the web is compact and the flange is noncompact, the shape is classified as noncompact. Table 5.3 has been extracted from AISC Table B5.1 and contains width–thickness ratios for hot-rolled I- and H-shaped cross sections.

TABLE 5.3 Width–Thickness Parameters*

Element	λ	λ_p	λ_r
Flange	$\dfrac{b_f}{2t_f}$	$\dfrac{65}{\sqrt{F_y}}$	$\dfrac{141}{\sqrt{F_y - 10}}$
Web	$\dfrac{h}{t_w}$	$\dfrac{640}{\sqrt{F_y}}$	$\dfrac{970}{\sqrt{F_y}}$

*For hot-rolled I- and H-shapes in flexure.

BENDING STRENGTH OF COMPACT SHAPES

A beam can fail by reaching M_p and becoming fully plastic, or it can fail by

1. lateral-torsional buckling (LTB), either elastically or inelastically;
2. flange local buckling (FLB), elastically or inelastically; or
3. web local buckling (WLB), elastically or inelastically.

If the maximum bending stress is less than the proportional limit when buckling occurs, the failure is said to be *elastic*. Otherwise, it is *inelastic*. (See the related discussion in Section 4.2, "Column Theory.")

For convenience we first categorize beams as compact, noncompact, or slender, and then determine the moment resistance based on the degree of lateral support. The discussion in this section applies to two types of beams: (1) hot-rolled I- and H-shapes bent about the strong axis and loaded in the plane of the weak axis; and (2) channels bent about the strong axis and either loaded through the shear center or restrained against twisting. (The shear center is the point on the cross section through which a transverse load must pass if the beam is to bend without twisting.) Emphasis will be on I- and H-shapes. Hybrid beams (those with different grades of steel in the web and flanges) will not be considered, and some of the AISC equations will be slightly modified to reflect this specialization; F_{yf} and F_{yw}, the yield strengths of the flange and web, will be replaced by F_y.

We begin with *compact shapes,* defined as those whose webs are continuously connected to the flanges and that satisfy the following width–thickness ratio requirements for the flange and the web:

$$\frac{b_f}{2t_f} \le \frac{65}{\sqrt{F_y}} \quad \text{and} \quad \frac{h}{t_w} \le \frac{640}{\sqrt{F_y}}$$

The web criterion is met by all standard hot-rolled shapes listed in the *Manual* (see Problems 5.4-1 and 5.4-2), so only the flange ratio need be checked. Most shapes will also satisfy the flange requirement and will therefore be classified as compact. If the beam is compact and has continuous lateral support, or if the unbraced length is very short, the nominal moment strength, M_n, is the full plastic moment capacity of the shape, M_p. For members with inadequate lateral support, the moment resistance is limited by the lateral-torsional buckling strength, either inelastic or elastic.

The first category, laterally supported compact beams, is quite common and is the simplest case. AISC F1.1 gives the nominal strength as

$$M_n = M_p \qquad \text{(AISC Equation F1-1)}$$

where

$$M_p = F_y Z \le 1.5 M_y$$

The limit of $1.5M_y$ for M_p is to prevent excessive working load deformations and is satisfied when

$$F_yZ \leq 1.5F_yS \quad \text{or} \quad \frac{Z}{S} \leq 1.5$$

For I- and H-shapes bent about the *strong* axis, Z/S will *always* be ≤ 1.5. (For I- and H-shapes bent about the *weak* axis, however, Z/S will *never* be ≤ 1.5.)

■ EXAMPLE 5.3

The beam shown in Figure 5.11 is a W16 × 31 of A36 steel. It supports a reinforced concrete floor slab that provides continuous lateral support of the compression flange. The service dead load is 450 lb/ft. This load is *superimposed* on the beam; it does not include the weight of the beam itself. The service live load is 550 lb/ft. Does this beam have adequate moment strength?

SOLUTION   The total service dead load, including the weight of the beam, is

$$w_D = 450 + 31 = 481 \text{ lb/ft}$$

For a simply supported, uniformly loaded beam, the maximum bending moment occurs at midspan and is equal to

$$M_{\max} = \frac{1}{8} wL^2$$

where w is the load in units of force per unit length, and L is the span length. Then

$$M_D = \frac{1}{8} w_D L^2 = \frac{0.481(30)^2}{8} = 54.11 \text{ ft-kips}$$

$$M_L = \frac{0.550(30)^2}{8} = 61.88 \text{ ft-kips}$$

■ **FIGURE 5.11**

$w_D = 450$ lb/ft
$w_L = 550$ lb/ft

|← 30' →|

The dead load is less than 8 times the live load, so load combination A4-2 controls:

$$M_u = 1.2M_D + 1.6M_L = 1.2(54.11) + 1.6(61.88) = 164 \text{ ft-kips}$$

Alternatively, the loads can be factored at the outset:

$$w_u = 1.2w_D + 1.6w_L = 1.2(0.431) + 1.6(0.550) = 1.457 \text{ kips/ft}$$

$$M_u = \frac{1}{8} w_u L^2 = \frac{1.457(30)^2}{8} = 164 \text{ ft-kips}$$

Check for compactness:

$$\frac{b_f}{2t_f} = 6.3 \qquad \text{(from Part 1 of the } \textit{Manual}\text{)}$$

$$\frac{65}{\sqrt{F_y}} = \frac{65}{\sqrt{36}} = 10.8 > 6.3 \qquad \therefore \text{ the flange is compact.}$$

$$\frac{h}{t_w} < \frac{640}{\sqrt{F_y}} \qquad \text{(for all shapes in the } \textit{Manual}\text{)}$$

$$\therefore \text{ a W16} \times 31 \text{ is compact for A36 steel.}$$

Because the beam is compact and laterally supported,

$$M_n = M_p = F_y Z_x = 36(54.0) = 1944 \text{ in.-kips} = 162.0 \text{ ft-kips}$$

Check for $M_p \le 1.5M_y$:

$$\frac{Z_x}{S_x} = \frac{54.0}{47.2} = 1.14 < 1.5 \qquad \text{(OK)}$$

$$\phi_b M_n = 0.90(162.0) = 146 \text{ ft-kips} < 164 \text{ ft-kips} \qquad \text{(N.G.)}$$

ANSWER   The design moment is less than the factored load moment, so the W16 × 31 is unsatisfactory. ■

Although a check for $M_p \le 1.5M_y$ was made in the preceding example, it is not necessary for I- and H-shapes bent about the strong axis and is not routinely done in this book.

The moment strength of compact shapes is a function of the unbraced length, L_b, defined as the distance between points of lateral support, or bracing. In this book, we indicate points of lateral support with an "×," as shown in Figure 5.12. The relationship between the nominal strength M_n and the unbraced length is shown in Figure 5.13. If the unbraced length is no greater than L_p, to be defined presently, the beam is considered to have full lateral support, and $M_n = M_p$. If L_b is greater than L_p but less than or equal to the parameter L_r, the strength is based on inelastic LTB. If L_b is greater than L_r, the strength is based on elastic LTB.

■ **FIGURE 5.12**

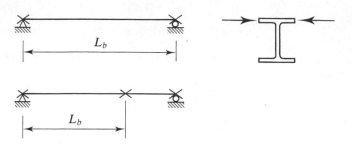

The equation for the theoretical elastic lateral-torsional buckling strength can be found in the *Theory of Elastic Stability* (Timoshenko and Gere, 1961) and, with some notational changes, is

$$M_n = \frac{\pi}{L_b} \sqrt{EI_y GJ + \left(\frac{\pi E}{L_b}\right)^2 I_y C_w} \tag{5.3}$$

■ **FIGURE 5.13**

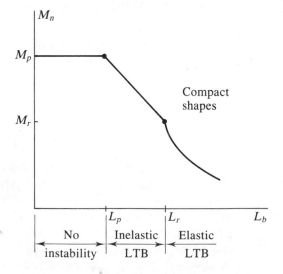

where

L_b = unbraced length (in.)

G = shear modulus = 11,200 ksi for structural steel

J = torsional constant (in.^{4})

C_w = warping constant (in.^{6})

If the moment when lateral-torsional buckling occurs is greater than the moment corresponding to first yield, the strength is based on inelastic behavior. The moment corresponding to first yield is

$$M_r = F_L S_x \qquad \text{(AISC Equation F1-7)}$$

where F_L is the smaller of $(F_{yf} - F_r)$ or F_{yw}. In this expression, the flange yield stress is reduced by F_r, the residual stress. For a nonhybrid member, $F_{yf} = F_{yw} = F_y$, and F_L will always equal $F_y - F_r$. In the remainder of this chapter, we replace F_L with $F_y - F_r$. For example, we write AISC Equation F1-7 as

$$M_r = (F_y - F_r)S_x \qquad \text{(AISC Equation F1-7)}$$

where the residual stress F_r is 10 ksi for rolled shapes and 16.5 ksi for welded built-up shapes. As shown in Figure 5.13, the boundary between elastic and inelastic behavior will be for an unbraced length of L_r, which is the value of L_b obtained from Equation 5.3 when M_n is set equal to M_r. The following equation is obtained:

$$L_r = \frac{r_y X_1}{(F_y - F_r)} \sqrt{1 + \sqrt{1 + X_2(F_y - F_r)^2}} \qquad \text{(AISC Equation F1-6)}$$

where

$$X_1 = \frac{\pi}{S_x} \sqrt{\frac{EGJA}{2}}$$

$$X_2 = \frac{4C_w}{I_y} \left(\frac{S_x}{GJ}\right)^2 \qquad \text{(AISC Equations F1-8 and F1-9)}$$

As with columns, inelastic behavior of beams is more complicated than elastic behavior, and empirical formulas are often used. With one minor modification, the following equation is used by AISC:

$$M_n = M_p - (M_p - M_r)\left(\frac{L_b - L_p}{L_r - L_p}\right) \qquad (5.4)$$

where

$$L_p = \frac{300 r_y}{\sqrt{F_y}} \qquad \text{(AISC Equation F1-4)}$$

The nominal bending strength of compact beams is completely described by Equations 5.3 and 5.4, subject to an upper limit of M_p for inelastic beams, provided that the applied moment is uniform over the unbraced length L_b. Otherwise there is a *moment gradient,* and Equations 5.3 and 5.4 must be modified by a factor C_b. This factor is given in AISC F1.2 as

$$C_b = \frac{12.5M_{\max}}{2.5M_{\max} + 3M_A + 4M_B + 3M_C} \qquad \text{(AISC Equation F1-3)}$$

where

$M_{\max}$ = absolute value of the maximum moment within the unbraced length (including the end points)

M_A = absolute value of the moment at the quarter point of the unbraced length

M_B = absolute value of the moment at the midpoint of the unbraced length

M_C = absolute value of the moment at the three-quarter point of the unbraced length

When the bending moment is uniform, the value of C_b is

$$C_b = \frac{12.5M}{2.5M + 3M + 4M + 3M} = 1.0$$

■ EXAMPLE 5.4

Determine C_b for a uniformly loaded, simply supported beam with lateral support at its ends only.

SOLUTION Because of symmetry, the maximum moment is at midspan, so

$$M_{\max} = M_B = \frac{1}{8} wL^2$$

Also because of symmetry, the moment at the quarter point equals the moment at the three-quarter point. From Figure 5.14,

$$M_A = M_C = \frac{wL}{2}\left(\frac{L}{4}\right) - \frac{wL}{4}\left(\frac{L}{8}\right) = \frac{wL^2}{8} - \frac{wL^2}{32} = \frac{3}{32} wL^2$$

$$C_b = \frac{12.5M_{\max}}{2.5M_{\max} + 3M_A + 4M_B + 3M_C}$$

$$= \frac{12.5(1/8)}{2.5(1/8) + 3(3/32) + 4(1/8) + 3(3/32)} = 1.14$$

■ **FIGURE 5.14**

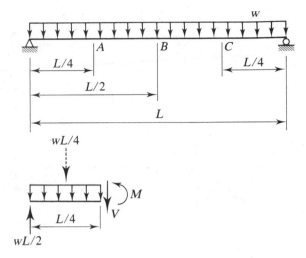

ANSWER $C_b = 1.14$.

Figure 5.15 shows the value of C_b for several common cases of loading and lateral support.

For unbraced cantilever beams, AISC specifies a value of C_b of 1.0. A value of 1.0 is always conservative, regardless of beam configuration or loading, but in some cases it may be excessively conservative. The complete specification of nominal moment strength for compact shapes can now be summarized.

For $L_b \leq L_p$,

$$M_n = M_p \leq 1.5M_y \qquad \text{(AISC Equation F1-1)}$$

For $L_p < L_b \leq L_r$,

$$M_n = C_b\left[M_p - (M_p - M_r)\left(\frac{L_b - L_p}{L_r - L_p}\right)\right] \leq M_p \qquad \text{(AISC Equation F1-2)}$$

For $L_b > L_r$,

$$M_n = M_{cr} \leq M_p \qquad \text{(AISC Equation F1-12)}$$

■ **FIGURE 5.15**

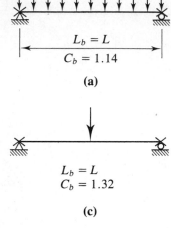

$L_b = L$
$C_b = 1.14$

(a)

$L_b = L/2$
$C_b = 1.30$

(b)

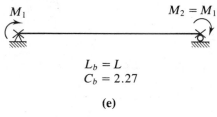

$L_b = L$
$C_b = 1.32$

(c)

$L_b = L/2$
$C_b = 1.67$

(d)

M_1

$M_2 = M_1$

$L_b = L$
$C_b = 2.27$

(e)

A B C D

a a

AB and CD : $C_b = 1.67$
BC : $C_b = 1.00$

(f)

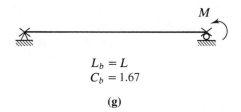

M

$L_b = L$
$C_b = 1.67$

(g)

where

$$M_{cr} = C_b \frac{\pi}{L_b} \sqrt{EI_y GJ + \left(\frac{\pi E}{L_b}\right)^2 I_y C_w} \qquad \text{(AISC Equation F1-13)}$$

$$= \frac{C_b S_x X_1 \sqrt{2}}{L_b / r_y} \sqrt{1 + \frac{X_1^2 X_2}{2(L_b / r_y)^2}}$$

■ **FIGURE 5.16**

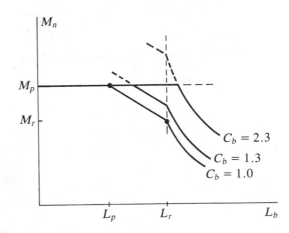

The constants X_1 and X_2 have been previously defined and are tabulated in the dimensions and properties tables in the *Manual*.

The effect of C_b on the nominal strength is illustrated in Figure 5.16. Although the strength is directly proportional to C_b, this graph clearly shows the importance of observing the upper limit of M_p, regardless of which equation is used for M_n.

■ **EXAMPLE 5.5**

Determine the design strength $\phi_b M_n$ for a W14 × 68 of A242 steel subject to

    a. continuous lateral support.
    b. unbraced length = 20 ft; C_b = 1.0.
    c. unbraced length = 20 ft; C_b = 1.75.

SOLUTION

    a. From Part 1 of the *Manual*, a W14 × 68 is in shape group 2 and is therefore available with a yield stress F_y of 50 ksi. Determine whether shape is compact, noncompact, or slender:

$$\frac{b_f}{2t_f} = 7.0 < \frac{65}{\sqrt{50}}$$

This shape is compact and thus

$$M_n = M_p = F_y Z_x = 50(115) = 5750 \text{ in.-kips} = 479.2 \text{ ft-kips}$$

W14×68

ANSWER $\phi_b M_n = 0.90(479.2) = 431$ ft-kips

b. $L_b = 20$ ft and $C_b = 1.0$. Compute L_p and L_r:

$$L_p = \frac{300 r_y}{\sqrt{F_y}} = \frac{300(2.46)}{\sqrt{50}} = 104.4 \text{ in.} = 8.700 \text{ ft}$$

From the torsion properties tables in Part 1 of the *Manual*,

$$J = 3.02 \text{ in.}^4 \quad \text{and} \quad C_w = 5380 \text{ in.}^6$$

Although X_1 and X_2 are tabulated in the dimensions and properties tables in Part 1 of the *Manual*, we compute them here for illustration:

$$X_1 = \frac{\pi}{S_x} \sqrt{\frac{EGJA}{2}} = \frac{\pi}{103} \sqrt{\frac{29,000(11,200)(3.02)(20)}{2}} = 3021 \text{ ksi}$$

$$X_2 = 4 \frac{C_w}{I_y} \left(\frac{S_x}{GJ}\right)^2 = 4\left(\frac{5380}{121}\right)\left(\frac{103}{11,200 \times 3.02}\right)^2 = 0.001649 \text{ (ksi)}^{-2}$$

$$L_r = \frac{r_y X_1}{(F_y - F_r)} \sqrt{1 + \sqrt{1 + X_2(F_y - F_r)^2}}$$

$$= \frac{2.46(3021)}{(50 - 10)} \sqrt{1 + \sqrt{1 + 0.001649(50 - 10)^2}} = 316.8 \text{ in.} = 26.40 \text{ ft}$$

Since $L_p < L_b < L_r$, the strength is based on inelastic LTB and

$$M_r = (F_y - F_r)S_x = \frac{(50 - 10)(103)}{12} = 343.3 \text{ ft-kips}$$

$$M_n = C_b \left[M_p - (M_p - M_r)\left(\frac{L_b - L_p}{L_r - L_p}\right) \right]$$

$$= 1.0 \left[479.2 - (479.2 - 343.3)\left(\frac{20 - 8.700}{26.40 - 8.700}\right) \right]$$

$$= 392.4 \text{ ft-kips} < M_p$$

ANSWER $\phi_b M_n = 0.90(392.4) = 353$ ft-kips

c. $L_b = 20$ ft and $C_b = 1.75$. The design strength for $C_b = 1.75$ is 1.75 times the design strength for $C_b = 1.0$. Therefore

$$M_n = 1.75(392.4) = 686.7 \text{ ft-kips} > M_p = 479.2 \text{ ft-kips}$$

The nominal strength cannot exceed M_p; hence use a nominal strength of $M_n = 479.2$ ft-kips:

ANSWER $\phi_b M_n = 0.90(479.2) = 431$ ft-kips   ■

Part 4 of the *Manual of Steel Construction,* "Beam and Girder Design," contains several useful tables and charts for the analysis and design of beams. For example, the Load Factor Design Selection Table lists shapes commonly used for beams, arranged in order of plastic section modulus, Z_x. Since $M_p = F_y Z_x$, the shapes are also in order of the design moment $\phi_b M_p$, which is also listed. Other useful constants tabulated include L_p and L_r (which is particularly tedious to compute). We cover additional design aids in Part 4 of the *Manual* in other sections of this chapter.

Plastic Analysis

In most cases, the maximum factored load moment M_u will be obtained by an elastic structural analysis using factored loads. Under certain conditions, the required strength for a statically indeterminate structure may be found by *plastic analysis*. AISC authorizes the use of plastic analysis if the shape is compact and if

$$L_b \leq L_{pd}$$

where

$$L_{pd} = \frac{3600 + 2200(M_1/M_2)}{F_y} r_y \qquad \text{(AISC Equation F1-17)}$$

M_1 = smaller of the two end moments for the unbraced segment

M_2 = larger of the two end moments for the unbraced segment

The ratio M_1/M_2 is positive when the moments cause reverse curvature in the unbraced segment. In this context, L_b is the unbraced length adjacent to a plastic hinge that is part of the failure mechanism. If plastic analysis is used, however, the nominal moment strength M_n adjacent to the last hinge to form and in regions not adjacent to plastic hinges must be determined in the same way as for beams analyzed by elastic methods, and it may be less than M_p. For a more complete discussion of plastic analysis, refer to Appendix A.

5.6 BENDING STRENGTH OF NONCOMPACT SHAPES

As previously noted, most standard W-, M-, and S-shapes are compact for $F_y = 36$ ksi and $F_y = 50$ ksi. A few are noncompact because of the flange width–thickness ratio, but none are slender. For these reasons, the AISC Specification deals with noncompact and slender flexural members in an appendix (Appendix F). In this book, we defer consideration of slender flexural members until Chapter 10, "Plate Girders."

In general, a beam may fail by lateral-torsional buckling, flange local buckling, or web local buckling. Any of these types of failure can be in either the elastic range or the inelastic range. The webs of all rolled shapes in the *Manual* are compact, so the noncompact shapes are subject only to the limit states of lateral-torsional buckling and flange local buckling. The strengths corresponding to both limit states must be computed and the smaller value will control. From AISC Appendix F, with

$$\lambda = \frac{b_f}{2t_f}$$

if $\lambda_p < \lambda \le \lambda_r$, the flange is noncompact, buckling will be inelastic, and

$$M_n = M_p - (M_p - M_r)\left(\frac{\lambda - \lambda_p}{\lambda_r - \lambda_p}\right) \qquad \text{(AISC Equation A-F1-3)}$$

where

$$\lambda_p = \frac{65}{\sqrt{F_y}}$$

$$\lambda_r = \frac{141}{\sqrt{F_y - F_r}}$$

$$M_r = (F_y - F_r)S_x$$

$$F_r = \text{residual stress} = 10 \text{ ksi for rolled shapes}$$

(These terms have been specialized for nonhybrid beams.)

■ **EXAMPLE 5.6**

A simply supported beam with a span length of 40 feet is laterally supported at its ends and is subjected to the following service loads:

Dead load = 400 lb/ft (including the weight of the beam)

Live load = 1000 lb/ft

If ASTM A572 Grade 50 steel is used, is a W14 × 90 adequate?

SOLUTION   The factored load and moment are

$$w_u = 1.2w_D + 1.6w_L = 1.2(0.400) + 1.6(1.000) = 2.080 \text{ kips/ft}$$

$$M_u = \frac{1}{8} w_u L^2 = \frac{2.080(40)^2}{8} = 416.0 \text{ ft-kips}$$

Determine whether shape is compact, noncompact, or slender:

$$\lambda = \frac{b_f}{2t_f} = 10.2$$

$$\lambda_p = \frac{65}{\sqrt{F_y}} = \frac{65}{\sqrt{50}} = 9.19$$

$$\lambda_r = \frac{141}{\sqrt{F_y - F_r}} = \frac{141}{\sqrt{50 - 10}} = 22.3$$

As $\lambda_p < \lambda < \lambda_r$, this shape is noncompact. Check the capacity based on the limit state of flange local buckling:

$$M_p = F_y Z_x = \frac{50(157)}{12} = 654.2 \text{ ft-kips}$$

$$M_r = (F_y - F_r)S_x = \frac{(50 - 10)(143)}{12} = 476.7 \text{ ft-kips}$$

$$M_n = M_p - (M_p - M_r)\left(\frac{\lambda - \lambda_p}{\lambda_r - \lambda_p}\right)$$

$$= 654.2 - (654.2 - 476.7)\left(\frac{10.2 - 9.19}{22.3 - 9.19}\right) = 640.5 \text{ ft-kips}$$

The design strength based on FLB is therefore

$$\phi_b M_n = 0.90(640.5) = 576 \text{ ft-kips}$$

Check the capacity based on the limit state of lateral-torsional buckling. From the Load Factor Design Selection Table,

$$L_p = 15 \text{ ft} \quad \text{and} \quad L_r = 38.4 \text{ ft}$$
$$L_b = 40 \text{ ft} > L_r \quad \therefore \quad \text{failure is by } \textit{elastic} \text{ LTB.}$$

From Part 1 of the *Manual,*

$$I_y = 362 \text{ in.}^4$$
$$J = 4.06 \text{ in.}^4$$
$$C_w = 16,000 \text{ in.}^6$$

For a uniformly loaded, simply supported beam with lateral support at the ends,

$$C_b = 1.14.$$

AISC Equation F1-13 gives

$$M_n = C_b \frac{\pi}{L_b} \sqrt{EI_y GJ + \left(\frac{\pi E}{L_b}\right)^2 I_y C_w} \leq M_p$$

$$= 1.14\left[\frac{\pi}{40(12)} \sqrt{29,000(362)(11,200)(4.06) + \left(\frac{\pi \times 29,000}{40 \times 12}\right)^2 (362)(16,000)}\right]$$

$$= 1.14(5421) = 6180 \text{ in.-kips} = 515.0 \text{ ft-kips}$$
$$M_p = 654.2 \text{ ft-kips} > 515.0 \text{ ft-kips} \quad \text{(OK)}$$

Because $515.0 < 640.5$, LTB controls, and

$$\phi_b M_n = 0.90(515.0) = 464 \text{ ft-kips} > M_u = 416 \text{ ft-kips} \qquad \text{(OK)}$$

ANSWER   Since $M_u < \phi_b M_n$, the beam has adequate moment strength.   ■

The identification of noncompact shapes is facilitated by the Load Factor Design Selection Table. Noncompact shapes are marked by footnotes that specify whether the shape is noncompact for either $F_y = 36$ ksi or $F_y = 50$ ksi. Noncompact shapes are also treated differently in the table in the following ways.

1. For noncompact shapes, the tabulated values of $\phi_b M_p$ *are actually values of the design strength based on flange local buckling*. In Example 5.6, we computed this value to be 576 ft-kips, the correct tabulated value, whereas the actual value of $\phi_b M_p$ is $0.90(654.2) = 589$ ft-kips.

2. The tabulated value of L_p is the value of unbraced length at which the nominal strength based on inelastic lateral torsional buckling equals the nominal strength based on flange local buckling; that is, the maximum unbraced length for which the nominal strength can be taken as the strength based on web local buckling. (Recall that L_p for compact shapes is the maximum unbraced length for which the nominal strength can be taken as the plastic moment.) For the shape in Example 5.6, equate the nominal strength based on FLB to the strength based on inelastic LTB (AISC Equation F1-2), with $C_b = 1.0$:

$$M_n = M_p - (M_p - M_r)\left(\frac{L_b - L_p}{L_r - L_p}\right) \tag{5.5}$$

The values of M_r and L_r were given in Example 5.6 and are unchanged. The value of L_p, however, must be computed from AISC Equation F1-4:

$$L_p = \frac{300 r_y}{\sqrt{F_y}} = \frac{300(3.70)}{\sqrt{50}} = 157.0 \text{ in.} = 13.08 \text{ ft.}$$

Returning to Equation 5.5, we obtain

$$640.5 = 654.2 - (654.2 - 476.7)\left(\frac{L_b - 13.08}{38.4 - 13.08}\right)$$

$$L_b = 15.0 \text{ ft}$$

This is the value tabulated as L_p for a W14 × 90 with $F_y = 50$ ksi. Note that

$$L_p = \frac{300 r_y}{\sqrt{F_y}}$$

could still be used for noncompact shapes. If doing so resulted in the equation for inelastic LTB being used when L_b was not really large enough, the strength based on FLB would control anyway.

5.7 SUMMARY OF MOMENT STRENGTH

The procedure for computation of nominal moment strength for I- and H-shaped sections bent about the x-axis will now be summarized. All terms in the following equations have been previously defined, and AISC equation numbers are not shown. This summary is for compact and noncompact shapes only (no slender shapes).

1. Determine whether the shape is compact.
2. If the shape is compact, check for lateral-torsional buckling as follows:

   If $L_b \leq L_p$, there is no LTB, and $M_n = M_p$.

   If $L_p < L_b \leq L_r$, there is inelastic LTB, and

   $$M_n = C_b \left[M_p - (M_p - M_r) \left(\frac{L_b - L_p}{L_r - L_p} \right) \right] \leq M_p$$

   If $L_b > L_r$, there is elastic LTB, and

   $$M_n = C_b \frac{\pi}{L_b} \sqrt{EI_y GJ + \left(\frac{\pi E}{L_b} \right)^2 I_y C_w} \leq M_p$$

3. If the shape is noncompact because of the flange, the web, or both, the nominal strength will be the smallest of the strengths corresponding to flange local buckling, web local buckling, and lateral-torsional buckling.

   a. Flange local buckling:

      If $\lambda \leq \lambda_p$, there is no FLB.

      If $\lambda_p < \lambda \leq \lambda_r$, the flange is noncompact, and

      $$M_n = M_p - (M_p - M_r) \left(\frac{\lambda - \lambda_p}{\lambda_r - \lambda_p} \right) \leq M_p$$

   b. Web local buckling:

      If $\lambda \leq \lambda_p$, there is no WLB.

      If $\lambda_p < \lambda \leq \lambda_r$, the web is noncompact, and

      $$M_n = M_p - (M_p - M_r) \left(\frac{\lambda - \lambda_p}{\lambda_r - \lambda_p} \right) \leq M_p$$

c. Lateral-torsional buckling:

If $L_b \leq L_p$, there is no LTB.

If $L_p < L_b \leq L_r$, there is inelastic LTB, and

$$M_n = C_b \left[M_p - (M_p - M_r) \left(\frac{L_b - L_p}{L_r - L_p} \right) \right] \leq M_p$$

If $L_b > L_r$, there is elastic LTB, and

$$M_n = C_b \frac{\pi}{L_b} \sqrt{EI_y GJ + \left(\frac{\pi E}{L_b} \right)^2 I_y C_w} \leq M_p$$

5.8   SHEAR STRENGTH

The shear strength of a beam must be sufficient to satisfy the relationship

$$V_u \leq \phi_v V_n$$

where

V_u = maximum shear based on the controlling combination of factored loads

ϕ_v = resistance factor for shear = 0.90

V_n = nominal shear strength

Consider the simple beam of Figure 5.17. At a distance x from the left end and at the neutral axis of the cross section, the state of stress is as shown in Figure 5.17d. Because this element is located at the neutral axis, it is not subjected to flexural stress. From elementary mechanics of materials, the shearing stress is

$$f_v = \frac{VQ}{Ib} \tag{5.6}$$

where

f_v = vertical and horizontal shearing stress at the point of interest

V = vertical shear force at the section under consideration

Q = first moment, about the neutral axis, of the area of the cross section between the point of interest and the top or bottom of the cross section

I = moment of inertia about the neutral axis

b = width of the cross section at the point of interest

Equation 5.6 is based on the assumption that the stress is constant across the width b, and it is therefore accurate only for small values of b. For a rectangular cross section of depth d and width b, the error for $d/b = 2$ is approximately 3%. For $d/b = 1$, the error is 12% and for $d/b = \frac{1}{4}$, it is 100% (Higdon, Ohlsen, and Stiles, 1960). For this

■ **FIGURE 5.17**

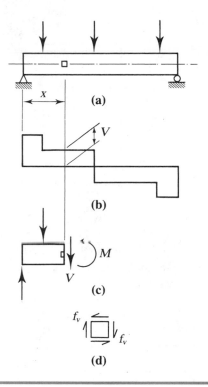

reason, Equation 5.6 cannot be applied to the flange of a W-shape in the same manner as for the web.

Figure 5.18 shows the shearing stress distribution for a W-shape. Superimposed on the actual distribution is the average stress in the web, V/A_w, which does not differ much from the maximum web stress. Clearly, the web will completely yield long before the flanges begin to yield. Because of this, yielding of the web represents one of the shear limit states. Taking the shear yield stress as 60% of the tensile yield stress, we can write the equation for the stress in the web at failure as

$$f_v = \frac{V_n}{A_w} = 0.60F_y$$

where A_w = area of the web. The nominal strength corresponding to this limit state is therefore

$$V_n = 0.6F_y A_w$$

and will be the nominal strength in shear provided that there is no shear buckling of the web. Whether that occurs will depend on h/t_w, the width–thickness ratio of the web. If this ratio is too large — that is, if the web is too slender — the web can buckle in shear,

■ **FIGURE 5.18**

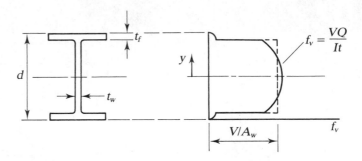

either inelastically or elastically. The relationship between shear strength and the width–thickness ratio is analogous to that between flexural strength and the width–thickness ratio (for FLB or WLB) and between flexural strength and the unbraced length (for LTB). This relationship is illustrated in Figure 5.19 and given in AISC F2.2 as follows:

For $h/t_w \leq 418/\sqrt{F_y}$, there is no web instability, and

$$V_n = 0.6F_yA_w \qquad \text{(AISC Equation F2-1)}$$

For $418/\sqrt{F_y} < h/t_w \leq 523/\sqrt{F_y}$, inelastic web buckling can occur, and

$$V_n = 0.6F_yA_w \frac{418/\sqrt{F_y}}{h/t_w} \qquad \text{(AISC Equation F2-2)}$$

■ **FIGURE 5.19**

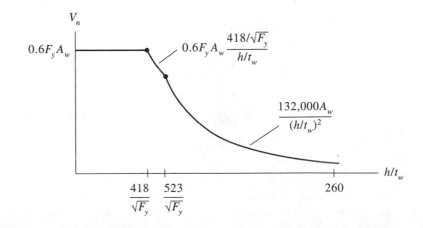

For $523/\sqrt{F_y} < h/t_w \leq 260,$ the limit state is elastic web buckling:

$$V_n = \frac{132,000 A_w}{(h/t_w)^2} \qquad \text{(AISC Equation F2-3)}$$

where

A_w = area of the web = dt_w

d = overall depth of the beam

If h/t_w is greater than 260, web stiffeners are required, and the provisions of Appendix F2 (or Appendix G for plate girders) must be consulted.

AISC Equation F2-3 is based on elastic stability theory, and Equation F2-2 is an empirical equation for the inelastic region, providing a transition between the limit states of web yielding and elastic web buckling.

Shear is rarely a problem in rolled steel beams; the usual practice is to design a beam for flexure and then to check it for shear.

■ **EXAMPLE 5.7**

Check the beam in Example 5.6 for shear.

SOLUTION   From Example 5.6, w_u = 2.080 kips/ft and L = 40 ft. A W14 × 90 with F_y = 50 ksi is used. For a simply supported, uniformly loaded beam, the maximum shear occurs at the support and is equal to the reaction

$$V_u = \frac{w_u L}{2} = \frac{2.080(40)}{2} = 41.6 \text{ kips}$$

From the dimensions and properties tables in Part 1 of the *Manual,* the web width–thickness ratio of a W14 × 90 is

$$\frac{h}{t_w} = 25.9$$

$$\frac{418}{\sqrt{F_y}} = \frac{418}{\sqrt{50}} = 59.11$$

Since h/t_w is less than $418/\sqrt{F_y}$, the strength is governed by shear yielding of the web:

$$V_n = 0.6F_y A_w = 0.6F_y(dt_w) = 0.6(50)(14.02)(0.440) = 185.1 \text{ kips}$$
$$\phi_v V_n = 0.90(185.1) = 167 \text{ kips} > 41.6 \text{ kips} \qquad \text{(OK)}$$

ANSWER   The shear design strength is greater than the factored load shear, so the beam is satisfactory.   ■

Values of $\phi_v V_n$ are tabulated in the factored uniform load tables in Part 4 of the *Manual,* so its computation is unnecessary for standard hot-rolled shapes.

Block Shear

Block shear, which was considered earlier in conjunction with tension member connections, can occur in certain types of beam connections. To facilitate the connection of beams to other beams so that the top flanges are at the same elevation, a short length of the top flange of one of the beams may be cut away, or *coped.* If a coped beam is connected with bolts as in Figure 5.20, segment *ABC* will tend to tear out. The applied load in this case will be the vertical beam reaction, so shear will occur along line *AB* and there will be tension along *BC.* Thus the block shear strength will be a limiting value of the reaction.

We covered the computation of block shear strength in Chapter 3, but we will review it here. Failure can occur by a combination of shear yielding and tension fracture or by shear fracture and tension yielding. AISC J4.3, "Block Shear Rupture Strength," gives two equations for the block shear design strength:

$$\phi R_n = \phi[0.6F_y A_{gv} + F_u A_{nt}] \qquad \text{(AISC Equation J4-3a)}$$

$$\phi R_n = \phi[0.6F_u A_{nv} + F_y A_{gt}] \qquad \text{(AISC Equation J4-3b)}$$

where

$\phi = 0.75$

A_{gv} = gross area in shear (in Figure 5.20, length *AB* times the web thickness)

A_{nv} = net area in shear

A_{gt} = gross area in tension (in Figure 5.20, length *BC* times the web thickness)

A_{nt} = net area in tension

The governing equation is the one that has the larger fracture term.

■ **FIGURE 5.20**

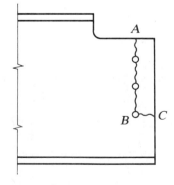

■ EXAMPLE 5.8

Determine the maximum factored load reaction, based on block shear, that can be resisted by the beam shown in Figure 5.21.

SOLUTION

The effective hole diameter is $\frac{3}{4} + \frac{1}{8} = \frac{7}{8}$ in. The gross and net shear areas are

$$A_{gv} = (2 + 3 + 3 + 3)t_w = 11(0.300) = 3.300 \text{ in.}^2$$

$$A_{nv} = \left(11 - 3.5 \times \frac{7}{8}\right)(0.300) = 2.381 \text{ in.}^2$$

The gross and net tension areas are

$$A_{gt} = 1.25t_w = 1.25(0.300) = 0.3750 \text{ in.}^2$$

$$A_{nt} = \left(1.25 - 0.5 \times \frac{7}{8}\right)(0.300) = 0.2438 \text{ in.}^2$$

AISC Equation J4-3a gives

$$\phi R_n = \phi[0.6F_y A_{gv} + F_u A_{nt}] = 0.75[0.6(36)(3.300) + 58(0.2438)]$$
$$= 0.75[71.28 + 14.14] = 64.1 \text{ kips}$$

AISC Equation J4-3b gives

$$\phi R_n = \phi[0.6F_u A_{nv} + F_y A_{gt}] = 0.75[0.6(58)(2.381) + 36(0.3750)]$$
$$= 0.75[82.86 + 13.50] = 72.3 \text{ kips}$$

The fracture term in AISC Equation J4-3b is the larger (that is, 82.86 > 14.14), so this equation governs.

ANSWER

The maximum factored load reaction based on block shear = 72.3 kips.

■ FIGURE 5.21

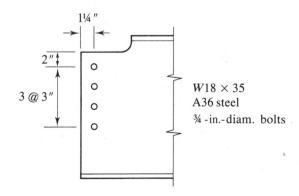

DEFLECTION

In addition to being safe, a structure must be *serviceable*. A serviceable structure is one that performs satisfactorily, not causing any discomfort or perceptions of unsafety for the occupants or users of the structure. For a beam, being serviceable usually means that the deformations, primarily the vertical sag, or deflection, must be limited. Excessive deflection is usually an indication of a very flexible beam, which can lead to problems with vibrations. The deflection itself can cause problems if elements attached to the beam can be damaged by small distortions. In addition, users of the structure may view large deflections negatively and wrongly assume that the structure is unsafe.

For the common case of a simply supported, uniformly loaded beam such as that in Figure 5.22, the maximum vertical deflection is

$$\Delta = \frac{5}{384} \frac{wL^4}{EI}$$

Deflection formulas for a variety of beams and loading conditions can be found in Part 4, "Beam and Girder Design," of the *Manual*. For more unusual situations, standard analytical methods such as the method of virtual work may be used. Deflection is a serviceability limit state, not one of strength, so deflections should always be computed with *service* loads.

The appropriate limit for the maximum deflection depends on the function of the beam and the likelihood of damage resulting from the deflection. The AISC Specification furnishes little guidance other than a statement in Chapter L, "Serviceability Design Considerations," that deflections should be checked. Appropriate limits for deflection can usually be found from the governing building code. The following values are typical maximum allowable total (service dead load plus service live load) deflections.

Plastered construction: $\dfrac{L}{360}$

Unplastered floor construction: $\dfrac{L}{240}$

Unplastered roof construction: $\dfrac{L}{180}$

where L is the span length.

▪ **FIGURE 5.22**

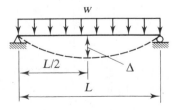

Sometimes the use of a numerical deflection limit, rather than a fraction of the span length, is necessary. Sometimes a limit is placed on the deflection caused only by the live loads, as dead load deflections can often be compensated for during fabrication and construction.

■ EXAMPLE 5.9

Check the deflection of the beam shown in Figure 5.23. The maximum permissible total deflection is L/240.

SOLUTION

$$\text{Maximum permissible deflection} = \frac{L}{240} = \frac{30(12)}{240} = 1.500 \text{ in.}$$

Total service load $= 500 + 550 = 1050 \text{ lb/ft} = 1.050 \text{ kips/ft}$

Expressing the deflection in inches rather than feet is more convenient, so we use units of inches in the following equation:

$$\text{Maximum total deflection} = \frac{5}{384}\frac{wL^4}{EI} = \frac{5}{384}\frac{(1.050/12)(30 \times 12)^4}{29,000(510)}$$
$$= 1.294 \text{ in.} < 1.500 \text{ in.} \quad (\text{OK})$$

ANSWER   The beam satisfies the deflection criterion.

■ **FIGURE 5.23**

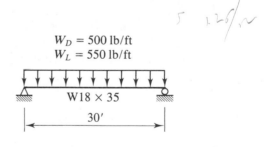

$W_D = 500 \text{ lb/ft}$
$W_L = 550 \text{ lb/ft}$

W18 × 35

30'

Ponding is one deflection problem that does affect the safety of a structure. It is a potential hazard for flat roof systems that can trap rainwater. If drains become clogged during a storm, the weight of the water will cause the roof to deflect, thus providing a reservoir for still more water. If this process proceeds unabated, collapse can occur. The AISC specification requires that the roof system have sufficient stiffness to prevent ponding, and it prescribes limits on moments of inertia and other parameters in Section K2, "Ponding."

5.10 **DESIGN**

Beam design entails the selection of a c⬛⬛⬛⬛⬛⬛pe that will have enough strength and that will meet serviceability r⬛⬛⬛⬛⬛⬛r as strength is concerned, flexure is almost always more critical thar⬛⬛⬛⬛⬛⬛usual practice is to design for flexure and then check shear. The design process can be outlined as follows.

1. Compute the factored load moment, M_u. It will be the same as the required design strength, $\phi_b M_n$. The weight of the beam is part of the dead load but is unknown at this point. A value may be assumed, or the weight may be ignored initially and checked after a shape has been selected.

2. Select a shape that satisfies this strength requirement. This can be done in one of two ways.

   a. Assume a shape, compute the design strength, and compare it with the factored load moment. Revise if necessary. The trial shape can be easily selected in only a limited number of situations (Example 5.10).

   b. Use the beam design charts in Part 4 of the *Manual*. This method is preferred, and we explain it following Example 5.10.

3. Check the shear strength.

4. Check the deflection.

■ **EXAMPLE 5.10**

Select a standard hot-rolled shape of A36 steel for the beam shown in Figure 5.24. The beam has continuous lateral support and must support a uniform service live load of 5 kips/ft. The maximum permissible *live* load deflection is $L/360$.

SOLUTION Assume a beam weight of 100 lb/ft. Then

$$w_u = 1.2w_D + 1.6w_L = 1.2(0.100) + 1.6(5.000) = 8.120 \text{ kips/ft}$$

$$M_u = \frac{1}{8} w_u L^2 = \frac{8.120(30)^2}{8}$$

$$= 913.5 \text{ ft-kips} = \text{required } \phi_b M_n$$

■ **FIGURE 5.24**

$M_u = 431.47$

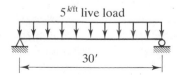

5$^{k/ft}$ live load

30'

Assume that the shape is compact. For a compact shape and continuous lateral support,

$$M_n = M_p = Z_x F_y$$

From $\phi_b M_n \geq M_u$,

$$\phi_b F_y Z_x \geq M_u$$

$$Z_x \geq \frac{M_u}{\phi_b F_y} = \frac{913.5(12)}{0.90(36)} = 338.3 \text{ in.}^3$$

$$\frac{435.44(12)}{0.90(36)} = 161.31$$

The Load Factor Design Selection Table lists rolled shapes normally used as beams in order of decreasing plastic section modulus. Furthermore, they are grouped so that the shape at the top of the group (in bold type) is the lightest one that has enough section modulus to satisfy a required section modulus falling within the group. In the current case, the shape that comes closest to meeting the section modulus requirement is a W27 × 114, with $Z_x = 343$ in.^{3}, but the lightest one is a W30 × 108, with $Z_x = 346$ in.^{3} Because section modulus is not directly proportional to area, having more section modulus with a smaller area, and hence less weight, is possible.

Try a W30 × 108. This shape is compact, as assumed (noncompact shapes are marked as such in the table); therefore $M_n = M_p$, as assumed.

The weight is slightly larger than that assumed, so the required strength will have to be recomputed, although the W30 × 108 has more capacity than originally required, and it will almost certainly be adequate.

$$w_u = 1.2(0.108) + 1.6(5.000) = 8.130 \text{ kips/ft}$$

$$M_u = \frac{8.130(30)^2}{8} = 914.6 \text{ ft-kips}$$

From the Load Factor Design Selection Table,

$$\phi_b M_p = \phi_b M_n = 934 \text{ ft-kips} > 914.6 \text{ ft-kips} \quad \text{(OK)}$$

In lieu of basing the search on the required section modulus, the design strength $\phi_b M_p$ could be used because it is directly proportional to Z_x and is also tabulated. Next, check the shear:

$$V_u = \frac{w_u L}{2} = \frac{8.130(30)}{2} = 122 \text{ kips}$$

From the factored uniform load tables,

$$\phi_v V_n = 316 \text{ kips} > 122 \text{ kips} \quad \text{(OK)}$$

Finally, check the deflection. The maximum permissible live load deflection is $L/360 = 30(12)/360 = 1$ in.

$$\Delta = \frac{5}{384} \frac{w_L L^4}{EI_x} = \frac{5}{384} \frac{(5.000/12)(30 \times 12)^4}{29,000(4470)} = 0.703 \text{ in.} < 1 \text{ in.} \quad \text{(OK)}$$

ANSWER   Use a W30 × 108.

Beam Design Charts

Many graphs, charts, and tables are available for the practicing engineer, and these aids can greatly simplify the design process. For the sake of efficiency, they are widely used in design offices, but you should approach their use with caution and not allow basic principles to become obscured. It is not our purpose to describe in this book all available design aids in detail, but some are worthy of note, particularly the curves of design moment versus unbraced length given in Part 4 of the *Manual*.

These curves will be described with reference to Figure 5.25, which shows a graph of design moment $\phi_b M_n$ as a function of unbraced length L_b for a particular compact shape. Such a graph can be constructed for any cross-sectional shape and specific values of F_y and C_b by using the appropriate equations for moment strength.

The *Manual* charts comprise a family of curves for various rolled shapes. All curves were generated with $C_b = 1.0$. For other values of C_b, simply multiply the design moments from the charts by C_b. Keep in mind, though, that $\phi_b M_n$ can never exceed $\phi_b M_p$ (or, for noncompact shapes, $\phi_b M_n$ based on local buckling). Use of the charts is illustrated in Figure 5.26, where two such curves are shown. Any point on this graph, such as the intersection of the two dashed lines, represents a design moment and an unbraced length. If the moment is a required moment capacity, then any curve above the point corresponds to a beam with a larger moment capacity. Any curve to the right is for a beam with exactly the required moment capacity, although for a larger unbraced length. In a design problem, therefore, if the charts are entered with a given unbraced length and a required design strength, curves above and to the right of the

■ **FIGURE 5.25**

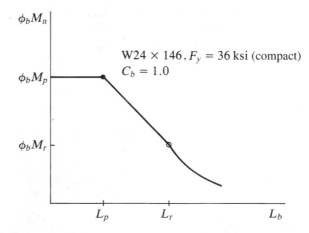

■ **FIGURE 5.26**

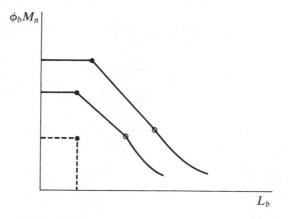

point correspond to acceptable beams. If a dashed portion of a curve is encountered, then a curve for a lighter shape lies above or to the right of the dashed curve. Points on the curves corresponding to L_p are indicated by a solid circle; L_r is represented by an open circle. Two sets of curves are available, one for $F_y = 36$ ksi and one for $F_y = 50$ ksi.

In Example 5.10, the required design strength (including an assumed beam weight) was 913.5 ft-kips, and there was continuous lateral support. For continuous lateral support, L_b can be taken as zero. From the charts for $F_y = 36$ ksi, the first solid curve above the 913.5 ft-kip mark is for a W30 × 108, the same as selected in Example 5.10. Although $L_b = 0$ is not on this particular chart, the smallest value of L_b shown is less than L_p for all shapes on that page.

The beam curve shown in Figure 5.25 is for a compact shape, so the value of $\phi_b M_n$ for sufficiently small values of L_b is $\phi_b M_p$. As discussed in Section 5.6, if the shape is noncompact, the maximum value of $\phi_b M_n$ will be based on flange local buckling. The maximum unbraced length for which this condition will be true will be different from the value of L_p obtained with AISC Equation F1-4. The moment strength of noncompact shapes is illustrated graphically in Figure 5.27, where the maximum design strength is denoted $\phi_b M_n'$, and the maximum unbraced length for which this strength is valid is denoted L_p'.

Although the charts for compact and noncompact shapes are similar in appearance, $\phi_b M_n$ and L_p are used for compact shapes, whereas $\phi_b M_n'$ and L_p' are used for noncompact shapes. (This notation is not used in the charts or in any of the other design aids in the *Manual,* but it is used in the discussion preceding the Load Factor Design Selection Table in Part 1 of the *Manual.*) Whether a shape is compact or noncompact is irrelevant to the *use* of the charts.

■ **FIGURE 5.27**

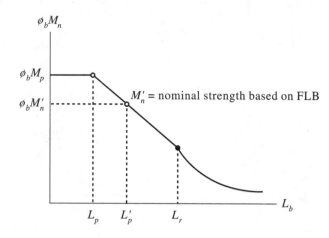

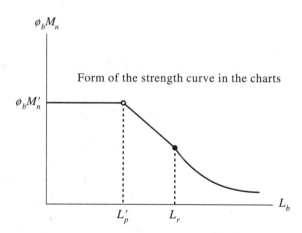

■ **EXAMPLE 5.11**

The beam shown in Figure 5.28 must support two concentrated *live* loads of 20 kips each at the quarter points. The maximum deflection must not exceed $L/240$. Lateral support is provided at the ends of the beam. Use A572 Grade 50 steel and select a rolled shape.

SOLUTION   If the weight of the beam is neglected, the central half of the beam is subjected to a uniform moment, and

$$M_A = M_B = M_C = M_{\max}, \quad \therefore C_b = 1.0$$

■ **FIGURE 5.28**

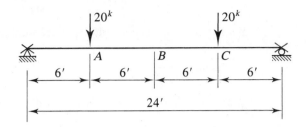

Even if the weight is included, it will be negligible compared to the concentrated loads, and C_b can still be taken as 1.0, permitting the charts to be used without modification. Temporarily ignoring the beam weight, we obtain

$$M_u = 6(1.6 \times 20) = 192 \text{ ft-kips}$$

From the charts, with $L_b = 24$ ft, **try W12 × 53:**

$$\phi_b M_n = 219 \text{ ft-kips} > 192 \text{ ft-kips} \quad \text{(OK)}$$

Now, we account for the beam weight:

$$M_u = 192 + \frac{1}{8} (1.2 \times 0.053)(24)^2 = 197 \text{ ft-kips} < 219 \text{ ft-kips} \quad \text{(OK)}$$

The shear is

$$V_u = 1.6(20) + \frac{1.2(0.053)(24)}{2} = 32.8 \text{ ft-kips}$$

From the factored uniform load tables,

$$\phi_v V_n = 112 \text{ kips} > 32.8 \text{ kips} \quad \text{(OK)}$$

The maximum permissible deflection is

$$\frac{L}{240} = \frac{24(12)}{240} = 1.200 \text{ in.}$$

From the Beam Diagrams and Formulas section in Part 4 of the *Manual,* the maximum deflection (at midspan) for two equal and symmetrically placed loads is

$$\Delta = \frac{Pa}{24EI} (3L^2 - 4a^2)$$

where
    P = magnitude of concentrated load
    a = distance from support to load
    L = span length

$$\Delta = \frac{20(6 \times 12)}{24EI}[3(24 \times 12)^2 - 4(6 \times 12)^2] = \frac{13.69 \times 10^6}{EI}$$

For the uniform beam weight, the maximum deflection is also at midspan, so

$$\Delta = \frac{5}{384}\frac{wL^4}{EI} = \frac{5}{384}\frac{(0.053/12)(24 \times 12)^4}{EI} = \frac{0.04 \times 10^6}{EI}$$

Both deflections occur at the same location, so they can be added:

$$\text{Total } \Delta = \frac{(13.69 + 0.04) \times 10^6}{EI}$$

$$= \frac{13.73 \times 10^6}{29{,}000(425)} = 1.114 \text{ in.} < 1.200 \text{ in.} \quad \text{(OK)}$$

ANSWER   Use a W12 × 53.   ■

Although the charts are based on $C_b = 1.0$, they can easily be used for design when C_b is not 1.0; simply divide the required design strength by C_b before entering the charts. We illustrate this technique in Example 5.12.

■ EXAMPLE 5.12

Use A36 steel and select a rolled shape for the beam in Figure 5.29. The concentrated load is a service live load, and the uniform load is 30% dead load and 70% live load. Lateral bracing is provided at the ends and at midspan. There is no restriction on deflection.

■ FIGURE 5.29

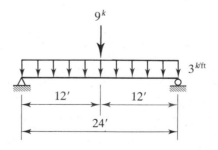

SOLUTION Assume a beam weight of 100 lb/ft. Then

$$w_D = 0.30(3) + 0.100 = 1.000 \text{ kips/ft}$$
$$w_u = 1.2(1.000) + 1.6(0.70 \times 3) = 4.560 \text{ kips/ft}$$
$$P_u = 1.6(9) = 14.40 \text{ kips}$$

The factored loads and reactions are shown in Figure 5.30.

Moments required for the computation of C_b: The bending moment at a distance x from the left end is

$$M = 61.92x - 4.560x\left(\frac{x}{2}\right) = 61.92x - 2.280x^2 \qquad \text{(for } x \leq 12 \text{ ft)}$$

For $x = 3$ ft, $M_A = 61.92(3) - 2.280(3)^2 = 165.2$ ft-kips
For $x = 6$ ft, $M_B = 61.92(6) - 2.280(6)^2 = 289.4$ ft-kips
For $x = 9$ ft, $M_C = 61.92(9) - 2.280(9)^2 = 372.6$ ft-kips
For $x = 12$ ft, $M_{max} = M_u = 61.92(12) - 2.280(12)^2 = 414.7$ ft-kips

$$C_b = \frac{12.5M_{max}}{2.5M_{max} + 3M_A + 4M_B + 3M_C}$$

$$= \frac{12.5(414.7)}{2.5(414.7) + 3(165.2) + 4(289.4) + 3(372.6)} = 1.36$$

Enter the charts with an unbraced length $L_b = 12$ ft and a bending moment of

$$\frac{M_u}{C_b} = \frac{414.7}{1.36} = 305 \text{ ft-kips}$$

Try W21 × 62:

$$\phi_b M_n = 343 \text{ ft-kips} \qquad \text{(for } C_b = 1)$$

■ **FIGURE 5.30**

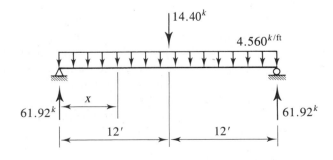

Since $C_b = 1.36$, the actual design strength is

$$\phi_b M_n = 1.36(343) = 466 \text{ ft-kips}$$

But the design strength cannot exceed $\phi_b M_p$, which is only 389 ft-kips (obtained from the chart), so the actual design strength must be taken as

$$\phi_b M_n = 389 \text{ ft-kips} < M_u = 414.7 \text{ ft-kips} \qquad \text{(N.G.)}$$

For the next trial shape, move up in the charts to the next solid curve and **try W21 × 68.** For $L_b = 12$ ft, the design strength from the chart is 385 ft-kips for $C_b = 1$. The strength for $C_b = 1.36$ is

$$\phi_b M_n = 1.36(385) = 524 \text{ ft-kips} > \phi_b M_p = 432 \text{ ft-kips}$$
$$\therefore \quad \phi_b M_n = \phi_b M_p = 432 \text{ ft-kips} > M_u = 414.7 \text{ ft-kips} \qquad \text{(OK)}$$

The beam weight is 68 lb/ft, which is less than the assumed weight of 100 lb/ft.   (OK)
The shear is

$$V_u = 61.92 \text{ kips}$$

(This result is slightly conservative because the beam weight is less than that assumed.) From the factored uniform load tables,

$$\phi_v V_n = 177 \text{ kips} > 61.92 \text{ kips} \qquad \text{(OK)}$$

ANSWER   Use a W21 × 68.   THEN CHECK FOR STIRRUP EDGE   ■

If deflection requirements control the design of a beam, a minimum required moment of inertia is computed, and the lightest shape having this value is sought. This task is greatly simplified by the moment of inertia selection tables in Part 4 of the *Manual*. We illustrate the use of these tables in Example 5.13, which also explains the design procedure for a beam in a typical floor or roof system.

■ EXAMPLE 5.13

Part of a floor framing system is shown in Figure 5.31. A 4-inch-thick reinforced concrete floor slab is supported by *floor beams* spaced at 7 feet. The floor beams are supported by *girders,* which in turn are supported by the columns. (The floor beams, which fill in the panels defined by the columns, are sometimes called *filler beams.*) In addition to the weight of the structure, loads consist of a uniform live load of 80 psf and movable partitions, to be accounted for by using a uniformly distributed load of 20 pounds per square foot of floor surface. The maximum total deflection must not exceed $\frac{1}{360}$ of the span length. Use A36 steel and design the floor beams. Assume that the slab provides continuous lateral support of the floor beams.

■ **FIGURE 5.31**

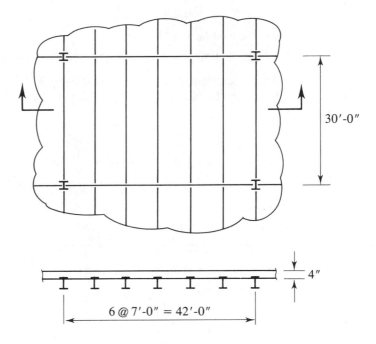

SOLUTION   For calculation of the dead load, use normal-weight reinforced concrete, which is usually considered to weigh 150 lb/ft³. The weight can be expressed as a load per square foot of floor surface by multiplying the unit weight by the volume of slab that has a surface area of one square foot; that is, multiply 150 lb/ft³ by the slab thickness in feet:

$$\text{Slab weight} = 150\left(\frac{4}{12}\right) = 50 \text{ psf}$$

Assume that each beam supports a 7-foot width (tributary width) of floor.

Slab: 50(7)	= 350 lb/ft
Partitions: 20(7)	= 140 lb/ft
Beam weight:	= 40 lb/ft (estimated)
Total:	= 530 lb/ft = service dead load

Although the partitions are movable, the national model building codes treat them as dead load (BOCA, 1996; ICBO, 1997; and SBCC, 1997). We also treat them as dead load in this book.

The live load is

80(7) = 560 lb/ft

and the total factored load is

$$w_u = 1.2w_D + 1.6w_L = 1.2(0.530) + 1.6(0.560) = 1.532 \text{ kips/ft}$$

The typical floor-beam connection will provide virtually no moment restraint, and the beams can be treated as simply supported. Hence

$$M_u = \frac{1}{8} w_u L^2 = \frac{1.532(30)^2}{8} = 172.4 \text{ ft-kips}$$

From the beam design charts, with $L_b = 0$, **try a W18 × 35:**

$$\phi_b M_n = 179.5 \text{ ft-kips} > 172.4 \text{ ft-kips} \qquad \text{(OK)}$$

The shear is

$$V_u \approx \frac{1.532(30)}{2} = 22.98 \text{ kips}$$

From the factored uniform load tables,

$$\phi_v V_n = 103 \text{ kips} > 22.98 \text{ kips} \qquad \text{(OK)}$$

The maximum permissible deflection is

$$\frac{L}{360} = \frac{30(12)}{360} = 1 \text{ in.}$$

$$\Delta = \frac{5}{384} \frac{wL^4}{EI} = \frac{5}{384} \frac{(0.350 + 0.140 + 0.035 + 0.560)(30)^4(12)^3}{29,000(510)}$$

$$= 1.3 \text{ in.} > 1 \text{ in.} \qquad \text{(N.G.)}$$

Solving the deflection equation for the required moment of inertia yields

$$I_{required} = \frac{5wL^4}{384E\Delta_{required}} = \frac{5(1.085)(30)^4(12)^3}{384(29,000)(1)} = 682 \text{ in.}^4$$

The Moment of Inertia Selection Table is organized in the same way as the Load Factor Design Selection Table, so selection of the lightest shape with sufficient moment of inertia is simple. From the I_x Table, try a W21 × 44:

$$I_x = 843 \text{ in.}^4 > 682 \text{ in.}^4 \qquad \text{(OK)}$$

$$\phi_b M_n = 257.5 \text{ ft-kips} > 172.4 \text{ ft-kips} \qquad \text{(OK)}$$

The weight of this shape is slightly more than the initial estimate, but the additional weight is insignificant compared to the excess moment capacity.

$$\phi_v V_n = 141 \text{ kips} > 22.98 \text{ kips} \qquad \text{(OK)}$$

ANSWER   Use a W21 × 44.   ■

HOLES IN BEAMS

If beam connections are made with bolts, holes will be punched or drilled in the beam web or flanges. In addition, relatively large holes are sometimes cut in beam webs to provide space for utilities such as electrical conduits and ventilation ducts. Ideally, holes should be placed in the web only at sections of low shear, and holes should be made in the flanges at points of low bending moment. That will not always be possible, so the effect of the holes must be accounted for.

For relatively small holes such as those for bolts, the effect will be small, particularly for flexure, for two reasons. First, the reduction in the cross section is usually small. Second, adjacent cross sections are not reduced, and the change in cross section is actually more of a minor discontinuity than a "weak link."

Hence AISC B10 permits bolt holes in flanges to be ignored when

$$0.75F_u A_{fn} \geq 0.9F_y A_{fg} \qquad \text{(AISC Equation B10-1)}$$

where

A_{fg} = gross flange area

A_{fn} = net flange area

If this condition is not met, the flexural properties must be based on an effective tension flange area of

$$A_{fe} = \frac{5}{6} \frac{F_u}{F_y} A_{fn} \qquad \text{(AISC Equation B10-3)}$$

■ **EXAMPLE 5.14**

Compute the reduced elastic section modulus S_x for the shape shown in Figure 5.32. A36 steel is used, and the holes are for 1-inch-diameter bolts.

■ **FIGURE 5.32**

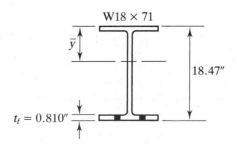

SOLUTION $A_{fg} = b_f t_f = 7.635(0.810) = 6.184$ in.^{2}

The effective hole diameter is

$$d_h = 1 + \frac{1}{8} = 1\frac{1}{8} \text{ in.}$$

and the net flange area is

$$A_{fn} = A_{fg} - \Sigma\, d_h t_f = 6.184 - 2(1.125)(0.810) = 4.362 \text{ in.}^2$$

From AISC Equation B10-1,

$$0.75 F_u A_{fn} = 0.75(58)(4.362) = 189.7 \text{ kips}$$

and

$$0.9 F_y A_{fg} = 0.9(36)(6.184) = 200.4 \text{ kips}$$

Since $0.75 F_u A_{fn} < 0.9 F_y A_{fg}$, we must account for the holes. Using AISC Equation B10-3 gives an effective flange area of

$$A_{fe} = \frac{5}{6} \frac{F_u}{F_y} A_{fn} = \frac{5}{6}\left(\frac{58}{36}\right)(4.362) = 5.856 \text{ in.}^2$$

This flange area corresponds to a reduction of $6.184 - 5.856 = 0.328$ in.^{2} The elastic neutral axis is located at a distance from the top of

$$\bar{y} = \frac{20.8(18.47/2) - 0.328(18.47 - 0.405)}{20.8 - 0.328} = 9.094 \text{ in.}$$

The reduced moment of inertia is

$$\bar{I}_x = 1170 + 20.8(9.094 - 9.235)^2 - 0.328(9.094 - 18.06)^2 = 1144 \text{ in.}^4$$

S_x referred to the top fiber is

$$S_x = \frac{I_x}{\bar{y}} = \frac{1144}{9.094} = 126 \text{ in.}^3$$

S_x referred to the bottom fiber is

$$S_x = \frac{I_x}{d - \bar{y}} = \frac{1144}{18.47 - 9.094} = 122 \text{ in.}^3$$

ANSWER The reduced elastic section modulus referred to the top is 126 in.^{3} and to the bottom is 122 in.^{3} ■

Beams with large holes in their webs will require special treatment and are beyond the scope of this book. *Design of Steel and Composite Beams with Web Openings* is a useful guide to this topic (Darwin, 1990).

5.12 OPEN-WEB STEEL JOISTS

Open-web steel joists are prefabricated trusses of the type shown in Figure 5.33. Many of the smaller ones use a continuous circular bar to form the web members and are commonly called *bar* joists. They are used in floor and roof systems in a wide variety of structures. For a given span, an open-web joist will be lighter in weight than a rolled shape, and the absence of a solid web allows for the easy passage of duct work and electrical conduits. Depending on the span length, open-web joists may be more economical than rolled shapes, although there are no general guidelines for making this determination.

Open-web joists are available in standard depths and load capacities from various manufacturers. Some open-web joists are designed to function as floor or roof joists, and others are designed to function as girders, supporting the concentrated reactions from joists. The AISC Specification does not cover open-web steel joists; a separate organization, the Steel Joist Institute (SJI), exists for this purpose. All aspects of steel joist usage, including their design and manufacture, are addressed in the publication *Standard Specifications, Load Tables, and Weight Tables for Steel Joists and Joist Girders* (SJI, 1994).

An open-web joist can be selected with the aid of the standard load tables (SJI, 1994), one of which is reproduced in Figure 5.34. For each combination of span and joist, a pair of load values is given. The top number is the total service load capacity in pounds per foot, and the bottom number is the service live load per foot that will produce a deflection of $\frac{1}{360}$ of the span length. (Although the loads in the tables are service load capacities, the tables can easily be adapted for use with the LRFD approach, which we demonstrate presently.) The first number in the designation is the nominal depth in inches. The table also gives the approximate weight in pounds per foot of length. Steel fabricators who furnish open-web steel joists must certify that a particular joist of a given designation, such as a 10K1 of span length 20 feet, will have a safe load capacity of at least the value given in the table. Different manufacturers' 10K1 joists may have different member cross sections, but they all must have a nominal depth of 10 inches and, for a span length of 20 feet, a service load capacity of at least 199 pounds per foot.

■ **FIGURE 5.33**

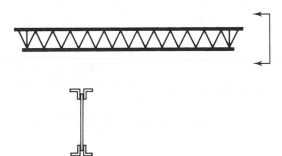

■ **FIGURE 5.34**

Joist Designation	8K1	10K1	12K1	12K3	12K5	14K1	14K3	14K4	14K6	16K2	16K3	16K4	16K5	16K6	16K7	16K9
Depth (In.)	8	10	12	12	12	14	14	14	14	16	16	16	16	16	16	16
Approx. Wt. (lbs./ft.)	5.1	5.0	5.0	5.7	7.1	5.2	6.0	6.7	7.7	5.5	6.3	7.0	7.5	8.1	8.6	10.0
Span (ft.) ↓																
8	550 / 550															
9	550 / 550															
10	550 / 480	550 / 550														
11	532 / 377	550 / 542														
12	444 / 288	550 / 455	550 / 550	550 / 550	550 / 550											
13	377 / 225	479 / 363	550 / 510	550 / 510	550 / 510											
14	324 / 179	412 / 289	500 / 425	550 / 463	550 / 463	550 / 550	550 / 550	550 / 550	550 / 550							
15	281 / 145	358 / 234	434 / 344	543 / 428	550 / 434	511 / 475	550 / 507	550 / 507	550 / 507							
16	246 / 119	313 / 192	380 / 282	476 / 351	550 / 396	448 / 390	550 / 467	550 / 467	550 / 467	550 / 550	550 / 550	550 / 550	550 / 550	550 / 550	550 / 550	550 / 550
17		277 / 159	336 / 234	420 / 291	550 / 366	395 / 324	495 / 404	550 / 443	550 / 443	512 / 488	550 / 526	550 / 526	550 / 526	550 / 526	550 / 526	550 / 526
18		246 / 134	299 / 197	374 / 245	507 / 317	352 / 272	441 / 339	530 / 397	550 / 408	456 / 409	508 / 456	550 / 490	550 / 490	550 / 490	550 / 490	550 / 490
19		221 / 113	268 / 167	335 / 207	454 / 269	315 / 230	395 / 287	475 / 336	550 / 383	408 / 347	455 / 386	547 / 452	550 / 455	550 / 455	550 / 455	550 / 455
20		199 / 97	241 / 142	302 / 177	409 / 230	284 / 197	356 / 246	428 / 287	525 / 347	368 / 297	410 / 330	493 / 386	550 / 426	550 / 426	550 / 426	550 / 426
21			218 / 123	273 / 153	370 / 198	257 / 170	322 / 212	388 / 248	475 / 299	333 / 255	371 / 285	447 / 333	503 / 373	548 / 405	550 / 406	550 / 406
22			199 / 106	249 / 132	337 / 172	234 / 147	293 / 184	353 / 215	432 / 259	303 / 222	337 / 247	406 / 289	458 / 323	498 / 351	550 / 385	550 / 385
23			181 / 93	227 / 116	308 / 150	214 / 128	268 / 160	322 / 188	395 / 226	277 / 194	308 / 216	371 / 252	418 / 282	455 / 307	507 / 339	550 / 363
24			166 / 81	208 / 101	282 / 132	196 / 113	245 / 141	295 / 165	362 / 199	254 / 170	283 / 189	340 / 221	384 / 248	418 / 269	465 / 298	550 / 346
25						180 / 100	226 / 124	272 / 145	334 / 175	234 / 150	260 / 167	313 / 195	353 / 219	384 / 238	428 / 263	514 / 311
26						166 / 88	209 / 110	251 / 129	308 / 156	216 / 133	240 / 148	289 / 173	326 / 194	355 / 211	395 / 233	474 / 276
27						154 / 79	193 / 98	233 / 115	285 / 139	200 / 119	223 / 132	268 / 155	302 / 173	329 / 188	366 / 208	439 / 246
28						143 / 70	180 / 88	216 / 103	265 / 124	186 / 106	207 / 118	249 / 138	281 / 155	306 / 168	340 / 186	408 / 220
29										173 / 95	193 / 106	232 / 124	261 / 139	285 / 151	317 / 167	380 / 198
30										161 / 86	180 / 96	216 / 112	244 / 126	266 / 137	296 / 151	355 / 178
31										151 / 78	168 / 87	203 / 101	228 / 114	249 / 124	277 / 137	332 / 161
32										142 / 71	158 / 79	190 / 92	214 / 103	233 / 112	259 / 124	311 / 147

Source: Standard Specifications, Load Tables, and Weight Tables for Steel Joists and Joist Girders.
Myrtle Beach, S.C.: Steel Joist Institute, 1994. Reprinted with permission.

The open-web steel joists that are designed to function as floor or roof joists (in contrast to girders) are available as open-web steel joists (K-series, both standard and KCS), longspan steel joists (LH-series), and deep longspan steel joists (DLH-series). Standard load tables are given for each of these categories (SJI, 1994). The higher you move up the series, the greater the available span lengths and load-carrying capacities become. At the lower end, an 8K1 is available with a span length of 8 feet and a capacity of 550 pounds per foot, whereas a 72DLH19 can support a load of 497 pounds per foot on a span of 144 feet.

With the exception of the KCS joists, all open-web steel joists are designed as simply supported trusses with uniformly distributed loads on the top chord. This loading subjects the top chord to bending as well as axial compression, so the top chord is designed as a beam-column (see Chapter 6). To ensure stability of the top chord, the floor or roof deck must be attached in such a way that continuous lateral support is provided.

KCS joists are designed to support both concentrated loads and distributed loads (including nonuniform distributions). To select a KCS joist, the engineer must compute a maximum moment and shear in the joist and enter the KCS tables with these values. (The KCS joists are designed to resist a uniform moment and a constant shear.) If concentrated loads must be supported by an LH or a DLH joist, a special analysis should be requested from the manufacturer.

Both top and bottom chord members of K-series joists must be made of steel with a yield stress of 50 ksi, and the web members may have a yield stress of either 36 ksi or 50 ksi. All members of LH- and DLH-series joists can be made with steel of any yield stress between 36 ksi and 50 ksi inclusive. The load capacity of K-series joists must be verified by the manufacturer by testing, and a minimum factor of safety of 1.65 must be demonstrated. No such testing program is required for LH- or DLH-series joists.

Joist girders are designed to support open-web steel joists (K-, LH-, and DLH-series). The engineer designates a joist girder by specifying its depth, the number of joist spaces, and the load in kips at each loaded top-chord panel point of the joist girder. A 52G9N10K, for example, is 52 inches deep, provides for 9 equal joist spaces on the top chord, and will support 10 kips at each joist location. The joist girder weight tables give the weight in pounds per linear foot for the specified joist girder for a specific span length.

A simple procedure for using the standard load tables in an LRFD context is given by the SJI (1994) and is shown here in slightly modified form. Consider the basic LRFD relationship, Equation 2.3:

$$\Sigma \gamma_i Q_i \leq \phi R_n$$

It can be written for a uniformly distributed load as

$$w_u \leq \phi w_n \tag{5.7}$$

where w_u is the factored uniform load and w_n is the nominal uniform load strength of the joist. If we use an average ratio of nominal strength to *allowable* strength of 1.65,* we can express the nominal strength as

$$w_n = 1.65 w_{sji}$$

where w_{sji} is the allowable strength (allowable load) given in the standard load tables. The design strength is

$$\phi w_n = 0.9(1.65 w_{sji}) = 1.485 w_{sji} \approx \tfrac{3}{2}\, w_{sji}$$

We can now write Equation 5.7 as

$$w_u \leq \tfrac{3}{2}\, w_{sji}$$

---

*Note that this is the factor of safety for K-series joists that must be demonstrated by manufacturers' testing.

For design purposes, we can express this relationship as

$$\text{Required } w_{sji} = \tfrac{2}{3} w_u$$

■ **EXAMPLE 5.15**

Use the load table given in Figure 5.34 to select an open-web steel joist for the following floor system and loads.

Joist spacing = 3 ft 0 in.

Span length = 20 ft 0 in.

The loads are:

3-in. floor slab

Other dead load: 20 psf

Live load: 50 psf

SOLUTION   For the dead loads of

$$\text{Slab: } 150\left(\frac{3}{12}\right) = 37.5 \text{ psf}$$

Other dead load	= 20 psf
Joist weight	= 3 psf (est.)
Total:	= 60.5 psf

$$w_D = 60.5(3) = 181.5 \text{ lb/ft}$$

For the live load of 50 psf,

$$w_L = 50(3) = 150 \text{ lb/ft}$$

The factored load is

$$w_u = 1.2w_D + 1.6w_L = 1.2\,(181.5) + 1.6(150) = 457.8 \text{ lb/ft}$$

Convert this load to a required service load:

$$\text{Required } w_{sji} = \frac{2}{3} w_u = \frac{2}{3}\,(457.8) = 305 \text{ lb/ft}$$

Figure 5.34 indicates that the following joists satisfy the load requirement: a 12K5, weighing approximately 7.1 lb/ft; a 14K3, weighing approximately 6.0 lb/ft; and a 16K2, weighing approximately 5.5 lb/ft. No restriction was placed on the depth, so we choose the lightest joist.

ANSWER   Use a 16K2.   ■

The standard load tables also include a K-series economy table, which facilitates the selection of the lightest joist for a given load.

5.13 BEAM BEARING PLATES AND COLUMN BASE PLATES

The design procedure for column base plates is similar to that for beam bearing plates, and for that reason we consider them together. In addition, the determination of the thickness of a column base plate requires consideration of flexure, so it logically belongs in this chapter rather than in Chapter 4. In both cases, the function of the plate is to distribute a concentrated load to the supporting material.

Two types of beam bearing plates are considered: one that transmits the beam reaction to a support such as a concrete wall and one that transmits a load to the top flange of a beam. Consider first the beam support shown in Figure 5.35. Although many beams are connected to columns or other beams, the type of support shown here is occasionally used, particularly at bridge abutments. The design of the bearing plate consists of three steps.

1. Determine dimension N so that web yielding and web crippling are prevented.
2. Determine dimension B so that the area $B \times N$ is sufficient to prevent the supporting material (usually concrete) from being crushed in bearing.
3. Determine the thickness t so that the plate has sufficient bending strength.

Web yielding and web crippling are addressed by AISC in Chapter K of the Specification, "Strength Design Considerations." Bearing strength of concrete is covered in Chapter J, "Connections, Joints, and Fasteners."

Web Yielding

Web yielding is the compressive crushing of a beam web caused by the application of a compressive force to the flange directly above or below the web. This force could be an end reaction from a support of the type shown in Figure 5.35, or it could be a load

---

■ **FIGURE 5.35**

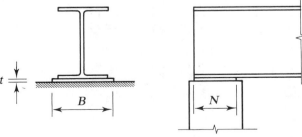

delivered to the top flange by a column or another beam. Yielding occurs when the compressive stress on a horizontal section through the web reaches the yield point. When the load is transmitted through a plate, web yielding is assumed to take place on the nearest section of width t_w. In a rolled shape, this section will be at the toe of the fillet, a distance k from the outside face of the flange (this dimension is tabulated in the dimensions and properties tables in the *Manual*). If the load is assumed to distribute itself at a slope of 1 : 2.5 as shown in Figure 5.36, the area at the support subject to yielding is $(2.5k + N)t_w$. Multiplying this area by the yield stress gives the nominal strength for web yielding at the support:

$$R_n = (2.5k + N)F_y t_w \qquad \text{(AISC Equation K1-3)}$$

The bearing length N at the support should not be less than k.

At the interior load, the length of the section subject to yielding is

$$2(2.5k) + N = 5k + N$$

and the nominal strength is

$$R_n = (5k + N)F_y t_w \qquad \text{(AISC Equation K1-2)}$$

The design strength is ϕR_n, where $\phi = 1.0$.

Web Crippling

Web crippling is buckling of the web caused by the compressive force delivered through the flange. For an interior load, the nominal strength for web crippling is

$$R_n = 135 t_w^2 \left[1 + 3\left(\frac{N}{d}\right)\left(\frac{t_w}{t_f}\right)^{1.5} \right] \sqrt{\frac{F_y t_f}{t_w}} \qquad \text{(AISC Equation K1-4)}$$

■ **FIGURE 5.36**

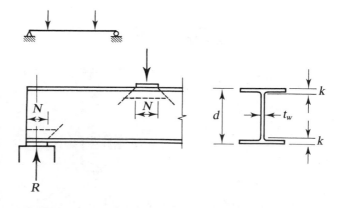

For a load at or near the support (no greater than half the beam depth from the end), the nominal strength is

$$R_n = 68t_w^2 \left[1 + 3\left(\frac{N}{d}\right)\left(\frac{t_w}{t_f}\right)^{1.5}\right] \sqrt{\frac{F_y t_f}{t_w}} \qquad \text{for } \frac{N}{d} \le 0.2 \quad \text{(AISC Equation K1-5a)}$$

or

$$R_n = 68t_w^2 \left[1 + \left(4\frac{N}{d} - 0.2\right)\left(\frac{t_w}{t_f}\right)^{1.5}\right] \sqrt{\frac{F_y t_f}{t_w}} \qquad \text{for } \frac{N}{d} > 0.2$$

$$\text{(AISC Equation K1-5b)}$$

The resistance factor for this limit state is $\phi = 0.75$.

Concrete Bearing Strength

The material used for a beam support can be concrete, brick, or some other material, but it usually will be concrete. This material must resist the bearing load applied by the steel plate. The nominal bearing strength specified in AISC J9 is the same as that given in the American Concrete Institute's Building Code (ACI, 1995) and may be used if no other building code requirements are in effect. If the plate covers the full area of the support, the nominal strength is

$$P_p = 0.85 f_c' A_1 \qquad \text{(AISC Equation J9-1)}$$

If the plate does not cover the full area of the support,

$$P_p = 0.85 f_c' A_1 \sqrt{A_2 / A_1} \qquad \text{(AISC Equation J9-2)}$$

where
  $f_c' = $ 28-day compressive strength of the concrete
  $A_1 = $ bearing area
  $A_2 = $ full area of the support

If area A_2 is not concentric with A_1, then A_2 should be taken as the largest concentric area that is geometrically similar to A_1, as illustrated in Figure 5.37. AISC also requires that

$$\sqrt{A_2 / A_1} \le 2$$

The design bearing strength is $\phi_c P_p$, where $\phi_c = 0.60$.

Plate Thickness

Once the length and width of the plate have been determined, the average bearing pressure is treated as a uniform load on the bottom of the plate, which is assumed to be

■ **FIGURE 5.37**

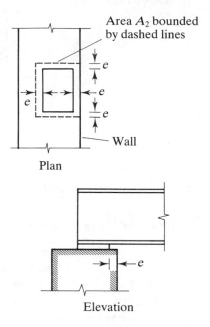

Area A_2 bounded by dashed lines

Wall

Plan

Elevation

supported at the top over a central width of $2k$ and length N, as shown in Figure 5.38. The plate is then considered to bend about an axis parallel to the beam span. Thus the plate is treated as a cantilever of span length $n = (B - 2k)/2$ and a width of N. For convenience, a 1-inch width is considered, with a uniform load in pounds per linear inch numerically equal to the bearing pressure in pounds per square inch.

From Figure 5.38, the maximum bending moment in the plate is

$$M_u = \frac{R_u}{BN} \times n \times \frac{n}{2} = \frac{R_u n^2}{2BN}$$

where R_u/BN is the average bearing pressure between the plate and the concrete. For a rectangular cross section bent about the minor axis, the nominal moment strength M_n is equal to the plastic moment capacity M_p. As illustrated in Figure 5.39, for a rectangular cross section of unit width and depth t, the plastic moment is

$$M_p = F_y\left(1 \times \frac{t}{2}\right)\left(\frac{t}{2}\right) = F_y \frac{t^2}{4}$$

■ **FIGURE 5.38**

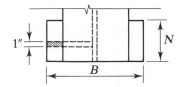

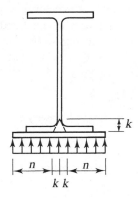

■ **FIGURE 5.39**

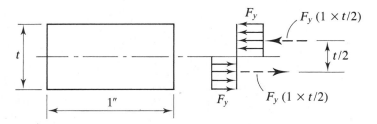

Since $\phi_b M_n$ must at least equal M_u,

$$\phi_b M_p \geq M_u$$

$$0.9 F_y \frac{t^2}{4} \geq \frac{R_u n^2}{2BN}$$

$$t \geq \sqrt{\frac{2R_u n^2}{0.9BNF_y}} \quad \text{or} \quad t \geq \sqrt{\frac{2.222R_u n^2}{BNF_y}} \qquad (5.8/5.9)$$

The foregoing procedure, which is demonstrated in Example 5.16, is essentially the same as that given in Part 11 of the *Manual* (Volume II), "Connections for Tension and Compression."

■ **EXAMPLE 5.16**

Design a bearing plate to distribute the reaction of a W21 × 68 with a span length of 15 feet 10 inches center-to-center of supports. The total service load, including the beam weight, is 9 kips/ft, with equal parts dead and live load. The beam is to be supported on reinforced concrete walls with $f'_c = 3500$ psi. Both the plate and the beam are A36 steel.

SOLUTION

The factored load is

$$w_u = 1.2w_D + 1.6w_L = 1.2(4.5) + 1.6(4.5) = 12.60 \text{ kips/ft}$$

and the reaction is

$$R_u = \frac{w_u L}{2} = \frac{12.60(15.83)}{2} = 99.73 \text{ kips}$$

Determine the length of bearing N required to prevent web yielding. From AISC Equation K1-3, the design strength for this limit state is

$$R_n = (2.5k + N)F_y t_w$$

For $\phi R_n \geq R_u$,

$$1.0[2.5(1.438) + N](36)(0.430) \geq 99.73$$
$$N \geq 2.85 \text{ in.}$$

Use AISC Equation K1-5 to determine the value of N required to prevent web crippling. Assume $N/d > 0.2$ and try the second form of the equation. For $\phi R_n \geq R_u$,

$$\phi 68 t_w^2 \left[1 + \left(4\frac{N}{d} - 0.2 \right)\left(\frac{t_w}{t_f} \right)^{1.5} \right] \sqrt{\frac{F_y t_f}{t_w}} \geq R_u$$

$$0.75(68)(0.430)^2 \left[1 + \left(\frac{4N}{21.13} - 0.2 \right)\left(\frac{0.430}{0.685} \right)^{1.5} \right] \sqrt{\frac{36(0.685)}{0.430}} \geq 99.73$$

$$N \geq 5.27 \text{ in.} \qquad \text{(controls)}$$

Check the assumption:

$$\frac{N}{d} = \frac{5.268}{21.13} = 0.25 > 0.2 \qquad \text{(OK)}$$

Try $N = 6$ in. Determine dimension B from a consideration of bearing strength. A conservative assumption is that the full area of the support is used. Then, the required plate area A_1 can be found as follows:

$$\phi_c(0.85)f_c'A_1 \geq R_u$$
$$0.6(0.85)(3.5)A_1 \geq 99.73$$
$$A_1 \geq 55.87 \text{ in.}^2$$

The minimum value of dimension B is

$$B = \frac{A_1}{N} = \frac{55.87}{6} = 9.31 \text{ in.}$$

The flange width of a W21 × 68 is 8.270 inches, making the plate slightly wider than the flange, which is desirable. Rounding up to the nearest inch, **try $B = 10$ in.**
Compute the required plate thickness:

$$n = \frac{B - 2k}{2} = \frac{10 - 2(1.438)}{2} = 3.562 \text{ in.}$$

From Equation (5.9),

$$t = \sqrt{\frac{2.222R_u n^2}{BNF_y}} = \sqrt{\frac{2.222(99.73)(3.562)^2}{10(6)(36)}} = 1.14 \text{ in.}$$

ANSWER   Use a PL $1\frac{1}{4} \times 6 \times 10$.

If the beam is not laterally braced at the load point (in such a way as to prevent relative lateral displacement between the loaded compression flange and the tension flange), the Specification requires that *sidesway web buckling* be investigated (AISC K1.5). When loads are applied to both flanges, *compression buckling* must be checked (AISC K1.6).

Column Base Plates

As with beam bearing plates, the design of column base plates requires consideration of bearing pressure on the supporting material and bending of the plate. A major difference is that bending in beam bearing plates is in one direction, whereas column base plates are subjected to two-way bending. Moreover, web crippling and web yielding are not factors in column base plate design.

Column base plates can be categorized as large or small, where small plates are those whose dimensions are approximately the same as the column dimensions. Furthermore, small plates behave differently when lightly loaded than when they are more heavily loaded.

The thickness of large plates is determined from consideration of bending of the portions of the plate that extend beyond the column outline. Bending is assumed to take place about axes at middepth of the plate near the edges of the column flanges. Two of the axes are parallel to the web and $0.80b_f$ apart, and two axes are parallel to the flanges and $0.95d$ apart. Of the two 1-inch cantilever strips, labeled m and n in Figure 5.40, the larger is used in place of n in Equation 5.8 to compute the plate thickness, or

$$t \geq \ell \sqrt{\frac{2P_u}{0.9BNF_y}} \tag{5.10}$$

where ℓ is the larger of m and n. This approach is referred to as the *cantilever method*.

Lightly loaded small base plates can be designed by using the Murray–Stockwell method (Murray, 1983). In this approach, the portion of the column load that falls within the confines of the column cross section — that is, over an area $b_f d$ — is assumed to be uniformly distributed over the H-shaped area shown in Figure 5.41. Thus the bearing pressure is concentrated near the column outline. The plate thickness is determined

---

■ **FIGURE 5.40**

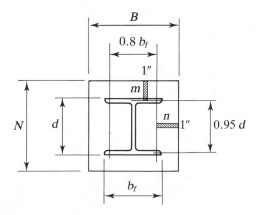

■ **FIGURE 5.41**

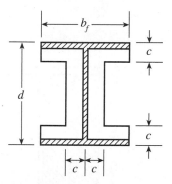

from a flexural analysis of a cantilever strip of unit width and of length c. This approach results in the equation

$$t \geq c\sqrt{\frac{2P_o}{0.9A_H F_y}}$$ (5.11)

where

$$P_o = \frac{P_u}{BN} \times b_f d$$

= load within the area $b_f d$

= load on H-shaped area

A_H = H-shaped area

c = dimension needed to give a stress of $\dfrac{P_o}{A_H}$ equal to

the design bearing stress of the supporting material

Note that Equation 5.11 has the same form as Equation 5.10 but with the stress P_u/BN replaced by P_o/A_H.

For more heavily loaded base plates (the boundary between lightly loaded and heavily loaded plates is not well defined), Thornton (1990a) proposed an analysis based on two-way bending of the portion of the plate between the web and the flanges. As shown in Figure 5.42, this plate segment is assumed to be fixed at the web, simply supported at the flanges, and free at the other edge. The required thickness is

$$t \geq n'\sqrt{\frac{2P_u}{0.9BNF_y}}$$

■ **FIGURE 5.42**

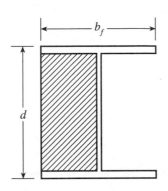 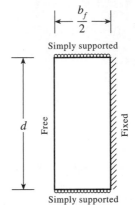

(a) Area of plate considered   (b) Approximated size and edge conditions

where

$$n' = \frac{1}{4}\sqrt{db_f} \qquad (5.12)$$

These three approaches were combined by Thornton (1990b), and a summary of the resulting unified procedure follows. The required plate thickness is

$$t \geq \ell\sqrt{\frac{2P_u}{0.9BNF_y}} \qquad (5.13)$$

where

$$\ell = \max(m, n, \lambda n')$$

$$\lambda = \frac{2\sqrt{X}}{1 - \sqrt{1 - X}} \leq 1$$

$$X = \left(\frac{4db_f}{(d + b_f)^2}\right)\frac{P_u}{\phi_c P_p}$$

$$n' = \frac{1}{4}\sqrt{db_f}$$

$$\phi_c = 0.60$$

P_p = nominal bearing strength from AISC Equation J9-1 or J9-2

There is no need to determine whether the plate is large or small, lightly loaded, or heavily loaded. As a simplification, λ can always be conservatively taken as 1.0 (Thornton, 1990b).

This procedure is the same as that given in Part 11 of the *Manual* (Volume II), "Connections for Tension and Compression."

■ **EXAMPLE 5.17**

A W10 × 49 is used as a column and is supported by a concrete pier as shown in Figure 5.43. The top surface of the pier is 18 inches by 18 inches. Design an A36 base plate for a column dead load of 98 kips and a live load of 145 kips. The concrete strength is $f'_c = 3000$ psi.

SOLUTION

The factored load is

$$P_u = 1.2D + 1.6L = 1.2(98) + 1.6(145) = 349.6 \text{ kips}$$

Compute the required bearing area:

$$\phi_c P_p \geq P_u$$

$$\phi_c (0.85) f'_c A_1 \sqrt{A_2 / A_1} \geq P_u$$

$$0.6(0.85)(3) A_1 \sqrt{18(18)/A_1} \geq 349.6$$

$$A_1 \geq 161.1 \text{ in.}^2$$

Check:

$$\sqrt{A_2 / A_1} = \sqrt{18(18)/161.1} = 1.41 < 2 \qquad \text{(OK)}$$

Also, the plate must be at least as large as the column, so

$$b_f d = 10.00(9.98) = 99.8 \text{ in.}^2 < 161.1 \text{ in.}^2 \qquad \text{(OK)}$$

■ **FIGURE 5.43**

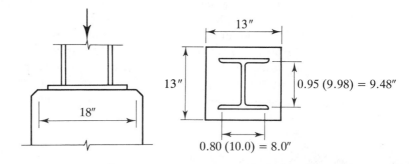

For $B = N = 13$ in., A_1 provided $= 13(13) = 169$ in.^{2}

The dimensions of the cantilever strips m and n can be determined from Figure 5.43, or

$$m = \frac{N - 0.95d}{2} = \frac{13 - 9.48}{2} = 1.76 \text{ in.}$$

$$n = \frac{B - 0.8b_f}{2} = \frac{13 - 8}{2} = 2.5 \text{ in.}$$

From Equation 5.12,

$$n' = \frac{1}{4} \sqrt{db_f} = \frac{1}{4} \sqrt{9.98(10)} = 2.497 \text{ in.}$$

As a conservative simplification, let $\lambda = 1.0$, giving

$$\ell = \max(m, n, \lambda n') = \max(1.76, 2.5, 2.497) = 2.5 \text{ in.}$$

From Equation 5.13, the required plate thickness is

$$t = \ell \sqrt{\frac{2P_u}{0.9BNF_y}} = 2.5 \sqrt{\frac{2(349.6)}{0.9(13)(13)(36)}} = 0.893 \text{ in.}$$

ANSWER   Use a PL $1 \times 13 \times 13$.   ■

5.14   BIAXIAL BENDING

Biaxial bending occurs when a beam is subjected to a loading condition that produces bending about both the major (strong) axis and the minor (weak) axis. Such a case is illustrated in Figure 5.44, where a single concentrated load acts normal to the longitudinal axis of the beam but is inclined with respect to each of the principal axes of the cross section. Although this loading is more general than those previously considered, it is still a special case: The load passes through the shear center of the cross section. The shear center is that point through which the loads must act if there is to be no twisting, or torsion, of the beam. The location of the shear center can be determined from elementary mechanics of materials by equating the internal resisting torsional moment, derived from the shear flow on the cross section, to the external torque.

The location of the shear center for several common cross sections is shown in Figure 5.45a, where the shear center is indicated by an "o." The value of e_o, which locates the shear center for channel shapes, is tabulated in the *Manual*. The shear center is always located on an axis of symmetry; thus the shear center will be at the centroid of a cross section with two axes of symmetry. Figure 5.45b shows the deflected position of two different beams when loads are applied through the shear center and when they are not.

■ **FIGURE 5.44**

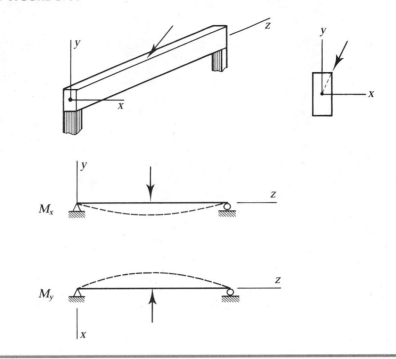

Case I: Loads Applied Through the Shear Center

If loads act through the shear center, the problem is one of simple bending in two per-pendicular directions. As illustrated in Figure 5.46, the load can be resolved into rec-tangular components in the x- and y-directions, each producing bending about a different axis.

To deal with this combined loading, we temporarily look ahead to Chapter H of the Specification, "Members Under Combined Forces and Torsion" (also the subject of Chapter 6 of this textbook, "Beam-Columns"). The Specification deals with com-bined loading primarily through the use of *interaction formulas,* which account for the relative importance of each load effect in relation to the strength corresponding to that effect. For example, if there is bending about the x-axis only,

$$M_{ux} \leq \phi_b M_{nx} \ \text{ or } \ \frac{M_{ux}}{\phi_b M_{nx}} \leq 1.0$$

where

$\quad M_{ux}$ = factored load bending moment about the x-axis

$\quad M_{nx}$ = nominal moment strength about the x-axis

■ **FIGURE 5.45**

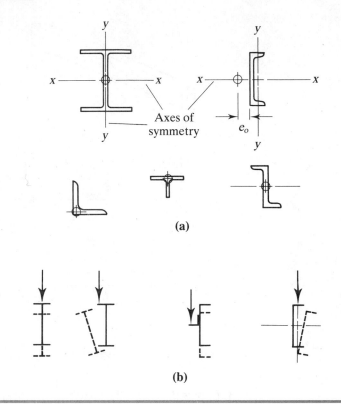

Axes of symmetry

(a)

(b)

■ **FIGURE 5.46**

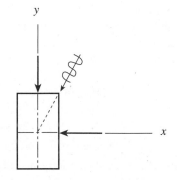

Similarly, if there were bending about the y-axis only, the requirement would be

$$\frac{M_{uy}}{\phi_b M_{ny}} \le 1.0$$

where

M_{uy} = factored load bending moment about the y-axis

M_{ny} = nominal moment strength about the y-axis

When both types of bending are present, the interaction formula approach requires that the sum of the two ratios be less than 1; that is,

$$\frac{M_{ux}}{\phi_b M_{nx}} + \frac{M_{uy}}{\phi_b M_{ny}} \le 1.0 \qquad\qquad (5.14)$$

In effect, this requirement allows the designer to allocate to one direction what has not been "used up" in the other direction. AISC Section H1 incorporates a comparable ratio for axial loads and gives two interaction formulas, one for small axial loads and one for large axial loads (we discuss the reason for this in Chapter 6). With biaxial bending and no axial load, the formula for small axial load is the appropriate one:

$$\frac{P_u}{2\phi P_n} + \left(\frac{M_{ux}}{\phi_b M_{nx}} + \frac{M_{uy}}{\phi_b M_{ny}} \right) \le 1.0 \qquad\qquad \text{(AISC Equation H1-1b)}$$

If the axial load P_u is zero, this equation reduces to Equation 5.14.

To this point, the strength of I- and H-shaped cross sections bent about the weak axis has not been considered. Doing so is relatively simple. Any shape bent about its weak axis cannot buckle in the other direction, so lateral-torsional buckling is not a limit state. If the shape is compact, then

$$M_{ny} = M_{py} = F_y Z_y \le 1.5 M_{yy}$$

where $M_{yy} = F_y S_y$ = yield moment for the y-axis. For I- and H-shaped sections bent about the minor axis, the upper limit of $1.5M_{yy}$ will always control (Z_y/S_y will always be > 1.5). If the shape is noncompact, the strength will be given by AISC Equation A-F1-3 for flange local buckling or web local buckling. (All standard shapes tabulated in the *Manual* have compact webs, so only flange local buckling will be a possibility.)

■ **EXAMPLE 5.18**

A W21 × 68 is used as a simply supported beam with a span length of 12 feet. Lateral support of the compression flange is provided only at the ends. Loads act through the shear center with factored load moments of M_{ux} = 200 ft-kips and M_{uy} = 25 ft-kips. If A36 steel is used, does this beam satisfy the provisions of the AISC Specification? Assume that both moments are uniform over the length of the beam.

SOLUTION   The following data for a W21 × 68 are obtained from the Load Factor Design Selection Table. The shape is compact (no footnote to indicate otherwise) and

$$L_p = 7.5 \text{ ft}, L_r = 22.8 \text{ ft}$$
$$\phi_b M_p = 432 \text{ ft-kips}$$
$$\phi_b M_r = 273 \text{ ft-kips}$$

The unbraced length $L_b = 12$ ft, so $L_p < L_b < L_r$, and the controlling limit state is inelastic lateral-torsional buckling. Then

$$\phi_b M_{nx} = \phi_b C_b \left[M_p - (M_p - M_r) \left(\frac{L_b - L_p}{L_r - L_p} \right) \right]$$

$$= C_b \left[\phi_b M_p - (\phi_b M_p - \phi_b M_r) \left(\frac{L_b - L_p}{L_r - L_p} \right) \right] \leq \phi_b M_p$$

Because the bending moment is uniform, $C_b = 1$, and

$$\phi_b M_{nx} = 1.0 \left[432 - (432 - 273) \left(\frac{12 - 7.5}{22.8 - 7.5} \right) \right] = 385.2 \text{ ft-kips}$$

Alternatively, $\phi_b M_{nx}$ can be obtained from the beam design charts.

Since the shape is compact, there is no flange local buckling and

$$\phi_b M_{ny} = \phi_b M_{py} = \phi_b Z_y F_y = 0.90(24.4)(36) = 790.6 \text{ in.-kips} = 65.88 \text{ ft-kips}$$

Check:

$$\frac{Z_y}{S_y} = \frac{24.4}{15.7} = 1.55 > 1.5$$

$$\therefore \text{ Use } M_{ny} = 1.5 M_{yy} = 1.5 F_y S_y = 1.5(36)(15.7) = 847.8 \text{ in.-kips}$$

$$= 70.75 \text{ ft-kips}$$

$$\phi_b M_{ny} = 0.90(70.65) = 63.59 \text{ ft-kips}$$

From Equation 5.11,

$$\frac{M_{ux}}{\phi_b M_{nx}} + \frac{M_{uy}}{\phi_b M_{ny}} = \frac{200}{385.2} + \frac{25}{63.59} = 0.912 < 1.0 \qquad \text{(OK)}$$

ANSWER   The W21 × 68 is satisfactory.   ■

Case II: Loads Not Applied Through the Shear Center

When loads are not applied through the shear center of a cross section, the result is flexure plus torsion. If possible, the structure or connection geometry should be modified to remove the eccentricity. The problem of torsion in rolled shapes is a complex one, and we resort to approximate, although conservative, methods for dealing with it. A more detailed treatment of this topic, complete with design aids, can be found in *Torsional Analysis of Structural Steel Members* (AISC, 1997). A typical loading condition that gives rise to torsion is shown in Figure 5.47a. The resultant load is applied to the center of the top flange, but its line of action does not pass through the shear center of the section. As far as equilibrium is concerned, the force can be moved to the shear center provided that a couple is added. The equivalent system thus obtained will consist of the given force acting through the shear center plus a twisting moment as shown. In Figure 5.47b, there is only one component of load to contend with, but the concept is the same.

Figure 5.48 illustrates a simplified way of treating these two cases. In Figure 5.48a, the top flange is assumed to provide the total resistance to the horizontal component of the load. In Figure 5.48b, the twisting moment *Pe* is resisted by a couple consisting of equal forces acting at each flange. As an approximation, each flange can be considered to resist each of these forces independently. Consequently, the problem is reduced to a case of bending of two shapes, each one loaded through its shear center. In each of the two situations depicted in Figure 5.48, only about half the cross section is considered to be effective with respect to its y-axis; therefore, when considering the strength of a single flange, use half the tabulated value of Z_y for the shape.

■ **FIGURE 5.47**

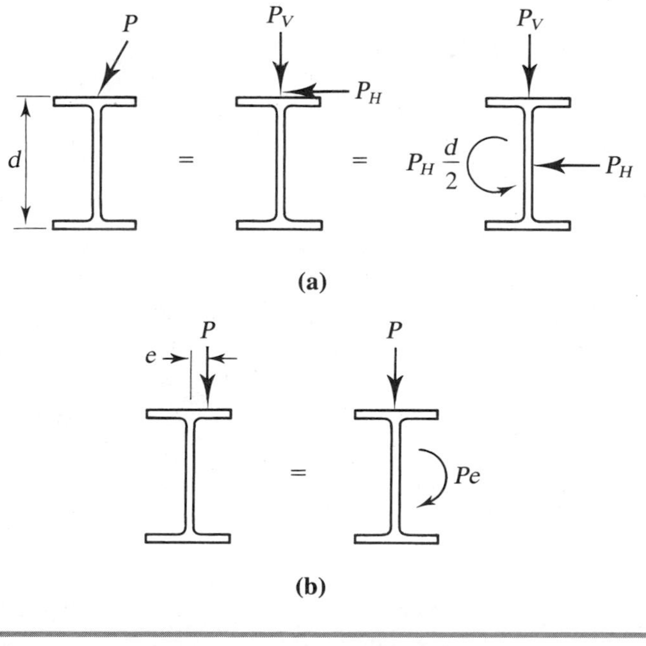

(a)

(b)

■ **FIGURE 5.48**

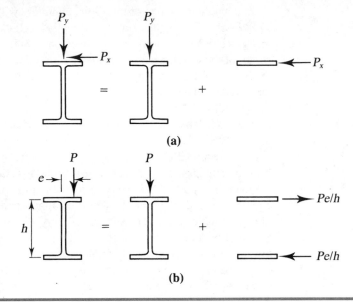

(a)

(b)

■ **FIGURE 5.49**

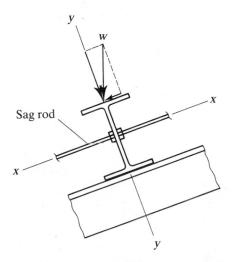

Design of Roof Purlins

Roof purlins that are part of a sloping roof system can be subjected to biaxial bending of the type just described. For the roof purlin shown in Figure 5.49, the load is vertical, but the axes of bending are inclined. This condition corresponds to the loading of Figure 5.48a. The component of load normal to the roof will cause bending about the x-axis, and the parallel component bends the beam about its y-axis. If the purlins are simply supported at the trusses (or rigid frame rafters), the maximum bending moment about each axis is $wL^2/8$, where w is the appropriate component of load. If sag rods are used, they will provide lateral support with respect to x-axis bending and will act as transverse supports for y-axis bending, requiring that the purlin be treated as a continuous beam. For uniform sag-rod spacings, the formulas for continuous beams in Part 4 of the *Manual* can be used.

■ **EXAMPLE 5.19**

A roof system consists of trusses of the type shown in Figure 5.50 spaced 15 feet apart. Purlins are to be placed at the joints and at the midpoint of each top chord member. Sag rods will be located at the center of each purlin. The total gravity load, including an estimated purlin weight, is 30 psf of roof surface, with a live load-to-dead load ratio of 1.0. Assuming that this is the critical loading condition, use A36 steel and select a 6-inch-deep W-shape for the purlins.

SOLUTION

For this loading condition, dead load plus a roof live load with no wind or snow, load combination (A4-3) will control:

$$w_u = 1.2w_D + 1.6L_r = 1.2(15) + 1.6(15) = 42 \text{ psf}$$

The width of roof surface tributary to each purlin is

$$\frac{15}{2} \frac{\sqrt{10}}{3} = 7.906 \text{ ft}$$

■ **FIGURE 5.50**

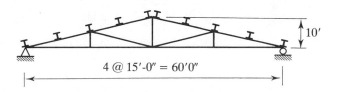

4 @ 15'-0" = 60'0"

10'

Then

$$\text{Purlin load} = 42(7.906) = 332.1 \text{ lb/ft}$$

$$\text{Normal component} = \frac{3}{\sqrt{10}} (332.1) = 315.1 \text{ lb/ft}$$

$$\text{Parallel component} = \frac{1}{\sqrt{10}} (332.1) = 105.0 \text{ lb/ft}$$

and

$$M_{ux} = \frac{1}{8} (0.3151)(15)^2 = 8.862 \text{ ft-kips}$$

With sag rods placed at the midpoint of each purlin, the purlins are two-span continuous beams with respect to weak axis bending. From the "Beam Diagrams and Formulas" section in Part 4 of the *Manual,* the bending moment at the interior support with only one span loaded is

$$M = \frac{1}{16} wL^2$$

where

w = uniform load intensity

L = span length (two equal spans)

With both spans loaded, the moment can be obtained by superposition:

$$M = M_{\max} = \frac{1}{16} wL^2(2) = \frac{1}{8} wL^2$$

$$\therefore M_{uy} = \frac{1}{8} (0.1050)(15/2)^2 = 0.7382 \text{ ft-kips}$$

To select a trial shape, use the beam design charts and choose a shape with a relatively large margin of strength with respect to major axis bending. For an unbraced length of $\frac{15}{2} = 7.5$ ft, **try a W6 × 9.**

For $C_b = 1.0$, $\phi_b M_{nx} = 14.0$ ft-kips. From Figure 5.15b, C_b is 1.30 for the load and lateral support conditions of this beam. Therefore

$$\phi_b M_{nx} = 1.30(14.0) = 18.20 \text{ ft-kips}$$

But

$$\phi_b M_{px} = 16.8 \text{ ft-kips} < 18.20$$
$$\therefore \quad \text{use } \phi_b M_{nx} = 16.8 \text{ ft-kips}$$

This shape is compact (no footnote in the factored uniform load tables), so

$$\phi_b M_{ny} = \phi_b M_{py} = \phi_b Z_y F_y = 0.90(1.72)(36) = 55.73 \text{ in.-kips} = 4.644 \text{ ft-kips}$$

but

$$\frac{Z_y}{S_y} = \frac{1.72}{1.11} = 1.55 > 1.5$$

$$\therefore \phi_b M_{ny} = \phi_b(1.5 M_{yy}) = \phi_b(1.5)(F_y S_y)$$
$$= 0.9(1.5)(36)(1.11) = 53.95 \text{ in.-kips} = 4.496 \text{ ft-kips}$$

Because the load is applied to the top flange, use only half this capacity to account for the torsional effects. From Equation 5.14,

$$\frac{M_{ux}}{\phi_b M_{nx}} + \frac{M_{uy}}{\phi_b M_{ny}} = \frac{8.862}{16.8} + \frac{0.7382}{4.496/2} = 0.856 < 1.0 \qquad (\text{OK})$$

The shear is

$$V_u = \frac{0.3151(15)}{2} = 2.363 \text{ kips}$$

From the factored uniform load tables,

$$\phi_v V_n = 19.5 \text{ kips} > 2.363 \text{ kips} \qquad (\text{OK})$$

ANSWER   Use a W6 × 9.   ■

5.15   BENDING STRENGTH OF VARIOUS SHAPES

W-, S-, and M-shapes are the most commonly used hot-rolled shapes for beams, and their bending strength has been covered in the preceding sections. Other shapes are sometimes used as flexural members, however, and this section provides a summary of the relevant AISC provisions. All equations are from Chapter F or Appendix F of the Specification. Nominal strength is given for compact and noncompact hot-rolled shapes, but not for slender shapes or shapes built up from plate elements. No numerical examples are given in this section, but Example 6.11 includes the computation of the flexural strength of a structural tee-shape.

As in previous coverage, equations are specialized for nonhybrid sections ($F_{yw} = F_{yf} = F_y$) and for the case of bending only (no axial load).

    I. Channels

        A. Width-thickness parameters for flexure

            1. Flange:

$$\lambda = \frac{b_f}{t_f}, \qquad \lambda_p = \frac{65}{\sqrt{F_y}}, \text{ and } \lambda_r = \frac{141}{\sqrt{F_y - 10}}$$

2. Web:

$$\lambda = \frac{h}{t_w}, \qquad \lambda_p = \frac{640}{\sqrt{F_y}}, \text{ and } \lambda_r = \frac{970}{\sqrt{F_y}}$$

B. Bending about the major axis [with either (1) the load applied through the shear center and in a plane parallel to the web or (2) restraint against twisting at the point of load application and at the supports]: M_n is the same as for I-shapes (see Sections 5.5 and 5.6).

C. Bending about the minor axis: M_n is the same as for I-shapes (see Section 5.14).

II. Rectangular Structural Tubes

A. Width–thickness parameters (see Figure 5.51)

1. Flange:

$$\lambda = \frac{b}{t}, \qquad \lambda_p = \frac{190}{\sqrt{F_y}}, \text{ and } \lambda_r = \frac{238}{\sqrt{F_y}}$$

2. Web:

$$\lambda = \frac{h}{t}, \qquad \lambda_p = \frac{640}{\sqrt{F_y}}, \text{ and } \lambda_r = \frac{970}{\sqrt{F_y}}$$

If the actual dimensions b and h, shown in Figure 5.51, are not known, they may be estimated as the total width or depth minus three times the thickness (properties of tubes listed in Part 1 of the *Manual* are based on an outside corner radius of $2t$).

■ **FIGURE 5.51**

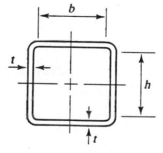

B. Bending about the major axis (loaded in the plane of symmetry)

   1. Compact shapes:

   For compact shapes, the strength will be based on the limit state of lateral-torsional buckling (LTB).

   For $L_b \leq L_p$,

   $$M_n = M_p \leq 1.5 M_y \qquad \text{(AISC Equation F1-1)}$$

   For $L_p < L_b \leq L_r$,

   $$M_n = C_b \left[M_p - (M_p - M_r) \left(\frac{L_b - L_p}{L_r - L_p} \right) \right] \leq M_i \qquad \text{(AISC Equation F1-2)}$$

   For $L_b > L_r$,

   $$M_n = M_{cr} \leq M_p \qquad \text{(AISC Equation F1-12)}$$

   where

   $$M_{cr} = \frac{57,000 C_b \sqrt{JA}}{L_b / r_y} \qquad \text{(AISC Equation F1-14)}$$

   $$L_p = \frac{3750 r_y}{M_p} \sqrt{JA} \qquad \text{(AISC Equation F1-5)}$$

   $$L_r = \frac{57,000 r_y \sqrt{JA}}{M_r} \qquad \text{(AISC Equation F1-10)}$$

   $$M_r = F_y S_x \qquad \text{(AISC Equation F1-11)}$$

   2. Noncompact shapes: The nominal strength M_n will be the smallest of the values computed for the following limit states: Lateral torsional buckling (LTB), flange local buckling (FLB), or web local buckling (WLB). For each of the two local buckling states, the strength is determined from the following equation:

   $$M_n = M_p - (M_p - M_r) \left(\frac{\lambda - \lambda_p}{\lambda_r - \lambda_p} \right) \qquad \text{(AISC Equation A-F1-3)}$$

C. Bending about the minor axis: There is no LTB limit state for any shape bent about its minor axis.

   1. Compact shapes:

   $$M_n = M_p \leq 1.5 M_y \qquad \text{(AISC Equation F1-1)}$$

   2. Noncompact shapes: Check FLB and WLB with AISC Equation A-F1-3.

III. Square Structural Tubes

    A. Width–thickness parameters

$$\lambda = \frac{b}{t}, \qquad \lambda_p = \frac{190}{\sqrt{F_y}}, \text{ and } \lambda_r = \frac{238}{\sqrt{F_y}}$$

    B. Nominal bending strength

    There is no LTB limit state for square (or circular) shapes.

        1. Compact shapes:

$$M_n = M_p \leq 1.5 M_y \qquad \text{(AISC Equation F1-1)}$$

        2. Noncompact shapes: The strength is limited by local buckling, either FLB or WLB, whichever yields the smaller value of M_n:

$$M_n = M_p - (M_p - M_r)\left(\frac{\lambda - \lambda_p}{\lambda_r - \lambda_p}\right) \qquad \text{(AISC Equation A-F1-3)}$$

IV. Solid Rectangular Bars

    For rectangular bars, the applicable limit state is LTB for major axis bending; local buckling is not a limit state for either major or minor axis bending.

    A. Bending about the major axis:

    For $L_b \leq L_p$,

$$M_n = M_p \leq 1.5 M_y \qquad \text{(AISC Equation F1-1)}$$

    For $L_p < L_b \leq L_r$,

$$M_n = C_b\left[M_p - (M_p - M_r)\left(\frac{L_b - L_p}{L_r - L_p}\right)\right] \leq M_p \qquad \text{(AISC Equation F1-2)}$$

    For $L_b > L_r$,

$$M_n = M_{cr} \leq M_p \qquad \text{(AISC Equation F1-12)}$$

$$M_{cr} = \frac{57{,}000 C_b \sqrt{JA}}{L_b / r_y} \qquad \text{(AISC Equation F1-14)}$$

    where

$$L_p = \frac{3750 r_y}{M_p} \sqrt{JA} \qquad \text{(AISC Equation F1-5)}$$

$$L_r = \frac{57{,}000 r_y \sqrt{JA}}{M_r} \qquad \text{(AISC Equation F1-10)}$$

$$M_r = F_y S_x \qquad \text{(AISC Equation F1-11)}$$

B. Bending about the minor axis:

$$M_n = M_p \leq 1.5M_y \qquad \text{(AISC Equation F1-1)}$$

V. Tees and Double-Angle Shapes

    A. Width–thickness parameters

        1. Tees

            a. Flange:

$$\lambda = \frac{b_f}{2t_f} \quad \text{and} \quad \lambda_r = \frac{95}{\sqrt{F_y}} \qquad (\lambda_p \text{ does not apply})$$

            b. Web:

$$\lambda = \frac{d}{t_w} \quad \text{and} \quad \lambda_r = \frac{127}{\sqrt{F_y}} \qquad (\lambda_p \text{ does not apply})$$

        2. Double angles with separators, either leg:

$$\lambda = \frac{b}{t} \quad \text{and} \quad \lambda_r = \frac{76}{\sqrt{F_y}} \qquad (\lambda_p \text{ does not apply})$$

        3. Double angles in continuous contact, outstanding leg:

$$\lambda = \frac{b}{t} \quad \text{and} \quad \lambda_r = \frac{95}{\sqrt{F_y}} \qquad (\lambda_p \text{ does not apply})$$

    B. With loading in the plane of symmetry:

$$M_n = M_{cr} = \frac{\pi\sqrt{EI_yGJ}}{L_b}\left[B + \sqrt{1 + B^2}\right] \qquad \text{(AISC Equation F1-15)}$$

    where

        $M_n \leq 1.5M_y$ for stems in tension

        $M_n \leq 1.0M_y$ for stems in compression

$$B = \pm 2.3\left(\frac{d}{L_b}\right)\sqrt{\frac{I_y}{J}} \qquad \text{(AISC Equation F1-16)}$$

        $M_y = F_yS_x$

    The positive sign is used for B when the stem is in tension, and the negative sign is used when the stem is in compression *anywhere* along the unbraced length.

    C. Bending about the minor axis:

    For nonslender shapes ($\lambda \leq \lambda_r$),

$$M_n = M_p \leq 1.5M_y$$

VI. Solid circular and square shapes:

$$M_n = M_p \leq 1.5M_y$$

VII. Hollow circular shapes

  A. Width–thickness parameters:

$$\lambda = \frac{D}{t}, \qquad \lambda_p = \frac{2070}{F_y}, \text{ and } \lambda_r = \frac{8970}{F_y}$$

  where D is the outer diameter.

  B. Nominal bending strength:

  There is no LTB limit state for circular (or square) shapes. The strength is limited by local buckling.

  For $\lambda \leq \lambda_p$,

$$M_n = M_p \leq 1.5M_y$$

  For $\lambda_p < \lambda \leq \lambda_r$,

$$M_n = \left(\frac{600}{D/t} + F_y \right) S \qquad \text{(AISC Appendix F, Table A-F1.1)}$$

■ PROBLEMS

Bending Stress and the Plastic Moment

5.2-1 A flexural member is fabricated from two flange plates $\frac{1}{2} \times 7\frac{1}{2}$ and a web plate $\frac{3}{8} \times 17$. The yield stress of the steel is 50 ksi.

  a. Compute the plastic moment M_p and the plastic section modulus Z with respect to the major principal axis.

  b. Compute the elastic section modulus S and the yield moment M_y with respect to the major principal axis.

  c. Would this shape be classified as a beam or as a plate girder by the AISC Specification?

5.2-2 An unsymmetrical flexural member consists of a $\frac{1}{2} \times 12$ top flange, a $\frac{1}{2} \times 7$ bottom flange, and a $\frac{3}{8} \times 16$ web.

  a. Compute the plastic section modulus Z_x with respect to the major principal axis.

  b. Compute the plastic section modulus Z_y with respect to the minor principal axis.

5.2-3   Verify the value of Z_x for a W18 × 50 that is tabulated in the properties tables in Part 1 of the *Manual*.

Classification of Shapes

5.4-1   For W-, M-, and S-shapes of A36 steel,

        a. List the shapes in Part 1 of the *Manual* that are noncompact (when used as flexural members). State whether they are noncompact because of the flange, the web, or both.

        b. List the shapes in Part 1 of the *Manual* that are slender. State whether they are slender because of the flange, the web, or both.

5.4-2   Repeat Problem 5.4-1 for steel with $F_y = 50$ ksi.

5.4-3   Determine the smallest value of yield stress F_y for which a W-, M-, or S-shape from Part 1 of the *Manual* will become slender. To which shape does this value apply? What conclusion can you draw from your answer?

Bending Strength of Compact Shapes

5.5-1   The beam shown in Figure P5.5-1 is a W21 × 73 of A36 steel with continuous lateral support. The uniform load is a superimposed dead load and the concentrated load is a live load. Is this beam adequate?

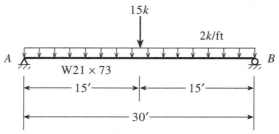

FIGURE P5.5-1

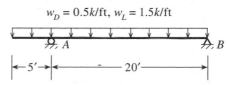

5.5-2   A 25-foot-long beam is pin-supported at the right end and roller-supported at a point 5 feet from the left end, as shown in Figure P5.5-2. There is continuous lateral support. The beam is subjected to a uniform load over its entire length consisting of a service dead load of 0.5 kip/ft (including the weight of the beam) and a service live load of 1.5 kips/ft. Is a W16 × 31 of A36 steel adequate?

FIGURE P5.5-2

5.5-3   A W16 × 40 is used as a simply supported, uniformly loaded beam with a span length of 50 feet and continuous lateral support. The yield stress F_y is 50 ksi. If the ratio of live load to dead load is 3.0, what is the maximum total service load, in kips/ft, that can be supported?

5.5-4   A W33 × 130 of A572 Grade 50 steel is used as a cantilever beam. It must support a uniform service dead load of 1.1 kips/ft (in addition to the beam weight) and a service live load of 2.6 kips/ft. The beam has continuous lateral support. What is the maximum permissible span length?

5.5-5   A W30 × 99 of A36 steel is subjected to the loading shown in Figure P5.5-5. There is continuous lateral support. The concentrated loads, P, are service live loads, and $w = P/10$ is a service dead load. Determine the maximum values of P and w that can be supported.

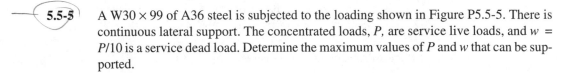

FIGURE P5.5-5

5.5-6   The beam shown in Figure P5.5-6 is a W36 × 182. It is laterally supported at A and B. The 300 kip load is a service live load.

   a. Compute C_b. Do not include the beam weight in the loading.

   b. Compute C_b. Include the beam weight in the loading.

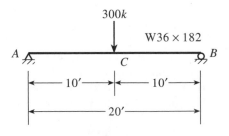

FIGURE P5.5-6

5.5-7   If the beam in Problem 5.5-6 is braced at A, B, and C, compute C_b for the unbraced length AC (same as C_b for unbraced length CB). Do not include the beam weight in the loading.

5.5-8 Compute L_p and L_r for a W36 × 160 of A572 Grade 60 steel.

5.5-9 A W21 × 44 of A572 Grade 50 steel is used as a simply supported beam. It is uniformly loaded and has lateral support only at its ends.

    a. Compute the design strength for a span length of 10 ft. Do not use the design aids in Part 4 of the *Manual*.

    b. Compute the design strength for a span length of 15 ft. Do not use the design aids in Part 4 of the *Manual*.

5.5-10 Repeat Problem 5.5-1 for lateral bracing only at the ends. Do not use the design aids in Part 4 of the *Manual*.

5.5-11 The simply supported beam shown in Figure P5.5-11 has lateral bracing at 10-foot intervals. What total service load can be supported if the load is equal parts dead load and live load? Do not use the design aids in Part 4 of the *Manual*. (Hint: $\phi_b M_n$ can be different for each unbraced segment.)

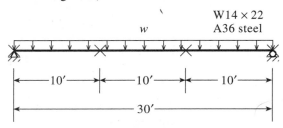

FIGURE P5.5-11

5.5-12 Repeat Problem 5.5-1 for lateral bracing only at the ends and at midspan. Do not use the design aids in Part 4 of the *Manual*.

5.5-13 Repeat Problem 5.5-2 for lateral bracing only at *A* and *B*. Do not use the design aids in Part 4 of the *Manual*.

5.5-14 Repeat Problem 5.5-5 for lateral bracing only at the ends and at the points where the concentrated loads are applied. Do not use the design aids in Part 4 of the *Manual*.

Bending Strength of Noncompact Shapes

5.6-1 A W6 × 15 of A36 steel is used as a simply supported, uniformly loaded beam with continuous lateral support. Use the AISC equations for M_n and compute the flexural design strength.

5.6-2 A W14 × 90 of A572 Grade 50 steel is used as a simply supported beam with lateral support only at the ends. The loading consists of equal couples applied at the ends, as shown in Figure P5.6-2. Use the AISC equations for M_n and compute the flexural design strength.

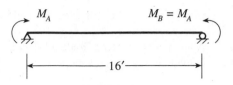

FIGURE P5.6-2

5.6-3 The beam shown in Figure P5.6-3 is a W12 × 65 with F_y = 50 ksi. The loading consists of a concentrated live load at midspan. Lateral support is provided at the ends and at midspan. Neglect the weight of the beam and determine the maximum permissible value of P, a service live load. Use the AISC equations for M_n.

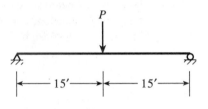

FIGURE P5.6-3

5.6-4 Verify the values of $\phi_b M_p$ and L_p in the Load Factor Design Selection Table for a W40 × 174 of A572 Grade 50 steel.

Shear Strength

5.8-1 Compute the design shear strength of an M12 × 10.8 of A572 Grade 65 steel.

5.8-2 Compute the design shear strength of a W16 × 26 of A572 Grade 60 steel.

5.8-3 The beam shown in Figure P5.8-3 is laterally supported only at the ends. The concentrated loads are service live loads. Is a W12 × 30 adequate if A572 grade 50 steel is used?

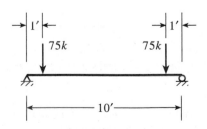

FIGURE P5.8-3

5.8-4 A W18 × 35 of A36 steel is used as a cantilever beam, as shown in Figure P5.8-4. The concentrated load is a service live load, and the uniform load is a service dead load that includes the weight of the beam. Lateral support is provided only at the fixed end. Is this beam adequate?

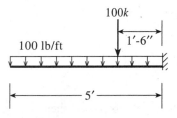

FIGURE P5.8-4

5.8-5 A W12 × 22 is used as a simply supported beam, as shown in Figure P5.8-5. Continuous lateral support is provided. The concentrated loads P are service live loads. Determine the maximum value of P if $F_y = 50$ ksi.

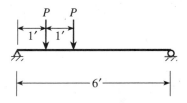

FIGURE P5.8-5

5.8-6 The beam shown in Figure P5.8-6 is a W16 × 40 of A36 steel. It is laterally supported only at A and B. The concentrated loads are service live loads. Is this beam adequate?

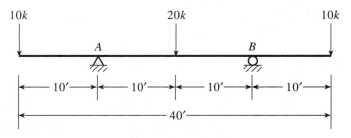

FIGURE P5.8-6

5.8-7 Compute the block shear strength of the beam shown in Figure P5.8-7, which is a W18 × 50 of A36 steel. The bolts are ¾-inch in diameter.

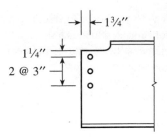

FIGURE P5.8-7

5.8-8 Compute the block shear strength of the beam shown in Figure P5.8-8, which is a W30 × 116 of A572 Grade 50 steel. The bolts are ⅞-inch in diameter.

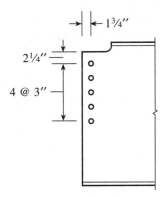

FIGURE P5.8-8

Design

5.10-1 The beam shown in Figure P5.10-1 has continuous lateral support. The concentrated load is a live load, and the uniform load is a dead load that does not include the weight of the beam. Use A36 steel and select a shape. The total deflection must not exceed *L*/240.

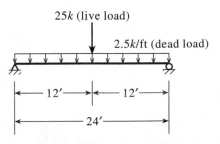

FIGURE P5.10-1

5.10-2   The beam shown in Figure P5.10-2 has continuous lateral support. The 20-kip concentrated load is a live load, and the uniform load consists of 6 kips/ft dead load (not
including the weight of the beam) and 2 kips/ft live load. Use A572 Grade 50 steel and
select a shape. There is no limit on the deflection.

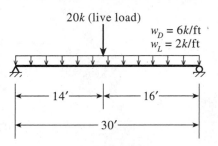

FIGURE P5.10-2

5.10-3   The beam shown in Figure P5.10-3 has continuous lateral support. The 3-kip concentrated load is a live load, and the uniform load is a 6-kip/ft dead load that does not include the weight of the beam. Use A572 Grade 50 steel and select a shape. There is no
limit on the deflection.

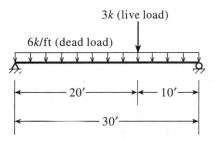

FIGURE P5.10-3

5.10-4   The beam shown in Figure P5.10-4 is a two-span beam with a pin (hinge) in the center of the left span, making the beam statically determinate. The concentrated loads are
service live loads. There is continuous lateral support. Use steel with a yield stress of
50 ksi and select a shape. There is no limit on the deflection.

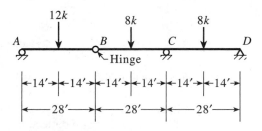

FIGURE P5.10-4

5.10-5 The beam shown in Figure P5.10-5 is laterally supported at the ends and at the points of load application. The 14 kip loads are service live loads. Use A36 steel and select a shape. The total deflection must not exceed $L/240$.

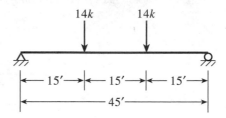

FIGURE P5.10-5

5.10-6 The cantilever beam shown in Figure P5.10-6 is laterally supported only at the fixed end. The concentrated load is a service live load. Use A572 Grade 50 steel and select a shape. There is no limit on deflection.

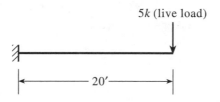

FIGURE P5.10-6

5.10-7 The beam shown in Figure P5.10-7 is part of a roof system. It is laterally supported only at its ends. The loading consists of 170 lb/ft dead load (not including the weight of the beam), 100 lb/ft roof live load, 280 lb/ft snow load, and 180 lb/ft of wind load acting *upward*. The dead, live, and snow loads are gravity loads and always act downward, whereas the wind load on the roof will always act upward. Use A572 Grade 50 steel and select a shape. The total deflection must not exceed $L/180$.

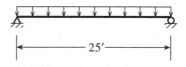

FIGURE P5.10-7

5.10-8 The beam shown in Figure P5.10-8 is laterally supported at *A, B,* and *C*. The concentrated load consists of 6 kips of dead load and 6 kips of live load. The uniform load consists of 2 kips/ft dead load and 2 kips/ft live load. Use A572 Grade 50 steel and select a shape. There is no limit on deflection.

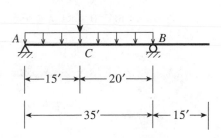

FIGURE P5.10-8

5.10-9 The beam shown in Figure P5.10-9 is laterally supported at *A*, *B*, and *C*. The concentrated load is a service live load and the uniform load is a service dead load. Use A36 steel and select a shape. There is no limit on deflection.

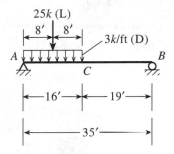

FIGURE P5.10-9

5.10-10 The beam shown in Figure P5.10-10 is a three-span beam with hinges in the outer spans, making the beam statically determinate. There is lateral bracing only at the supports *A*, *B*, *C*, and *D*. The uniform load consists of 1 kip/ft dead load (not including the weight of the beam) and 3 kips/ft live load. Use steel with a yield stress of 50 ksi and select a shape. There is no limit on deflection.

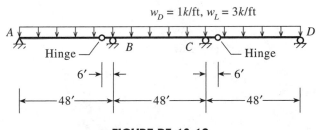

FIGURE P5.10-10

5.10-11 The beam shown in Figure P5.10-11 is laterally supported at *A*, *B*, *C*, and *D*. The load consists of 1.5 kips/ft dead load and 4 kips/ft live load, on both segments *AB* and *CD*.

Use A572 Grade 50 steel and select a shape. The total deflection must not exceed *L*/240.

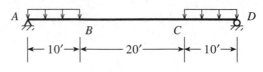

FIGURE P5.10-11

5.10-12 A partial floor plan for a storage facility is shown in Figure P5.10-12. The floor beams are 30 feet long and are spaced 12 feet apart. The only superimposed dead load is a 6-inch-thick reinforced concrete floor slab of normal-weight concrete that provides continuous lateral support of the compression flanges of the beams. The live load is 250 psf. A36 steel will be used, and there is no deflection limit.

    a. Select a typical interior floor beam *AB*.

    b. Select an exterior floor beam *CD*.

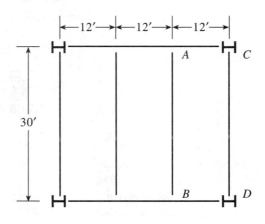

FIGURE P5.10-12

5.10-13 The beams in a roof framing system span 35 feet and are spaced at 6 feet 6 inches. The loads are as follows.

Roof live load:	12 psf
Snow:	20 psf
Metal deck:	2.5 psf
Rigid insulation:	1.5 psf
Membrane roof system:	12 psf
Suspending ceiling:	5 psf

Assume that the deck provides adequate lateral support of the compression flange (that is, $L_b \leq L_p$). Use A572 Grade 50 steel and select a shape for an interior beam. The total deflection must not exceed $L/180$.

5.10-14 Beams in a floor system are spaced 8 feet apart and have a span length of 40 feet. They must support a 5-inch-thick reinforced concrete slab of lightweight concrete (use 115 pcf). Other loads include floor tiles (1 psf), a suspended ceiling with electrical and mechanical services (5 psf), and partitions (20 psf). The live load is 40 psf. Assume that the slab provides continuous lateral support. Use A572 Grade 50 steel and select a shape for an interior beam. The deflection must not exceed $L/240$.

5.10-15 Design an interior floor beam of A572 Grade 50 steel for a span length of 25 feet and a beam spacing of 7 feet. There is a 4-inch-thick reinforced concrete floor slab of normal weight concrete. Assume that the slab provides continuous lateral support. Other loads include flooring (4 psf) and a suspended ceiling (5 psf). Use a live load of either a uniformly distributed load of 60 psf or a concentrated load of 1000 pounds, whichever causes the worst effect (the concentrated load is moveable, and should be considered to act where it causes the worst effect). The maximum deflection should not exceed ¾ inch.

5.10-16 A floor system consists of 45-foot-long girders supporting W24 × 55 floor beams spaced at 9 feet. A partial floor plan is shown in Figure P5.10-16. The loads consist of a 5-inch-thick reinforced concrete floor slab of normal-weight concrete, a partition load of 20 psf, and miscellaneous dead loads of 6 psf. The live load is 125 psf. Assume continuous lateral support.

   a. Use A572 Grade 50 steel and design girder *AB*. For the girder loading, use both a uniform load and concentrated loads. The uniform load consists of only the girder weight, and the concentrated loads are the beam reactions. There is no deflection limit.

   b. To compute the maximum bending moment in the girder, convert all loads, including the weight of the floor beams, into a uniform load acting on the girder. Then compare this moment with the one computed in part (a), but *do not* redesign the girder.

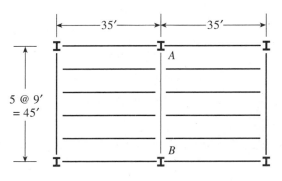

FIGURE P5.10-16

Holes in Beams

5.11-1 A W36 × 160 beam is connected with $1\frac{1}{8}$-inch-diameter bolts. There are two bolts in each flange. A572 Grade 50 steel is used.

   a. Determine the elastic section modulus S_x.

   b. Determine the plastic section modulus Z_x.

5.11-2 A beam is built up from a W21 × 93 and two 8 × $\frac{1}{2}$ flange cover plates, as shown in Figure P5.11-2. The bolts are 1 inch in diameter and A572 Grade 50 steel is used.

   a. Determine the elastic section modulus S_x.

   b. Determine the plastic section modulus Z_x.

FIGURE P5.11-2

5.11-3 A W14 × 34 is connected through the top flange with two $\frac{7}{8}$-inch-diameter bolts. There are no bolts in the bottom flange. A36 steel is used.

   a. Determine the elastic section modulus S_x for both the top and bottom fibers of the beam.

   b. Determine the plastic section modulus Z_x.

Open-Web Steel Joists

5.12-1 A floor system uses open-web steel joists spaced at 4 feet with a span length of 18 feet. The joists must support a 4-inch-thick reinforced concrete slab of normal weight concrete, 4 psf of flooring, and a ceiling load of 5 psf. The live load is 75 psf. The slab provides continuous lateral support of the top chord.

   a. Select a K-series joist from the table in Figure 5.34.

   b. Is there likely to be a live load deflection problem with your selection? Why or why not?

5.12-2 Select a K-series open-web steel joist from the table in Figure 5.34 to satisfy the following conditions:

span length = 30 ft and joist spacing = 3 ft;

3-inch-thick reinforced concrete floor slab (provides continuous lateral support);

20 psf movable partition load;

6 psf other dead load; and

40 psf live load.

Beam Bearing Plates and Column Base Plates

5.13-1   The beam in Figure P5.13-1 has continuous lateral support. The uniform load is a dead load, and the concentrated load is a live load.

    a. Use A36 steel and design the beam. The total deflection is limited to $\frac{1}{240}$ of the span length.

    b. Design bearing plates for the supports and the concentrated load. Assume that the bearing plate at the support rests on concrete with a surface area larger than the bearing area by an amount equal to 1 inch of concrete on all sides of the plate. Use A36 steel and a concrete strength of $f'_c = 3500$ psi.

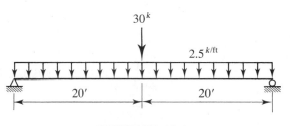

FIGURE P5.13-1

5.13-2   Design a bearing plate to support a concentrated factored load of 150 kips on a W18 × 50. The beam is A572 Grade 50 steel, but use A36 for the bearing plate.

5.13-3   Design a base plate for a W14 × 82 column supporting a factored axial load of 500 kips. The support will be an 18 inch by 18 inch concrete pier. Use A36 steel and $f'_c = 3500$ psi.

5.13-4   Design a column base plate for a W10 × 49 with a factored axial load of 220 kips. The column is supported by a 12 inch by 12 inch concrete pier. Use A36 steel and $f'_c = 3000$ psi.

Biaxial Bending

5.14-1   The 40 kip concentrated load shown in Figure P5.14-1 is a live load. Neglect the weight of the beam and determine whether the beam satisfies the AISC Specification if A572 Grade 50 steel is used. Lateral support is provided at the ends only.

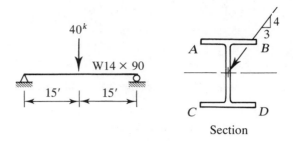

FIGURE P5.14-1

5.14-2 The beam shown in Figure P5.14-2 is a W21 × 68 of A36 steel and has lateral support only at the ends. Check it for compliance with the AISC Specification.

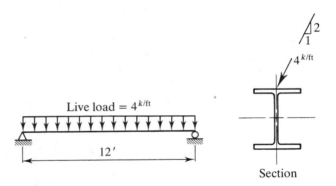

FIGURE P5.14-2

5.14-3 Horizontal live loads of 12 kips each are applied at B and C as shown in Figure P5.14-3. Lateral support is provided only at the ends. Use A572 Grade 50 steel and select a W-shape.

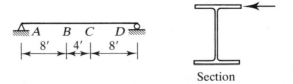

FIGURE P5.14-3

5.14-4 The beam shown in Figure P5.14-4 is simply supported and has lateral support only at its ends. Neglect the beam weight and determine whether it is satisfactory for each of the loading conditions shown. A572 Grade 50 steel is used, and the 1.0 kip/ft is a live load.

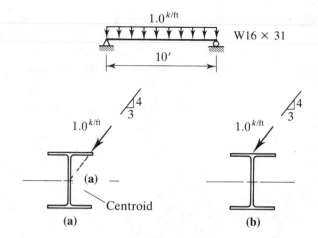

FIGURE P5.14-4

5.14-5   Check the beam shown in Figure P5.14-5 for compliance with the AISC Specification. Lateral support is provided only at the ends, and A36 steel is used. The 10 kip service loads are 30% dead load and 70% live load.

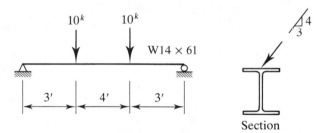

FIGURE P5.14-5

5.14-6   The truss shown in Figure P5.14-6 is part of a roof system supporting a total gravity load of 35 psf of roof surface, half dead load and half snow. Trusses are spaced at 12 feet on centers. Assume that wind load is not a factor and investigate the adequacy of a W6 × 16 of A36 steel for use as a purlin. No sag rods are used. Assume lateral support at the ends only.

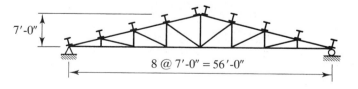

FIGURE P5.14-6

5.14-7   The truss shown in Figure P5.14-7 is one of several roof trusses spaced 16 feet apart. Purlins are located at the joints and halfway between the joints. The weight of the roofing materials is 15 psf, and the snow load is 20 psf of horizontal projection of the roof surface. Use A572 Grade 50 steel and select W-shapes for the purlins in the following cases:

a. Sag rods midway between the trusses.

b. Sag rods at the third points.

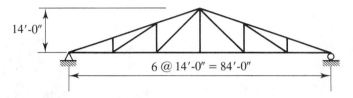

FIGURE P5.14-7

6 Beam-Columns

6.1 DEFINITION

While many structural members can be treated as axially loaded columns or as beams with only flexural loading, most beams and columns are subjected to some degree of both bending and axial load. This is especially true of statically indeterminate structures. Even the roller support of a simple beam can experience friction that restrains the beam longitudinally, inducing axial tension when transverse loads are applied. In this particular case, however, the secondary effects are usually small and can be neglected. Many columns can be treated as pure compression members with negligible error. If the column is a one-story member and can be treated as pinned at both ends, the only bending will result from minor accidental eccentricity of the load.

For many structural members, however, there will be a significant amount of both effects, and such members are called *beam-columns*. Consider the rigid frame in Figure 6.1. For the given loading condition, the horizontal member AB must not only support the vertical uniform load but must also assist the vertical members in resisting the concentrated lateral load P_1. Member CD is a more critical case, because it must resist the load $P_1 + P_2$ without any assistance from the vertical members. The reason is that the x-bracing, indicated by dashed lines, prevents sidesway in the lower story. For the direction of P_2 shown, member ED will be in tension and member CF will be slack, provided that the bracing elements have been designed to resist only tension. For this condition to occur, however, member CD must transmit the load $P_1 + P_2$ from C to D.

■ **FIGURE 6.1**

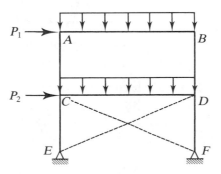

The vertical members of this frame must also be treated as beam-columns. In the upper story, members *AC* and *BD* will bend under the influence of P_1. In addition, at *A* and *B,* bending moments are transmitted from the horizontal member through the rigid joints. This transmission of moments also takes place at *C* and *D* and is true in any rigid frame, although these moments are usually smaller than those resulting from lateral loads. Most columns in rigid frames are actually beam-columns, and the effects of bending should not be ignored. However, many isolated one-story columns can be realistically treated as axially loaded compression members.

Another example of beam-columns can sometimes be found in roof trusses. Although the top chord is normally treated as an axially loaded compression member, if purlins are placed between the joints, their reactions will cause bending, which must be accounted for. We discuss methods for handling this problem later in this chapter.

6.2 INTERACTION FORMULAS

The inequality of Equation 2.3 may be written in the following form:

$$\frac{\sum \gamma_i Q_i}{\phi R_n} \leq 1.0 \tag{6.1}$$

or

$$\frac{\sum \text{ load effects}}{\text{resistance}} \leq 1.0$$

If more than one type of resistance is involved, Equation 6.1 can be used to form the basis of an interaction formula. As we discussed in Chapter 5 in conjunction with biaxial bending, the sum of the load-to-resistance ratios must be limited to unity. For example, if both bending and axial compression are acting, the interaction formula would be

$$\frac{P_u}{\phi_c P_n} + \frac{M_u}{\phi_b M_n} \leq 1.0 \tag{6.2}$$

where

$$P_u = \text{factored axial compressive load}$$
$$\phi_c P_n = \text{compressive design strength}$$
$$M_u = \text{factored bending moment}$$
$$\phi_b M_n = \text{design moment}$$

For biaxial bending, there will be two bending ratios:

$$\frac{P_u}{\phi_c P_n} + \left(\frac{M_{ux}}{\phi_b M_{nx}} + \frac{M_{uy}}{\phi_b M_{ny}} \right) \leq 1.0 \tag{6.3}$$

where the *x* and *y* subscripts refer to bending about the *x*- and *y*-axes.

Equation 6.3 is the basis for the AISC formulas for members subject to bending plus axial compressive load. Two formulas are given in the Specification: one for small axial load and one for large axial load. If the axial load is small, the axial load term is reduced. For large axial load, the bending term is slightly reduced. The AISC requirements are given in Chapter H, "Members Under Combined Forces and Torsion," and are summarized as follows:

For $\dfrac{P_u}{\phi_c P_n} \geq 0.2$,

$$\frac{P_u}{\phi_c P_n} + \frac{8}{9}\left(\frac{M_{ux}}{\phi_b M_{nx}} + \frac{M_{uy}}{\phi_b M_{ny}}\right) \leq 1.0 \qquad \text{(AISC Equation H1-1a)}$$

For $\dfrac{P_u}{\phi_c P_n} < 0.2$,

$$\frac{P_u}{2\phi_c P_n} + \left(\frac{M_{ux}}{\phi_b M_{nx}} + \frac{M_{uy}}{\phi_b M_{ny}}\right) \leq 1.0 \qquad \text{(AISC Equation H1-1b)}$$

Example 6.1 illustrates the application of these equations.

■ EXAMPLE 6.1

The beam-column shown in Figure 6.2 is pinned at both ends and is subjected to the factored loads shown. Bending is about the strong axis. Determine whether this member satisfies the appropriate AISC Specification interaction equation.

■ FIGURE 6.2

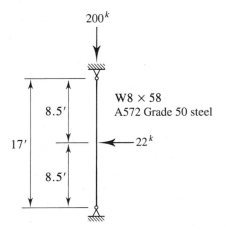

SOLUTION   As we demonstrate in Section 6.3, the applied moments in AISC Equations H1-1a and b will sometimes be increased by *moment amplification*. The purpose of this example is to illustrate how interaction formulas work.

From the column load tables, the axial compressive design strength of a W8 × 58 with $F_y = 50$ ksi and an effective length of $K_y L = 1.0 \times 17 = 17$ feet is

$$\phi_c P_n = 365 \text{ kips}$$

Since bending is about the strong axis, the design moment, $\phi_b M_n$, for $C_b = 1.0$ can be obtained from the beam design charts in Part 4 of the *Manual*.

For an unbraced length $L_b = 17$ ft,

$$\phi_b M_n = 202 \text{ ft-kips}$$

For the end conditions and loading of this problem, $C_b = 1.32$ (see Figure 5.15c).

For $C_b = 1.32$,

$$\phi_b M_n = 1.32(202) = 267 \text{ ft-kips}$$

But this moment is larger than $\phi_b M_p = 224$ ft-kips (also obtained from the beam design charts), so the design moment must be limited to $\phi_b M_p$. Therefore

$$\phi_b M_n = 224 \text{ ft-kips}$$

The maximum bending moment occurs at midheight, so

$$M_u = \frac{22(17)}{4} = 93.5 \text{ ft-kips}$$

Determine which interaction equation controls:

$$\frac{P_u}{\phi_c P_n} = \frac{200}{365} = 0.5479 > 0.2 \qquad \therefore \text{ use AISC Eq. H1-1a.}$$

$$\frac{P_u}{\phi_c P_n} + \frac{8}{9}\left(\frac{M_{ux}}{\phi_b M_{nx}} + \frac{M_{uy}}{\phi_b M_{ny}} \right) = \frac{200}{365} + \frac{8}{9}\left(\frac{93.5}{224} + 0 \right) = 0.919 < 1.0 \quad \text{(OK)}$$

ANSWER   This member satisfies the AISC Specification.   ■

6.3   ## MOMENT AMPLIFICATION

The foregoing approach to the analysis of members subjected to both bending and axial load is satisfactory so long as the axial load is not too large. The presence of the axial load produces secondary moments, and unless the axial load is relatively small, these additional moments must be accounted for. For an explanation, refer to Figure 6.3, which

■ **FIGURE 6.3**

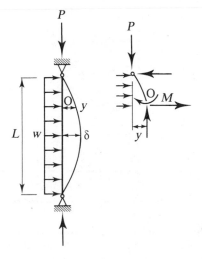

shows a beam-column with an axial load and a transverse uniform load. At an arbitrary point O, there is a bending moment caused by the uniform load and an additional moment Py, caused by the axial load acting at an eccentricity from the longitudinal axis of the member. This secondary moment is largest where the deflection is largest — in this case, at the centerline, where the total moment is $wL^2/8 + P\delta$. Of course, the additional moment causes an additional deflection over and above that resulting from the transverse load. Because the total deflection cannot be found directly, this problem is nonlinear, and without knowing the deflection, we cannot compute the moment.

Ordinary structural analysis methods that do not take the displaced geometry into account are referred to as *first-order* methods. Iterative numerical techniques, called *second-order* methods, can be used to find the deflections and secondary moments, but these methods are impractical for manual calculations and are usually implemented with a computer program. Most current design codes and specifications, including the AISC Specification, permit the use of either a second-order analysis or the *moment amplification method*. This method entails computing the maximum bending moment resulting from flexural loading (transverse loads or member end moments) by a first-order analysis, then multiplying by a *moment amplification factor* to account for the secondary moment. An expression for this factor will now be developed.

Figure 6.4 shows a simply supported member with an axial load and an initial out-of-straightness. This initial crookedness can be approximated by

$$y_0 = e \sin \frac{\pi x}{L}$$

■ **FIGURE 6.4**

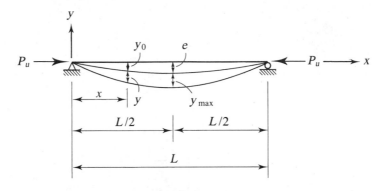

where e is the maximum initial displacement, occurring at midspan. For the coordinate system shown, the moment – curvature relationship can be written as

$$\frac{d^2y}{dx^2} = -\frac{M}{EI}$$

The bending moment M is caused by the eccentricity of the axial load P_u with respect to the axis of the member. This eccentricity consists of the initial crookedness y_0 plus additional deflection y resulting from bending. At any location, the moment is

$$M = P_u(y_0 + y)$$

Substituting this equation into the differential equation, we obtain

$$\frac{d^2y}{dx^2} = -\frac{P_u}{EI}\left(e \sin\frac{\pi x}{L} + y\right)$$

Rearranging gives

$$\frac{d^2y}{dx^2} + \frac{P_u}{EI}y = -\frac{P_u e}{EI} \sin\frac{\pi x}{L}$$

which is an ordinary, nonhomogeneous differential equation. Because it is a second-order equation, there are two boundary conditions. For the support conditions shown, the boundary conditions are

At $x = 0$, $y = 0$ and at $x = L$, $y = 0$

That is, the displacement is zero at each end. A function that satisfies both the differential equation and the boundary conditions is

$$y = B \sin\frac{\pi x}{L}$$

where B is a constant. Substituting into the differential equation, we get

$$-\frac{\pi^2}{L^2} B \sin \frac{\pi x}{L} + \frac{P_u}{EI} B \sin \frac{\pi x}{L} = -\frac{P_u e}{EI} \sin \frac{\pi x}{L}$$

Solving for the constant gives

$$B = \frac{-\dfrac{P_u e}{EI}}{\dfrac{P_u}{EI} - \dfrac{\pi^2}{L^2}} = \frac{-e}{1 - \dfrac{\pi^2 EI}{P_u L^2}} = \frac{e}{\dfrac{P_e}{P_u} - 1}$$

where

$$P_e = \frac{\pi^2 EI}{L^2} = \text{ the Euler buckling load}$$

$$\therefore \quad y = B \sin \frac{\pi x}{L} - \left[\frac{e}{(P_e/P_u) - 1} \right] \sin \frac{\pi x}{L}$$

$$M = P_u(y_0 + y)$$

$$= P_u \left\{ e \sin \frac{\pi x}{L} + \left[\frac{e}{(P_e/P_u) - 1} \right] \sin \frac{\pi x}{L} \right\}$$

The maximum moment occurs at $x = L/2$:

$$M_{\max} = P_u \left[e + \frac{e}{(P_e/P_u) - 1} \right]$$

$$= P_u e \left[\frac{(P_e/P_u) - 1 + 1}{(P_e/P_u) - 1} \right]$$

$$= M_0 \left[\frac{1}{1 - (P_u/P_e)} \right]$$

where M_0 is the unamplified maximum moment. In this case, it results from initial crookedness, but in general it can be the result of transverse loads or end moments. The moment amplification factor is therefore

$$\frac{1}{1 - (P_u/P_e)} \tag{6.4}$$

As we describe later, the exact form of the AISC moment amplification factor can be slightly different from that shown in Expression 6.4.

■ **EXAMPLE 6.2**

Use Expression 6.4 to compute the amplification factor for the beam-column of Example 6.1.

SOLUTION Since the Euler load P_e is part of an amplification factor for a *moment,* it must be computed for the axis of bending, which in this case is the x-axis. In terms of effective length and slenderness ratio, the Euler load can be written as

$$P_e = \frac{\pi^2 E A_g}{(KL/r)^2}$$

(See Chapter 4, Equation 4.6a.) For the axis of bending,

$$\frac{KL}{r} = \frac{K_x L}{r_x} = \frac{1.0(17)(12)}{3.65} = 55.89$$

$$P_e = \frac{\pi^2 E A_g}{(KL/r)^2} = \frac{\pi^2 (29,000)(17.1)}{(55.89)^2} = 1567 \text{ kips}$$

From Expression 6.4,

$$\frac{1}{1 - (P_u/P_e)} = \frac{1}{1 - (200/1567)} = 1.15$$

which represents a 15% increase in bending moment. The amplified primary moment is

$$1.15 \times M_u = 1.15(93.5) = 107.5 \text{ ft-kips}$$

ANSWER Amplification factor = 1.15

6.4 **WEB LOCAL BUCKLING IN BEAM-COLUMNS**

The determination of the design moment requires that the cross section be checked for compactness. The web is compact for all tabulated shapes so long as there is no axial load. In the presence of axial load, the web may not be compact. When we use the notation $\lambda = h/t_w$,

if $\lambda \leq \lambda_p$, the shape is compact;

if $\lambda_p < \lambda \leq \lambda_r$, the shape is noncompact; and

if $\lambda > \lambda_r$, the shape is slender.

AISC B5, in Table B5.1, prescribes the following limits:

$$\text{For } \frac{P_u}{\phi_b P_y} \leq 0.125, \ \lambda_p = \frac{640}{\sqrt{F_y}} \left(1 - \frac{2.75 P_u}{\phi_b P_y} \right)$$

$$\text{For } \frac{P_u}{\phi_b P_y} > 0.125, \ \lambda_p = \frac{191}{\sqrt{F_y}} \left(2.33 - \frac{P_u}{\phi_b P_y} \right) \geq \frac{253}{\sqrt{F_y}}$$

$$\text{For any value of } \frac{P_u}{\phi_b P_y}, \ \lambda_r = \frac{970}{\sqrt{F_y}} \left(1 - 0.74 \frac{P_u}{\phi_b P_y} \right)$$

where $P_y = A_g F_y$, the axial load required to reach the limit state of yielding.

Because P_u is variable, compactness of the web cannot be checked and tabulated in advance. Some rolled shapes satisfy the worst case limit of $253/\sqrt{F_y}$, however, which means that these shapes have compact webs no matter what the axial load. Shapes listed in the column load tables in Part 3 of the *Manual* that do not satisfy this criterion are marked, and only these shapes need be checked for compactness of the web. Shapes whose flanges are not compact are also marked, so if there is no indication to the contrary, shapes found in the column load tables are compact.

■ **EXAMPLE 6.3**

A W12 × 65 of A36 steel is subjected to a bending moment and a factored axial load of 300 kips. Check the web for compactness.

SOLUTION

This shape is compact for any value of axial load because there is no footnote in the applicable column load table to indicate otherwise. For illustration, however, we check the web width–thickness ratio here:

$$\frac{P_u}{\phi_b P_y} = \frac{P_u}{\phi_b (A_g F_y)} = \frac{300}{0.90(19.1)(36)} = 0.4848 > 0.125$$

$$\therefore \lambda_p = \frac{191}{\sqrt{F_y}} \left(2.33 - \frac{P_u}{\phi_b P_y} \right) = \frac{191}{\sqrt{36}} (2.33 - 0.4848) = 58.74$$

$$\frac{253}{\sqrt{F_y}} = \frac{253}{\sqrt{36}} = 42.17 < 58.74 \qquad \therefore \lambda_p = 58.74$$

From the dimensions and properties tables,

$$\lambda = \frac{h}{t_w} = 24.9 < \lambda_p$$

The web is therefore compact. Note that, for any available F_y, h/t_w will be less than $253/\sqrt{F_y}$, the smallest possible value of λ_p, so the web of a W12 × 65 will always be compact. ■

<h2>6.5   BRACED VERSUS UNBRACED FRAMES</h2>

The AISC Specification covers moment amplification in Chapter C, "Frames and Other Structures." Two amplification factors are used in LRFD: one to account for amplification resulting from the member deflection and one to account for the effect of sway when the member is part of an unbraced frame. This approach is the same as the one used in the ACI Building Code for reinforced concrete (ACI, 1995). Figure 6.5 illustrates these two components. In Figure 6.5a, the member is restrained against sidesway, and

■ **FIGURE 6.5**

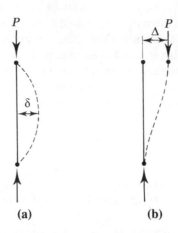

(a)   (b)

the maximum secondary moment is $P\delta$, which is added to the maximum moment within the member. If the frame is actually unbraced, there is an additional component of the secondary moment, shown in Figure 6.5b, that is caused by sidesway. This secondary moment has a maximum value of $P\Delta$, which represents an amplification of the *end* moment.

To approximate these two effects, two amplification factors, B_1 and B_2, are used for the two types of moments. The amplified moment to be used in design is computed from the factored loads and moments as follows (x and y subscripts are not used here; amplified moments must be computed in the following manner for each axis about which there are moments):

$$M_u = B_1 M_{nt} + B_2 M_{\ell t} \qquad \text{(AISC Equation C1-1)}$$

where

M_{nt} = maximum moment assuming that no sidesway occurs, whether the frame is actually braced or not (the subscript "*nt*" is for "no translation").

$M_{\ell t}$ = maximum moment caused by sidesway (the subscript "*ℓt*" is for "lateral translation"). This moment can be caused by lateral loads or by unbalanced gravity loads. Gravity loads can produce sidesway if the frame is unsymmetrical or if the gravity loads are unsymmetrically placed. $M_{\ell t}$ will be zero if the frame is actually braced.

B_1 = amplification factor for the moments occurring in the member when it is braced against sidesway.

B_2 = amplification factor for the moments resulting from sidesway.

We cover the evaluation of B_1 and B_2 in the following sections.

MEMBERS IN BRACED FRAMES

The amplification factor given by Expression 6.4 was derived for a member braced against sidesway — that is, one whose ends cannot translate with respect to each other. Figure 6.6 shows a member of this type subjected to equal end moments producing *single-curvature bending* (bending that produces tension or compression on one side throughout the length of the member). Maximum moment amplification occurs at the center, where the deflection is largest. For equal end moments, the moment is constant throughout the length of the member, so the maximum primary moment also occurs at the center. Thus the maximum secondary moment and maximum primary moment are additive. Even if the end moments are not equal, as long as one is clockwise and the other is counterclockwise there will be single-curvature bending, and the maximum primary and secondary moments will occur near each other.

That is not the case if applied end moments produce reverse-curvature bending as shown in Figure 6.7. Here the maximum primary moment is at one of the ends, and maximum moment amplification occurs between the ends. Depending on the value of the axial load P_u, the amplified moment can be either larger or smaller than the end moment.

The maximum moment in a beam-column therefore depends on the distribution of bending moment within the member. This distribution is accounted for by a factor, C_m, applied to the amplification factor B_1. The amplification factor given by Expression 6.4 was derived for the worst case, so C_m will never be greater than 1.0. The final form of the amplification factor is

$$B_1 = \frac{C_m}{1 - (P_u / P_{e1})} \geq 1 \qquad \text{(AISC Equation C1-2)}$$

■ FIGURE 6.6

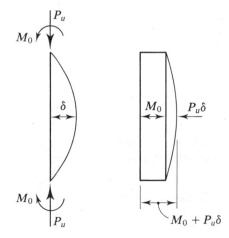

■ **FIGURE 6.7**

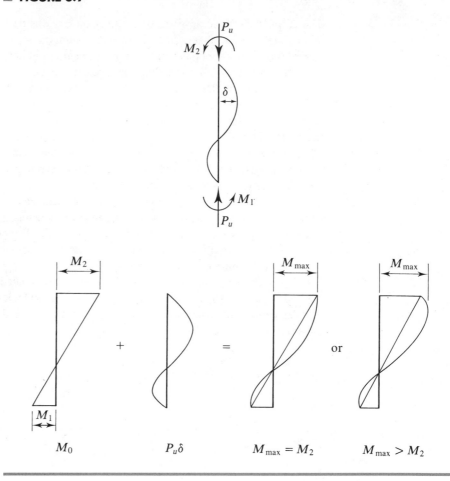

where

$$P_{e1} = \frac{A_g F_y}{\lambda_c^2} = \frac{\pi^2 E A_g}{(KL/r)^2}$$

When computing P_{e1}, use KL/r for the axis of bending and an effective length factor K less than or equal to 1.0 (corresponding to the braced condition).

Evaluation of C_m

The factor C_m applies only to the braced condition. There are two categories of members: those with transverse loads applied between the ends and those with no transverse loads.

Figure 6.8b and c illustrate these two cases (member AB is the beam-column under consideration).

1. If there are no transverse loads acting on the member,

$$C_m = 0.6 - 0.4\left(\frac{M_1}{M_2}\right)$$ (AISC Equation C1-3)

M_1/M_2 is a ratio of the bending moments at the ends of the member. M_1 is the end moment that is smaller in absolute value, M_2 is the larger, and the ratio is positive for members bent in reverse curvature and negative for single-curvature bending (Figure 6.9). Reverse curvature (a positive ratio) occurs when M_1 and M_2 are both clockwise or both counterclockwise.

2. For transversely loaded members, C_m can be taken as 0.85 if the ends are restrained against rotation and 1.0 if the ends are unrestrained against rotation (pinned). End restraint will usually result from the stiffness of members connected to the beam-column. The pinned end condition is the one used in the derivation of the amplification factor; hence there is no reduction for this case, which corresponds to $C_m = 1.0$. Although the actual end condition may lie between full fixity and a frictionless pin, use of one of the two values given here will give satisfactory results.

■ **FIGURE 6.8**

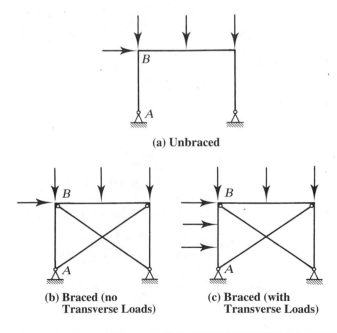

(a) Unbraced

(b) Braced (no
Transverse Loads)

(c) Braced (with
Transverse Loads)

■ **FIGURE 6.9**

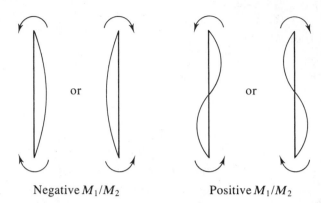

Negative M_1/M_2 Positive M_1/M_2

A more refined procedure for transversely loaded members (the second case) is provided in Section C1 of the Commentary to the Specification. The reduction factor is

$$C_m = 1 + \Psi \frac{P_u}{P_{e1}}$$

For simply supported members,

$$\Psi = \frac{\pi^2 \delta_0 EI}{M_0 L^2} - 1$$

where δ_0 is the maximum deflection resulting from transverse loading, and M_0 is the maximum moment between supports resulting from the transverse loads. The factor Ψ has been evaluated for several common situations and is given in Commentary Table C-C1.1.

■ **EXAMPLE 6.4**

The member shown in Figure 6.10 is part of a braced frame. The load and moments have been computed with factored loads, and bending is about the strong axis. If A572 Grade 50 steel is used, is this member adequate? $K_x L = K_y L = 14$ feet.

SOLUTION   Determine which interaction formula to apply:

$$\text{Maximum } \frac{KL}{r} = \frac{K_y L}{r_y} = \frac{14(12)}{3.02} = 55.63$$

From AISC Table 3-50, $\phi_c F_{cr} = 33.89$ ksi, so

$$\phi_c P_n = A_g(\phi_c F_{cr}) = 19.1(33.89) = 647.4 \text{ kips}$$

$$\frac{P_u}{\phi_c P_n} = \frac{420}{647.4} = 0.6487 > 0.2 \qquad \therefore \text{ use AISC Equation H1-1a.}$$

■ **FIGURE 6.10**

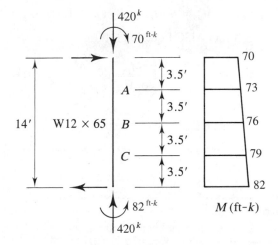

In the plane of bending,

$$\frac{KL}{r} = \frac{K_x L}{r_x} = \frac{14(12)}{5.28} = 31.82$$

$$P_{e1} = \frac{A_g F_y}{\lambda_c^2} = \frac{\pi^2 E A_g}{(K_x L / r_x)^2} = \frac{\pi^2 (29,000)(19.1)}{(31.82)^2} = 5399 \text{ kips}$$

$$C_m = 0.6 - 0.4\left(\frac{M_1}{M_2}\right) = 0.6 - 0.4\left(-\frac{70}{82}\right) = 0.9415$$

$$B_1 = \frac{C_m}{1 - (P_u / P_{e1})} = \frac{0.9415}{1 - (420/5399)} = 1.021$$

From the Beam Design charts, with $C_b = 1.0$ and $L_b = 14$ feet, the moment strength is

$$\phi_b M_n = 347 \text{ ft-kips}$$

For the actual value of C_b, refer to the moment diagram shown in Figure 6.10:

$$C_b = \frac{12.5 M_{\max}}{2.5 M_{\max} + 3M_A + 4M_B + 3M_C} = \frac{12.5(82)}{2.5(82) + 3(73) + 4(76) + 3(79)}$$

$$= 1.06$$

$$\therefore \ \phi_b M_n = C_b(347) = 1.06(347) = 368 \text{ ft-kips}$$

But

$$\phi_b M_p = 358 \text{ ft-kips (from the charts)} < 368 \text{ ft-kips}$$

$$\therefore \quad \text{use } \phi_b M_n = 358 \text{ ft-kips}$$

The factored load moments are

$$M_{nt} = 82 \text{ ft-kips} \qquad M_{\ell t} = 0$$

From AISC Equation C1-1,

$$M_u = B_1 M_{nt} + B_2 M_{\ell t} = 1.021(82) + 0 = 83.72 \text{ ft-kips} = M_{ux}$$

From AISC Equation H1-1a,

$$\frac{P_u}{\phi_c P_n} + \frac{8}{9} \left(\frac{M_{ux}}{\phi_b M_{nx}} + \frac{M_{uy}}{\phi_b M_{ny}} \right) = 0.6487 + \frac{8}{9} \left(\frac{83.72}{358} \right) = 0.857 < 1.0 \qquad \text{(OK}$$

ANSWER This member is satisfactory. ■

■ EXAMPLE 6.5

The horizontal beam-column shown in Figure 6.11 is subjected to the service live loads shown. This member is laterally braced at its ends, and bending is about the x-axis. Check for compliance with the AISC Specification.

SOLUTION The factored load is

$$P_u = 1.6(20) = 32.0 \text{ kips}$$

and the maximum moment is

$$M_{nt} = \frac{(1.6 \times 20)(10)}{4} + \frac{(1.2 \times 0.035)(10)^2}{8} = 80.52 \text{ ft-kips}$$

This member is braced against end translation, so $M_{\ell t} = 0$.

Compute the moment amplification factor:

For a member braced against sidesway, transversely loaded, and with unrestrained ends, C_m can be taken as 1.0. A more accurate value from AISC Commentary Table C-C1.1 is

$$C_m = 1 - 0.2 \frac{P_u}{P_{e1}}$$

■ FIGURE 6.11

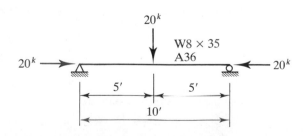

For the axis of bending,

$$\frac{KL}{r} = \frac{K_x L}{r_x} = \frac{1.0(10)(12)}{3.51} = 34.19$$

$$P_{e1} = \frac{\pi^2 E A_g}{(KL/r)^2} = \frac{\pi^2 (29,000)(10.3)}{(34.19)^2} = 2522 \text{ kips}$$

$$C_m = 1 - 0.2\left(\frac{32.0}{2522}\right) = 0.9975$$

The amplification factor is

$$B_1 = \frac{C_m}{1 - (P_u/P_{e1})} = \frac{0.9975}{1 - (32.0/2522)} = 1.010 > 1.0$$

For the axis of bending,

$$M_u = B_1 M_{nt} + B_2 M_{\ell t} = 1.010(80.52) + 0 = 81.33 \text{ ft-kips}$$

To obtain the design strengths, first go to the column load tables in Part 3 of the *Manual,* which give

$$\phi_c P_n = 262 \text{ kips}$$

From the beam design charts in Part 4 of the *Manual,* for $L_b = 10$ feet and $C_b = 1.0$,

$$\phi_b M_n = 91.8 \text{ ft-kips}$$

Because the beam weight is very small in relation to the concentrated live load, C_b may be taken from Figure 5.13c as 1.32. This value results in a design moment of

$$\phi_b M_n = 1.32(91.8) = 121 \text{ ft-kips}$$

This moment is greater than $\phi_b M_p = 93.6$ ft-kips, also taken from the beam design charts, so the design strength must be limited to this value. Therefore

$$\phi_b M_n = 93.6 \text{ ft-kips}$$

Check the interaction formula:

$$\frac{P_u}{\phi_c P_n} = \frac{32.0}{262} = 0.1221 < 0.2 \qquad \therefore \text{ use AISC Equation H1-1b:}$$

$$\frac{P_u}{2\phi_c P_n} + \left(\frac{M_{ux}}{\phi_b M_{nx}} + \frac{M_{uy}}{\phi_b M_{ny}}\right) = \frac{0.1221}{2} + \left(\frac{81.33}{93.6} + 0\right)$$

$$= 0.930 < 1.0 \qquad \text{(OK)}$$

ANSWER   A W8 × 35 is adequate

■ EXAMPLE 6.6

The member shown in Figure 6.12 is a W12 × 65 of A242 steel and must support a factored axial compressive load of 300 kips. One end is pinned, and the other is subjected to factored load moments of 135 ft-kips about the strong axis and 30 ft-kips about the weak axis. Use $K_x = K_y = 1.0$, and investigate for compliance with the AISC Specification.

SOLUTION   First, determine the yield stress F_y. From Table 1-2, Part 1 of the *Manual,* a W12 × 65 is a Group 2 shape. From Table 1-1, Group 2 shapes are available in only one strength in A242 steel, with $F_y = 50$ ksi.

Next, find the compressive strength. For $KL = 1.0(15) = 15$ feet, the axial compressive design strength from the column load tables is

$$\phi_c P_n = 626 \text{ kips}$$

Note that the table indicates that the flange of a W12 × 65 is noncompact for $F_y = 50$ ksi. Compute the strong axis bending moments.

$$C_{mx} = 0.6 - 0.4 \frac{M_1}{M_2} = 0.6 - 0.4(0) = 0.6$$

$$\frac{K_x L}{r_x} = \frac{15(12)}{5.28} = 34.09$$

$$P_{e1x} = \frac{\pi^2 E A_g}{(K_x L / r_x)^2} = \frac{\pi^2 (29{,}000)(19.1)}{(34.09)^2} = 4704 \text{ kips}$$

$$B_{1x} = \frac{C_{mx}}{1 - (P_u / P_{e1x})} = \frac{0.6}{1 - (300/4704)} = 0.641 < 1.0 \qquad \therefore \text{ use } B_{1x} = 1.0.$$

$$M_{ux} = B_{1x} M_{ntx} + B_{2x} M_{\ell tx} = 1.0(135) + 0 = 135 \text{ ft-kips}$$

■ FIGURE 6.12

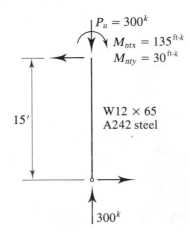

$P_u = 300^k$

$M_{ntx} = 135^{\text{ft-}k}$
$M_{nty} = 30^{\text{ft-}k}$

15′

W12 × 65
A242 steel

300^k

From the beam design charts with $L_b = 15$ ft, $\phi_b M_{nx} = 342$ ft-kips for $C_b = 1.0$, and $\phi_b M_{px} = 357.8$ ft-kips. From Figure 5.15g, $C_b = 1.67$, and

$$C_b \times (\phi_b M_{nx} \text{ for } C_b = 1.0) = 1.67(342) = 571 \text{ ft-kips}$$

This result is larger than $\phi_b M_{px}$; therefore use $\phi_b M_{nx} = \phi_b M_{px} = 357.8$ ft-kips. Compute the weak axis bending moments.

$$C_{my} = 0.6 - 0.4\,\frac{M_1}{M_2} = 0.6 - 0.4(0) = 0.6$$

$$\frac{K_y L}{r_y} = \frac{15(12)}{3.02} = 59.60$$

$$P_{ely} = \frac{\pi^2 E A_g}{(K_y L/r_y)^2} = \frac{\pi^2 (29{,}000)(19.1)}{(59.60)^2} = 1539 \text{ kips}$$

$$B_{1y} = \frac{C_{my}}{1 - (P_u/P_{ey})} = \frac{0.6}{1 - (300/1539)} = 0.745 < 1.0 \qquad \therefore \text{ use } B_{1y} = 1.0.$$

$$M_{uy} = B_{1y} M_{nty} + B_{2y} M_{\ell ty} = 1.0(30) + 0 = 30 \text{ ft-kips}$$

Because the flange of this shape is noncompact, the weak axis bending strength is limited by FLB.

$$\lambda = \frac{b_f}{2t_f} = 9.9$$

$$\lambda_p = \frac{65}{\sqrt{F_y}} = \frac{65}{\sqrt{50}} = 9.192$$

$$\lambda_r = \frac{141}{\sqrt{F_y - 10}} = \frac{141}{\sqrt{50 - 10}} = 22.29$$

Since $\lambda_p < \lambda < \lambda_r$,

$$M_n = M_p - (M_p - M_r)\left(\frac{\lambda - \lambda_p}{\lambda_r - \lambda_p}\right) \qquad \text{(AISC Equation A-F1-3)}$$

$$M_p = M_{py} = F_y Z_y = \frac{50(44.1)}{12} = 183.8 \text{ ft-kips}$$

$$M_r = M_{ry} = (F_y - F_r)S_y = (50 - 10)(29.1) = 1164 \text{ in.-kips} = 97.00 \text{ ft-kips}$$

Substituting into AISC Equation A-F1-3, we get

$$M_n = M_{ny} = 183.8 - (183.8 - 97.0)\left(\frac{9.9 - 9.192}{22.29 - 9.192}\right) = 179.1 \text{ ft-kips}$$

$$\phi_b M_{ny} = 0.90(179.1) = 161.2 \text{ ft-kips}$$

The interaction formula gives

$$\frac{P_u}{\phi_c P_n} = \frac{300}{626} = 0.4792 > 0.2 \qquad \therefore \text{ use AISC Equation H1-1a:}$$

$$\frac{P_u}{\phi P_n} + \frac{8}{9}\left(\frac{M_{ux}}{\phi_b M_{nx}} + \frac{M_{uy}}{\phi_b M_{ny}}\right) = 0.4792 + \frac{8}{9}\left(\frac{135}{357.8} + \frac{30}{161.2}\right)$$

$$= 0.980 < 1.0 \qquad \text{(OK)}$$

ANSWER   The W12 × 65 is satisfactory.   ■

6.7 MEMBERS IN UNBRACED FRAMES

In a beam-column whose ends are free to translate, the maximum primary moment resulting from the sidesway is almost always at one end. As was illustrated in Figure 6.5, the maximum secondary moment from the sidesway is *always* at the end. As a consequence of this condition, the maximum primary and secondary moments are usually additive and there is no need for the factor C_m; in effect, $C_m = 1.0$. Even when there is a reduction, it will be slight and can be neglected. Consider the beam-column shown in Figure 6.13. Here the equal end moments are caused by the sidesway (from the horizontal load). The axial load, which partly results from loads not causing the sidesway, is carried along and amplifies the end moment.

The amplification factor for the sidesway moments, B_2, is given by two equations. Either may be used; the choice is usually one of convenience:

$$B_2 = \frac{1}{1 - \Sigma P_u(\Delta_{oh}/\Sigma\, HL)} \qquad \text{(AISC Equation C1-4)}$$

or

$$B_2 = \frac{1}{1 - (\Sigma P_u/\Sigma P_{e2})} \qquad \text{(AISC Equation C1-5)}$$

where

ΣP_u = sum of factored loads on all columns in the story under consideration

Δ_{oh} = drift (sidesway displacement) of the story under consideration

ΣH = sum of all horizontal forces causing Δ_{oh}

L = story height

ΣP_{e2} = sum of the Euler loads for all columns in the story (when computing P_{e2}, use KL/r for the axis of bending and a value of K corresponding to the unbraced condition)

The summations for P_u and P_{e2} apply to all columns that are in the same story as the column under consideration. The rationale for using the summations is that B_2 applies

■ **FIGURE 6.13**

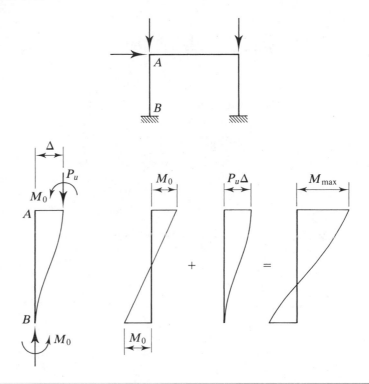

to unbraced frames, and if sidesway is going to occur, all columns in the story must sway simultaneously. In most cases, the structure will be made up of plane frames, so ΣP_u and ΣP_{e2} are for the columns within a story of the frame, and the lateral loads H are the lateral loads acting on the frame at and above the story. With Δ_{oh} caused by ΣH, the ratio $\Delta_{oh}/\Sigma H$ can be based on either factored or unfactored loads. The alternative form of B_2 given by AISC Equation C1-5 is almost the same as the equation for B_1 except for the summations.

AISC Equations C1-4 and C1-5 were derived by using two different methods, but in most cases they will give almost identical results (Yura, 1988). In those cases where the two values of B_2 would be significantly different, the axial load term of the interaction formula will be dominant, and the end results will not differ by much. As mentioned earlier, the choice is based on convenience; it depends on which terms in the equations are readily available.

In situations where M_{nt} and $M_{\ell t}$ act at two different points on the member, as in Figure 6.5, AISC Equation C1-1 will produce conservative results.

Figure 6.14 further illustrates the superposition concept. Figure 6.14a shows an unbraced frame subject to both gravity and lateral loads. The moment M_{nt} in member AB

■ **FIGURE 6.14**

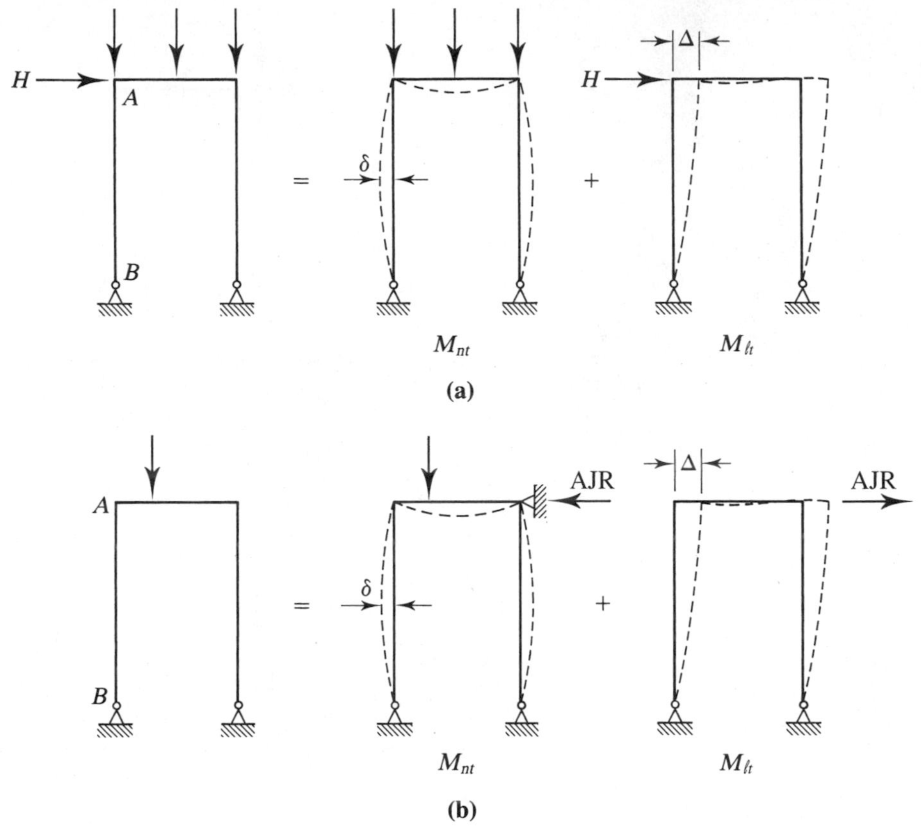

(a)

(b)

is computed by using only the gravity loads. Because of symmetry, no bracing is needed to prevent sidesway from these loads. This moment is amplified with the factor B_1 to account for the $P\delta$ effect. $M_{\ell t}$, the moment corresponding to the sway (caused by the horizontal load H), will be amplified by B_2 to account for the $P\Delta$ effect.

In Figure 6.14b, the unbraced frame supports only a vertical load. Because of the unsymmetrical placement of this load, there will be a small amount of sidesway. The moment M_{nt} is computed by considering the frame to be braced — in this case, by a fictitious horizontal support and corresponding reaction called an artificial joint restraint (AJR). To compute the sidesway moment, the fictitious support is removed, and a force equal to the artificial joint restraint, but opposite in direction, is applied to the frame. In cases such as this one, the secondary moment $P\Delta$ will be very small, and $M_{\ell t}$ can usually be neglected.

If both lateral loads and unsymmetrical gravity loads are present, the AJR force should be added to the actual lateral loads when $M_{\ell t}$ is computed.

■ **EXAMPLE 6.7**

A W12 × 65 of A572 Grade 50 steel, 15 feet long, is to be investigated for use as a column in an unbraced frame. The axial load and end moments obtained from a first-order analysis of the gravity loads (dead load and live load) are shown in Figure 6.15a. The frame is symmetrical, and the gravity loads are symmetrically placed. Figure 6.15b shows the wind load moments obtained from a first-order analysis. All bending moments are about the strong axis. Effective length factors are $K_x = 1.2$ for the sway case, $K_x = 1.0$ for the nonsway case, and $K_y = 1.0$. Determine whether this member is in compliance with the AISC Specification.

SOLUTION

All the load combinations given in AISC A4.1 involve dead load, and except for the first one, all combinations also involve live load or wind load, or both. If load types not present in this example (E, L_r, S, and R) are omitted, the load conditions can be summarized as

$$1.4D \qquad\qquad\qquad\qquad\qquad\qquad\qquad\qquad\qquad (A4\text{-}1)$$

$$1.2D + 1.6L \qquad\qquad\qquad\qquad\qquad\qquad\qquad (A4\text{-}2)$$

$$1.2D + (0.5L \text{ or } 0.8W) \qquad\qquad\qquad\qquad (A4\text{-}3)$$

$$1.2D + 1.3W + 0.5L \qquad\qquad\qquad\qquad\quad (A4\text{-}4)$$

$$1.2D + 0.5L \qquad\qquad\qquad\qquad\qquad\qquad\quad (A4\text{-}5)$$

$$0.9D \pm 1.3W \qquad\qquad\qquad\qquad\qquad\qquad\quad (A4\text{-}6)$$

The dead load is less than eight times the live load, so combination (A4-1) can be ruled out. Load combination (A4-4) will be more critical than (A4-3), so (A4-3) can be eliminated. Combination (A4-5) can be eliminated because it will be less critical than

■ **FIGURE 6.15**

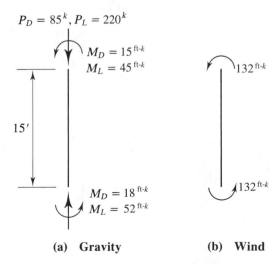

(a) **Gravity** (b) **Wind**

(A4-2). Finally, combination (A4-6) will not be as critical as (A4-4) and can be removed from consideration, leaving only two load combinations to be investigated, (A4-2) and (A4-4):

$$1.2D + 1.6L \quad \text{and} \quad 1.2D + 1.3W + 0.5L$$

Figure 6.16 shows the axial loads and bending moments calculated for these two combinations.

Determine the critical axis for axial compressive strength:

$$K_y L = 15 \text{ ft}$$

$$\frac{K_x L}{r_x / r_y} = \frac{1.2(15)}{1.75} = 10.29 \text{ ft} < 15 \text{ ft} \qquad \therefore \text{ use } KL = 15 \text{ ft}$$

■ **FIGURE 6.16**

454^k

90$^{ft\text{-}k}$

104.8$^{ft\text{-}k}$

(a) Load Combination A4-2 (1.2D + 1.6L)

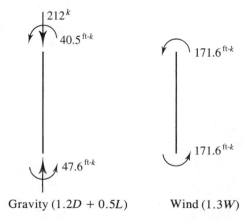

212^k

40.5$^{ft\text{-}k}$

171.6$^{ft\text{-}k}$

47.6$^{ft\text{-}k}$

171.6$^{ft\text{-}k}$

Gravity (1.2D + 0.5L)   Wind (1.3W)

(b) Load Combination A4-4

From the column load tables, with $KL = 15$ ft, $\phi_c P_n = 626$ kips.

For load condition (A4-2), $P_u = 454$ kips, $M_{nt} = 104.8$ ft-kips, and $M_{\ell t} = 0$ (because of symmetry, there are no sidesway moments). The bending factor is

$$C_m = 0.6 - 0.4\left(\frac{M_1}{M_2}\right) = 0.6 - 0.4\left(\frac{90}{104.8}\right) = 0.2565$$

For the axis of bending,

$$\frac{KL}{r} = \frac{K_x L}{r_x} = \frac{1.0(15)(12)}{5.28} = 34.09$$

(This case involves no sidesway, so K_x for the braced condition is used.) Then

$$P_{e1} = \frac{\pi^2 EA_g}{(KL/r)^2} = \frac{\pi^2(29,000)(19.1)}{(34.09)^2} = 4704 \text{ kips}$$

The amplification factor for nonsway moments is

$$B_1 = \frac{C_m}{1 - (P_u/P_{e1})} = \frac{0.2565}{1 - (454/4704)} = 0.284 < 1.0 \qquad \therefore \text{ use } B_1 = 1.0$$

$$M_u = B_1 M_{nt} + B_2 M_{\ell t} = 1.0(104.8) + 0 = 104.8 \text{ ft-kips}$$

From the beam design charts, with $L_b = 15$ feet,

$$\phi_b M_n = 343 \text{ ft-kips (for } C_b = 1.0)$$
$$\phi_b M_p = 358 \text{ ft-kips}$$

Figure 6.17 shows the bending moment diagram for the gravity load moments. (The computation of C_b is based on absolute values, so a sign convention for the diagram is not necessary.) Hence

$$C_b = \frac{12.5 M_{max}}{2.5 M_{max} + 3M_A + 4M_B + 3M_C}$$

$$= \frac{12.5(104.8)}{2.5(104.8) + 3(41.30) + 4(7.400) + 3(56.10)} = 2.24$$

For $C_b = 2.24$,

$$\phi_b M_n = 2.24(343) > \phi_b M_p = 358 \text{ kips} \qquad \therefore \text{ use } \phi_b M_n = 358 \text{ ft-kips}$$

Determine the appropriate interaction equation:

$$\frac{P_u}{\phi_c P_n} = \frac{454}{626} = 0.7252 > 0.2 \qquad \therefore \text{ use AISC Equation H1-1a.}$$

$$\frac{P_u}{\phi_c P_n} + \frac{8}{9}\left(\frac{M_{ux}}{\phi_b M_{nx}} + \frac{M_{uy}}{\phi_b M_{ny}}\right) = 0.7252 + \frac{8}{9}\left(\frac{104.8}{358} + 0\right) = 0.985 < 1.0 \quad \text{(OK)}$$

■ **FIGURE 6.17**

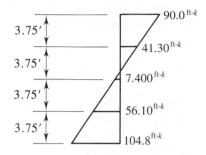

For load condition (A4-4), P_u = 212 kips, M_{nt} = 47.6 ft-kips, and $M_{\ell t}$ = 171.6 ft-kips. For the braced condition,

$$C_m = 0.6 - 0.4\left(\frac{M_1}{M_2}\right) = 0.6 - 0.4\left(\frac{40.5}{47.6}\right) = 0.2597$$

$$P_{e1} = 4704 \text{ kips} \qquad (P_{e1} \text{ is independent of the loading condition})$$

$$B_1 = \frac{C_m}{1 - (P_u/P_{e1})} = \frac{0.2597}{1 - (212/4704)} = 0.272 < 1.0 \qquad \therefore \text{ use } B_1 = 1.0$$

Not enough data are available to compute accurately the amplification factor for sidesway moments, B_2, from either AISC Equation C1-4 or C1-5. If we assume that the ratio of applied axial load to Euler load capacity is the same for all the columns in the story as for the column under consideration, we can use Equation C1-5:

$$B_2 = \frac{1}{1 - (\Sigma P_u/\Sigma P_{e2})} \approx \frac{1}{1 - (P_u/P_{e2})}$$

For P_{e2}, use K_x corresponding to the unbraced condition:

$$\frac{KL}{r} = \frac{K_x L}{r_x} = \frac{1.2(15)(12)}{5.28} = 40.91$$

$$P_{e2} = \frac{\pi^2 EA_g}{(KL/r)^2} = \frac{\pi^2(29,000)(19.1)}{(40.91)^2} = 3266 \text{ kips}$$

From AISC Equation C1-5,

$$B_2 \approx \frac{1}{1 - (P_u/P_{e2})} = \frac{1}{1 - (212/3266)} = 1.069$$

The total amplified moment is

$$M_u = B_1 M_{nt} + B_2 M_{\ell t} = 1.0(47.6) + 1.069(171.6) = 231.0 \text{ ft-kips}$$

Although the moments M_{nt} and $M_{\ell t}$ are different, they are distributed similarly, and C_b will be roughly the same; at any rate, they are large enough that $\phi_b M_p = 358$ ft-kips will be the design strength regardless of which moment is considered.

$$\frac{P_u}{\phi_c P_n} = \frac{212}{626} = 0.3387 > 0.2 \qquad \therefore \text{ use AISC Equation H1-1a:}$$

$$\frac{P_u}{\phi_c P_n} + \frac{8}{9}\left(\frac{M_{ux}}{\phi_b M_{nx}} + \frac{M_{uy}}{\phi_b M_{ny}}\right) = 0.3387 + \frac{8}{9}\left(\frac{231.0}{358} + 0\right) = 0.912 < 1.0 \quad \text{(OK)}$$

ANSWER   This member satisfies the AISC Specification requirements.

6.8   DESIGN OF BEAM-COLUMNS

Because of the many variables in the interaction formulas, the design of beam-columns is essentially a trial-and-error process. A trial shape is selected and then reviewed for satisfaction of the governing interaction formula. Obviously, the closer the trial shape is to the final selection, the better. A very efficient procedure for choosing a trial shape, originally developed for allowable stress design (Burgett, 1973), has been adapted for LRFD and is given in Part 3 of the *Manual,* "Column Design." The essence of this method is to "convert" the bending moments to equivalent axial loads. These fictitious loads are added to the actual loads, and a shape that will support the total load is selected from the column load tables. This selection must then be investigated with AISC Equation H1-1a or H1-1b. The total effective axial load is given by

$$P_{u\,eq} = P_u + M_{ux}m + M_{uy}mu$$

where

P_u = actual factored axial load (kips)

M_{ux} = factored moment about the *x*-axis (ft-kips)

M_{uy} = factored moment about the *y*-axis (ft-kips)

m = a tabulated constant

u = a tabulated constant

The basis of this procedure can be examined by rewriting Equation 6.3 as follows. First, multiply both sides by $\phi_c P_n$:

$$P_u + \frac{\phi_c P_n M_{ux}}{\phi_b M_{nx}} + \frac{\phi_c P_n M_{uy}}{\phi_b M_{ny}} \leq \phi_c P_n$$

or

$$P_u + (M_{ux} \times \text{a constant}) + (M_{uy} \times \text{a constant}) \leq \phi_c P_n$$

The right side of this inequality is the design strength of a member under consideration, and the left side can be thought of as the applied factored load to be resisted. Each of the three terms on the left must have units of force, so the constants "convert" the bending moments M_{ux} and M_{uy} to axial load components.

Average values of the constant m have been computed for different groups of W-shapes and are given in Table 3-2 in Part 3 of the *Manual*. Values of u are given in the column load tables for each shape listed. To select a trial shape for a member with axial load and bending about both axes, proceed as follows.

1. Select a trial value of m based on the effective length KL. Let $u = 2$.

2. Compute an effective axial compressive load:

$$P_{u\,eq} = P_u + M_{ux}m + M_{uy}mu$$

Use this load to select a shape from the column load tables.

3. Use the value of u given in the column load tables and an improved value of m from Table 3-2 to compute an improved value of $P_{u\,eq}$. Select another shape.

4. Repeat until there is no change in $P_{u\,eq}$.

■ EXAMPLE 6.8

A certain structural member in a braced frame must support a factored axial compressive load of 150 kips and factored end moments of 75 ft-kips about the strong axis and 30 ft-kips about the weak axis. Both of these moments occur at one end; the other end is pinned. The effective length with respect to each axis is 15 feet. There are no transverse loads on the member. Use A36 steel and select the lightest W-shape.

SOLUTION The amplification factor B_1 can be estimated as 1.0 for purposes of making a trial selection. For each of the two axes,

$$M_{ux} = B_1 M_{ntx} \approx 1.0(75) = 75 \text{ ft-kips}$$
$$M_{uy} = B_1 M_{nty} \approx 1.0(30) = 30 \text{ ft-kips}$$

From Table 3-2, Part 3 of the *Manual, m* = 1.75 by interpolation.

Use an initial value of $u = 2.0$.

$$P_{u\,eq} = P_u + M_{ux}m + M_{uy}mu = 150 + 75(1.75) + 30(1.75)(2.0) = 386 \text{ kips}$$

Beginning with the smaller shapes in the column load tables, **try W8 × 67** ($\phi_c P_n$ = 412 kips, u = 2.03):

$$m = 2.1$$
$$P_{u\,eq} = 150 + 75(2.1) + 30(2.1)(2.03) = 435 \text{ kips}$$

This value is larger than the design strength of 412 kips, so another shape must be tried. **Try W10 × 60** ($\phi_c P_n$ = 416 kips, u = 2.00):

$$m = 1.85$$
$$P_{u\,eq} = 150 + 75(1.85) + 30(1.85)(2.00) = 400 \text{ kips} < 416 \text{ kips} \qquad \text{(OK)}$$

The W10 × 60 is therefore a potential trial shape. Check the W12s and W14s for other possibilities. **Try W12 × 58** ($\phi_c P_n = 397$ kips, $u = 2.41$):

$$m = 1.55$$
$$P_{u\,eq} = 150 + 75(1.55) + 30(1.55)(2.41) = 378 \text{ kips} < 397 \text{ kips} \qquad \text{(OK)}$$

A W12 × 58 is therefore a potential trial shape. The lightest W14 with a chance of working is a W14 × 61, and it is heavier than the W12 × 58. Therefore use a W12 × 58 as a trial shape:

$$\frac{P_u}{\phi_c P_n} = \frac{150}{397} = 0.3778 > 0.2 \qquad \therefore \text{ use AISC Equation H1-1a.}$$

Compute the x-axis bending moments.

$$\frac{K_x L}{r_x} = \frac{15(12)}{5.28} = 34.09$$

$$P_{e1} = \frac{\pi^2 E A_g}{(KL/r)^2} = \frac{\pi^2 (29,000)(17.0)}{(34.09)^2} = 4187 \text{ kips}$$

$$C_m = 0.6 - 0.4(M_1/M_2) = 0.6 - 0.4(0/M_2) = 0.6 \text{ (for both axes)}$$

$$B_1 = \frac{C_m}{1 - (P_u/P_{e1})} = \frac{0.6}{1 - (150/4187)} = 0.622 < 1.0 \qquad \therefore \text{ use } B_1 = 1.0.$$

$$M_{ux} = B_1 M_{ntx} = 1.0(75) = 75 \text{ ft-kips}$$

Next, determine the design strength. From the beam design curves, for $C_b = 1$ and $L_b = 15$ ft, $\phi_b M_n = 220$ ft-kips. From Figure 5.15g, $C_b = 1.67$. For $C_b = 1.67$, the design strength is

$$C_b \times 220 = 1.67(220) = 367 \text{ ft-kips}$$

This moment is greater than $\phi_b M_p = 233$ ft-kips $\quad \therefore$ use $\phi_b M_n = 233$ ft-kips.

Compute the y-axis bending moments.

$$\frac{K_y L}{r_y} = \frac{15(12)}{2.51} = 71.71$$

$$P_{e1} = \frac{\pi^2 E A_g}{(KL/r)^2} = \frac{\pi^2 (29,000)(17.0)}{(71.71)^2} = 946.2 \text{ kips}$$

$$B_1 = \frac{C_m}{1 - (P_u/P_{e1})} = \frac{0.6}{1 - (150/946.2)} = 0.713 < 1.0 \qquad \therefore \text{ use } B_1 = 1.0.$$

$$M_{uy} = B_1 M_{nty} = 1.0(30) = 30 \text{ ft-kips}$$

A W12 × 58 is compact for any value of P_u (there is no footnote in the column load tables), so the nominal strength is $M_{py} \leq 1.5\ M_{yy}$. The design strength is

$$\phi_b M_{ny} = \phi_b M_{py} = \phi_b Z_y F_y = 0.90(32.5)(36) = 1053\ \text{in.-kips}$$
$$= 87.75\ \text{ft-kips}$$

But $Z_y/S_y = 32.5/21.4 = 1.52 > 1.5$, which means that $\phi_b M_{ny}$ should be taken as

$$\phi_b(1.5 M_{yy}) = \phi_b(1.5 F_y S_y) = 0.90(1.5)(36)(21.4) = 1040\ \text{in.-kips} = 86.67\ \text{ft-kips}$$

From AISC Equation H1-1a,

$$\frac{P_u}{\phi_c P_n} + \frac{8}{9}\left(\frac{M_{ux}}{\phi_b M_{nx}} + \frac{M_{uy}}{\phi_b M_{ny}} \right) = 0.3778 + \frac{8}{9}\left(\frac{75}{233} + \frac{30}{86.67} \right)$$
$$= 0.972 < 1.0 \quad \text{(OK)}$$

ANSWER   Use a W12 × 58.   ■

Although the method just presented for selecting a trial shape converges quickly, a somewhat simpler approach is suggested by Yura (1988). The equivalent axial load to be used is

$$P_{\text{equiv}} = P + \frac{2M_x}{d} + \frac{7.5M_y}{b} \tag{6.5}$$

where
  P = factored axial load
  M_x = factored moment about x-axis
  M_y = factored moment about y-axis
  d = beam depth
  b = beam width

All terms in Equation 6.5 must have consistent units.

■ EXAMPLE 6.9

Use Yura's method to select a trial W12 shape for the beam-column of Example 6.8.

SOLUTION   From Equation 6.5, the equivalent axial load is

$$P_{\text{equiv}} = P + \frac{2M_x}{d} + \frac{7.5M_y}{b} = 150 + \frac{2(75 \times 12)}{12} + \frac{7.5(30 \times 12)}{12} = 525\ \text{kips}$$

where the width b is assumed to be 12 inches. From the column load tables, **try a W12 × 72** ($\phi_c P_n = 537$ kips).

The flange width is 12 inches as assumed, so no iteration is required. For this particular problem, Yura's method produces a more conservative trial shape than the *Manual* method, although that may not always be true. ■

When the bending terms dominate (i.e., the member is more a beam than a column), Yura recommends that the axial load be converted to an equivalent bending moment about the *x*-axis. A trial shape can then be selected from the beam design charts in Part 3 of the *Manual*. This equivalent moment is

$$M_{\text{equiv}} = M_x + P\frac{d}{2}$$

Design of Unbraced Beam-Columns

The preliminary design of beam columns in braced frames has been illustrated. The amplification factor B_1 was assumed to be equal to 1.0 for purposes of selecting a trial shape; B_1 could then be evaluated for this trial shape. For beam-columns subject to sidesway, the amplification factor B_2 is based on several quantities that may not be known until all columns in the frame have been selected. If AISC Equation C1-4 is used for B_2, the sidesway deflection Δ_{oh} may not be available for a preliminary design. If AISC Equation C1-5 is used, ΣP_{e2} may not be known. The following methods are suggested for evaluating B_2.

Method 1. Assume that $B_2 = 1.0$. After a trial shape has been selected, compute B_2 from AISC Equation C1-5 by assuming that $\Sigma P_u/\Sigma P_{e2}$ is the same as P_u/P_{e2} for the member under consideration (as in Example 6.7).

Method 2. Use a predetermined limit for the *drift index* Δ_{oh}/L, the ratio of story drift to story height. The use of a maximum permissible drift index is a serviceability requirement that is similar to a limit on beam deflection. Although no building code or other standard used in the United States contains a limit on the drift index, values of $\frac{1}{500}$ to $\frac{1}{200}$ are commonly used (Ad Hoc Committee on Serviceability, 1986). Remember that Δ_{oh} is the drift caused by ΣH, so if the drift index is based on service loads, then the lateral loads H must also be service loads. Use of a prescribed drift index enables the designer to determine the final value of B_2 at the outset.

■ EXAMPLE 6.10

Figure 6.18 shows a single-story unbraced frame subjected to dead load, roof live load, and wind. The service gravity loads are shown in Figure 6.18a, and the service wind load (including an uplift, or suction, on the roof) is shown in Figure 6.18b. Use A572 Grade

■ **FIGURE 6.18**

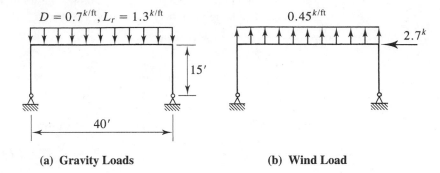

$D = 0.7^{k/ft}, L_r = 1.3^{k/ft}$

$0.45^{k/ft}$

2.7^k

15'

40'

(a) Gravity Loads

(b) Wind Load

50 steel and select a W12 shape for the columns (vertical members). Design for a drift index of ¼₀₀ based on service wind load. Bending is about the strong axis, and each column is laterally braced at the top and bottom.

SOLUTION This frame is statically indeterminate to the first degree. The analysis of statically indeterminate structures is not a prerequisite for the use of this book, so the frame will not be analyzed here. The results of an approximate analysis, which is adequate for the early stages of a structural design, are summarized in Figure 6.19. The axial load and end moment are given separately for dead load, live load, wind acting on the roof, and lateral wind load. All vertical loads are symmetrically placed and contribute only to the M_{nt} moments. The lateral load produces an $M_{\ell t}$ moment.

■ **FIGURE 6.19**

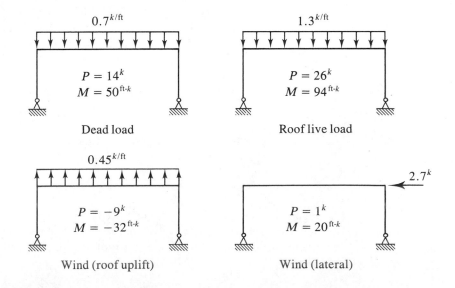

$0.7^{k/ft}$

$P = 14^k$
$M = 50^{ft\text{-}k}$

Dead load

$1.3^{k/ft}$

$P = 26^k$
$M = 94^{ft\text{-}k}$

Roof live load

$0.45^{k/ft}$

$P = -9^k$
$M = -32^{ft\text{-}k}$

Wind (roof uplift)

2.7^k

$P = 1^k$
$M = 20^{ft\text{-}k}$

Wind (lateral)

Load combinations involving dead load, D, roof live load, L_r, and wind, W, are as follows.

A4-2: $1.2D + 0.5L_r$

$P_u = 1.2(14) + 0.5(26) = 29.8$ kips

$M_{nt} = 1.2(50) + 0.5(94) = 107$ ft-kips

$M_{\ell t} = 0$

A4-3: $1.2D + 1.6L_r + 0.8W$

$P_u = 1.2(14) + 1.6(26) + 0.8(-9 + 1) = 52.0$ kips

$M_{nt} = 1.2(50) + 1.6(94) + 0.8(-32) = 184.8$ ft-kips

$M_{\ell t} = 0.8(20) = 16.0$ ft-kips

A4-4: $1.2D + 1.3W + 0.5L_r$

$P_u = 1.2(14) + 1.3(-9 + 1) + 0.5(26) = 19.4$ kips

$M_{nt} = 1.2(50) + 1.3(-32) + 0.5(94) = 65.4$ ft-kips

$M_{\ell t} = 1.3(20) = 26$ ft-kips

Load Combination A4-3 will obviously govern. It produces the largest axial load and the largest total moment. (Combination A4-4 could not control unless B_2, the amplification factor for $M_{\ell t}$, were unrealistically large.)

For purposes of selecting a trial shape, assume that $B_1 = 1.0$. The value of B_2 can be computed with AISC Equation C1-4 and the design drift index:

$$B_2 = \frac{1}{1 - \Sigma P_u(\Delta_{oh}/\Sigma HL)} = \frac{1}{1 - (\Sigma P_u/\Sigma H)(\Delta_{oh}/L)}$$

$$= \frac{1}{1 - [2(52.0)/2.7](1/400)} = 1.107$$

The unfactored horizontal load ΣH is used because the drift index is based on the maximum drift caused by *service* loads. Therefore

$$M_u = B_1 M_{nt} + B_2 M_{\ell t} = 1.0(184.8) + 1.107(16) = 202.5 \text{ ft-kips}$$

Without knowing the frame member sizes, the alignment chart for the effective length factor cannot be used. Table C-C2.1 in the Commentary to the Specification reveals that case (f) corresponds most closely to the end conditions for the sidesway case of this example and that $K_x = 2.0$.

For the braced condition, $K_x = 1.0$ will be used. Because the member is braced in the out-of-plane direction, $K_y = 1.0$ will be used. Then, a trial selection can be made by using the method given in Part 3 of the *Manual*. From Table 3-2, the bending factor m is 1.5 for W12 shapes with $KL = 15$ feet.

$$P_{u\,eq} = P_u + M_{ux}m + M_{uy}mu = 52.0 + 202.5(1.5) + 0 = 356 \text{ kips}$$

For $KL = K_y L = 15$ ft, a W12 × 53 has a design strength of $\phi_c P_n = 451$ kips. For the x-axis,

$$\frac{K_x L}{r_x/r_y} = \frac{2.0(15)}{2.11} = 14.2 \text{ ft} < 15 \text{ ft} \qquad \therefore \; KL = 15 \text{ ft controls}$$

Try a W12 × 53. For the braced condition,

$$\frac{K_x L}{r_x} = \frac{1.0(15)(12)}{5.23} = 34.42$$

$$P_{e1x} = \frac{\pi^2 E A_g}{(K_x L / r_x)^2} = \frac{\pi^2 (29{,}000)(15.6)}{(34.42)^2} = 3769 \text{ kips}$$

$$C_m = 0.6 - 0.4\left(\frac{M_1}{M_2}\right) = 0.6 - 0.4\left(\frac{0}{M_2}\right) = 0.6$$

From AISC Equation C1-2,

$$B_1 = \frac{C_m}{1 - (P_u / P_{e1})} = \frac{0.6}{1 - (52.0/3769)} = 0.608 < 1.0 \qquad \therefore \text{ use } B_1 = 1.0$$

Note that $B_1 = 1.0$ is the value originally assumed, and since B_2 will not change, the previously computed value of $M_u = 202.5$ ft-kips is unchanged.

From the beam design charts in Part 4 of the *Manual* with $L_b = 15$ feet, the design moment for a W12 × 53 with $C_b = 1.0$ is

$$\phi_b M_n = 262 \text{ ft-kips}$$

For a bending moment that varies linearly from zero at one end to a maximum at the other, the value of C_b is 1.67 (see Figure 5.15g). The corrected value of the design moment is therefore

$$\phi_b M_n = 1.67(262) = 438 \text{ ft-kips}$$

This moment, however, is larger than the plastic moment capacity of $\phi_b M_p = 292$ ft-kips, which can also be read from the charts. Therefore the design strength must be limited to

$$\phi_b M_n = \phi_b M_p = 292 \text{ ft-kips}$$

Determine the appropriate interaction formula:

$$\frac{P_u}{\phi_c P_n} = \frac{52.0}{451} = 0.1153 < 0.2 \qquad \therefore \text{ use AISC Equation H1-1b:}$$

$$\frac{P_u}{2\phi_c P_n} + \left(\frac{M_{ux}}{\phi_b M_{nx}} + \frac{M_{uy}}{\phi_b M_{ny}}\right) = \frac{0.1153}{2} + \left(\frac{202.5}{292} + 0\right) = 0.751 < 1.0 \quad \text{(OK}$$

This result is significantly smaller than 1.0, so try a shape two sizes smaller.

Try a W12 × 45. For $KL = K_y L = 15$ ft, $\phi_c P_n = 299$ kips. For the x-axis,

$$\frac{K_x L}{r_x / r_y} = \frac{2.0(15)}{2.65} = 11.3 \text{ ft} < 15 \text{ ft} \qquad \therefore KL = 15 \text{ ft controls}$$

For the braced condition,

$$\frac{K_x L}{r_x} = \frac{1.0(15)(12)}{5.15} = 34.95$$

$$P_{e1x} = \frac{\pi^2 E A_g}{(K_x L / r_x)^2} = \frac{\pi^2 (29{,}000)(13.2)}{(34.95)^2} = 3093 \text{ kips}$$

From AISC Equation C1-2,

$$B_1 = \frac{C_m}{1 - (P_u/P_{e1})} = \frac{0.6}{1 - (52.0/3093)} = 0.610 < 1.0 \qquad \therefore \text{ use } B_1 = 1.0$$

From the beam design charts with $L_b = 15$ feet, the design moment for a W12 × 45 with $C_b = 1.0$ is

$$\phi_b M_n = 201 \text{ ft-kips}$$

For $C_b = 1.67$,

$$\phi_b M_n = 1.67(201) = 336 \text{ ft-kips} > \phi_b M_p = 242.5 \text{ ft-kips}$$

The design strength is therefore

$$\phi_b M_n = \phi_b M_p = 242.5 \text{ ft-kips}$$

Determine the appropriate interaction formula:

$$\frac{P_u}{\phi_c P_n} = \frac{52.0}{299} = 0.1739 < 0.2 \qquad \therefore \text{ use AISC Equation H1-1b:}$$

$$\frac{P_u}{2\phi_c P_n} + \left(\frac{M_{ux}}{\phi_b M_{nx}} + \frac{M_{uy}}{\phi_b M_{ny}} \right) = \frac{0.1739}{2} + \left(\frac{202.5}{242.5} + 0 \right) = 0.922 < 1.0 \quad \text{(OK}$$

ANSWER Use a W12 × 45. ■

In Example 6.10, limiting the drift index was a design criterion, and there was no freedom to choose the method of computing the amplification factor B_2. If there had been no imposed drift index, a different value of B_2 could have been computed from AISC Equation C1-5, as follows (using properties of the W12 × 45):

$$\frac{K_x L}{r_x} = \frac{2.0(15)(12)}{5.15} = 69.90$$

$$P_{e2x} = \frac{\pi^2 EA_g}{(K_x L/r_x)^2} = \frac{\pi^2 (29,000)(13.2)}{(69.90)^2} = 773.2 \text{ kips}$$

$$B_2 = \frac{1}{1 - (\Sigma P_u / \Sigma P_{e2})} = \frac{1}{1 - [2(52.0)]/[2(773.2)]} = 1.072$$

6.9 ## TRUSSES WITH TOP CHORD LOADS BETWEEN JOINTS

If a compression member in a truss must support transverse load between its ends, it will be subjected to bending as well as axial compression and is therefore a beam-column. This condition can occur in the top chord of a roof truss with purlins located between the joints. The top chord of an open-web steel joist must also be designed as a beam-column because an open-web steel joist must support uniformly distributed gravity loads on its top chord. To account for loadings of this nature, a truss can be modeled as an assembly of continuous chord members and pin-connected web members. The axial loads and bending moments can then be found by using a method of structural analysis such as the stiffness method. The magnitude of the moments involved, however, does not usually warrant this degree of sophistication, and in most cases an approximate analysis will suffice. The following procedure is recommended.

1. Consider each member of the top chord to be a fixed-end beam. Use the fixed-end moment as the maximum bending moment in the member. The top chord is actually one continuous member rather than a series of individual pin-connected members, so this approximation is more accurate than treating each member as a simple beam.

2. Add the reactions from this fixed-end beam to the actual joint loads to obtain total joint loads.

3. Analyze the truss with these total joint loads acting. The resulting axial load in the top chord member is the axial compressive load to be used in the design.

This method is represented schematically in Figure 6.20. Alternatively, the bending moments and beam reactions can be found by treating the top chord as a continuous beam with supports at the panel points.

■ FIGURE 6.20

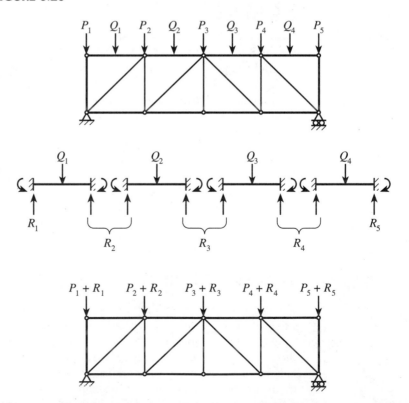

■ EXAMPLE 6.11

The parallel-chord roof truss shown in Figure 6.21 supports purlins at the top chord panel points and midway between the panel points. The *factored* loads transmitted by the purlins are as shown. Design the top chord. Use A36 steel and select a structural tee cut from a W-shape.

SOLUTION

The bending moments and panel point forces caused by the loads acting between the joints will be found by treating each top chord member as a fixed-end beam. From Part 4 of the *Manual,* "Beam and Girder Design," the fixed-end moment for each top chord member is

$$M = M_{nt} = \frac{PL}{8} = \frac{2.4(10)}{8} = 3.0 \text{ ft-kips}$$

These end moments and the corresponding reactions are shown in Figure 6.22a. When the reactions are added to the loads that are directly applied to the joints, the loading condition shown in Figure 6.22b is obtained. The maximum axial compressive force will occur in member *DE* (and in the adjacent member, to the right of the center of the span) and can be found by considering the equilibrium of a free body of the portion of the truss to the left of section *a–a:*

$$\Sigma M_I = (19.2 - 2.4)(30) - 4.8(10 + 20) + F_{DE}(4) = 0$$
$$F_{DE} = -90 \text{ kips (compression)}$$

Design for an axial load of 90 kips and a bending moment of 3.0 ft-kips.

Table 3-2 in Part 3 of the *Manual* does not provide bending factors for structural tees. Although Yura's method (Yura, 1988) was probably developed for I- and H-shaped members, it will be tried here. An examination of the column load tables shows that a small shape will be needed because the axial load is small and the moment is small relative to the axial load. If a 6-inch-deep tee is used,

$$P_{equiv} = P + \frac{2M_x}{d} + \frac{7.5M_y}{b} = 90 + \frac{2(3)(12)}{6} + 0 = 102 \text{ kips}$$

■ FIGURE 6.21

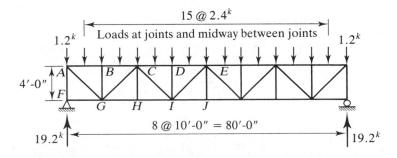

■ **FIGURE 6.22**

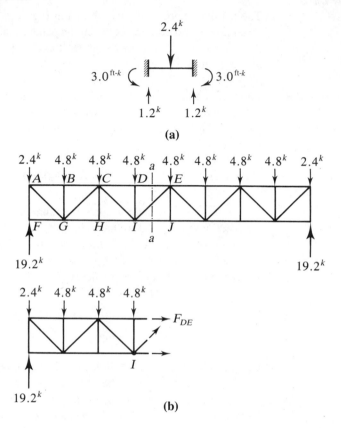

From the column load tables, with $K_x L = 10$ ft and $K_y L = 5$ ft, **try a WT6 × 17.5** ($\phi_c P_n$ = 124 kips). Bending is about the x-axis, and the member is braced against sidesway:

$$M_{nt} = 3.0 \text{ ft-kips}, \quad M_{\ell t} = 0$$

Because there is a transverse load on the member, and the ends are restrained, $C_m = 0.85$ (the Commentary approach will not be used here). Compute B_1:

$$\frac{KL}{r} = \frac{K_x L}{r_x} = \frac{10(12)}{1.76} = 68.18$$

$$P_{e1} = \frac{\pi^2 E A_g}{(KL/r)^2} = \frac{\pi^2 (29,000)(5.17)}{(68.18)^2} = 318.3 \text{ kips}$$

$$B_1 = \frac{C_m}{1 - (P_u/P_{e1})} = \frac{0.85}{1 - (90/318.3)} = 1.185$$

The amplified moment is

$$M_u = B_1 M_{nt} + B_2 M_{\ell t} = 1.185(3.0) + 0 = 3.555 \text{ ft-kips}$$

The nominal moment strength of a structural tee will be based on local buckling if the cross section is classified as slender; otherwise, it will be based on lateral-torsional buckling (see AISC F1.2c and Section 5.14 of this book). For the flange,

$$\lambda = \frac{b_f}{2t_f} = \frac{6.560}{2(0.520)} = 6.308$$

$$\lambda_r = \frac{95}{\sqrt{F_y}} = \frac{95}{\sqrt{36}} = 15.83 > \lambda$$

For the web,

$$\lambda = \frac{d}{t_w} = \frac{6.25}{0.300} = 20.83$$

$$\lambda_r = \frac{127}{\sqrt{F_y}} = \frac{127}{\sqrt{36}} = 21.17 > \lambda$$

Since $\lambda < \lambda_r$ for both the flange and the web, the shape is not slender, and lateral-torsional buckling controls. From AISC Equation F1-15,

$$M_n = M_{cr} = \frac{\pi\sqrt{EI_y GJ}}{L_b}\left(B + \sqrt{1 + B^2}\right)$$

$$\leq 1.5 M_y \text{ for stems in tension}$$

$$\leq 1.0 M_y \text{ for stems in compression}$$

(AISC Equation F1-15)

From AISC Equation F1-16,

$$B = \pm 2.3(d/L_b)\sqrt{I_y/J} = \pm 2.3\left[\frac{6.25}{5(12)}\right]\sqrt{\frac{12.2}{0.369}} = \pm 1.378$$

and the nominal strength from AISC Equation F1-15 is

$$M_n = \frac{\pi\sqrt{29,000(12.2)(11,200)(0.369)}}{5(12)}\left(\pm 1.378 + \sqrt{1 + (1.378)^2}\right)$$

$$= 2002(\pm 1.378 + 1.703) = 6168 \text{ in.-kips or } 650.5 \text{ in.-kips}$$

The positive value of B corresponds to tension in the stem of the tee, and the negative sign is used to obtain the strength when the stem is in compression. For the loading in this problem, the maximum moment occurs at both the fixed ends and at midspan; thus the strength is controlled by compression in the stem, and

$$M_n = 650.5 \text{ in.-kips} = 54.21 \text{ ft-kips}$$

subject to a maximum of

$$1.0 M_y = 1.0 F_y S_x = \frac{1.0(36)(3.23)}{12} = 9.690 \text{ ft-kips} < 54.21 \text{ ft-kips}$$

$$\therefore \text{ use } M_n = 9.690 \text{ ft-kips}$$

$$\phi_b M_n = 0.90(9.690) = 8.721 \text{ ft-kips}$$

Determine which interaction formula to use:

$$\frac{P_u}{\phi_c P_n} = \frac{90}{124} = 0.7258 > 0.2 \quad \therefore \text{ use AISC Equation H1-1a:}$$

$$\frac{P_u}{\phi_c P_n} + \frac{8}{9}\left(\frac{M_{ux}}{\phi_b M_{nx}} + \frac{M_{uy}}{\phi_b M_{ny}}\right) = 0.7258 + \frac{8}{9}\left(\frac{3.555}{8.721} + 0\right)$$

$$= 0.7258 + 0.3623 = 1.09 > 1.0 \quad \text{(N.G.)}$$

The bending moment is small in this example, but so is the bending strength, and the bending term of the interaction formula is not insignificant. In searching for a better shape, the designer must be sure that the bending strength, as well as the axial compressive strength, is larger. The next shape in the column load tables is a WT6×20, with an axial compressive design strength of 133 kips. An inspection of the dimensions and properties tables shows that we are entering a group of shapes in which the x-axis is the weak axis. Hence bending is now about the weak axis, and there is no lateral-torsional buckling limit state. Further, if the shape is nonslender, the nominal strength will be based on yielding and is equal to the plastic moment capacity, subject to a limit of $1.5 M_y$.

Therefore **try a WT6 × 20** ($\phi_c P_n = 133$ kips). First, compute B_1:

$$\frac{KL}{r} = \frac{K_x L}{r_x} = \frac{10(12)}{1.57} = 76.43$$

$$P_{e1} = \frac{\pi^2 E A_g}{(KL/r)^2} = \frac{\pi^2 (29,000)(5.89)}{(76.43)^2} = 288.6 \text{ kips}$$

$$B_1 = \frac{C_m}{1 - (P_u/P_{e1})} = \frac{0.85}{1 - (90/288.6)} = 1.235$$

The amplified moment is

$$M_u = B_1 M_{nt} = 1.235(3.0) = 3.705 \text{ ft-kips}$$

Check the slenderness parameters. For the flange:

$$\lambda = \frac{b_f}{2 t_f} = \frac{8.005}{2(0.515)} = 7.772 < \lambda_r = 15.83$$

For the web:

$$\lambda = \frac{d}{t_w} = \frac{5.970}{0.295} = 20.2 < \lambda_r = 21.17$$

Because bending is about the weak axis,

$$M_n = M_p = Z_x F_y = \frac{5.30(36)}{12} = 15.9 \text{ ft-kips}$$

subject to a maximum of

$$1.5M_y = 1.5F_y S_x = \frac{1.5(36)(2.95)}{12} = 13.28 \text{ ft-kips}$$

Since $M_p > 1.5M_y$,

$$\phi_b M_n = \phi_b(1.5M_y) = 0.90(13.28) = 11.95 \text{ ft-kips}$$

Determine which interaction formula to use:

$$\frac{P_u}{\phi_c P_n} = \frac{90}{133} = 0.6767 > 0.2 \qquad \therefore \text{ use AISC Equation H1-1a:}$$

$$\frac{P_u}{\phi_c P_n} + \frac{8}{9}\left(\frac{M_{ux}}{\phi_b M_{nx}} + \frac{M_{uy}}{\phi_b M_{ny}}\right) = 0.6767 + \frac{8}{9}\left(\frac{3.705}{11.95} + 0\right)$$

$$= 0.952 < 1.0 \qquad \text{(OK)}$$

ANSWER   Use a WT6 × 20.   ■

■ PROBLEMS

Interaction Formulas

6.2-1   Determine whether the beam-column shown in Figure P6.2-1 satisfies the appropriate AISC interaction equation. The load and moment shown are factored (the end shears are not shown), and bending is about the strong axis.

6.2-2   Determine whether the beam-column shown in Figure P6.2-2 satisfies the appropriate AISC interaction equation. The load and moment shown are factored (the end shears are not shown), and bending is about the strong axis.

Moment Amplification

6.3-1   Repeat Problem 6.2-1, but incorporate the amplification factor of Expression 6.4.

6.3-2   Repeat Problem 6.2-2, but incorporate the amplification factor of Expression 6.4.

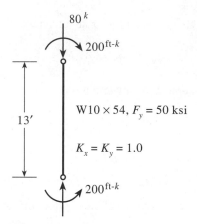

FIGURE P6.2-1

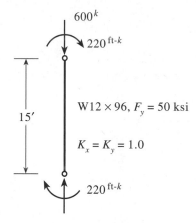

FIGURE P6.2-2

Web Local Buckling in Beam-Columns

6.4-1 A W14 × 43 of A572 Grade 50 steel is used as a beam-column. If the factored axial load is 65 kips, is the web compact?

6.4-2 If the beam-column in Problem 6.4-1 is subjected to an axial load of 480 kips, is the web compact?

Members in Braced Frames

6.6-1 A W12 × 58 of A572 Grade 50 steel is loaded as shown in Figure P6.6-1. The axial load and bending moments are computed from factored loads (the end shears are not shown).

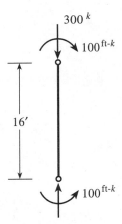

FIGURE P6.6-1

Bending is about the strong axis, and $K_x = K_y = 1.0$. Lateral support is provided only at the ends. Does this member satisfy the provisions of the AISC Specification?

6.6-2 A W10 × 77 of A572 Grade 50 steel is subjected to a factored axial load of 140 kips, as shown in Figure P6.6-2. If $K_x = 0.8$ and $K_y = 1.0$, what is the maximum factored moment M_{ntx} that can be applied about the strong axis at the top end of the member? Lateral support is provided only at the ends, and the bottom end is pinned.

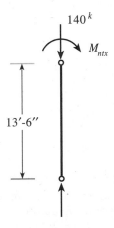

FIGURE P6.6-2

6.6-3 A W10 × 39 of A572 Grade 50 steel is loaded as shown in Figure P6.6-3. The loads are factored. Bending is about the strong axis, and $K_x = K_y = 1.0$. Lateral support is provided only at the ends. Does this member satisfy the provisions of the AISC Specification?

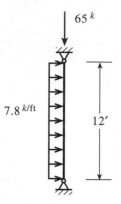

FIGURE P6.6-3

6.6-4 A W8 × 67 of A36 steel is loaded as shown in Figure P6.6-4. Bending is about the strong axis, and $K_x = K_y = 1.0$. Lateral support is provided only at the ends. What is the maximum factored axial load P_u that can be applied?

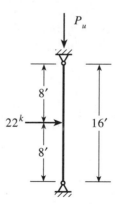

FIGURE P6.6-4

6.6-5 A W12 × 252 must support the service loads shown in Figure P6.6-5. Bending is about the strong axis, and lateral support is provided at *A, B,* and *C*. A572 Grade 50 steel is used. Investigate this member for compliance with the AISC Specification.

6.6-6 A W10 × 88 of A572 Grade 50 steel supports an axial compressive load and end moments that cause bending about both axes of the member, as shown in Figure P6.6-6. The axial load and bending moments are computed from factored loads (the end shears are not shown). $K_x = K_y = 1.0$. Investigate this member for compliance with the AISC Specification.

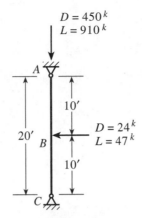

FIGURE P6.6-5

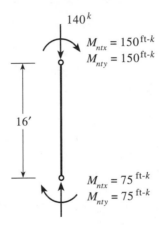

FIGURE P6.6-6

6.6-7 An $8 \times 8 \times \frac{1}{2}$ structural tube of A500 Grade B steel is used as a beam-column, as shown in Figure P6.6-7. The loads are factored. Lateral support is provided only at the ends, and $K_x = K_y = 1.0$. (See Section 5.15 for bending strength of structural tubes.) Investigate this member for compliance with the AISC Specification.

6.6-8 A W12 × 152 of A36 steel supports an axial compressive load and end moments at the top that cause bending about both axes of the member, as shown in Figure P6.6-8. The axial load and bending moments shown are computed from factored loads (the end shears are not shown). Lateral support is provided only at the ends, and $K_x = K_y = 1.0$. Determine the maximum factored moment M_{nty} that can be applied about the weak axis.

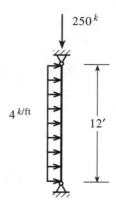

FIGURE P6.6-7

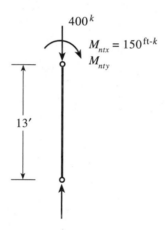

FIGURE P6.6-8

6.6-9   The horizontal member shown in Figure P6.6-9 supports a uniformly distributed transverse load and an axial compressive force. Both loads are factored, and the uniform load includes the weight of the beam. Bending is about the strong axis, and lateral support is provided at A, B, and C. Does this member satisfy the provisions of the AISC Specification?

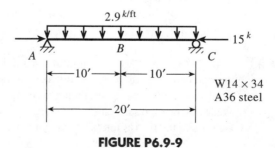

FIGURE P6.9-9

6.6-10 The fixed-end member shown in Figure P6.6-10 is a W14 × 90 of A572 Grade 50 steel. Lateral support is provided only at the ends, and $K_y = 1.0$. The loads are factored, and bending is about the strong axis. Does this member comply with the provisions of the AISC Specification? (Moments can be computed from the equations given in Part 4 of the *Manual*.)

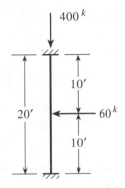

FIGURE P6.6-10

6.6-11 A W18 × 55 is loaded as shown in Figure P6.6-11, with an axial compressive force and forces at midspan that cause bending about both the strong and weak axes. The loads are factored. Determine whether the AISC Specification is satisfied. The steel is A572 Grade 50. Lateral bracing is provided only at the ends.

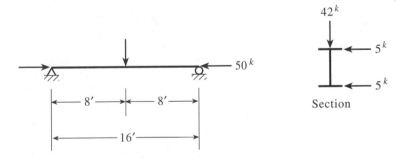

FIGURE P6.6-11

Members in Unbraced Frames

6.7-1 A W14 × 109 of A572 Grade 50 steel is to be investigated for use as a beam-column in an unbraced frame. The length is 14 feet, and there are no transverse loads on the member. First-order analyses of the frame were performed for both the sway and the nonsway cases. The factored loads and moments corresponding to one of the load combinations to be investigated are given for this member in Table P6.7-1.

TABLE P6.7-1

Type of Analysis	P_u (kips)	M_{top} (ft-kips)	M_{bot} (ft-kips)
Nonsway	300	62	35
Sway	—	54	135

Bending is about the strong axis, and all moments cause double-curvature bending (all end moments are in the same direction; that is, all clockwise or all counterclockwise). The following values are also available from the results of a preliminary design:

$$\Sigma P_{e2} = 54{,}000 \text{ kips and } \Sigma P_u = 8200 \text{ kips}$$

Use $K_x = 1.0$ (nonsway case), $K_x = 1.7$ (sway case), and $K_y = 1.0$. Determine whether this member satisfies the provisions of the AISC Specification.

6.7-2 A W14 × 61 of A572 Grade 50 steel, 16 feet long, is to be investigated for use as a column in an unbraced frame. The axial load and end moments obtained from a first-order analysis of the gravity loads (dead load and live load) are shown in Figure P6.7-2a. The frame is symmetric, and the gravity loads are symmetrically placed. Figure P6.7-2b shows the wind load moments obtained from a first-order analysis. All loads and moments are based on service loads, and all bending moments are about the strong axis. Effective length factors are $K_x = 0.85$ (braced case), $K_x = 1.2$ (unbraced case), and $K_y = 1.0$. Determine whether this member is in compliance with the AISC Specification.

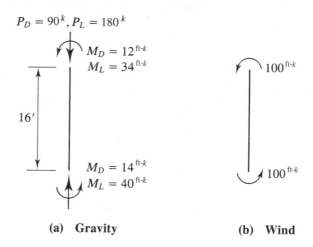

(a) **Gravity**   (b) **Wind**

FIGURE P6.7.2

Design of Beam-Columns

6.8-1 Use A36 steel and select the lightest W-shape for the beam-column shown in Figure P6.8-1. The axial load and bending moment are computed from factored loads (the end

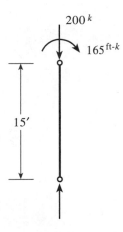

FIGURE P6.8-1

shears are not shown). Bending is about the strong axis, and $K_x = K_y = 1.0$. The member is laterally braced only at the ends.

6.8-2   Use A572 Grade 50 steel and select the lightest W-shape for the beam-column shown in Figure P6.8-2. Use $K_x = 0.85$ and $K_y = 1.0$. Lateral support is provided only at the ends.

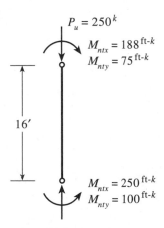

FIGURE P6.8-2

6.8-3   Use A572 grade 50 steel and select the lightest W-shape for the beam-column shown in Figure P6.8-3. The axial load and bending moments are computed from factored loads (the end shears are not shown). Bending is about the strong axis, and $K_x = K_y = 1.0$. The member is laterally braced only at the ends.

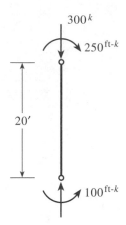

FIGURE P6.8-3

6.8-4   Use A572 grade 50 steel and select the lightest W-shape for the member shown in Figure P6.8-4. The axial load and bending moments are computed from factored loads. Bending is about the strong axis, and lateral bracing is provided at A, B, and C.

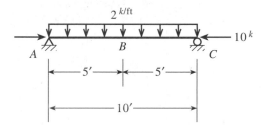

FIGURE P6.8-4

6.8-5   Use A36 steel and select the lightest W-shape for a 14-foot-long beam-column with K_x = 0.9 and K_y = 1.0. The member is braced against sidesway, and lateral support is provided only at the ends. The ends are restrained, and there are transverse loads between the ends. The member is subjected to the following factored loads: P_u = 300 kips, M_{ntx} = 180 ft-kips, and M_{nty} = 60 ft-kips. Use C_b = 1.6.

6.8-6   Use A572 Grade 50 steel and select the lightest W-shape for the beam-column shown in Figure P6.8-6. The loads are factored, and the lateral loading causes bending about the weak axis. Lateral bracing is provided only at the ends.

6.8-7   Select the lightest square structural tube for the beam-column shown in Figure P6.8-7. Use A500 Grade B steel (see Section 5.15 for bending strength of tubes). Lateral support is provided only at the ends. The loads are factored.

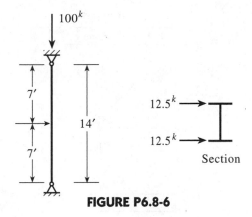

FIGURE P6.8-6

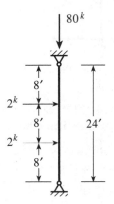

FIGURE P6.8-7

6.8-8 Use A36 steel and select the lightest W12 shape to be used as a beam-column in an *unbraced* frame. The member length is 16 feet, and the effective length factors are $K_x = 1.0$ (nonsway), $K_x = 2.0$ (sway), and $K_y = 1.0$. Factored loads and moments based on first-order analyses are $P_u = 50$ kips, $M_{nt} = 180$ ft-kips, and $M_{\ell t} = 20$ ft-kips. Use $C_m = 0.6$ and $C_b = 1.67$. Bending is about the strong axis.

6.8-9 The single-story unbraced frame shown in Figure P6.8-9 is subjected to dead load, roof live load, and wind. The results of an approximate analysis are summarized in the figure. The axial load and end moment are given separately for dead load, roof live load, wind uplift on the roof, and lateral wind load. All vertical loads are symmetrically placed and contribute only to the M_{nt} moments. The lateral load produces $M_{\ell t}$ moments. Use A572 Grade 50 steel and select a W12 shape for the columns (vertical members). Design for a drift index of $\frac{1}{400}$ based on service wind load. Bending is about the strong axis, and each column is laterally braced at the top and bottom.

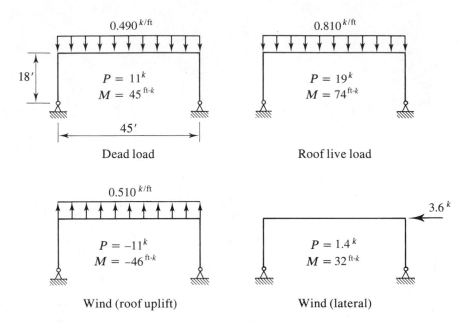

FIGURE P6.8-9

Trusses with Top Chord Loads Between Joints

6.9-1 Use A36 steel and select a structural tee-shape for the top chord of the truss shown in Figure P6.9-1. The trusses are spaced at 18 feet and are subjected to the following loads.

Purlins:	M6 × 4.4, located at joints and midway between joints
Snow:	20 psf of horizontal roof surface projection
Metal deck:	2 psf
Roofing:	4 psf
Insulation:	3 psf

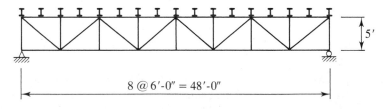

FIGURE P6.9-1

6.9-2 Use A572 Grade 50 steel and select a structural tee shape for the top chord of the truss shown in Figure P6.9-2. This is the truss in Example 3.13. The trusses are spaced at 20 feet on centers and support W6 × 12 purlins at the joints and midway between the joints. The other pertinent data are summarized as follows.

Metal deck: 2 psf

Built-up roof: 5 psf

Snow: 18 psf of horizontal roof surface projection

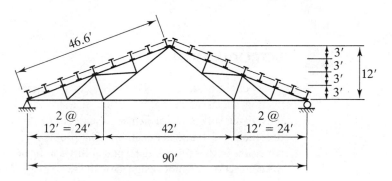

FIGURE P6.9-2

7 Simple Connections

7.1 ## INTRODUCTION

Connections of structural steel members are of critical importance. An inadequate connection, which can be the "weak link" in a structure, has been the cause of numerous failures. Failure of structural *members* is rare; most structural failures are the result of poorly designed or detailed connections. The problem is compounded by the confusion that sometimes exists regarding responsibility for the design of connections. In many cases, the connections are not designed by the same engineer who designs the rest of the structure, but by someone associated with the steel fabricator who furnishes the material for the project. The structural engineer responsible for the production of the design drawings, however, is responsible for the complete design, including the connections. It is therefore incumbent upon the engineer to be proficient in connection design, if only for the purpose of validating a connection designed by someone else.

Modern steel structures are connected by welding or bolting (either high-strength or "common" bolts) or by a combination of both. Until fairly recently, connections were either welded or riveted. In 1947, the Research Council of Riveted and Bolted Structural Joints was formed, and its first specification was issued in 1951. This document authorized the substitution of high-strength bolts for rivets on a one-for-one basis. Since that time, high-strength bolting has rapidly gained in popularity, and today the widespread use of high-strength bolts has rendered the rivet obsolete in civil engineering structures. There are several reasons for this change. Two relatively unskilled workers can install high-strength bolts, whereas four skilled workers are required for riveting. In addition, the riveting operation is noisy and somewhat dangerous because of the practice of tossing the heated rivet from the point of heating to the point of installation. Riveted construction is still covered by the AISC Specification and the *Manual of Steel Construction,* however; many existing structures contain riveted joints, and an understanding of their behavior is essential for the strength evaluation and rehabilitation of older structures. The design and analysis of riveted connections is essentially the same as for connections with common bolts; only the material properties are different.

Welding has several advantages over bolting. A welded connection is often simpler in concept and requires few, if any, holes (sometimes erection bolts may be required to hold the members in position for the welding operation). Connections that are extremely complex with fasteners can become very simple when welds are used. A case in point is the plate girder shown in Figure 7.1. Before welding became widely used, this type of built-up shape was fabricated by riveting. To attach the flange plates to the web plate, angle shapes were used to transfer load between the two elements. If cover plates were added, the finished product became even more complicated. The welded

■ **FIGURE 7.1**

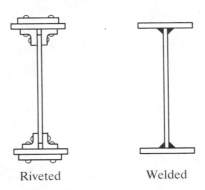

Riveted   Welded

version, however, is elegant in its simplicity. On the negative side, skilled workers are required for welding, and inspection can be difficult and costly. This last disadvantage can be partially overcome by using shop welding instead of field welding whenever possible. Quality welding can be more easily ensured under the controlled conditions of a fabricating shop. When a connection is made with a combination of welds and bolts, welding can be done in the shop and bolting in the field. In the single-plate beam-to-column connection shown in Figure 7.2, the plate is shop welded to the column flange and field bolted to the beam web.

In considering the behavior of different types of connections, it is convenient to categorize them according to the type of loading. The tension member lap splice shown in Figure 7.3a subjects the fasteners to forces that tend to shear the shank of the fastener. Similarly, the weld shown in Figure 7.3b must resist shearing forces. The connection of a bracket to a column flange, as in Figure 7.3c, whether by fasteners or welds, subjects the connection to shear when loaded as shown. The hanger connection shown in Figure 7.3d puts the fasteners in tension. The connection shown in Figure 7.3e produces both shear and tension in the upper row of fasteners. The strength of a fastener depends on whether it is subjected to shear or tension, or both. Welds are weak in shear and are usually assumed to fail in shear, regardless of the direction of loading.

■ **FIGURE 7.2**

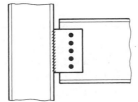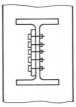

■ **FIGURE 7.3**

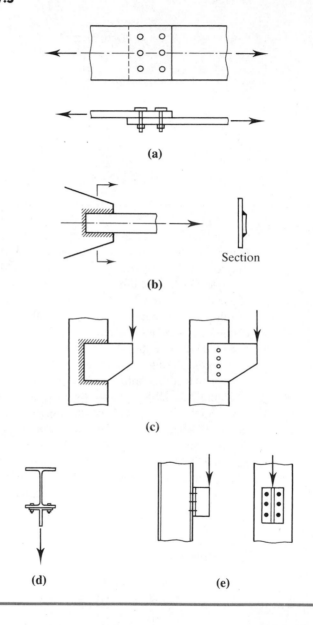

(a)

(b)

(c)

(d)   (e)

Once the force per fastener or force per unit length of weld has been determined, it is a simple matter to evaluate the adequacy of the connection. This determination is the basis for the two major categories of connections. If the line of action of the resultant force to be resisted passes through the center of gravity of the connection, each part of the connection is assumed to resist an equal share of the load, and the connection is called a *simple* connection. In such connections, illustrated in Figure 7.3a and b, each fastener or each unit length of weld will resist an equal amount of force.* The load capacity of the connection can then be found by multiplying the capacity of each fastener or inch of weld by the total number of fasteners or the total length of weld. This chapter is devoted to simple connections. Eccentrically loaded connections, covered in Chapter 8, are those in which the line of action of the load does not act through the center of gravity of the connection. The connections shown in Figure 7.3c and e are of this type. In these cases, the load is not resisted equally by each fastener or each segment of weld, and the determination of the distribution of the load is the complicating factor in the design of this type of connection.

The AISC Specification deals with connections in Chapter J, "Connections, Joints and Fasteners." Bolts, rivets, and welds are covered, although we do not consider rivets in this book.

BOLTED SHEAR CONNECTIONS: FAILURE MODES

Before considering the strength of specific grades of bolts, we need to examine the various modes of failure that are possible in connections with fasteners subjected to shear. There are two broad categories of failure: failure of the fastener and failure of the parts being connected. Consider the lap joint shown in Figure 7.4a. Failure of the fastener can be assumed to occur as shown. The average shearing stress in this case will be

$$f_v = \frac{P}{A} = \frac{P}{\pi d^2 / 4}$$

where P is the load acting on an individual fastener, A is the cross-sectional area of the fastener, and d is its diameter. The load can then be written as

$$P = f_v A$$

Although the loading in this case is not perfectly concentric, the eccentricity is small and can be neglected. The connection in Figure 7.4b is similar, but an analysis of free-body diagrams of portions of the fastener shank shows that each cross-sectional area is subjected to half the total load, or, equivalently, two cross sections are effective in resisting the total load. In either case, the load is $P = 2f_v A$, and this loading is called *double shear*. The bolt loading in the connection in Figure 7.4a, with only one shear plane, is called *single shear*. The addition of more thicknesses of material to the connection will increase the number of shear planes and further reduce the load on each plane. However, that will also increase the length of the fastener and could subject it to bending.

Other modes of failure in shear connections involve failure of the parts being connected and fall into two general categories.

---

*There is actually a small eccentricity in each of these two connections, but it is usually neglected.

■ **FIGURE 7.4**

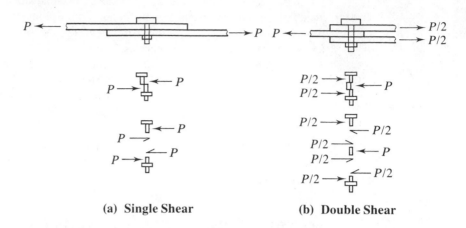

           (a) **Single Shear**   (b) **Double Shear**

1. **Failure resulting from excessive tension, shear, or bending in the parts being connected.** If a tension member is being connected, tension on both the gross area and effective net area must be investigated. Depending on the configuration of the connection, block shear might also need to be considered. Block shear must also be examined in beam-to-column connections in which the top flange of the beam is coped. (We covered block shear in Chapters 3 and 5, and it is also described in AISC J4.3.) Depending on the type of connection and loading, connection fittings such as gusset plates and framing angles may require an analysis for shear, tension, bending, or block shear. The design of a tension member connection will usually be done in parallel with the design of the member itself because the two processes are interdependent.

2. **Failure of the connected part because of bearing exerted by the fasteners.** If the hole is slightly larger than the fastener and the fastener is assumed to be placed loosely in the hole, contact between the fastener and the connected part will exist over approximately half the circumference of the fastener when a load is applied. This condition is illustrated in Figure 7.5. The stress will vary from a maximum at *A* to zero at *B;* for simplicity, an average stress, computed as the applied force divided by the projected area of contact, is used.

    Thus the bearing stress would be computed as $f_p = P/(dt)$, where P is the force applied to the fastener, d is the fastener diameter, and t is the thickness of the part subjected to the bearing. The bearing load is therefore $P = f_p dt$.

    The bearing problem can be complicated by the presence of a nearby bolt or the proximity of an edge in the direction of the load, as shown in Figure 7.6. The bolt spacing and edge distance will have an effect on the bearing strength.

■ **FIGURE 7.5**

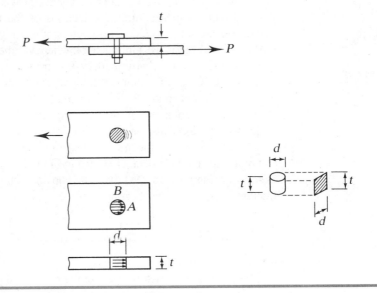

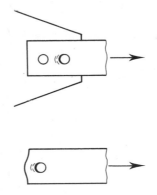

■ **FIGURE 7.6**

7.3 BEARING STRENGTH, SPACING, AND EDGE-DISTANCE REQUIREMENTS

Bearing strength is independent of the type of fastener because the stress under consideration is on the part being connected rather than on the fastener. For this reason, bearing strength, as well as spacing and edge-distance requirements, which also are independent of the type of fastener, will be considered before bolt shear and tensile strength.

The AISC Specification provisions for bearing strength, as well as all the requirements for high-strength bolts, are based on the provisions of the specification of the Research Council on Structural Connections of the Engineering Foundation (RCSC, 1994). The current edition of this document is not yet part of the AISC Specification (AISC, 1993a) but is used in this book whenever there is a conflict between it and the AISC Specification. Otherwise, the AISC provisions will be used. When an equation from the RCSC specification is displayed, the numbering scheme from that document will be used (for example, RCSC Equation LRFD 4.3). The following discussion, which is based on the commentary that accompanies the RCSC specification, explains the basis of the RCSC equations for bearing strength.

A possible failure mode resulting from excessive bearing is shear tear-out at the end of a connected element, as shown in Figure 7.7a. If the failure surface is idealized as shown in Figure 7.7b, the failure load on one of the two surfaces is equal to the shear fracture stress times the shear area, or

$$\frac{R_n}{2} = 0.6F_u L_c t$$

where

$0.6F_u$ = shear fracture stress of the connected part

L_c = distance from edge of hole to edge of connected part

t = thickness of connected part

The total strength is

$$R_n = 2(0.6F_u L_c t) = 1.2F_u L_c t \tag{7.1}$$

■ **FIGURE 7.7**

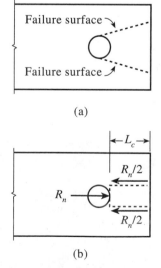

(a)

(b)

This tear-out can take place at the edge of a connected part, as shown, or between two holes in the direction of the bearing load. To prevent excessive elongation of the hole, an upper limit is placed on the bearing load given by Equation 7.1. This upper limit is proportional to the fracture stress times the projected bearing area, or

$$R_n = C \times F_u \times \text{bearing area} = CF_u dt \tag{7.2}$$

where

C = a constant

d = bolt diameter

t = thickness of the connected part

The RCSC Specification uses Equation 7.1 for the bearing strength, subject to an upper limit given by Equation 7.2. If deformation is not a concern, the constant C can be taken as 3.0. If excessive deformation is a concern, and it usually is, C is taken as 2.4. This value corresponds to a hole elongation of about ¼ inch (RCSC, 1994). In this book, we consider deformation to be a design consideration. The RCSC bearing strength for a single bolt therefore can be expressed as ϕR_n, where

$$\phi = 0.75$$

and

$$R_n = 1.2L_c tF_u \le 2.4 dtF_u \tag{RCSC Equation LRFD 4.3}$$

where

L_c = clear distance, in the direction parallel to the applied load, from the edge of the bolt hole to the edge of the adjacent hole or to the edge of the material

t = thickness of the connected part

d = bolt diameter (*not* the hole diameter)

F_u = ultimate tensile stress of the connected part (*not* the bolt)

Figure 7.8 further illustrates the distance L_c. When computing the bearing strength for a bolt, consider the distance from that bolt to the adjacent bolt or edge in the direction of the bearing load on the connected part. For the case shown, the bearing load would

---

■ FIGURE 7.8

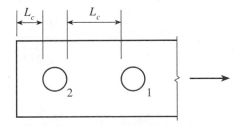

be on the left side of each hole. Thus the strength for bolt 1 is calculated with L_c measured to the edge of bolt 2, and the strength for bolt 2 is calculated with L_c measured to the edge of the connected part.

RCSC Equation LRFD 4.3 is valid for standard, oversized, short-slotted, and long-slotted holes with the slot parallel to the load. We use only standard holes in this book (holes $\frac{1}{16}$ inch larger than the bolt diameter).

When computing the distance L_c, use the actual hole diameter and do not add the $\frac{1}{16}$ inch as required in AISC B.2 for computing the net area of a tension member. In other words, use a hole diameter of $d + \frac{1}{16}$ inch, *not* $d + \frac{1}{8}$ inch. If h denotes the hole diameter, then

$$h = d + \frac{1}{16} \text{ in.}$$

The computation of bearing strength from RCSC Equation LRFD 4.3 can be simplified somewhat as follows. The upper limit will become effective when

$$1.2L_c t F_u = 2.4 dt F_u$$

or, after simplification, when

$$L_c = 2d$$

This relationship can be used to determine when the upper limit of $2.4 dt F_u$ controls:

If $L_c \leq 2d$, use $R_n = 1.2 L_c t F_u$.
If $L_c > 2d$, use $R_n = 2.4 dt F_u$.

Spacing and Edge-Distance Requirements

To maintain clearances between bolt nuts and to provide room for wrench sockets, AISC J3.3 requires that center-to-center spacing of fasteners (in any direction) be no less than $2\frac{2}{3}d$ and preferably no less than $3d$, where d is the fastener diameter. Minimum edge distances (in any direction), measured from the center of the hole, are given in AISC Table J3.4 as a function of bolt size and type of edge — sheared, rolled, or gas cut. The spacing and edge distance, denoted s and L_e, are illustrated in Figure 7.9.

■ **FIGURE 7.9**

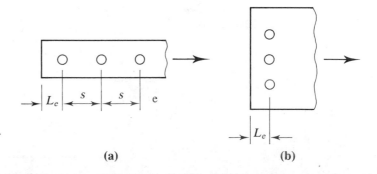

(a)   (b)

Summary of Bearing Strength, Spacing, and Edge-Distance Requirements (Standard Holes)

a. Bearing strength:

$$\phi R_n = 0.75(1.2L_c t F_u) \leq 0.75(2.4 dt F_u) \qquad \text{(RCSC Equation LRFD 4.3)}$$

or, equivalently,

If $L_c \leq 2d$, $\phi R_n = 0.75(1.2L_c t F_u)$.

If $L_c > 2d$, $\phi R_n = 0.75(2.4 dt F_u)$.

b. Minimum spacing and edge distance: In any direction, both in the line of force and transverse to the line of force,

$$s \geq 2\tfrac{2}{3}d \qquad \text{(preferably } 3d\text{)}$$

$L_e \geq$ value from AISC Table J3.4

For single- and double-angle shapes, the usual gage distances given in Part 9 of the *Manual*, Volume II (see Section 3.6), may be used in lieu of these minimums.

■ **EXAMPLE 7.1**

Check bearing, bolt spacing, and edge distances for the connection shown in Figure 7.10.

SOLUTION From AISC J3.3, the minimum spacing in any direction is

$$2\tfrac{2}{3}d = 2.667\left(\frac{3}{4}\right) = 2.00 \text{ in.}$$

Actual spacing $= 2.50$ in. > 2.00 in.   (OK)

The minimum edge distance in any direction is obtained from AISC Table J3.4. If we assume sheared edges (the worst case), the minimum edge distance is $1\frac{1}{4}$ in., so

$$\text{Actual edge distance} = 1\frac{1}{4} \text{ in.} \qquad \text{(OK)}$$

■ **FIGURE 7.10**

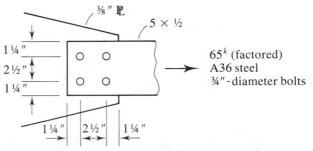

$\frac{3}{8}$" ℙ

$5 \times \frac{1}{2}$

$1\frac{1}{4}$"

$2\frac{1}{2}$"

$1\frac{1}{4}$"

$1\frac{1}{4}$" $2\frac{1}{2}$" $1\frac{1}{4}$"

65^k (factored)
A36 steel
$\frac{3}{4}$"-diameter bolts

$(3.4.5)(.707)(.1875)$

For computation of the bearing strength, use a hole diameter of

$$h = d + \frac{1}{16} = \frac{3}{4} + \frac{1}{16} = \frac{13}{16} \text{ in.}$$

Check bearing on both the tension member and the gusset plate. For the tension member and the holes nearest the edge of the member,

$$L_c = L_e - \frac{h}{2} = 1.25 - \frac{13/16}{2} = 0.8438 \text{ in.}$$

$$\phi R_n = \phi(1.2L_c t F_u) \le \phi(2.4 d t F_u)$$

$$\phi(1.2L_c t F_u) = 0.75(1.2)(0.8438)\left(\frac{1}{2}\right)(58) = 22.02 \text{ kips}$$

$$\phi(2.4 d t F_u) = 0.75(2.4)\left(\frac{3}{4}\right)\left(\frac{1}{2}\right)(58)$$

$$= 39.15 \text{ kips} > 22.02 \text{ kips} \qquad \therefore \text{ use } \phi R_n = 22.02 \text{ kips/bolt}$$

For the other holes,

$$L_c = s - h = 2.5 - \frac{13}{16} = 1.688 \text{ in.}$$

$$\phi R_n = \phi(1.2L_c t F_u) \le (\phi 2.4 d t F_u)$$

$$\phi(1.2L_c t F_u) = 0.75(1.2)(1.688)\left(\frac{1}{2}\right)(58) = 44.06 \text{ kips}$$

$$\phi(2.4 d t F_u) = 39.15 \text{ kips} < 44.06 \text{ kips} \qquad \therefore \text{ use } \phi R_n = 39.15 \text{ kips/bolt}$$

The bearing strength for the tension member is

$$\phi R_n = 2(22.02) + 2(39.15) = 122 \text{ kips}$$

For the gusset plate and the holes nearest the edge of the plate,

$$L_c = L_e - \frac{h}{2} = 1.25 - \frac{13/16}{2} = 0.8438 \text{ in.}$$

$$\phi R_n = \phi(1.2L_c t F_u) \le \phi(2.4 d t F_u)$$

$$\phi(1.2L_c t F_u) = 0.75(1.2)(0.8438)\left(\frac{3}{8}\right)(58) = 16.52 \text{ kips}$$

$$\phi(2.4 d t F_u) = 0.75(2.4)\left(\frac{3}{4}\right)\left(\frac{3}{8}\right)(58)$$

$$= 29.36 \text{ kips} > 16.52 \text{ kips} \qquad \therefore \text{ use } \phi R_n = 16.52 \text{ kips/bolt}$$

For the other holes,

$$L_c = s - h = 2.5 - \frac{13}{16} = 1.688 \text{ in.}$$

$$\phi R_n = \phi(1.2L_c t F_u) \le (\phi 2.4 d t F_u)$$

$$\phi(1.2L_c t F_u) = 0.75(1.2)(1.688)\left(\frac{3}{8}\right)(58) = 33.04 \text{ kips}$$

$$\phi(2.4 d t F_u) = 29.36 \text{ kips} < 33.04 \text{ kips} \qquad \therefore \text{ use } \phi R_n = 29.36 \text{ kips/bolt}$$

The bearing strength for the gusset plate is

$$\phi R_n = 2(16.52) + 2(29.36) = 91.8 \text{ kips}$$

The gusset plate controls. The bearing strength for the connection is therefore

$$\phi R_n = 91.8 \text{ kips} > 65 \text{ kips} \quad \text{(OK)}$$

ANSWER   Bearing strength, spacing, and edge-distance requirements are satisfied.   ■

The bolt spacing and edge distances in Example 7.1 are the same for both the tension member and the gusset plate. Only the thicknesses are different, so the gusset plate will control. In cases such as this one, only the thinner component need be checked. If the edge distances are different, however, both the tension member and the gusset plate should be checked.

7.4   COMMON BOLTS

We begin our coverage of fastener strength with *common bolts,* which differ from high-strength bolts not only in material properties, but also in that we do not account for the clamping force from the tightening of the bolt. Common bolts, also known as *unfinished bolts,* are designated as ASTM A307.

The design shear strength of A307 bolts is ϕR_n, where the resistance factor ϕ is 0.75, and the nominal shear strength is

$$R_n = F_v A_b$$

where

F_v = ultimate shearing stress

A_b = cross-sectional area of the unthreaded part of the bolt (also known as the *nominal bolt area* or *nominal body area*)

The ultimate shearing stress is given in AISC Table J3.2 as 24 ksi, giving a nominal strength of

$$R_n = F_v A_b = 24 A_b$$

■ **EXAMPLE 7.2**

Determine the design strength of the connection shown in Figure 7.11 based on shear and bearing.

SOLUTION   The connection can be classified as a simple connection, and each fastener can be considered to resist an equal share of the load. It will be convenient in most cases to determine the strength corresponding to one fastener and then to multiply by the total number of fasteners.

■ **FIGURE 7.11**

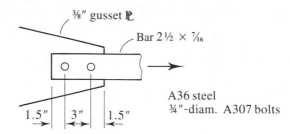

A36 steel
¾"-diam. A307 bolts

Shear strength: This is a case of single shear, and the design shear strength of one bolt is

$$\phi R_n = \phi F_v A_b = 0.75(24 A_b)$$

The nominal bolt area is

$$A_b = \frac{\pi d^2}{4} = \frac{\pi (3/4)^2}{4} = 0.4418 \text{ in.}^2$$

The design shear strength per bolt is therefore

$$\phi R_n = 0.75(24)(0.4418) = 7.952 \text{ kips}$$

For two bolts, it is

$$\phi R_n = 2(7.952) = 15.9 \text{ kips}$$

Bearing strength: Since the edge distances are the same for both the tension member and the gusset plate, the bearing strength of the gusset plate will control, because it is thinner than the tension member. For bearing strength computation, use a hole diameter of

$$h = d + \frac{1}{16} = \frac{3}{4} + \frac{1}{16} = \frac{13}{16} \text{ in.}$$

For the hole nearest the edge of the gusset plate,

$$L_c = L_e - \frac{h}{2} = 1.5 - \frac{13/16}{2} = 1.094 \text{ in.}$$

$$2d = 2\left(\frac{3}{4}\right) = 1.5 \text{ in.}$$

Because $L_c < 2d$,

$$\phi R_n = \phi(1.2 L_c t F_u) = 0.75(1.2)(1.094)\left(\frac{3}{8}\right)(58) = 21.42 \text{ kips}$$

For the other hole,

$$L_c = s - h = 3 - \frac{13}{16} = 2.188 \text{ in.} > 2d$$

$$\therefore \ \phi R_n = \phi(2.4 dt F_u) = 0.75(2.4)\left(\frac{3}{4}\right)\left(\frac{3}{8}\right)(58) = 29.36 \text{ kips}$$

The bearing strength for the connection is

$$\phi R_n = 21.42 + 29.36 = 50.8 \text{ kips}$$

This strength is larger than the shear strength of 15.9 kips, so shear controls and the strength of the connection is

$$\phi R_n = 15.9 \text{ kips}$$

Note that all spacing and edge-distance requirements are satisfied. For a sheared edge, the minimum edge distance required by AISC Table J3.4 is $1\frac{1}{4}$ in., and this requirement is satisfied in both the longitudinal and transverse directions. The bolt spacing s is 3 in., which is greater than $2\frac{2}{3}d = 2.667(\frac{3}{4}) = 2$ in.

ANSWER   Based on shear and bearing, the design strength of the connection is 15.9 kips. (Note that some other limit state that has not been checked, such as tension on the net area of the bar, may in fact govern the design strength.) ■

■ EXAMPLE 7.3

A bar $4 \times \frac{3}{8}$ is used as a tension member to resist a service dead load of 8 kips and a service live load of 22 kips. This member was designed under the assumption that a single line of $\frac{3}{4}$-inch-diameter A307 bolts would be used to connect the member to a $\frac{3}{8}$-inch gusset plate. A36 steel is used for both the tension member and the gusset plate. How many bolts are required?

SOLUTION   The factored load is

$$P_u = 1.2D + 1.6L = 1.2(8) + 1.6(22) = 44.80 \text{ kips}$$

Compute the capacity of one bolt. From Example 7.2, the shear strength is

$$\phi R_n = 7.952 \text{ kips per bolt}$$

For the bearing strength, the spacing and edge distances are not known. We assume that the upper limit of $\phi 2.4\, dt F_u$ will control, giving a bearing strength of

$$\phi R_n = 0.75(2.4)\left(\frac{3}{4}\right)\left(\frac{3}{8}\right)(58) = 29.36 \text{ kips per bolt}$$

The actual bearing strength for this connection will depend on the value of L_c for each bolt. When this value has been determined in the final design, the bearing strength should be rechecked, but shear is still likely to control.

The number of bolts required is

$$\frac{44.80 \text{ kips}}{7.952 \text{ kips/bolt}} = 5.63 \text{ bolts}$$

ANSWER   Use six ¾-inch-diameter A307 bolts.   ■

7.5   HIGH-STRENGTH BOLTS

High-strength bolts for structural joints are available in two grades: ASTM A325 and ASTM A490. The AISC provisions for high-strength bolts are based in part on the provisions of the specification of the Research Council on Structural Connections of the Engineering Foundation (RCSC, 1994).

A490 bolts have a higher ultimate tensile strength than A325 bolts and are assigned a higher nominal strength. They were introduced long after A325 bolts had been in general use, primarily for use with high-strength steels (Bethlehem Steel, 1969). A490 bolts are more expensive than A325 bolts, but usually fewer are required.

In certain cases, A325 and A490 bolts are installed to such a degree of tightness that they are subjected to extremely large tensile forces. For example, the initial tension in a ⅝-inch-diameter A325 bolt can be as high as 19 kips. A complete list of minimum tension values, for those connections in which a minimum tension is required, is given in AISC Table J3.1, Minimum Bolt Tension. Each value is equal to 70% of the minimum tensile strength of the bolt. The purpose of such a large tensile force is to achieve the clamping force illustrated in Figure 7.12. Such bolts are said to be *fully tensioned*.

As a nut is turned and advanced along the threads of a bolt, the connected parts undergo compression and the bolt elongates. The free-body diagrams in Figure 7.12a show that the total compressive force acting on the connected part is numerically equal to the tension in the bolt. If an external load P is applied, a friction force will develop between the connected parts. The maximum possible value of this force is

$$F = \mu N$$

where μ is the coefficient of static friction between the connected parts, and N is the normal compressive force acting on the inner surfaces. The value of μ will depend on the surface condition of the steel — for example, whether it is painted or whether rust is present. Thus each bolt in the connection is capable of resisting a load of $P = F$, even if the bolt shank does not bear on the connected part. As long as this frictional force is not exceeded, there is no bearing or shear. If P is greater than F and slippage occurs, shear and bearing will then exist and will affect the capacity of the connection.

■ **FIGURE 7.12**

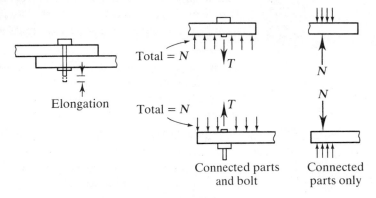

Total = N

Elongation

Total = N

Connected parts
and bolt

Connected
parts only

(a) No External Loads

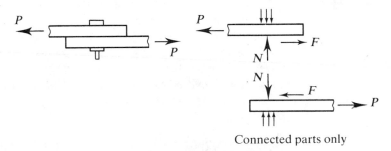

Connected parts only

(b) External Load Applied

Installation

How is this high tension accurately achieved? There are currently four authorized procedures for the installation of high-strength bolts (RCSC, 1994).

1. **Turn-of-the-nut method.** This procedure is based on the load-deformation characteristics of the fastener and the connected parts. One full turn of a nut corresponds to a fixed length of travel along the bolt threads, which can be correlated to the elongation of the bolt. The stress–strain relationship for the bolt material can then be used to compute the tension in the bolt.

   For any size and type of bolt, therefore, the number of turns of the nut required to produce a given tensile force can be computed. Table 5 in the high-strength bolt specification (RCSC, 1994) gives the required nut rotation for various bolt sizes in terms of the ratio of length to diameter. The specified rotation is from a snug position, with *snug* being defined as the tightness after a few impacts with an impact wrench or after the full effort of one worker with an ordinary spud wrench, with all parts of the connection in firm contact. In spite of all the uncertainties and variables involved, the turn-of-the-nut method has proved to be reliable and surprisingly accurate.

2. **Calibrated wrench tightening.** Torque wrenches are used for this purpose. The torque required to attain a specified tension in a bolt of a given size and grade is determined by tightening this bolt in a tension-indicating device. This calibration must be done daily during construction for bolts of each size and grade.

3. **Alternate design bolts.** These are specially designed A325 and A490 bolts whose tips twist off when the proper tension has been achieved. Special wrenches are required for their installation. Inspection of this type of bolt installation is particularly easy.

4. **Direct tension indicators.** The most common of these devices is a washer with protrusions on its surface. When the bolt is tightened, the protrusions are compressed in proportion to the tension in the bolt. A prescribed amount of deformation can be established for any bolt, and when that amount has been achieved, the bolt will have the proper tension. The deformation can be determined by measuring the gap between the nut or bolt head and the undeformed part of the washer surface. Inspection of the bolt installation is also simplified when this type of direct tension indicator is used, as only a feeler gage is required.

7.6 SHEAR STRENGTH OF HIGH-STRENGTH BOLTS

The design shear strength of both A325 and A490 bolts is ϕR_n, where the resistance factor ϕ is 0.75. As with common bolts, the nominal shear strength of high-strength bolts is given by the ultimate shearing stress times the nominal bolt area. Unlike A307 bolts, however, the shear strength of A325 and A490 bolts depends on whether the threads are in a plane of shear. Rather than use a reduced cross-sectional area when the threaded portion is subject to shear, the ultimate shearing stress is multiplied by a factor of 0.75, the approximate ratio of threaded area to unthreaded area. The strengths are given in AISC Table J3.2 and are summarized in Table 7.1. AISC Table J3.2 refers to threads

TABLE 7.1

Fastener	Nominal Shear Strength $R_n = F_v A_b$
A325, threads in plane of shear	$48A_b$
A325, threads not in plane of shear	$60A_b$
A490, threads in plane of shear	$60A_b$
A490, threads not in plane of shear	$75A_b$

in a plane of shear as "*not* excluded from shear planes" and refers to threads not in a plane of shear as "excluded from shear planes." The first category, threads included in the shear plane, is sometimes referred to as connection type "N," and an A325 bolt of this type can be denoted as an A325-N bolt. The designation "X" can be used to indicate that threads are excluded from the plane of shear — for example, an A325-X bolt.

■ EXAMPLE 7.4

Determine the design strength of the connection shown in Figure 7.13. Investigate bolt shear, bearing, and the tensile strength of the member. The bolts are ⅞-inch-diameter A325 with the threads *not* in the plane of shear. A572 Grade 50 steel is used.

SOLUTION

Shear strength: For one bolt,

$$A_b = \frac{\pi(7/8)^2}{4} = 0.6013 \text{ in.}^2$$

$$\phi R_n = \phi F_v A_b = 0.75(60)(0.6013) = 27.06 \text{ kips}$$

For three bolts,

$$\phi R_n = 3(27.06) = 81.2 \text{ kips}$$

Bearing strength: For computation of the bearing strength, use a hole diameter of

$$h = d + \frac{1}{16} = \frac{7}{8} + \frac{1}{16} = \frac{15}{16} \text{ in.}$$

Check the bearing on both the tension member and the gusset plate. For the tension member and the hole nearest the edge of the member,

$$L_c = L_e - \frac{h}{2} = 1.25 - \frac{15/16}{2} = 0.7812 \text{ in.}$$

$$2d = 2(7/8) = 1.75 \text{ in.}$$

■ FIGURE 7.13

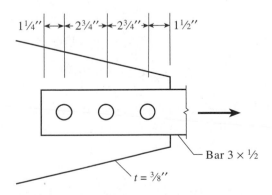

Because $L_c < 2d,$

$$\phi R_n = \phi(1.2L_c t F_u) = 0.75(1.2)(0.7812)\left(\frac{1}{2}\right)(65) = 22.85 \text{ kips}$$

For the other holes,

$$L_c = s - h = 2.75 - \frac{15}{16} = 1.812 \text{ in.} > 2d$$

$$\therefore \ \phi R_n = \phi(2.4dtF_u) = 0.75(2.4)\left(\frac{7}{8}\right)\left(\frac{1}{2}\right)(65) = 51.19 \text{ kips}$$

The bearing strength for the tension member is

$$\phi R_n = 22.85 + 2(51.19) = 125 \text{ kips}$$

Compute the bearing strength of the gusset plate. For the hole nearest the edge of the gusset plate,

$$L_c = L_e - \frac{h}{2} = 1.5 - \frac{15/16}{2} = 1.031 \text{ in.} < 2d$$

$$\therefore \ \phi R_n = \phi(1.2L_c t F_u) = 0.75(1.2)(1.031)\left(\frac{3}{8}\right)(65) = 22.62 \text{ kips}$$

For the other holes,

$$L_c = s - h = 2.75 - \frac{15}{16} = 1.812 \text{ in.} > 2d$$

$$\therefore \ \phi R_n = \phi(2.4dtF_u) = 0.75(2.4)\left(\frac{7}{8}\right)\left(\frac{3}{8}\right)(65) = 38.39 \text{ kips}$$

The bearing strength for the gusset plate is

$$\phi R_n = 22.62 + 2(38.39) = 99.4 \text{ kips}$$

The gusset plate controls. The bearing strength for the connection is therefore

$$\phi R_n = 99.4 \text{ kips}$$

Check the tensile strength of the tension member. **Tension on the gross area:**

$$\phi_t P_n = \phi_t F_y A_g = 0.90(50)\left(3 \times \frac{1}{2}\right) = 67.5 \text{ kips}$$

Tension on the net area: All elements of the cross section are connected, so shear lag is not a factor and $A_e = A_n$. For the hole diameter, use

$$h = d + \frac{1}{8} = \frac{7}{8} + \frac{1}{8} = 1.0 \text{ in.}$$

The design strength is

$$\phi_t P_n = \phi_t F_u A_e = \phi_t F_u t\left(w_g - \Sigma\, h\right) = 0.75(65)\left(\frac{1}{2}\right)[3 - 1(1.0)] = 48.8 \text{ kips}$$

Tension on the net section controls.

ANSWER The design strength of the connection is 48.8 kips ■

7.7   SLIP-CRITICAL CONNECTIONS

A connection with high-strength bolts is classified as either a *slip-critical* connection or a *bearing-type* connection. A slip-critical connection is one in which no slippage is permitted — that is, the friction force must not be exceeded. In a bearing-type connection, slip is acceptable, and shear and bearing actually occur. In some types of structures, notably bridges, the load on connections can undergo many cycles of reversal. In such cases, fatigue of the fasteners can become critical if the connection is allowed to slip with each reversal, and a slip-critical connection is advisable. In most structures, however, slip is perfectly acceptable, and a bearing-type connection is adequate. (A307 bolts are used only in bearing-type connections.) Proper installation and achievement of the prescribed initial tension is necessary for slip-critical connections. Specific situations in which high-strength bolts must be fully tensioned are listed in AISC J1.11. In bearing-type connections, the only practical requirement for the installation of the bolts is that they be tensioned enough so that the surfaces of contact in the connection firmly bear on one another. This installation produces the snug-tight condition referred to earlier in the discussion of the turn-of-the-nut method.

Although slip-critical connections theoretically are not subject to shear and bearing, they must have sufficient shear and bearing strength in the event of an overload that may cause slip to occur.

To prevent slip, the load must be limited, which can be accomplished by limiting either the service load or the factored load. Although preventing slip is essentially a serviceability requirement, the AISC Specification permits the slip-critical strength to be based on either service loads or factored loads. For uniformity, it is more convenient to use factored loads, and that is the practice in this book.

As discussed earlier, the resistance to slip will be a function of the product of the coefficient of static friction and the normal force between the connected parts. This relationship is reflected in the provisions of the RCSC specification, which we use here for slip-critical connections (RCSC, 1994). The slip-critical strength of a connection is given by ϕR_{str}, where $\phi = 1.0$ for standard holes, and

$$R_{str} = 1.13\mu T_m N_b N_s \qquad \text{(RCSC Equation LRFD 5.3)}$$

where

μ = mean slip coefficient (coefficient of static friction) = 0.33 for Class A surfaces

T_m = minimum fastener tension from AISC Table J3.1 or RCSC Table 4

N_b = number of bolts in the connection

N_s = number of slip planes (shear planes)

A class A surface is one with clean mill scale (mill scale is an iron oxide that forms on the steel when it is produced). The specification covers other surfaces, but in this book we conservatively use class A surfaces, which are assigned the smallest slip coefficient. With this stipulation, the slip-critical design strength for one bolt in single shear is

$$\phi R_{str} = \phi(1.13\mu T_m N_b N_s) = 1.0(1.13)(0.33)T_m(1)(1)$$
$$= 0.373T_m \text{ kips} \tag{7.3}$$

for the bolt tension T_m in kips.

■ **EXAMPLE 7.5**

The connection shown in Figure 7.14a uses ¾-inch-diameter A325 bolts with the threads in the shear plane. No slip is permitted. Both the tension member and the gusset plate are of A36 steel. Determine the design strength.

SOLUTION   **Shear strength:** For one bolt

$$A_b = \frac{\pi(3/4)^2}{4} = 0.4418 \text{ in.}^2$$

$$\phi R_n = \phi F_v A_b = 0.75(48)(0.4418) = 15.90 \text{ kips}$$

■ **FIGURE 7.14**

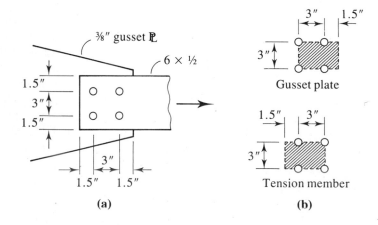

(a)

Gusset plate

Tension member

(b)

For four bolts,

$$\phi R_n = 4(15.90) = 63.6 \text{ kips}$$

Slip-critical strength: Because no slippage is permitted, this connection is classified as slip-critical. From AISC Table J3.1, the minimum bolt tension is $T_m = 28$ kips. From Equation 7.3,

$$\phi R_{str} = 0.373 T_m = 0.373(28) = 10.4 \text{ kips/bolt}$$

For four bolts,

$$\phi R_{str} = 4(10.4) = 41.6 \text{ kips}$$

Bearing strength: Since both edge distances are the same, and the gusset plate is thinner than the bar, the gusset plate thickness of $\frac{3}{8}$-in. will be used. For bearing strength computation, use a hole diameter of

$$h = d + \frac{1}{16} = \frac{3}{4} + \frac{1}{16} = \frac{13}{16} \text{ in.}$$

For the holes nearest the edge of the gusset plate,

$$L_c = L_e - \frac{h}{2} = 1.5 - \frac{13/16}{2} = 1.094 \text{ in.}$$

$$2d = 2\left(\frac{3}{4}\right) = 1.5 \text{ in.}$$

Because $L_c < 2d$,

$$\phi R_n = \phi(1.2 L_c t F_u) = 0.75(1.2)(1.094)\left(\frac{3}{8}\right)(58) = 21.42 \text{ kips/bolt}$$

For the other holes,

$$L_c = s - h = 3 - \frac{13}{16} = 2.188 \text{ in.} > 2d$$

$$\therefore \ \phi R_n = \phi(2.4 d t F_u) = 29.36 \text{ kips/bolt}$$

The bearing strength for the connection is

$$\phi R_n = 2(21.42) + 2(29.36) = 102 \text{ kips}$$

Check the tensile strength of the tension member. **Tension on the gross area:**

$$\phi_t P_n = \phi_t F_y A_g = 0.90(36)\left(6 \times \frac{1}{2}\right) = 97.2 \text{ kips}$$

Tension on the net area: All elements of the cross section are connected, so shear lag is not a factor and $A_e = A_n$. For the hole diameter, use

$$h = d + \frac{1}{8} = \frac{3}{4} + \frac{1}{8} = \frac{7}{8} \text{ in.}$$

The design strength is

$$\phi_t P_n \;=\; \phi_t F_u A_e \;=\; \phi_t F_u t\!\left(w_g - \Sigma\, h\right) \;=\; 0.75(58)\!\left(\frac{1}{2}\right)\!\left[6 - 2\!\left(\frac{7}{8}\right)\right] \;=\; 92.4 \text{ kips}$$

Block shear strength: The failure block for the gusset plate has the same dimensions as the block for the tension member except for the thickness (Figure 7.14b). The gusset plate, which is the thinner element, will control. There are two potential shear-failure planes:

$$A_{gv} \;=\; 2 \times \frac{3}{8}\,(3 + 1.5) \;=\; 3.375 \text{ in.}^2$$

and since there are 1.5 hole diameters per horizontal line of bolts,

$$A_{nv} \;=\; 2 \times \frac{3}{8}\!\left[(3 + 1.5) - (1.5)\!\left(\frac{7}{8}\right)\right] \;=\; 2.391 \text{ in.}^2$$

For the tension areas,

$$A_{gt} \;=\; \frac{3}{8}\,(3) \;=\; 1.125 \text{ in.}^2$$

$$A_{nt} \;=\; \frac{3}{8}\!\left(3 - \frac{7}{8}\right) \;=\; 0.7969 \text{ in.}^2$$

AISC Equation J4-3a gives

$$\phi R_n \;=\; \phi[0.6 F_y A_{gv} + F_u A_{nt}] \;=\; 0.75[0.6(36)(3.375) + 58(0.7969)]$$
$$= 0.75[72.90 + 46.22] \;=\; 89.3 \text{ kips}$$

AISC Equation J4-3b gives

$$\phi R_n \;=\; \phi[0.6 F_u A_{nv} + F_y A_{gt}] \;=\; 0.75[0.6(58)(2.391) + 36(1.125)]$$
$$= 0.75[83.21 + 40.50] \;=\; 92.8 \text{ kips}$$

The fracture term (the one involving F_u) in the second equation is larger than the one in the first equation; therefore AISC Equation J4-3b governs.

Design strength for block shear = 92.8 kips

Of all the limit states investigated, the strength corresponding to slip is the smallest.

ANSWER The design strength of the connection is 41.6 kips. ■

■ EXAMPLE 7.6

A ⅝-inch-thick tension member is connected to two ¼-inch splice plates, as shown in Figure 7.15. The loads shown are service loads. A36 steel and ⅝-inch-diameter A325 bolts will be used. If slip *is* permissible, how many bolts are required? Each bolt centerline shown represents a row of bolts in the direction of the width of the plates.

■ **FIGURE 7.15**

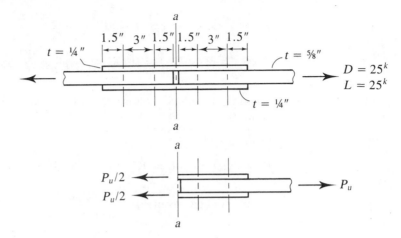

SOLUTION **Shear:** For shear, the nominal bolt area is

$$A_b = \frac{\pi(5/8)^2}{4} = 0.3068 \text{ in.}^2$$

Assume that the bolt threads are in the planes of shear. Then, the design strength for one bolt is

$$\phi R_n = \phi F_v A_b \times 2 \text{ planes of shear} = 0.75(48)(0.3068)(2) = 22.09 \text{ kips}$$

Bearing: The bearing force on the ⅝-inch-thick tension member will be twice as large as the bearing force on each of the ¼-inch splice plates. Because the total load on the splice plates is the same as the load on the tension member, for the splice plates to be critical the total splice plate thickness must be less than the thickness of the tension member — and it is. Use a hole diameter of

$$h = d + \frac{1}{16} = \frac{5}{8} + \frac{1}{16} = \frac{11}{16} \text{ in.}$$

For the holes nearest the edge of the plate,

$$L_c = L_e - \frac{h}{2} = 1.5 - \frac{11/16}{2} = 1.156 \text{ in.}$$

$$2d = 2\left(\frac{5}{8}\right) = 1.25 \text{ in.}$$

Because $L_c < 2d$, the bearing strength is

$$\phi R_n = \phi(1.2L_c t F_u) = 0.75(1.2)(1.156)\left(\frac{1}{4} + \frac{1}{4}\right)(58) = 30.17 \text{ kips/bolt}$$

For the other holes,

$$L_c = s - h = 3 - \frac{11}{16} = 2.312 \text{ in.} > 2d$$

$$\therefore \phi R_n = \phi(2.4dtF_u) = 0.75(2.4)\left(\frac{5}{8}\right)\left(\frac{1}{4} + \frac{1}{4}\right)(58) = 32.62 \text{ kips/bolt}$$

The shearing strength per bolt is smaller than both of these bearing values and therefore controls.

The factored load is

$$P_u = 1.2D + 1.6L = 1.2(25) + 1.6(25) = 70 \text{ kips}$$

$$\text{Number of bolts required} = \frac{\text{total load}}{\text{load per bolt}}$$

$$= \frac{70}{22.09} = 3.17 \text{ bolts}$$

ANSWER   Use four bolts, two per line, on each side of the splice. A total of eight bolts will be required for the connection. ■

■ **EXAMPLE 7.7**

The C6 × 13 shown in Figure 7.16 has been selected to resist a factored tensile load of 108 kips. It is to be attached to a ⅜-inch gusset plate with ⅞-inch-diameter A325 bolts. Assume that the threads are in the plane of shear and that slip of the connection is permissible. Determine the number and required layout of bolts such that the length of connection h is a minimum. A36 steel is used.

■ **FIGURE 7.16**

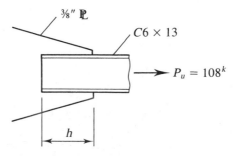

SOLUTION   Determine the capacity of a single bolt:

Shear:

$$A_b = \frac{\pi(7/8)^2}{4} = 0.6013 \text{ in.}^2$$

$$\phi R_n = \phi F_v A_b = 0.75(48)(0.6013) = 21.65 \text{ kips}$$

Bearing: The gusset plate is thinner than the web of the channel and will control. Assume that along a line parallel to the force the length L_c is greater than $2d$ for all bolts. Then

$$\phi R_n = \phi(2.4dtF_u) = 0.75(2.4)\left(\frac{7}{8}\right)\left(\frac{3}{8}\right)(58) = 34.26 \text{ kips}$$

and shear controls. Hence the

$$\text{Number of bolts required} = \frac{108}{21.65} = 4.99$$

Although five bolts will furnish enough capacity, try six bolts so that a symmetrical layout with two gage lines of three bolts each can be used, as shown in Figure 7.17. (Two gage lines are used to minimize the length of the connection.) We do not know whether the design of this tension member was based on the assumption of one line or two lines of fasteners; the tensile capacity of the channel with two lines of bolts must be checked before proceeding.

The tension on the gross area:

$$\phi_t P_n = 0.90 F_y A_g = 0.90(36)(3.83) = 124 \text{ kips}$$

Tension on the effective net area:

$$A_n = 3.83 - 2(1.0)(0.437) = 2.96 \text{ in.}^2$$

The exact length of the connection is not yet known, so we must use the average value of U from the Commentary.

$$A_e = U A_n = 0.85(2.96) = 2.51 \text{ in.}^2$$

$$\phi_t P_n = 0.75 F_u A_e = 0.75(58)(2.51) = 109 \text{ kips} \qquad \text{(controls)}$$

■ **FIGURE 7.17**

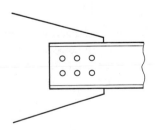

The member capacity is therefore based on two gage lines of bolts.

Check the spacing and edge-distance transverse to the load. From AISC J3.3,

$$\text{Minimum spacing} = 2.667\left(\frac{7}{8}\right) = 2.33 \text{ in.}$$

From AISC Table J3.4,

Minimum edge distance $= 1\frac{1}{8}$ in.

A spacing of 3 inches and an edge distance of $1\frac{1}{2}$ inches will be used transverse to the load.

The minimum length of the connection can be established by using the minimum permissible spacing and edge distances in the longitudinal direction. The minimum spacing in any direction is $2\frac{2}{3}\, d = 2.33$ in. **Try 2½ in.** The minimum edge distance in any direction is $1\frac{1}{8}$ in. These minimum distances will now be used to check the bearing strength of the connection. For the bearing strength computation, use a hole diameter of

$$h = d + \frac{1}{16} = \frac{7}{8} + \frac{1}{16} = \frac{15}{16} \text{ in.}$$

For the holes nearest the edge of the gusset plate,

$$L_c = L_e - \frac{h}{2} = 1.125 - \frac{15/16}{2} = 0.6562 \text{ in.}$$

$$2d = 2(7/8) = 1.75 \text{ in.}$$

Because $L_c < 2d$, the bearing strength is

$$\phi R_n = \phi(1.2L_ctF_u) = 0.75(1.2)(0.6562)\left(\frac{3}{8}\right)(58) = 12.85 \text{ kips/bolt}$$

For the other holes,

$$L_c = s - h = 2.5 - \frac{15}{16} = 1.562 \text{ in.} < 2d$$

$$\therefore \phi R_n = \phi(1.2L_ctF_u) = 0.75(1.2)(1.562)\left(\frac{3}{8}\right)(58) = 30.58 \text{ kips/bolt}$$

The total bearing strength for the connection is

$$\phi R_n = 2(12.85) + 4(30.58) = 148 \text{ kips} > P_u = 108 \text{ kips} \qquad \text{(OK)}$$

The tentative connection design is shown in Figure 7.18 and will now be checked for block shear in the gusset plate (the geometry of the failure block in the channel is identical, but the gusset plate is thinner).

Shear areas:

$$A_{gv} = \frac{3}{8}(2.5 + 2.5 + 1.125)(2) = 4.594 \text{ in.}^2$$

■ **FIGURE 7.18**

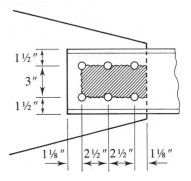

and, as there are 2½ hole diameters along each of the shear surfaces,

$$A_{nv} = \frac{3}{8}[6.125 - 2.5(1.0)](2) = 2.719 \text{ in.}^2$$

Tension areas:

$$A_{gt} = \frac{3}{8}(3) = 1.125 \text{ in.}^2 \quad \text{and} \quad A_{nt} = \frac{3}{8}(3 - 1.0) = 0.75 \text{ in.}^2$$

Check for tension yield and shear fracture with AISC Equation J4-3a:

$$\phi R_n = \phi[0.6F_y A_{gv} + F_u A_{nt}]$$
$$= 0.75[0.6(36)(4.594) + 58(0.75)] = 0.75[99.23 + 43.50] = 107.0 \text{ kips}$$

Check for tensile fracture and shear yield with AISC Equation J4-3b:

$$\phi R_n = \phi[0.6F_u A_{nv} + F_y A_{gt}]$$
$$= 0.75[0.6(58)(2.719) + 36(1.125)] = 0.75[94.62 + 40.50] = 101.3 \text{ kips}$$

The fracture term (the one involving F_u) in the second equation is larger than the one in the first equation; therefore, the second equation governs. Hence the

Design strength for block shear = 101.3 kips < 108 kips   (N.G.)

The simplest way to increase the block shear strength for this connection is to increase the shear areas by increasing the bolt spacing. If the spacing is increased, AISC Equation J4-3b will still control. Although the required spacing can be determined by trial and error, it can be solved for directly, which we do here. From AISC Equation J4-3b, let

$$0.75[0.6(58)A_{nv} + 40.50] = 108 \text{ kips}$$

Required $A_{nv} = 2.974 \text{ in.}^2$

$$A_{nv} = \frac{3}{8}(2s + 1.125 - 2.5)(2) = 2.974 \text{ in.}^2$$

Required $s = 2.67$ in.   use $s = 3.0$ in.

With a spacing of 3.0 in., the net shear area is

$$A_{nv} = \frac{3}{8}(3 + 3 + 1.125 - 2.5)(2) = 3.469 \text{ in.}^2$$

and the block shear strength from AISC Equation J4-3b is

$$\phi R_n = \phi[0.6F_u A_{nv} + F_y A_{gt}]$$
$$= 0.75[0.6(58)(3.469) + 36(1.125)] = 0.75[120.7 + 40.50] = 120.9 \text{ kips}$$

Using the spacing and edge distances determined, the minimum length is therefore

h = 1.125 in. at the end of the channel

+ 2 spaces @ 3.0 in.

+ 1.125 in. at the end of the gusset plate

= 8.25 in. total

ANSWER   Use the connection detail as shown in Figure 7.19.

■ **FIGURE 7.19**

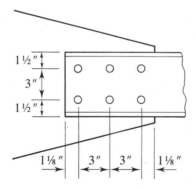

The bolt layout in Example 7.7 is symmetrical with respect to the longitudinal centroidal axis of the member. Consequently, the resultant resisting force provided by the fasteners also acts along this line, and the geometry is consistent with the definition of a simple connection. If an odd number of bolts had been required and two rows had been used, the symmetry would not exist and the connection would be eccentric. In such cases, the designer has several choices: (1) ignore the eccentricity, assuming that the effects are negligible; (2) account for the eccentricity; (3) use a staggered pattern of fasteners that would preserve the symmetry; or (4) add an extra bolt and remove the eccentricity. Most engineers would probably choose the last alternative.

■ **EXAMPLE 7.8**

A 13-foot-long tension member and its connection must be designed for a service dead load of 8 kips and a service live load of 22 kips. No slip of the connection is permitted. The connection will be to a ⅜-inch-thick gusset plate, as shown in Figure 7.20. Use a single angle for the tension member. Use A325 bolts and A572 Grade 50 steel for both the tension member and the gusset plate.

SOLUTION

The factored load to be resisted is

$$P_u = 1.2D + 1.6L = 1.2(8) + 1.6(22) = 44.8 \text{ kips}$$

Because the bolt size and layout will affect the net area of the tension member, we will begin with selection of the bolts. The strategy will be to select a bolt size for trial, determine the number required, and then try a different size if the number is too large or too small. Bolt diameters typically range from ½ inch to 1½ inches in ⅛-inch increments.

   Try ⅝-inch bolts. The nominal bolt area is

$$A_b = \frac{\pi(5/8)^2}{4} = 0.3068 \text{ in.}^2$$

The shear strength is

$$\phi R_n = \phi F_v A_b = 0.75(48)A_b = 0.75(48)(0.3068)$$
$$= 11.04 \text{ kips/bolt} \quad \text{(assuming that the threads are in the shear plane)}$$

No slip is permitted, so this connection is slip-critical. We will assume class A surfaces, and for a ⅝-inch-diameter A325 bolt, the minimum tension is $T_m = 19$ kips (from AISC Table J3.1). From RCSC Equation LRFD 5.3, the slip-critical strength for one bolt is

$$\phi R_{str} = \phi(1.13\mu T_m N_b N_s) = 1.0(1.13)(0.33)(19)(1)(1) = 7.085 \text{ kips/bolt}$$

■ **FIGURE 7.20**

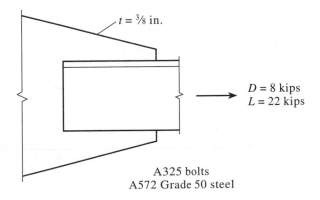

$t = $ ⅜ in.

$D = 8$ kips
$L = 22$ kips

A325 bolts
A572 Grade 50 steel

The slip-critical strength controls. We will determine the number of bolts based on this strength and check bearing after selecting the member (because the bearing strength cannot be computed until the member thickness is known). Hence the

$$\text{Number of bolts} = \frac{\text{total load}}{\text{load per bolt}} = \frac{44.8}{7.085} = 6.3 \text{ bolts}$$

A minimum of seven bolts will be required. If two rows are used, an extra bolt could be added to maintain symmetry. Figure 7.21 shows several potential bolt layouts. Although any of these arrangements could be used, the connection length can be decreased by using a larger bolt size and fewer bolts.

Try ⅞-inch bolts. The nominal bolt area is

$$A_b = \frac{\pi(7/8)^2}{4} = 0.6013 \text{ in.}^2$$

The shear strength is

$$\phi R_n = 0.75(48)A_b = 0.75(48)(0.6013)$$
$$= 21.65 \text{ kips/bolt} \qquad \text{(assuming that the threads are in the shear plane)}$$

The minimum tension for a ⅞-inch A325 bolt is $T_m = 39$ kips, so the slip-critical strength is

$$\phi R_{str} = \phi(1.13\mu T_m N_b N_s) = 1.0(1.13)(0.33)(39)(1)(1) = 14.54 \text{ kips/bolt} \qquad \text{(controls)}$$

The number of ⅞-inch bolts required is

$$\frac{44.8}{14.54} = 3.1 \text{ bolts}$$

■ **FIGURE 7.21**

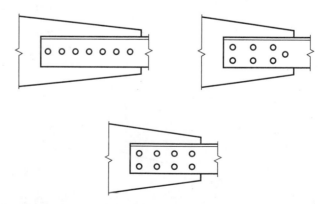

Four $\frac{7}{8}$-inch-diameter A325 bolts will be used. From AISC J3.3, the minimum spacing is

$$s = 2.667d = 2.667\left(\frac{7}{8}\right) = 2.33 \text{ in.} \quad \text{(or, preferably, } 3d = 3\left(\frac{7}{8}\right) = 2.62 \text{ in.)}$$

From AISC Table J3.4, the minimum edge distance is

$$L_e = 1.5 \text{ in.} \quad \text{(assuming sheared edges)}$$

Try the layout shown in Figure 7.22 and select a tension member. The required gross area is

$$A_g \geq \frac{P_u}{0.9F_y} = \frac{44.8}{0.9(50)} = 0.996 \text{ in.}^2$$

and the required effective net area is

$$A_e \geq \frac{P_u}{0.75F_u} = \frac{44.8}{0.75(65)} = 0.9190 \text{ in.}^2$$

Since the effective net area is $A_e = UA_n$, the required net area is

$$A_n \geq \frac{\text{required } A_e}{U}$$

For the layout shown in Figure 7.22, with more than two bolts in the line of force, the average value of U from the Commentary to the AISC Specification is 0.85. (Once a member has been selected, U can be computed with AISC Equation B3-2). Therefore

$$A_n \geq \frac{0.9190}{0.85} = 1.08 \text{ in.}^2$$

■ **FIGURE 7.22**

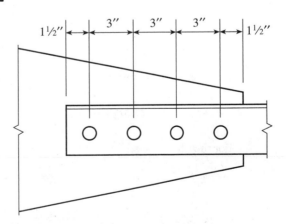

Note that this value is greater than the required gross area. The required minimum radius of gyration is

$$r_{min} = \frac{L}{300} = \frac{13(12)}{300} = 0.52 \text{ in.}$$

Try an L3½ × 2½ × ¼.

$$A_g = 1.44 \text{ in.}^2 > 0.996 \text{ in.}^2 \quad \text{(OK)}$$
$$r_{min} = r_z = 0.544 \text{ in.} > 0.52 \text{ in.} \quad \text{(OK)}$$

For net area computation, use a hole diameter of ⅞ + ⅛ = 1.0 in.

$$A_n = A_g - A_{hole} = 1.44 - 1.0\left(\frac{1}{4}\right) = 1.190 \text{ in.}^2 > 1.08 \text{ in.}^2 \quad \text{(OK)}$$

Compute U with AISC Equation B3-2:

$$U = 1 - \frac{\bar{x}}{L} \le 0.9$$
$$= 1 - \frac{0.785}{9} = 0.913$$

This value is greater than 0.9, so use $U = 0.9$. The effective net area is

$$A_e = UA_n = 0.9(1.190) = 1.071 \text{ in.}^2 > 0.9190 \text{ in.}^2 \quad \text{(OK)}$$

Now check the bearing strength. The edge distance for the angle is the same as the edge distance for the gusset plate and the angle is thinner than the gusset plate, so the angle thickness of ¼-inch will be used. For bearing strength computation, use a hole diameter of

$$h = d + \frac{1}{16} = \frac{7}{8} + \frac{1}{16} = \frac{15}{16} \text{ in.}$$

For the hole nearest the edge of the member,

$$L_c = L_e - \frac{h}{2} = 1.5 - \frac{15/16}{2} = 1.031 \text{ in.}$$
$$2d = 2(7/8) = 1.75 \text{ in.}$$

Because $L_c < 2d$, the bearing strength is

$$\phi R_n = \phi(1.2 L_c t F_u) = 0.75(1.2)(1.031)\left(\frac{1}{4}\right)(65) = 15.08 \text{ kips/bolt}$$

For the other holes,

$$L_c = s - h = 3 - \frac{15}{16} = 2.062 \text{ in.} > 2d$$
$$\therefore \phi R_n = \phi(2.4 dt F_u) = 0.75(2.4)\left(\frac{7}{8}\right)\left(\frac{1}{4}\right)(65) = 25.59 \text{ kips/bolt}$$

The total bearing strength for the connection is

$$\phi R_n = 15.08 + 3(25.59) = 91.9 \text{ kips} > P_u = 44.8 \text{ kips} \quad \text{(OK)}$$

Now check block shear. With the bolts placed in the long leg at the usual gage distance (see Chapter 3, Figure 3.22), the failure block is as shown in Figure 7.23. The shear areas are

$$A_{gv} = \frac{1}{4}(1.5 + 9) = 2.625 \text{ in.}^2$$

$$A_{nv} = \frac{1}{4}[1.5 + 9 - 3.5(1.0)] = 1.750 \text{ in.}^2 \quad \text{(3.5 hole diameters)}$$

The tension areas are

$$A_{gt} = \frac{1}{4}(1.5) = 0.3750 \text{ in.}^2$$

$$A_{nt} = \frac{1}{4}[1.5 \quad 0.5(1.0)] = 0.2500 \text{ in.}^2 \quad \text{(0.5 hole diameter)}$$

AISC Equation J4-3a gives

$$\phi R_n = \phi[0.6F_yA_{gv} + F_uA_{nt}]$$
$$= 0.75[0.6(50)(2.625) + 65(0.2500)] = 0.75(78.75 + 16.25) = 71.2 \text{ kips}$$

AISC Equation J4-3b gives

$$\phi R_n = \phi[0.6F_uA_{nv} + F_yA_{gt}]$$
$$= 0.75[0.6(65)(1.750) + 50(0.3750)] = 0.75(68.25 + 18.75) = 65.2 \text{ kips}$$

Equation J4-3b has the larger fracture term, so it controls. The block shear strength is therefore

$$\phi R_n = 65.2 \text{ kips} > P_u = 44.8 \text{ kips} \quad \text{(OK)}$$

■ **FIGURE 7.23**

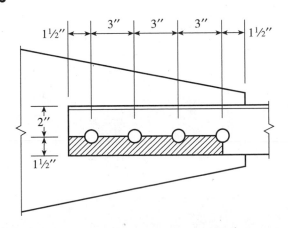

■ **FIGURE 7.24**

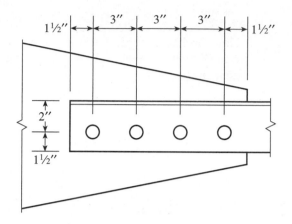

ANSWER   Use an L3½ × 2½ × ¼ with the long leg connected. Use four ⅞-inch-diameter A325 bolts, as shown in Figure 7.24.   ■

7.8   ## HIGH-STRENGTH BOLTS IN TENSION

When a tensile load is applied to a bolt with no initial tension, the tensile force in the bolt is equal to the applied load. If the bolt is pretensioned, however, a large part of the applied load is used to relieve the compression, or clamping forces, on the connected parts, as determined by Kulak, Fisher, and Struik (1987) and demonstrated here. Figure 7.25 shows a hanger connection consisting of a structural tee shape bolted to the bottom flange of a W-shape and subjected to a tensile load. A single bolt and a portion of the connected parts will be isolated and examined both before and after loading.

■ **FIGURE 7.25**

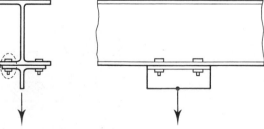

Before loading, all forces are internal, and a free-body diagram of the assembly is as shown in Figure 7.26a. For simplicity, all forces will be assumed to be symmetrical with respect to the axis of the bolt, and any eccentricity will be neglected. If the connected parts are considered as separate free bodies, the forces consist of the bolt tension T_0 and the normal clamping force N_0, shown here as uniformly distributed. Equilibrium requires that T_0 equal N_0. When the external tensile load is applied, the forces on the assembly are as shown in Figure 7.26b, with F representing the total tensile force applied to one bolt (again, the actual distribution of the applied force per bolt has been idealized for simplicity). Figure 7.26c shows the forces acting on a free-body diagram of the segment of the structural tee flange and the corresponding segment of the bolt. Summing forces in the direction of the bolt axis gives

$$T = F + N$$

The application of force F will increase the bolt tension and cause it to elongate by an amount δ_b. Compression in the flange of the structural tee will be reduced, resulting in a distortion $\delta_{f\ell}$ in the same sense as δ_b. A relationship between the applied force and the change in bolt tension can be approximated as follows.

■ **FIGURE 7.26**

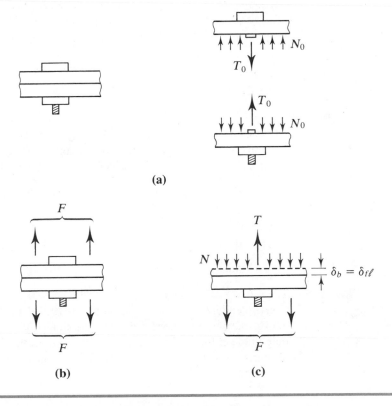

(a)

(b)

(c)

From elementary mechanics of materials, the axial deformation of an axially loaded uniform member is

$$\delta = \frac{PL}{AE} \qquad (7.4)$$

where
P = applied axial force
L = original, undeformed length
A = cross-sectional area
E = modulus of elasticity

Equation 7.4 can be solved for the load:

$$P = \frac{AE\delta}{L} \qquad (7.5)$$

The change in bolt force corresponding to a given axial displacement δ_b therefore is

$$\Delta T = \frac{A_b E_b \delta_b}{L_b} \qquad (7.6)$$

where the subscript indicates a property or dimension of the bolt. Application of Equation 7.5 to the compression flange requires a somewhat more liberal interpretation of the load distribution in that N must be treated as if it were uniformly applied over a surface area, $A_{f\ell}$. The change in the force N is then obtained from Equation 7.5 as

$$\Delta N = \frac{A_{f\ell} E_{f\ell} \delta_{f\ell}}{L_{f\ell}} \qquad (7.7)$$

where $L_{f\ell}$ is the flange thickness. As long as the connected parts (the two flanges) remain in contact, the bolt deformation, δ_b, and the flange deformation, $\delta_{f\ell}$, will be equal. Because $E_{f\ell}$ approximately equals E_b (Bickford, 1981), and $A_{f\ell}$ is much larger than A_b,

$$\frac{A_{f\ell} E_{f\ell} \delta_{f\ell}}{L_{f\ell}} \gg \frac{A_b E_b \delta_b}{L_b}$$

and therefore

$$\Delta N \gg \Delta T$$

The ratio of ΔT to ΔN is in the range of 0.05 to 0.1 (Kulak, Fisher, and Struik, 1987). Consequently, ΔT will be no greater than $0.1\Delta N$, demonstrating that most of the applied load is devoted to relieving the compression of the connected parts. To estimate the magnitude of the load required to overcome completely the clamping effect and cause the parts to separate, consider the free-body diagram in Figure 7.27. When the parts have separated,

$$T = F$$

■ **FIGURE 7.27**

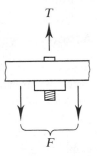

or

$$T_0 + \Delta T = F \tag{7.8}$$

At the point of impending separation, the bolt elongation and plate decompression are the same, and

$$\Delta T = \frac{A_b E_b}{L_b} \delta_b = \frac{A_b E_b}{L_b} \delta_{f\ell} \tag{7.9}$$

where $\delta_{f\ell}$ is the elongation corresponding to the release of the initial compression force N_0. From Equation 7.4,

$$\delta_{f\ell} = \frac{N_0 L_{f\ell}}{A_{f\ell} E_{f\ell}}$$

Substituting into Equation 7.9 yields

$$\Delta T = \left(\frac{A_b E_b}{L_b} \right) \left(\frac{N_0 L_{f\ell}}{A_{f\ell} E_{f\ell}} \right) = \left(\frac{A_b E_b / L_b}{A_{f\ell} E_{f\ell} / L_{f\ell}} \right) N_0 = \left(\frac{A_b E_b / L_b}{A_{f\ell} E_{f\ell} / L_{f\ell}} \right) T_0 \approx 0.1 T_0$$

From Equation 7.8,

$$T_0 + 0.1 T_0 = F \quad \text{or} \quad F = 1.1 T_0$$

Therefore, at the instant of separation, the bolt tension is approximately 10% larger than its initial value at installation. Once the connected parts separate, however, any increase in external load will be resisted entirely by a corresponding increase in bolt tension. If the bolt tension is assumed to be equal to the externally applied force (as if there were no initial tension) and the connection is loaded until the connected parts separate, the bolt tension will be underestimated by less than 10%. This discrepancy has been accounted for in the nominal strength values given in AISC Table J3.2. These values depend on the connected parts not separating, so the initial tension must not be overcome. For this reason, high-strength bolts subject to direct tension must be pretensioned to the values prescribed in AISC Table J3.1, whether the connection is slip-critical or not.

In summary, the tensile force in the bolt should be computed without considering the initial tension.

Prying Action

In most connections in which fasteners are subjected to tension forces, the flexibility of the connected parts can lead to deformations that increase the tension applied to the fasteners. A hanger connection of the type used in the preceding discussion is subject to this type of behavior. The additional tension is called a *prying force* and is illustrated in Figure 7.28, which shows the forces on a free body of the hanger. Before the external load is applied, the normal compressive force N_0 is centered on the bolt. As the load is applied, if the flange is flexible enough to deform as shown, the compressive forces will migrate toward the edges of the flange. This redistribution will change the relationship between all forces, and the bolt tension will increase. If the connected parts are sufficiently rigid, however, this shifting of forces will not occur, and there will be no prying action. The maximum value of the prying force will be reached when only the corners of the flange remain in contact with the other connected part.

In connections of this type, bending caused by the prying action will usually control the design of the connected part. AISC J3.6 requires that prying action be included in the computation of tensile loads applied to fasteners.

■ FIGURE 7.28

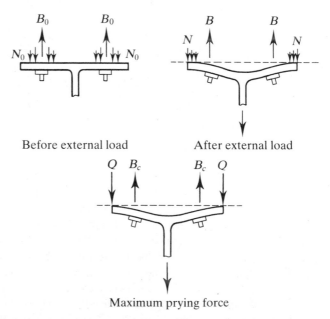

Before external load   After external load

Maximum prying force

A procedure for the determination of prying forces, based on research reported in *Guide to Design Criteria for Bolted and Riveted Joints* (Kulak, Fisher, and Struick, 1987), is given in the *Manual* in Part 11, "Connections for Tension and Compression" (Volume II). The specific case examined is the connection of a segment of a structural tee shape, but a pair of back-to-back angles would be treated in exactly the same way. The method is presented here in a somewhat different form but gives the same results.

The method used is based on the model shown in Figure 7.29. All forces are for one fastener. Thus T is the external factored tension force applied to one bolt, Q is the prying force corresponding to one bolt, and B_c is the total bolt force. The prying force has shifted to the tip of the flange and is at its maximum value.

The equations that follow are derived from consideration of equilibrium based on the free-body diagrams in Figure 7.29. From the summation of moments at *b–b* in Figure 7.29b,

$$Tb - M_{a-a} = Qa \qquad (7.10)$$

■ **FIGURE 7.29**

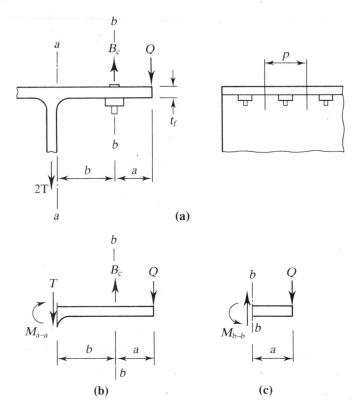

From Figure 7.29c,

$$M_{b-b} = Qa \tag{7.11}$$

Finally, equilibrium of forces requires that

$$B_c = T + Q \tag{7.12}$$

These three equilibrium equations can be combined to obtain a single equation for the total bolt force, which includes the effects of prying action. We first define the variable α as the ratio of the moment per unit length along the bolt line to the moment per unit length at the face of the stem. For the bolt line, the length is a net length, so

$$\alpha = \frac{M_{b-b}/(p - d')}{M_{a-a}/p} = \frac{M_{b-b}}{M_{a-a}}\left(\frac{1}{1 - d'/p}\right) = \frac{M_{b-b}}{\delta M_{a-a}} \tag{7.13}$$

where

$$p = \text{length of flange tributary to one bolt} \quad \text{(see Figure 7.29a)}$$

$$d' = \text{diameter of bolt hole}$$

$$\delta = 1 - \frac{d'}{p} = \frac{\text{net area at bolt line}}{\text{gross area at web face}}$$

$$M_{a-a} = \text{design strength at } a-a = \phi_b M_p = \phi_b(pt_f^2 F_y/4)$$

(The numerical evaluation of α will require the use of another equation, which we develop shortly.) With this notation, we can combine the three equilibrium equations 7.10–7.12 to obtain the total bolt force, B_c:

$$B_c = T\left[1 + \frac{\delta\alpha}{(1 + \delta\alpha)}\frac{b}{a}\right] \tag{7.14}$$

At this level of loading, deformations are so large that the resultant of tensile stresses in the bolt does not coincide with the axis of the bolt. Consequently, the bolt force predicted by Equation 7.14 is conservative and does not quite agree with test results. Much better agreement is obtained if the force B_c is shifted toward the stem of the tee by an amount $d/2$, where d is the bolt diameter. The modified values of b and a are therefore defined as

$$b' = b - \frac{d}{2} \quad \text{and} \quad a' = a + \frac{d}{2}$$

(For best agreement with test results, the value of a should be no greater than $1.25b$.) With this modification, we can write Equation 7.14 as

$$B_c = T\left[1 + \frac{\delta\alpha}{(1 + \delta\alpha)}\frac{b'}{a'}\right] \tag{7.15}$$

We can evaluate α from Equation 7.15 by setting the bolt force B_c equal to the bolt design tensile strength, which we denote B. Doing so results in

$$\alpha = \frac{[(B/T) - 1](a'/b')}{\delta\{1 - [(B/T) - 1](a'/b')\}} \tag{7.16}$$

Two limit states are possible: tensile failure of the bolts and bending failure of the tee. Failure of the tee is assumed to occur when plastic hinges form at section a–a, the face of the stem of the tee, and at b–b, the bolt line, thereby creating a beam mechanism. The moment of each of these locations will equal M_p, the plastic moment capacity of a length of flange tributary to one bolt. If the absolute value of α, obtained from Equation 7.16, is less than 1.0, the moment at the bolt line is less than the moment at the face of the stem, indicating that the beam mechanism has not formed, and the controlling limit state will be tensile failure of the bolt. The bolt force B_c in this case will equal the design strength B. If the absolute value of α is equal to or greater than 1.0, plastic hinges have formed at both a–a and b–b, and the controlling limit state is flexural failure of the tee flange. Since the moments at these two locations are limited to the plastic moment M_p, α should be set equal to 1.0.

The three equilibrium Equations 7.10–7.12 can also be combined into a single equation for the required flange thickness, t_f. From Equations 7.10 and 7.11, we can write

$$Tb' - M_{a-a} = M_{b-b}$$

where b' has been substituted for b. From Equation 7.13,

$$Tb' - M_{a-a} = \delta\alpha M_{a-a} \tag{7.17}$$

Setting M_{a-a} equal to the design strength gives

$$M_{a-a} = \phi_b M_p = \phi_b \frac{pt_f^2 F_y}{4}$$

where t_f is the required flange thickness. Substituting into Equation 7.17, we obtain

$$t_f = \sqrt{\frac{4Tb'}{\phi_b p F_y (1 + \delta\alpha)}}$$

With $\phi_b = 0.90$,

$$\text{Required } t_f = \sqrt{\frac{4.444 Tb'}{p F_y (1 + \delta\alpha)}} \tag{7.18}$$

The design of connections subjected to prying is essentially a trial-and-error process. When selecting the size or number of bolts, we must make an allowance for the prying force. The selection of the tee flange thickness is more difficult in that it is a function of the bolt selection and tee dimensions. The Preliminary Hanger Connection Selection

Table provided in Part 11 of the *Manual* can be helpful in choosing a trial shape. Once the trial shape has been selected and the number of bolts and their layout estimated, Equations 7.15 and 7.18 can be used to verify or disprove the choices.

If the actual flange thickness is different from the required value, the actual values of α and B_c will be different from those previously calculated. If the actual bolt force, which includes the prying force Q, is desired, α will need to be recomputed as follows.

First, combine Equations 7.10 and 7.11, using b' instead of b:

$$M_{b-b} = Tb' - M_{a-a}$$

From Equation 7.13,

$$\alpha = \frac{M_{b-b}}{\delta M_{a-a}}$$

$$= \frac{Tb' - M_{a-a}}{\delta M_{a-a}} = \frac{Tb'/M_{a-a} - 1}{\delta}$$

Setting M_{a-a} equal to the design moment gives

$$M_{a-a} = \phi_b M_p = 0.90 \left(\frac{pt_f^2 F_y}{4} \right)$$

then

$$\alpha = \frac{\dfrac{Tb'}{0.90\, pt_f^2 F_y/4} - 1}{\delta} = \frac{1}{\delta}\left(\frac{4.444 Tb'}{pt_f^2 F_y} - 1 \right) \tag{7.19}$$

The total bolt force can the be found from Equation 7.15.

■ EXAMPLE 7.9

An 8-inch-long WT10.5 × 66 is attached to the bottom flange of a beam, as shown in Figure 7.30. This hanger must support a factored load of 90 kips. Determine the number of ⅞-inch-diameter A325 bolts required and investigate the adequacy of the tee. A36 steel is used.

SOLUTION The bolt area is

$$A_b = \frac{\pi (7/8)^2}{4} = 0.6013 \text{ in.}^2$$

and the design strength of one bolt is

$$B = \phi R_n = \phi F_t A_b = 0.75(90.0)(0.6013) = 40.59 \text{ kips}$$

The number of bolts required is $90/40.59 = 2.22$. A minimum of four bolts will be needed to maintain symmetry, so **try four bolts.** From the dimensions shown in Figure 7.30,

■ **FIGURE 7.30**

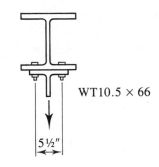

WT10.5 × 66

5½"

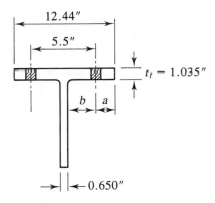

12.44"

5.5"

$t_f = 1.035''$

b | a

0.650"

$$b = \frac{(5.5 - 0.650)}{2} = 2.425 \text{ in.}$$

$$a = \frac{(12.44 - 5.5)}{2} = 3.470 \text{ in.}$$

$$1.25b = 1.25(2.425) = 3.031 \text{ in.} < 3.470 \text{ in.} \qquad \text{use } a = 3.031 \text{ in.}$$

$$b' = b - \frac{d}{2} = 2.425 - \frac{7/8}{2} = 1.988 \text{ in.}$$

$$a' = a + \frac{d}{2} = 3.031 + \frac{7/8}{2} = 3.468 \text{ in.}$$

The factored external load per bolt, excluding prying force is $T = 90/4 = 22.5$ kips. Compute δ:

$$d' = d + \frac{1}{8} = \frac{7}{8} + \frac{1}{8} = 1 \text{ in.}$$

$$p = \frac{8}{2} = 4 \text{ in.}$$

$$\delta = 1 - \frac{d'}{p} = 1 - \frac{1}{4} = 0.75$$

compute α:

$$\frac{B}{T} - 1 = \frac{40.59}{22.5} - 1 = 0.8040$$

$$\frac{a'}{b'} = \frac{3.468}{1.988} = 1.744$$

From Equation 7.16,

$$\alpha = \frac{[(B/T) - 1](a'/b')}{\delta\{1 - [(B/T) - 1](a'/b')\}} = \frac{0.8040(1.744)}{0.75[1 - 0.8040(1.744)]} = -4.65$$

Because $|\alpha| > 1.0$, use $\alpha = 1.0$. From Equation 7.18,

$$\text{Required } t_f = \sqrt{\frac{4.444Tb'}{pF_y(1 + \delta\alpha)}} = \sqrt{\frac{4.444(22.5)(1.988)}{4(36)(1 + 0.75)}}$$

$$= 0.888 \text{ in.} < 1.035 \text{ in.} \quad \text{(OK)}$$

Both the number of bolts selected and the flange thickness are adequate, and no further computations are required. To illustrate the procedure, however, we compute the prying force, using Equations 7.19 and 7.15. From Equation 7.19,

$$\alpha = \frac{1}{\delta}\left(\frac{4.444Tb'}{pt_f^2 F_y} - 1\right) = \frac{1}{0.75}\left[\frac{4.444(22.5)(1.988)}{4(1.035)^2(36)} - 1\right] = 0.3848$$

From Equation 7.15, the total bolt force, including prying, is

$$B_c = T\left[1 + \frac{\delta\alpha}{(1 + \delta\alpha)}\frac{b'}{a'}\right]$$

$$= 22.5\left[1 + \frac{0.75(0.3848)}{1 + 0.75(0.3848)}\left(\frac{1.988}{3.468}\right)\right] = 25.39 \text{ kips}$$

The prying force is

$$Q = B_c - T = 25.39 - 22.5 = 2.89 \text{ kips}$$

ANSWER A WT10.5 × 66 is satisfactory. Use four ⅞-inch-diameter A325 bolts. ■

If the flange thickness had proved to be inadequate, the alternatives would include trying a larger tee shape or using more bolts to reduce T, the external load per bolt. The prying force in Example 7.9 adds approximately 13% to the externally applied load. Neglect of this additional tension could have serious consequences. You should be alert for situations in which it may arise and, when feasible, minimize it by avoiding overly flexible connection elements.

7.9 COMBINED SHEAR AND TENSION IN FASTENERS

In most of the situations in which a bolt is subjected to both shear and tension, the connection is loaded eccentrically and falls within the realm of Chapter 8. However, in some simple connections the fasteners are in a state of combined loading. Figure 7.31 shows a structural tee segment connected to the flange of a column for the purpose of attaching a bracing member. This bracing member is oriented in such a way that the line of action of the member force passes through the center of gravity of the connection. The vertical component of the load will put the fasteners in shear, and the horizontal component will cause tension (with the possible inclusion of prying forces). Since the line of action of the load acts through the center of gravity of the connection, each fastener can be assumed to take an equal share of each component.

As in other cases of combined loading, an interaction formula approach can be used. The shear and tensile strengths for bearing-type bolts are based on test results and can be taken from the elliptical interaction curve shown in Figure 7.32. The equation of this curve is

$$\left[\frac{P_u}{(\phi R_n)_t} \right]^2 + \left[\frac{V_u}{(\phi R_n)_v} \right]^2 = 1.0$$

where

$$P_u = \text{factored tensile load on the bolt}$$
$$(\phi R_n)_t = \text{design strength of the bolt in tension}$$
$$V_u = \text{factored shear load on the bolt}$$
$$(\phi R_n)_v = \text{design strength of the bolt in shear}$$

■ **FIGURE 7.31**

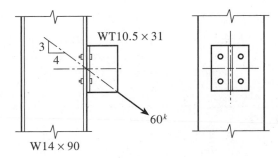

■ **FIGURE 7.32**

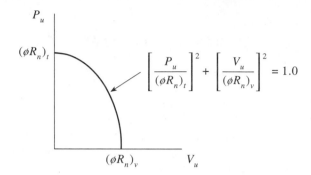

An acceptable combination of shear and tension is one that lies under this curve. This fact leads to the requirement from the RCSC specification that

$$\left[\frac{P_u}{\left(\phi R_n\right)_t}\right]^2 + \left[\frac{V_u}{\left(\phi R_n\right)_v}\right]^2 \le 1.0 \qquad \text{(RCSC Equation LRFD 4.2)}$$

For slip-critical connections in which the bolts are subject to both shear and tension, the effect of the applied tensile force is to relieve some of the clamping force, thereby reducing the available friction force. The AISC Specification reduces the slip-critical shear strength for this case. From AISC Appendix J, the slip-critical shear strength is to be multiplied by the factor

$$\left[1 - \frac{T_u}{1.13T_m N_b}\right] \qquad \text{(AISC Equation A-J3-2)}$$

where

T_u = factored tensile load on the connection

T_m = prescribed initial bolt tension from AISC Table J3.1

N_b = number of bolts in the connection

Note that RCSC Equation LRFD 4.2 has been presented here as applying to a single bolt, whereas AISC Equation A-J3-2 has been presented as applying to the entire connection. Each equation could be modified to apply in the alternative way.

■ **EXAMPLE 7.10**

A WT10.5 × 31 is used as a bracket to transmit a 60-kip service load to a W14 × 90 column, as previously shown in Figure 7.31. The load consists of 15 kips dead load and 45 kips live load. Four ⅞-inch-diameter A325 bolts are used. Both the column and bracket are of A36 steel. Assume that all spacing and edge-distance requirements are satisfied, including those necessary for the use of the maximum design strength in bearing (i.e., $\phi[2.4dtF_u]$), and determine the adequacy of the bolts for the following types of connections: (a) bearing-type connection with the threads in shear and (b) slip-critical connection with the threads in shear.

SOLUTION   The factored load is

$$1.2D + 1.6L = 1.2(15) + 1.6(45) = 90 \text{ kips}$$

a. For the bearing-type connection, with threads in the shear plane, the total shear force is

$$\frac{3}{5}(90) = 54 \text{ kips}$$

The shear force per bolt is

$$V_u = \frac{54}{4} = 13.5 \text{ kips}$$

and

$$A_b = \frac{\pi(7/8)^2}{4} = 0.6013 \text{ in.}^2$$

$$(\phi R_n)_v = \phi F_v A_b = 0.75(48)(0.6013)$$
$$= 21.65 \text{ kips} > 13.5 \text{ kips} \qquad \text{(OK)}$$

The bearing strength (flange of tee controls) is

$$\phi R_n = \phi(2.4dtF_u) = 0.75(2.4)\left(\frac{7}{8}\right)(0.615)(58)$$

$$= 56.18 \text{ kips} > 13.5 \text{ kips} \quad \text{(OK)}$$

The total tensile load is

$$\frac{4}{5}(90) = 72 \text{ kips}$$

and the tensile load per bolt is

$$P_u = \frac{72}{4} = 18 \text{ kips}$$

From AISC Table J3.2,

$$(\phi R_n)_t = \phi F_t A_b = 0.75(90)(0.6013) = 40.59 \text{ kips} > 18 \text{ kips} \qquad \text{(OK)}$$

From RCSC Equation LRFD 4.2,

$$\left[\frac{P_u}{(\phi R_n)_t}\right]^2 + \left[\frac{V_u}{(\phi R_n)_v}\right]^2 = \left(\frac{18}{40.59}\right)^2 + \left(\frac{13.5}{21.65}\right)^2$$

$$= 0.585 < 1.0 \qquad \text{(OK)}$$

ANSWER   The connection is adequate as a bearing-type connection. (In order not to obscure the combined loading features of this example, prying action has not been included in the analysis.)

b. For the slip-critical connection, with threads in the shear plane, from part (a), the shear, bearing, and tension strengths are satisfactory. From RCSC Equation LRFD 5.3, the slip-critical design strength is

$$\phi R_{str} = \phi(1.13\mu T_m N_b N_s)$$

From AISC Table J3.1, the prescribed tension for a ⅞-inch-diameter A325 bolt is

$$T_m = 39 \text{ kips}$$

If we assume Class A surfaces, the slip coefficient is $\mu = 0.33$, and for four bolts,

$$\phi R_{str} = \phi(1.13\mu T_m N_b N_s) = 1.0(1.13)(0.33)(39)(4)(1) = 58.17 \text{ kips}$$

Since there is a tensile load on the bolts, the slip-critical strength load must be reduced by a factor of

$$\left(1 - \frac{T_u}{1.13 T_m N_b}\right) = \left[1 - \frac{72}{1.13(39)(4)}\right] - 0.5916$$

The reduced strength is therefore

$$\phi R_{str} = 0.5916(58.17) = 34.4 \text{ kips} < 54 \text{ kips} \qquad \text{(N.G.)}$$

ANSWER   The connection is inadequate as a slip-critical connection.   ■

Bolted connections subjected to shear and tension can be designed directly. RCSC Equation LRFD 4.2 can be solved for the required bolt size as follows. First, let

$$\left[\frac{P_u}{(\phi R_n)_t}\right]^2 + \left[\frac{V_u}{(\phi R_n)_v}\right]^2 = \left(\frac{P_u}{\phi F_t \ \Sigma A_b}\right)^2 + \left(\frac{V_u}{\phi F_v \ \Sigma A_b}\right)^2$$

$$= \left(\frac{P_u}{\phi F_t}\right)^2 \frac{1}{\left(\Sigma A_b\right)^2} + \left(\frac{V_u}{\phi F_v}\right)^2 \frac{1}{\left(\Sigma A_b\right)^2}$$

where

P_u = total tensile load on the connection

F_t = ultimate tensile stress of bolt

V_u = total shear load on the connection

F_v = ultimate shear stress of bolt

ΣA_b = total bolt area

Substituting into RCSC Equation LRFD 4.2, we get

$$\left(\frac{P_u}{\phi F_t}\right)^2 \frac{1}{\left(\sum A_b\right)^2} + \left(\frac{V_u}{\phi F_v}\right)^2 \frac{1}{\left(\sum A_b\right)^2} \leq 1.0$$

or

$$\sum A_b \geq \sqrt{\left(\frac{P_u}{\phi F_t}\right)^2 + \left(\frac{V_u}{\phi F_v}\right)^2} \qquad (7.20)$$

where $\sum A_b$ is the required *total* bolt area.

■ EXAMPLE 7.11

A concentrically loaded connection is subjected to a service load shear force of 50 kips and a service load tensile force of 100 kips. The loads are 25% dead load and 75% live load. The fasteners will be in single shear, and bearing strength will be controlled by a ⁵⁄₁₆-inch-thick connected part. Assume that all spacing and edge distances are satisfactory, including those that permit the maximum bearing strength of $\phi(2.4 dt F_u)$ to be used. Determine the required number of ¾-inch-diameter A325 bolts for the following cases: (a) bearing-type connection with threads in the plane of shear and (b) slip-critical connection with threads in the plane of shear. All contact surfaces have clean mill scale.

Consider this design to be preliminary so that no consideration of prying action is necessary.

SOLUTION

Factored load shear = 1.2[0.25(50)] + 1.6[0.75(50)] = 75 kips

Factored load tension = 1.2[0.25(100)] + 1.6[0.75(100)] = 150 kips

a. For the bearing-type connection, with threads in the shear plane, Equation 7.20 gives

$$\sum A_b \geq \sqrt{\left(\frac{P_u}{\phi F_t}\right)^2 + \left(\frac{V_u}{\phi F_v}\right)^2} = \sqrt{\left[\frac{150}{0.75(90)}\right]^2 + \left[\frac{75}{0.75(48)}\right]^2} = 3.046 \text{ in.}^2$$

The area of one bolt is

$$A_b = \frac{\pi(3/4)^2}{4} = 0.4418 \text{ in.}^2$$

so the number of bolts required is

$$\frac{\sum A_b}{A_b} = \frac{3.046}{0.4418} = 6.89$$

Try 7 bolts and check the bearing:

$$\phi R_n = \phi(2.4dtF_u) \times 7 \text{ bolts}$$

$$= 0.75(2.4)\left(\frac{7}{8}\right)\left(\frac{5}{16}\right)(58)(7) = 171 \text{ kips} > 75 \text{ kips} \qquad \text{(OK)}$$

(Spacing and edge-distance requirements must be checked in the final design.)

ANSWER   Use seven bolts. (If the bolts are placed in two rows, use eight bolts for symmetry.)

b. For the slip-critical connection, the slip-critical strength will be given by RCSC Equation LRFD 5.3 multiplied by the reduction factor of AISC Equation A-J3-2:

$$\phi R_{str} = \phi(1.13\mu T_m N_b N_s)\left(1 - \frac{T_u}{1.13T_m N_b}\right) \qquad (7.21)$$

From AISC Table J3.1, for a ¾-inch-diameter A325 bolt, $T_m = 28$ kips. Substituting into Equation 7.21, we get

$$\phi R_{str} = \phi(1.13\mu T_m N_b N_s)\left(1 - \frac{T_u}{1.13T_m N_b}\right)$$

$$= 1.0(1.13)(0.33)(28)(N_b)(1)\left[1 - \frac{150}{1.13(28)N_b}\right]$$

$$= 10.44N_b\left(1 - \frac{4.741}{N_b}\right) = 10.44(N_b - 4.741)$$

Equating this result to the applied shear load, we can solve for the number of bolts required to prevent slip:

$$10.44(N_b - 4.741) = 75 \text{ kips} \quad \text{or} \quad N_b = 11.9 \text{ bolts}$$

Since seven bolts are adequate for shear, bearing, and tension, these limit states will not have to be checked.

ANSWER   Use twelve ¾-inch-diameter A325 bolts.   ■

7.10   ## WELDED CONNECTIONS

Structural welding is a process whereby the parts to be connected are heated and fused, with supplementary molten metal added to the joint. For example, the tension member lap joint shown in Figure 7.33a can be constructed by welding across the ends of both connected parts. A relatively small depth of material will become molten, and upon

■ **FIGURE 7.33**

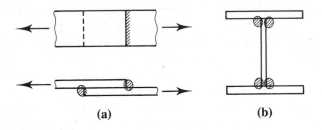

(a)   (b)

cooling, the structural steel and the weld metal will act as one continuous part where they are joined. The additional metal is deposited from a special electrode, which is part of an electrical circuit that includes the connected part, or base metal. In the shielded metal arc welding (SMAW) process, shown schematically in Figure 7.34, current arcs across a gap between the electrode and base metal, heating the connected parts and do positing part of the electrode into the molten base metal. A special coating on the electrode vaporizes and forms a protective gaseous shield, preventing the molten weld metal from oxidizing before it solidifies. The electrode is moved across the joint, and a weld bead is deposited, its size depending on the rate of travel of the electrode. As the weld cools, impurities rise to the surface, forming a coating called *slag* that must be removed before the member is painted or another pass is made with the electrode.

Shielded metal arc welding is normally done manually and is the process universally used for field welds. For shop welding, an automatic or semiautomatic process is usually used. Foremost among these processes is submerged arc welding (SAW). In this process, the end of the electrode and the arc are submerged in a granular flux that melts and forms a gaseous shield. There is more penetration into the base metal than with shielded metal arc welding, and higher strength results. Other commonly used

■ **FIGURE 7.34**

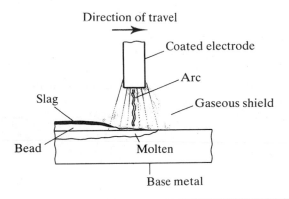

Direction of travel

Coated electrode

Arc

Gaseous shield

Slag

Bead

Molten

Base metal

processes for shop welding include gas shielded metal arc, flux cored arc, and electroslag welding.

Quality control of welded connections is particularly difficult, because defects below the surface, or even minor flaws at the surface, will escape visual detection. Welders must be properly certified, and for critical work, special inspection techniques such as radiography or ultrasonic testing must be used.

The two most common types of welds are the *fillet weld* and the *groove weld.* The lap joint illustrated in Figure 7.33a is made with fillet welds, which are defined as those placed in a corner formed by two parts in contact. Fillet welds can also be used in a tee joint, as shown in Figure 7.33b. Groove welds are those deposited in a gap, or groove, between two parts to be connected. They are most frequently used for butt, tee, and corner joints. In most cases, one or both of the connected parts will have beveled edges, called *prepared edges,* as shown in Figure 7.35a, although relatively thin material can be groove welded with no edge preparation. The welds shown in Figure 7.35a are complete penetration welds and can be made from one side, sometimes with the aid of a backing bar. Partial penetration groove welds can be made from one or both sides, with or without edge preparation (Figure 7.35b).

Figure 7.36 shows the plug or slot weld, which sometimes is used when more weld is needed than length of edge is available. A circular or slotted hole is cut in one of the parts to be connected and is filled with the weld metal.

Of the two major types of welds, fillet welds are the most common and are considered here in some detail. The design of complete penetration groove welds is a trivial exercise in that the weld will have the same strength as the base metal and the connected parts can be treated as completely continuous at the joint. The strength of a partial penetration groove weld will depend on the amount of penetration; once that has been determined, the design procedure will be essentially the same as that for a fillet weld.

■ **FIGURE 7.35**

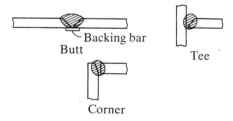

Backing bar
Butt
Tee
Corner

(a) Complete Penetration Groove Welds

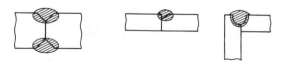

(b) Partial Penetration Groove Welds

■ **FIGURE 7.36**

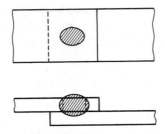

7.11 FILLET WELDS

The design and analysis of fillet welds is based on the assumption that the cross section of the weld is a 45° right triangle, as shown in Figure 7.37. Any reinforcement (buildup outside the hypotenuse of the triangle) or penetration is neglected. The size of a fillet weld is denoted w and is the length of one of the two equal sides of this idealized cross section. Standard weld sizes are specified in increments of $\frac{1}{16}$ inch. Although a length of weld can be loaded in any direction in shear, compression, or tension, a fillet weld is weakest in shear and is always assumed to fail in this mode. Specifically, failure is assumed to occur in shear on a plane through the throat of the weld. For fillet welds made with the shielded metal arc process, the throat is the perpendicular distance from

■ **FIGURE 7.37**

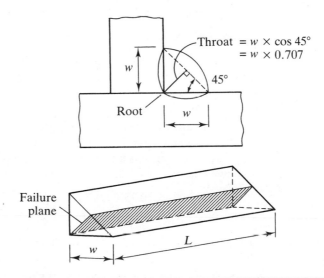

the corner, or root, of the weld to the hypotenuse and is equal to 0.707 times the size of the weld. (The effective throat thickness for a weld made with the submerged arc welding process is larger. In this book, we conservatively assume that the shielded metal arc welding process is used.) Thus, for a given length of weld L subjected to a load of P, the critical shearing stress is

$$f_v = \frac{P}{0.707 \times w \times L}$$

where w is the weld size.

If the weld ultimate shearing stress, F_W, is used in this equation, the nominal load capacity of the weld can be written as

$$R_n = 0.707 \times w \times L \times F_W$$

and the nominal design strength is

$$\phi R_n = 0.707 \times w \times L \times \phi F_W \tag{7.22}$$

The strength of a fillet weld depends on the weld metal used — that is, it is a function of the type of electrode. The strength of the electrode is defined as its ultimate tensile strength, with strengths of 60, 70, 80, 90, 100, 110, and 120 kips per square inch available for the shielded metal arc welding process. The standard notation for specifying an electrode is the letter E followed by two or three digits indicating the tensile strength in kips per square inch and two digits specifying the type of coating. As strength is the property of primary concern to the design engineer, the last two digits are usually represented by XX, and a typical designation would be E70XX or just E70, indicating an electrode with an ultimate tensile strength of 70 ksi. Electrodes should be selected to match the base metal. For the commonly used grades of steel, only two electrodes need be considered:

Use E70XX electrodes with steels that have a yield stress less than 60 ksi.

Use E80XX electrodes with steels that have a yield stress of 60 ksi or 65 ksi.

The notation for designating electrodes and all provisions of the AISC Specification dealing with welds have been taken from the *Structural Welding Code* of the American Welding Society (AWS, 1996). Criteria not covered in the AISC Specification can be found in the AWS Code.

The design strengths of welds are given in AISC Table J2.5. The ultimate shearing stress F_W in a fillet weld is 0.6 times the tensile strength of the weld metal, denoted F_{EXX}. The design stress is therefore ϕF_W, where $\phi = 0.75$ and $F_W = 0.60 F_{EXX}$. For the usual two electrodes, the design strengths (stresses) are as follows:

E70XX: $\phi F_W = 0.75[0.60(70)] = 31.5$ ksi

E80XX: $\phi F_W = 0.75[0.60(80)] = 36$ ksi

An additional requirement is that the factored load shear on the base metal shall not produce a stress in excess of ϕF_{BM}, where F_{BM} is the nominal shear strength of the connected material. Thus the factored load on the connection is subject to a limit of

$$\phi R_n = \phi F_{BM} \times \text{area of base metal subject to shear} \tag{7.23}$$

AISC J5, "Connecting Elements," gives the shear yielding strength as ϕR_n, where

$$\phi = 0.90$$

$$R_n = 0.60 A_g F_y \tag{AISC Equation J5-3}$$

and A_g is the area subjected to shear. The shear strength of the base metal can therefore be taken as

$$\phi F_{BM} = 0.90(0.60)F_y = 0.54 F_y$$

Consequently, when the load is in the same direction as the axis of the weld, the base metal must also be investigated by using the relationship of Equation 7.23. This requirement can be explained by an examination of the welded bracket connection shown in Figure 7.38. Assume that the load is close enough to the welded end that any eccentricity is negligible. If the two welds are the same size, the design strength of *each* weld per inch of length can be found from Equation 7.22 as

$$0.707 \times w \times \phi F_W$$

whereas from Equation 7.23, the capacity of the bracket plate in shear per inch of length is

$$t \times \phi F_{BM}$$

Obviously, more load cannot be resisted by the weld than the base metal (the bracket in this case) is capable of carrying. This investigation must always be made when the base metal is subjected to shear.

In most welded connection problems, whether analysis or design, it is advantageous to work with the strength per unit length of weld, which will be either the strength of the weld itself or the strength of the base metal, whichever is smaller. We illustrate this approach in the following examples.

■ **FIGURE 7.38**

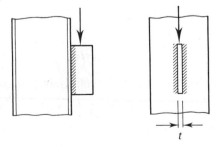

■ **EXAMPLE 7.12**

A flat bar used as a tension member is connected to a gusset plate, as shown in Figure 7.39. The welds are $\frac{3}{16}$-inch fillet welds made with E70XX electrodes. The connected parts are of A36 steel. Assume that the tensile strength of the member is adequate and determine the design strength of the welded connection.

■ **FIGURE 7.39**

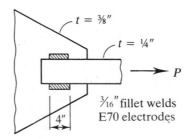

SOLUTION

Because the welds are placed symmetrically about the axis of the member, this connection qualifies as a simple connection, and each part of the weld will resist an equal share of the load. Only the total length of weld matters, and so long as the resultant applied load passes through the center of gravity of the weld (ignoring slight eccentricities), the location and orientation of the individual weld segments are irrelevant.

The strength of the weld metal for E70 electrodes is

$$\phi F_W = 31.5 \text{ ksi}$$

$$0.707 \times w \times \phi F_W = 0.707\left(\frac{3}{16}\right)(31.5) = 4.176 \text{ kips/in.}$$

Check the strength of the base metal (the smallest thickness controls). From Equation 7.23,

$$\phi R_n = \phi F_{BM} \times \text{area subject to shear}$$

$$= \phi F_{BM} \times t = 0.54 F_y t = 0.54(36)\left(\frac{1}{4}\right)$$

$$= 4.86 \text{ kips/in.}$$

The weld strength controls. For the connection,

$$\phi R_n = 4.176 \text{ kips/in.} \times (4 + 4) \text{ in.} = 33.4 \text{ kips}$$

ANSWER

Design strength of the weld is 33.4 kips.

■ EXAMPLE 7.13

A connection of the type used in Example 7.12 must resist a factored load of 40 kips. What total length of $\frac{3}{16}$-inch fillet weld, E70XX electrodes, is required?

SOLUTION

The capacity per inch of weld from Example 7.12 is

$$\phi R_n/\text{in.} = 4.176 \text{ kips/in.}$$

The total length required is

$$\frac{40 \text{ kips}}{4.176 \text{ kips/in.}} = 9.58 \text{ in.}$$

ANSWER

Use 10 inches total, 5 inches on each side.   ■

Practical design of welded connections requires a consideration of such details as maximum and minimum weld sizes and lengths. The requirements for fillet welds are found in AISC J2.2b and are summarized here.

Minimum Size

The minimum size permitted is a function of the thickness of the thicker connected part and is given in AISC Table J2.4.

Maximum Size

Along the edge of a part less than $\frac{1}{4}$-inch thick, the maximum fillet weld size is equal to the thickness of the part. For thicker parts, the maximum size is $t - \frac{1}{16}$ inch, where t is the thickness of the part.

Minimum Length

The minimum permissible length of a fillet weld is four times its size. This limitation is certainly not severe, but if this length is not available, a shorter length can be used if the effective size of the weld is taken as one fourth its length. Connections of the type shown in Figure 7.40, similar to those in the preceding examples, are in the special case category for shear lag in welded connections, covered in Chapter 3. AISC B3 states that the length of the welds in this case may not be less than the distance between them — that is, $L \geq W$.

End Returns

When a weld extends to the corner of a member, it must be continued around the corner, as shown in Figure 7.41. The reason for this continuation, called an *end return,* is primarily to avoid stress concentrations and ensure that the weld size is maintained over

■ **FIGURE 7.40**

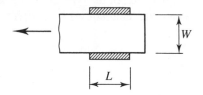

■ **FIGURE 7.41**

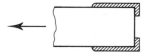

the full length of the weld. The end return should be at least two times the weld size. The length of an end return may be included in the computation of load capacity, or it may be conservatively neglected.

Small welds are generally cheaper than large welds. The maximum size that can be made with a single pass of the electrode is approximately ⁵⁄₁₆ inch, and multiple passes will add to the cost. In addition, for a given load capacity, although a small weld must be made longer, a larger and shorter weld will require more volume of weld metal.

■ **EXAMPLE 7.14**

A bar 4 × ½ of A36 steel is used as a tension member to carry a service dead load of 6 kips and a service live load of 18 kips. It is to be attached to a ⅜-inch gusset plate, as shown in Figure 7.42. Design a welded connection.

SOLUTION The base metal in this connection is A36 steel, so E70XX electrodes will be used. No restrictions on connection length have been stipulated, so weld length will not be limited, and the smallest permissible size will be used:

$$\text{Minimum size} = \frac{3}{16} \text{ in.} \quad \text{(AISC Table J2.4)}$$

Try a ³⁄₁₆-in. fillet weld, E70XX electrodes. The capacity per inch is

$$0.707w(\phi F_W) = 0.707\left(\frac{3}{16}\right)(31.5) = 4.176 \text{ kips/in.}$$

and the capacity of the base metal is

$$0.54F_y t = 0.54(36)\left(\frac{3}{8}\right) = 7.29 \text{ kips/in.}$$

■ **FIGURE 7.42**

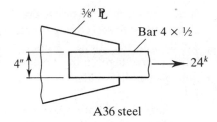

A36 steel

The weld strength of 4.176 kips/in. governs.

The factored load is

$$P_u = 1.2D + 1.6L = 1.2(6) + 1.6(18) = 36 \text{ kips}$$

and

$$\text{Required length} = \frac{36}{4.176} = 8.62 \text{ in.}$$

$$\text{Minimum length} = 4\left(\frac{3}{16}\right) = 0.75 \text{ in.} < 8.62 \text{ in.} \quad (\text{OK})$$

For the end returns,

$$\text{Minimum length} = 2\left(\frac{3}{16}\right) = 0.375 \text{ in.} \quad \text{use 1 in.}$$

For this type of connection, the length of the side welds must at least equal the transverse distance between them, or 4 inches in this case. The total length of weld provided, including end returns, will be

$$2(4 + 1) = 10 \text{ in.} > 8.62 \text{ in. required}$$

ANSWER   Use a ³⁄₁₆-inch fillet weld, E70XX electrodes, with a length of 10 inches, as shown in Figure 7.43.

■ **FIGURE 7.43**

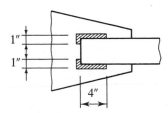

Weld Symbols

Welds are specified on design drawings by standard symbols, which provide a convenient method for describing the required weld configuration. Details are given in Part 8 of the *Manual,* "Bolts, Welds, and Connected Elements" (Volume II), and so are not fully covered here. In this book we provide only a brief introduction to the standard symbols for fillet welds. The following discussion refers to the symbols shown in Figure 7.44.

The basic symbol is a horizontal line (reference line) containing information on the type, size, and length of weld and an inclined arrow pointing to the weld. A right triangle with the vertical leg on the left side is used to indicate a fillet weld. If the symbol for the type of weld is below the reference line, the weld is on the arrow side of the joint, that is, the part of the joint that the arrow is touching. If the symbol is above the line, the weld is on the other side of the joint, which may or may not be hidden from view in the

---

■ **FIGURE 7.44**

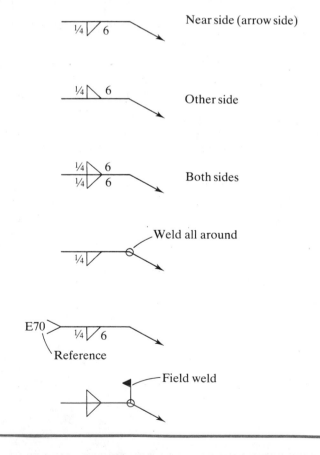

drawing. The numbers on the reference line, from left to right, are the weld size and length. They should always be shown in this order. If both the front and back sides of the joint are to be welded, all information should be shown on each side of the reference line. A circle at the bend in the reference line is an instruction to weld all around the joint. To specify the process used or to furnish other information, a tail can be placed at the end of the reference line and the desired notation placed beside it. If no such reference is to be provided, the tail is omitted. Finally, a flag placed at the bend in the reference line indicates a field weld.

■ **EXAMPLE 7.15**

A plate $\frac{3}{4} \times 8$ of A36 steel is used as a tension member and is to be connected to a $\frac{3}{8}$-inch-thick gusset plate, as shown in Figure 7.45. The length of the connection cannot exceed 8 inches, and all welding must be done on the near side. Design a weld to develop the full tensile capacity of the member.

SOLUTION

The design strength of the member based on its gross area is

$$\phi_t P_n = 0.90 F_y A_g = 0.90(36)\left(\frac{3}{4}\right)(8) = 194.4 \text{ kips}$$

Next compute the design strength of the member based on its net area. For a flat bar connection, if the welds are along the sides only, $A_e = UA_g$. If there is also a transverse weld at the end, then $A_e = A_g$. Assuming the latter, we have

$$\phi_t P_n = 0.75 F_u A_e = 0.75(58)\left(\frac{3}{4}\right)(8) = 261 \text{ kips}$$

Design for a factored load of 194.4 kips and use E70 electrodes:

$$\phi F_W = 31.5 \text{ ksi}$$

■ **FIGURE 7.45**

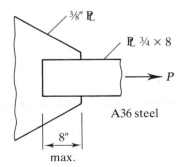

$\frac{3}{8}''$ ℙ

ℙ $\frac{3}{4} \times 8$

P

A36 steel

8"
max.

From AISC Table J2.4, the minimum weld size is ¼ inch. **Try ¼-inch E70 fillet weld:**

$$\text{Design strength per inch of weld} = 0.707\left(\frac{1}{4}\right)(31.5) = 5.568 \text{ kips/in.}$$

$$\text{Strength of base metal} = 0.54F_yt = 0.54(36)\left(\frac{3}{8}\right) = 7.290 \text{ kips/in.}$$

$$\text{Required length of weld} = \frac{194.4}{5.568} = 34.9 \text{ in.}$$

For the given constraints, the available length is 8 + 8 + 8 = 24 inches, so a ¼-inch weld cannot be used. If the length is set at 24 inches, the required weld size from Equation 7.22 is

$$\frac{\phi R_n}{0.707 \times L \times \phi F_W} = \frac{194.4}{0.707(24)(31.5)} = 0.364 \text{ in.}$$

Try ⅜ in.:

$$\text{Maximum weld size} = \frac{3}{4} - \frac{1}{16} = \frac{11}{16} \text{ in.} > \frac{3}{8} \text{ in.} \quad \text{(OK)}$$

$$\text{Minimum length} = 4\left(\frac{3}{8}\right) = 1.5 \text{ in.} < 24 \text{ in.} \quad \text{(OK)}$$

ANSWER   Use the weld shown in Figure 7.46.

■ **FIGURE 7.46**

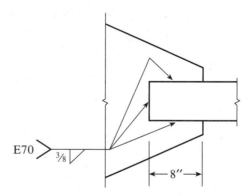

■ PROBLEMS

Bearing Strength, Spacing, and Edge-Distance Requirements

7.3-1   A 3-inch × ⅜-inch tension member is connected with ⅞-inch-diameter bolts to a ⅜-inch-thick gusset plate, as shown in Figure P7.3-1. A572 Grade 50 steel is used for both the tension member and the gusset plate.

    a. Check the spacing and edge distances for compliance with the AISC Specification.

    b. Compute the bearing strength of the connection.

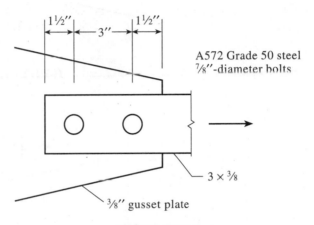

FIGURE P7.3-1

7.3-2   The tension member shown in Figure P7.3-2 is connected with 1-inch-diameter bolts. A36 steel is used for both the tension member and the gusset plate.

    a. Check the spacing and edge distances for compliance with the AISC Specification.

    b. Compute the bearing strength of the connection.

Common Bolts

7.4-1   Determine the design strength for the connection shown in Figure P7.4-1, based on shear and bearing. The bolts are ¾-inch-diameter A307 bolts, and A36 steel is used for both the member and the gusset plate.

7.4-2   Determine the design strength of the connection shown in Figure P7.4-2, based on shear and bearing. The bolts are 1¼-inch-diameter A307 bolts and A572 Grade 50 steel is used for both the member and the gusset plate.

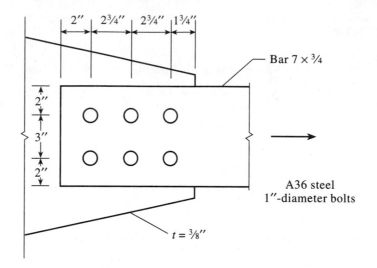

FIGURE P7.3-2

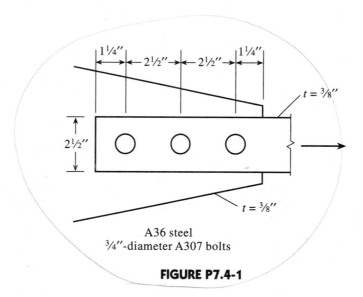

FIGURE P7.4-1

7.4-3 The tension member shown in Figure P7.4-3 was designed to accommodate one line of 1¼-inch-diameter bolts. A572 Grade 50 steel is used. How many A307 bolts are required for a service dead load of 28 kips and a service live load of 84 kips?

7.4-4 **a.** How many 1-inch-diameter A307 bolts are required for the connection shown in Figure P7.4-4? The bolts will be placed in two lines as shown.

b. What is the minimum splice length *L*?

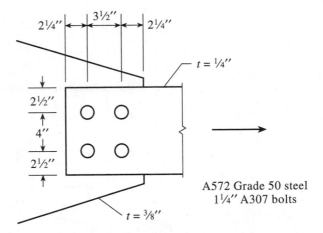

FIGURE P7.4-2

A572 Grade 50 steel
1¼″ A307 bolts

$t = \frac{1}{4}''$

$t = \frac{3}{8}''$

2¼″ 3½″ 2¼″

2½″ 4″ 2½″

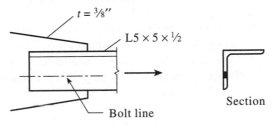

$t = \frac{3}{8}''$

L5 × 5 × ½

Section

Bolt line

FIGURE P7.4-3

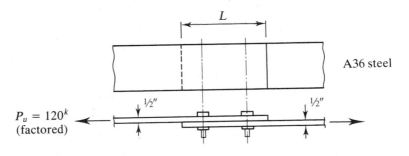

L

A36 steel

½″ ½″

$P_u = 120^k$
(factored)

FIGURE P7.4-4

Shear Strength of High-Strength Bolts

7.6-1 Determine the design strength of the connection shown in Figure P7.6-1, based on bearing and bolt shear. The bolts are ¾-inch-diameter A325 bolts and A572 Grade 50 steel is used for both the member and the gusset plate.

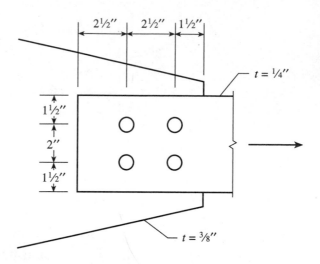

FIGURE P7.6-1

7.6-2 A double-angle tension member is connected to a ⅜-inch-thick gusset plate, as shown in Figure P7.6-2. The angles are $5 \times 3 \times \frac{5}{16}$ with the long legs back-to-back. The tension member is A572 Grade 50 steel and the gusset plate is A36 steel. The bolts are ⅞-inch-diameter A325. Compute the design strength of the connection based on bearing and bolt shear.

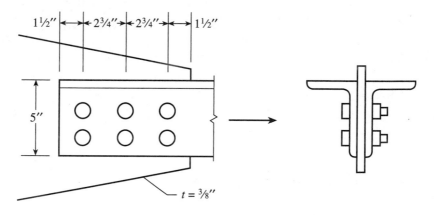

FIGURE P7.6-2

7.6-3 A ⅞-inch-thick tension member is spliced with two ⅜-inch-thick splice plates, as shown in Figure P7.6-3. The connection is made with four ⅝-inch-diameter A325 bolts. All steel is A36. Compute the design strength of the connection based on bearing and bolt shear.

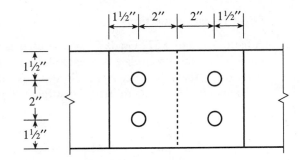

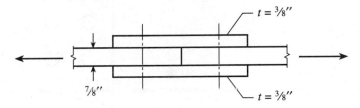

FIGURE P7.6-3

7.6-4   A WT7 × 19 of A572 Grade 50 steel is used as a tension member. It will be connected to a ⅜-inch-thick gusset plate, also of A572 Grade 50 steel, with ⅞-inch-diameter bolts, as shown in Figure P7.6-4. The connection is through the flange of the tee and is a bearing-type connection. The connection must resist a service dead load of 45 kips and a service live load of 90 kips. Assume that the nominal bearing strength will be $2.4dtF_u$, and answer the following questions.

a. How many A307 bolts are required?

b. How many A325 bolts are required?

c. How many A490 bolts are required?

d. If the relative costs of A307, A325, and A490 bolts are in the ratio 1.0:1.6:2.8, which type of bolt will be the most economical for this connection?

FIGURE P7.6-4

Slip-Critical Connections

7.7-1 A bar $6\frac{1}{2} \times \frac{1}{2}$ of A572 Grade 50 steel is used as a tension member, as shown in Figure P7.7-1. The gusset plate is $\frac{5}{8}$-inch-thick and is also of A572 Grade 50 steel. The bolts are $1\frac{1}{8}$-inch-diameter A325. No slip is permitted. Determine the maximum factored load P_u that can be resisted. *Investigate all possible failure modes.*

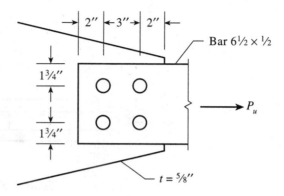

FIGURE P7.7-1

7.7-2 A C9 × 20 is used as a tension member and is connected to a $\frac{1}{2}$-inch gusset plate, as shown in Figure P7.7-2. A242 steel is used for both the tension member and the gusset plate. The member has been designed to resist a factored load of 175 kips.

a. If the connection is to be slip-critical, how many $1\frac{3}{8}$-inch-diameter A325 bolts are needed?

b. Show a sketch with a possible layout. Assume that the member tensile and block shear strengths will be adequate.

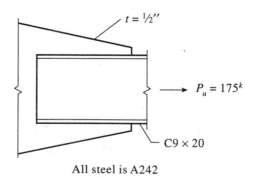

All steel is A242

FIGURE P7.7-2

7.7-3   A bar 6 × ½ tension member is spliced with two plates, as shown in Figure P7.7-3. Both the bar and the tension member are of A572 Grade 50 steel.

    a. Determine the number of 1-inch-diameter A325 slip-critical bolts required to resist the service loads shown.

    b. Determine the dimensions of the splice plates and show the arrangement of the bolts in a sketch.

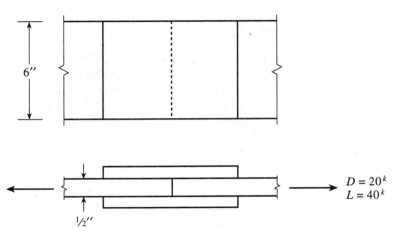

6″

½″

$D = 20^k$
$L = 40^k$

FIGURE P7.7-3

7.7-4   **a.** Prepare a table showing values of A325 bolt shear strength (both single and double-shear) and slip-critical strength for bolt diameters of ½ inch to 1½ inches in increments of ⅛ inch. Assume that threads are in the plane(s) of shear. Your table should be like Table P7.7-4.

TABLE P7.7-4

Bolt Diameter (in.)	Single-shear Design Strength (kips)	Double-shear Design Strength (kips)	Slip-critical Strength (kips)
½	7.068	14.14	4.476
.	.	.	.
.	.	.	.
.	.	.	.

    b. What conclusion can you draw from your table?

7.7-5   Design a 9-foot-long single-angle tension member and its connection for the following loads: service dead load = 20 kips, service live load = 60 kips, and service wind load = 20 kips. The member will be connected to a ⅜-inch-thick gusset plate. Use A572

Grade 50 steel for the angle and A36 steel for the gusset plate. Use A325 slip-critical bolts. Provide a sketch of the connection.

7.7-6 Design a 30-foot-long double-angle tension member and its connection for a service dead load of 34 kips and a service live load of 102 kips. The member will be connected to a $\frac{3}{8}$-inch-thick gusset plate. Use A325 bearing-type bolts and A572 Grade 50 steel for both the member and the gusset plate. Provide a sketch of the connection.

High-Strength Bolts in Tension

7.8-1 A WT9 × 38 of A572 Grade 50 steel is used in a hanger connection as shown in Figure P7.8-1. The tee section is 9 inches long and is connected with six $\frac{3}{4}$-inch-diameter A325 bolts. The total applied factored load is 90 kips. Investigate the adequacy of the bolts and the thickness of the tee. Include the effects of prying.

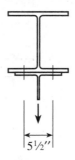

$5\frac{1}{2}''$

FIGURE P7.8-1

7.8-2 The hanger connection shown in Figure P7.8-2 consists of a pair of angles, $2\frac{1}{2} \times 2\frac{1}{2} \times \frac{1}{2}$, and a $\frac{3}{8}$-inch-thick plate. A36 steel is used. Determine the number of $\frac{7}{8}$-inch-diameter A325 bolts needed to resist a factored tensile load of 150 kips. The bolts in the vertical legs of the angles are bearing-type. The bolts will be placed at the usual

150^k

FIGURE P7.8-2

gage distances for angles (see Section 3.6). Along the length of the angles, use bolt edge distances of 1½ inches and spacings of 3 inches. Include the effects of prying.

Combined Shear and Tension in Fasteners

7.9-1   A single-angle tension member is connected to a W10 × 68 column flange by means of a bracket cut from a W10 × 68, as shown in Figure P7.9-1. The tension member is subjected to a factored load of 115 kips. A572 Grade 50 steel is used throughout, and the bolts in the column flange are ¾-inch-diameter A325 bearing-type. Determine whether these bolts are adequate.

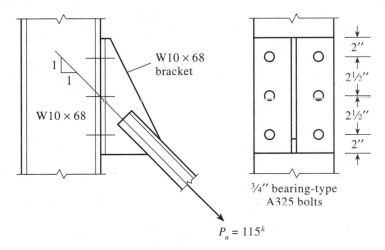

W10 × 68 bracket

W10 × 68

¾″ bearing-type A325 bolts

2″
2½″
2½″
2″

$P_u = 115^k$

FIGURE P7.9-1

7.9-2   A tension member is connected to a column by means of a bracket cut from a W-shape as shown in Figure P7.9-2. Eight ⅞-inch-diameter A325 slip-critical bolts are used. What is the maximum factored load P_u that can be applied? Assume that the bearing strength is adequate.

7.9-3   A connection similar to the one in Problem 7.9-2 is subjected to a factored load of 150 kips. The line of action of the load makes an angle of 30° with the horizontal. Assume that the bearing strength is adequate and determine the number of 1-inch-diameter A325 bearing-type bolts required. Use an even number for symmetry.

7.9-4   A connection similar to the one in Problem 7.9-2 is subjected to a factored shear load of 26 kips and a factored tension load of 72 kips. Assume that the bearing strength is adequate and determine the number of ¾-inch-diameter A325 slip-critical bolts required. Use an even number for symmetry.

7.9-5   A double-angle tension member is attached to a ⅞-inch gusset plate, which in turn is connected to a column flange via another pair of angles, as shown in Figure P7.9-5. The

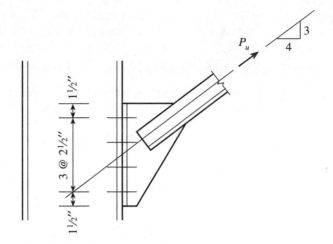

FIGURE P7.9-2

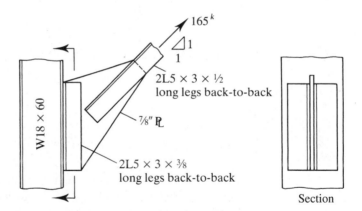

FIGURE P7.9-5

load is a service load consisting of 50% dead load and 50% live load. All connections are to be made with ¾-inch-diameter A490 slip-critical bolts. Assume that the threads are in shear. Determine the required number of bolts and show their location on a sketch. A572 Grade 50 steel is used throughout.

Fillet Welds

7.11-1 Determine the maximum factored load P_u that can be applied to the joint shown in Figure P7.11-1. A36 steel is used for both the tension member and the gusset plate. The weld is a ¼-inch fillet weld, E70 electrode.

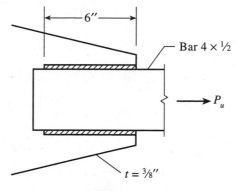

FIGURE P7.11-1

7.11-2 Determine the maximum factored load P_u that can be applied to the joint shown in Figure P7.11-2. Each component is a bar $7 \times \frac{3}{4}$ of A36 steel. The weld is a $\frac{1}{2}$-inch fillet weld, E70 electrode.

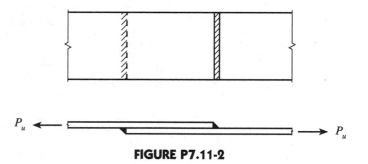

FIGURE P7.11-2

7.11-3 A single-angle tension member, an $L2 \times 2 \times \frac{5}{16}$, is connected to a $WT6 \times 8$ with $\frac{3}{16}$-inch fillet welds, as shown in Figure 7.11-3. Determine the maximum factored load P_u that can be applied. A572 Grade 50 steel and E70 electrodes are used. Each weld segment is 4 inches long.

7.11-4 A tension member splice is made with $\frac{1}{4}$-inch E70 fillet welds, as shown in Figure P7.11-4. Each side of the splice is welded as shown. The inner member is a bar $6 \times \frac{1}{2}$ and each outer member is a bar $3 \times \frac{5}{16}$. All steel is A572 Grade 50. Determine the maximum factored load P_u that can be applied.

7.11-5 Use the minimum weld size and determine the dimension L shown in Figure P7.11-5 to the nearest $\frac{1}{2}$ inch. A572 Grade 50 steel and E70 electrodes are used. Assume that the member tensile strength is adequate.

7.11-6 Use the minimum fillet weld size and determine the dimension L shown in Figure P7.11-6 to the nearest $\frac{1}{2}$ inch. The steel is A572 Grade 50 and the electrodes are E70.

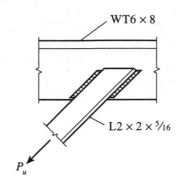

FIGURE P7.11-3

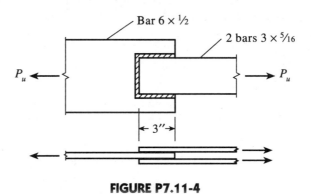

FIGURE P7.11-4

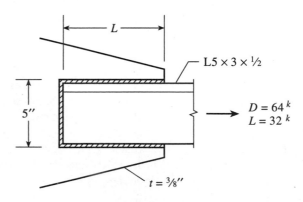

FIGURE P7.11-5

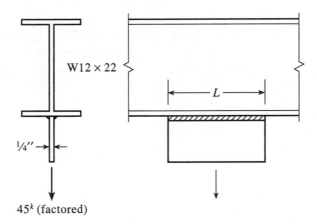

FIGURE P7.11-6

7.11-7 Design a welded connection for the tension member in Problem 7.7-2. Provide a sketch of your design.

7.11-8 Design a welded connection for the 2L5 × 3 × ½ tension member in Problem 7.9-5. Provide a sketch of your design.

7.11-9 Design a welded connection to resist 50% of the design strength of the tension member shown in Figure P7.11-9.

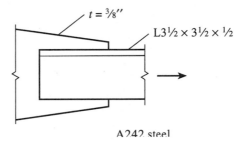

FIGURE P7.11-9

7.11-10 Design a tension member and a welded connection to resist a service dead load of 40 kips and a service live load of 120 kips. The member should have a rectangular cross section (bar or plate). The member is 5 feet long. Use A572 Grade 60 steel for the tension member. The gusset plate is ⅜-inch-thick A36 steel.

7.11-11 Design a WT shape of A36 steel with a nominal depth of 8 inches to resist a factored tensile load of 230 kips. The member will be 30 feet long. The flange will be welded to a ⅜-inch-thick gusset plate. Design the connection and show the details on a sketch.

7.11-12   Use A36 steel and select a Miscellaneous Channel Shape for use as an 18-foot-long tension member, and design a welded connection to a $\frac{3}{8}$-inch-thick gusset plate. The service dead load is 20 kips, the service live load is 42 kips, and the service wind load is 42 kips. Show the connection details on a sketch.

8

Eccentric Connections

8.1 EXAMPLES OF ECCENTRIC CONNECTIONS

An eccentric connection is one in which the resultant of the applied loads does not pass through the center of gravity of the fasteners or welds. If the connection has a plane of symmetry, the centroid of the shear area of the fasteners or welds may be used as the reference point, and the perpendicular distance from the line of action of the load to the centroid is called the *eccentricity*. Although a majority of connections are probably loaded eccentrically, in many cases the eccentricity is small and may be neglected.

The *framed beam* connection shown in Figure 8.1a is a typical eccentric connection. This connection, in either bolted or welded form, is commonly used to connect beams to columns. Although the eccentricities in this type of connection are usually negligible, they do exist and are used here for illustration. Two different connections are actually

FIGURE 8.1

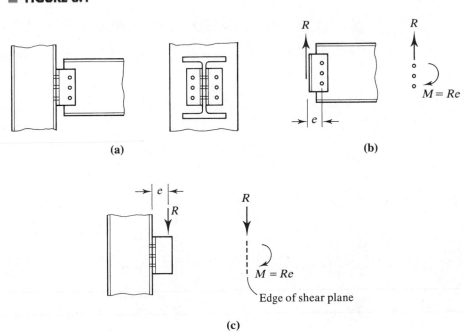

(a)

(b)

(c)

involved: the attachment of the beam to the framing angles and the attachment of the angles to the column. These connections illustrate the two basic categories of eccentric connections: those causing only shear in the fasteners or welds and those causing both shear and tension.

If the beam and angles are considered separately from the column, as shown in Figure 8.1b, it is clear that the reaction R acts at an eccentricity e from the centroid of the areas of the fasteners in the beam web. These fasteners are thus subjected to both a shearing force and a couple that lies in the plane of the connection and causes torsional shearing stress.

If the column and the angles are isolated from the beam, as shown in Figure 8.1c, it is clear that the fasteners in the column flange are subjected to the reaction R acting at an eccentricity e from the plane of the fasteners, producing the same couple as before. In this case, however, the load is not in the plane of the fasteners, so the couple will tend to put the upper part of the connection in tension and compress the lower part. The fasteners at the top of the connection will therefore be subjected to both shear and tension.

Although we used a bolted connection here for illustration, welded connections can be similarly categorized as either shear only or shear plus tension.

Maximum factored load reactions for various framed beam connections are given in Tables 9-2 through 9-12 in Part 9 of the *Manual,* "Simple Shear and PR Moment Connections" (Volume II). The relatively small eccentricity in these connections is neglected, and only the direct shear is considered.

8.2 ECCENTRIC BOLTED CONNECTIONS: SHEAR ONLY

The column bracket connection shown in Figure 8.2 is an example of a bolted connection subjected to eccentric shear. Two approaches exist for the solution of this problem: the traditional elastic analysis and the more accurate (but more complex) ultimate strength analysis. Both will be illustrated.

■ **FIGURE 8.2**

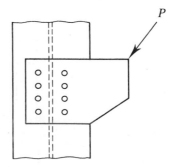

Elastic Analysis

In Figure 8.3a, the fastener shear areas and the load are shown separate from the column and bracket plate. The eccentric load P can be replaced with the same load acting at the centroid plus the couple, $M = Pe$, where e is the eccentricity. If this replacement is made, the load will be concentric, and each fastener can be assumed to resist an equal share of the load, given by $p_c = P/n$, where n is the number of fasteners. The fastener forces resulting from the couple can be found by considering the shearing stress in the fasteners to be the result of torsion of a cross section made up of the cross-sectional areas of the fasteners. If such an assumption is made, the shearing stress in each fastener can be found from the torsion formula

$$f_v = \frac{Md}{J} \tag{8.1}$$

where

d = distance from the centroid of the area to the point where the stress is being computed

J = polar moment of inertia of the area about the centroid

and the stress f_v is perpendicular to d. Although the torsion formula is applicable only to right circular cylinders, its use here is conservative, yielding stresses that are somewhat larger than the actual stresses.

■ **FIGURE 8.3**

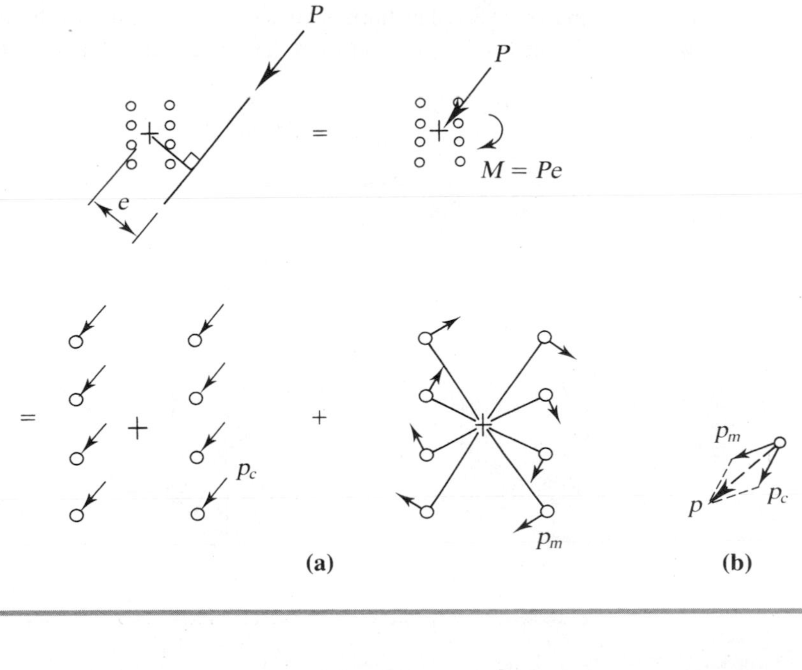

(a) (b)

If the parallel-axis theorem is used and the polar moment of inertia of each circular area about its own centroid is neglected, J for the total area can be approximated as

$$J = \Sigma Ad^2 = A \Sigma d^2$$

provided all fasteners have the same area, A. Equation 8.1 can then be written as

$$f_v = \frac{Md}{A \Sigma d^2}$$

and the shear force in each fastener caused by the couple is

$$p_m = Af_v = A \frac{Md}{A \Sigma d^2} = \frac{Md}{\Sigma d^2}$$

The two components of shear force thus determined can be added vectorially to obtain the resultant force, p, as shown in Figure 8.3b, where the lower right-hand fastener is used as an example. When the largest resultant is determined, the fastener size is selected so as to resist this force. The critical fastener cannot always be found by inspection, and several force calculations may be necessary.

It is generally more convenient to work with rectangular components of forces. For each fastener, the horizontal and vertical components of force resulting from direct shear are

$$p_{cx} = \frac{P_x}{n} \quad \text{and} \quad p_{cy} = \frac{P_y}{n}$$

where P_x and P_y are the x- and y-components of the total connection load, P, as shown in Figure 8.4. The horizontal and vertical components caused by the eccentricity can be found as follows. In terms of the x- and y-coordinates of the centers of the fastener areas,

$$\Sigma d^2 = \Sigma(x^2 + y^2)$$

■ **FIGURE 8.4**

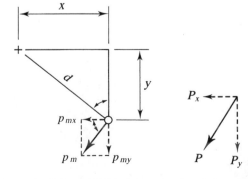

where the origin of the coordinate system is at the centroid of the total fastener shear area. The x-component of p_m is

$$p_{mx} = \frac{y}{d} \, p_m = \frac{y}{d} \frac{Md}{\Sigma \, d^2} = \frac{y}{d} \frac{Md}{\Sigma(x^2 + y^2)} = \frac{My}{\Sigma(x^2 + y^2)}$$

Similarly,

$$p_{my} = \frac{Mx}{\Sigma(x^2 + y^2)}$$

and the total fastener force is

$$p = \sqrt{(\Sigma \, p_x)^2 + (\Sigma \, p_y)^2}$$

where

$$\Sigma \, p_x = p_{cx} + p_{mx}$$
$$\Sigma \, p_y = p_{cy} + p_{my}$$

If P, the load applied to the connection, is a factored load, then force p on the fastener is the factored load to be resisted in shear and bearing — that is, the required design strength.

■ EXAMPLE 8.1

Determine the critical fastener force in the bracket connection shown in Figure 8.5.

■ FIGURE 8.5

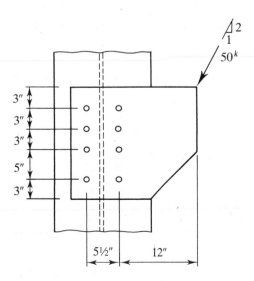

SOLUTION   The centroid of the fastener group can be found by using a horizontal axis through the lower row and applying the principal of moments:

$$\bar{y} = \frac{2(5) + 2(8) + 2(11)}{8} = 6 \text{ in.}$$

The horizontal and vertical components of the load are

$$P_x = \frac{1}{\sqrt{5}}(50) = 22.36 \text{ kips} \leftarrow \text{ and } p_y = \frac{2}{\sqrt{5}}(50) = 44.72 \text{ kips} \downarrow$$

Referring to Figure 8.6a, we can compute the moment of the load about the centroid:

$$M = 44.72(12 + 2.75) - 22.36(14 - 6) = 480.7 \text{ in.-kips (clockwise)}$$

■ **FIGURE 8.6**

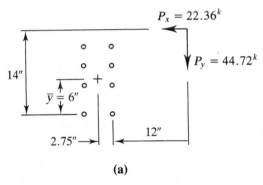

(a)

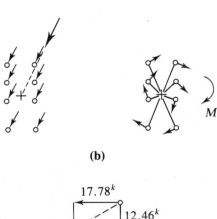

(b)

(c)

Figure 8.6b shows the directions of all component bolt forces and the relative magnitudes of the components caused by the couple. Using these directions and relative magnitudes as a guide and bearing in mind that forces add by the parallelogram law, we can conclude that the lower right-hand fastener will have the largest resultant force.

The horizontal and vertical components of force in each bolt resulting from the concentric load are

$$p_{cx} = \frac{22.36}{8} = 2.795 \text{ kips} \leftarrow \text{ and } p_{cy} = \frac{44.72}{8} = 5.590 \text{ kips} \downarrow$$

For the couple,

$$\Sigma(x^2 + y^2) = 8(2.75)^2 + 2[(6)^2 + (1)^2 + (2)^2 + (5)^2] = 192.5 \text{ in.}^2$$

$$p_{mx} = \frac{My}{\Sigma(x^2 + y^2)} = \frac{480.7(6)}{192.5} = 14.98 \text{ kips} \leftarrow$$

$$p_{my} = \frac{Mx}{\Sigma(x^2 + y^2)} = \frac{480.7(2.75)}{192.5} = 6.867 \text{ kips} \downarrow$$

$$\Sigma p_x = 2.795 + 14.98 = 17.78 \text{ kips} \leftarrow$$

$$\Sigma p_y = 5.590 + 6.867 = 12.46 \text{ kips} \downarrow$$

$$p = \sqrt{(17.78)^2 + (12.46)^2} = 21.7 \text{ kips} \quad \text{(see Figure 8.6c)}$$

ANSWER The critical fastener force is 21.7 kips. Inspection of the magnitudes and directions of the horizontal and vertical components of the forces confirms the earlier conclusion that the fastener selected is indeed the critical one. ■

Ultimate Strength Analysis

The foregoing procedure is relatively easy to apply but is inaccurate — on the conservative side. The major flaw in the analysis is the implied assumption that the fastener load–deformation relationship is linear and that the yield stress is not exceeded. Experimental evidence shows that this is not the case and that individual fasteners do not have a well-defined shear yield stress. The procedure to be described here determines the ultimate strength of the connection by using an experimentally determined nonlinear load–deformation relationship for the individual fasteners.

The experimental study reported in Crawford and Kulak (1971) used ¾-inch-diameter A325 bearing-type bolts and A36 steel plates, but the results can be used with little error for A325 bolts of different sizes and steels of other grades. The procedure gives conservative results when used with slip-critical bolts and with A490 bolts (AISC, 1994).

The bolt force R corresponding to a deformation Δ is

$$R = R_{ult}(1 - e^{-\mu\Delta})^\lambda$$
$$= 74(1 - e^{-10\Delta})^{0.55} \tag{8.2}$$

where

$$R_{ult} = \text{bolt shear force at failure} = 74 \text{ kips}$$
$$e = \text{base of natural logarithms}$$
$$\mu = \text{a regression coefficient} = 10$$
$$\lambda = \text{a regression coefficient} = 0.55$$

The ultimate strength of the connection is based on the following assumptions.

1. At failure, the fastener group rotates about an instantaneous center (IC).
2. The deformation of each fastener is proportional to its distance from the IC and acts perpendicularly to the radius of rotation.
3. The capacity of the connection is reached when the ultimate strength of the fastener farthest from the IC is reached. (Figure 8.7 shows the bolt forces as resisting forces acting to oppose the applied load.)
4. The connected parts remain rigid.

As a consequence of the second assumption, the deformation of an individual fastener is

$$\Delta = \frac{r}{r_{max}} \Delta_{max} = \frac{r}{r_{max}} (0.34) \tag{8.34}$$

where

$$r = \text{distance from the IC to the fastener}$$
$$r_{max} = \text{distance to the farthest fastener}$$
$$\Delta_{max} = \text{deformation of the farthest fastener at ultimate} = 0.34 \text{ in. (determined experimentally)}$$

■ **FIGURE 8.7**

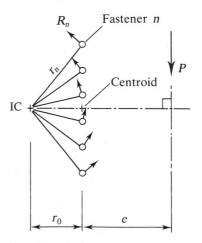

As with the elastic analysis, it is more convenient to work with rectangular components of forces, or

$$R_y = \frac{x}{r} R \text{ and } R_x = \frac{y}{r} R$$

where x and y are the horizontal and vertical distances from the instantaneous center to the fastener. At the instant of failure, equilibrium must be maintained, and the following three equations of equilibrium will be applied to the fastener group (refer to Figure 8.7):

$$\Sigma F_x = \sum_{n=1}^{m} (R_x)_n - P_x = 0 \tag{8.3}$$

$$M_{IC} = P(r_0 + e) - \sum_{n=1}^{m} (r_n \times R_n) = 0 \tag{8.4}$$

and

$$\Sigma F_y = \sum_{n=1}^{m} (R_y)_n - P_y = 0 \tag{8.5}$$

where the subscript n identifies an individual fastener and m is the total number of fasteners. The general procedure is to assume the location of the instantaneous center, then determine if the corresponding value of P satisfies the equilibrium equations. If so, this location is correct, and P is the capacity of the connection. The specific procedure is as follows.

1. Assume a value for r_0.
2. Solve for P from Equation 8.4.
3. Substitute r_0 and P into Equations 8.3 and 8.5.
4. If these equations are satisfied within an acceptable tolerance, the analysis is complete. Otherwise, a new trial value of r_0 must be selected and the process repeated.

For the usual case of vertical loading, Equation 8.3 will automatically be satisfied. For simplicity and without loss of generality, we consider only this case. Even with this assumption, however, the computations for even the most trivial problems are overwhelming, and computer assistance is needed. Part (b) of Example 8.2 was worked with the aid of a standard spreadsheet program for personal computers.

■ EXAMPLE 8.2

The bracket connection shown in Figure 8.8 must support an eccentric factored load of 53 kips. The connection was designed to have two vertical rows of four bolts, but one bolt was inadvertently omitted. If ⅞-inch-diameter A325 bearing-type bolts are used, is the connection adequate? Assume that the bolt threads are in the plane of shear. Use A36 steel and perform the following analyses: (a) elastic analysis; (b) ultimate strength analysis.

■ **FIGURE 8.8**

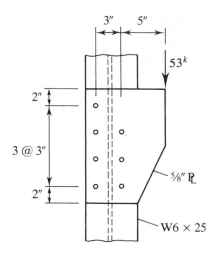

SOLUTION

a. Elastic analysis. For an xy-coordinate system with the origin at the center of the lower left bolt (Figure 8.9),

$$\bar{y} = \frac{2(3) + 2(6) + 1(9)}{7} = 3.857 \text{ in.}$$

$$\bar{x} = \frac{3(3)}{7} = 1.286 \text{ in.}$$

$$\Sigma(x^2 + y^2) = 4(1.286)^2 + 3(1.714)^2 + 2(3.857)^2 + 2(0.857)^2$$
$$+ 2(2.143)^2 + 1(5.143)^2 = 82.29 \text{ in.}^2$$

$$e = 3 + 5 - 1.286 = 6.714 \text{ in.}$$

$$M = Pe = 53(6.714) = 355.8 \text{ in.-kips} \qquad \text{(clockwise)}$$

$$p_{cy} = \frac{53}{7} = 7.571 \text{ kips } \downarrow \qquad p_{cx} = 0$$

From the directions and relative magnitudes shown in Figure 8.9, the lower right bolt is judged to be critical, so

$$p_{mx} = \frac{My}{\Sigma(x^2 + y^2)} = \frac{355.8(3.857)}{82.29} = 16.68 \text{ kips } \leftarrow$$

$$p_{my} = \frac{Mx}{\Sigma(x^2 + y^2)} = \frac{355.8(1.714)}{82.29} = 7.411 \text{ kips } \downarrow$$

$$\Sigma\,p_x = 16.68 \text{ kips}$$

$$\Sigma\,p_y = 7.571 + 7.411 = 14.98 \text{ kips}$$

$$p = \sqrt{(16.68)^2 + (14.98)^2} = 22.4 \text{ kips}$$

■ **FIGURE 8.9**

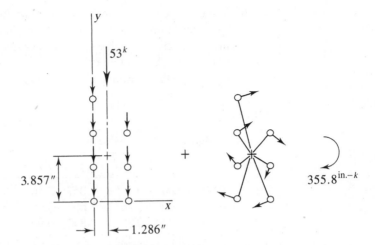

To determine the bolt design strength in bearing, use a hole diameter of

$$h = d + \frac{1}{16} = \frac{7}{8} + \frac{1}{16} = \frac{15}{16} \text{ in.}$$

For the holes nearest the edge, use $L_e = 2$ in. Then,

$$L_c = L_e - \frac{h}{2} = 2 - \frac{15/16}{2} = 1.531 \text{ in.}$$

$$2d = 2\left(\frac{7}{8}\right) = 1.75 \text{ in.}$$

Because $L_c < 2d$, the bearing strength is

$$\phi R_n = \phi(1.2L_c t F_u) = 0.75(1.2)(1.531)(0.455)(58) = 36.4 \text{ kips/bolt}$$

For the other holes, use $s = 3$ in. Then,

$$L_c = s - h = 3 - \frac{15}{16} = 2.062 \text{ in.} > 2d$$

$$\therefore \ \phi R_n = \phi(2.4 d t F_u) = 0.75(2.4)\left(\frac{7}{8}\right)(0.455)(58) = 41.56 \text{ kips/bolt}$$

Both bearing values are larger than the applied force per bolt, so the bearing strength is adequate.

For shear,

$$A_b = \frac{\pi(7/8)^2}{4} = 0.6013 \text{ in.}^2$$

$$\phi R_n = \phi F_v A_b = 0.75(48)(0.6013) = 21.6 \text{ kips} < 22.4 \text{ kips} \quad \text{(N.G.)}$$

ANSWER The connection is unsatisfactory by elastic analysis.

> **b.** The ultimate strength analysis will be performed with the aid of standard spreadsheet software. The results of the final trial value of $r_o = 1.57104$ inches are given in Table 8.1. The coordinate system and bolt numbering scheme are shown in Figure 8.10. (Values shown in the table have been rounded to three decimal places for presentation purposes.)

TABLE 8.1

| Fastener | Origin at Bolt 1 | | Origin at IC | | | | | | |
	x'	y'	x	y	r	Δ	R	rR	Ry
1	0.000	0.000	0.285	−3.857	3.868	0.255	70.774	273.731	5.221
2	3.000	0.000	3.285	−3.857	5.067	0.334	72.553	367.598	47.045
3	0.000	3.000	0.285	−0.857	0.903	0.060	47.649	43.046	15.050
4	3.000	3.000	3.285	−0.857	3.395	0.224	69.563	236.188	67.310
5	0.000	6.000	0.285	2.143	2.162	0.143	63.631	137.555	8.398
6	3.000	6.000	3.285	2.143	3.922	0.259	70.891	278.061	59.377
7	0.000	9.000	0.285	5.143	5.151	0.340	72.631	374.107	4.023
Sum								1710.287	206.424

From Equation 8.4,

$$P(r_0 + e) = \Sigma\, rR$$

$$P = \frac{\Sigma\, rR}{r_0 + e} = \frac{1710.29}{1.57104 + 6.71429} = 206.424 \text{ kips}$$

■ **FIGURE 8.10**

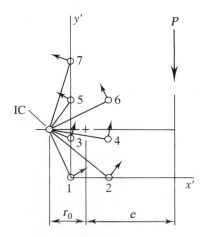

where e has been expressed to six significant figures for consistency.
From Equation 8.5,

$$\Sigma\, F_y \;=\; \Sigma\, R_y \;-\; P \;=\; 206.424 \;-\; 206.424 \;=\; 0.000$$

The applied load has no horizontal component, so Equation 8.3 is automatically satisfied.

The load of 206.424 kips just computed is the failure load for the connection and is based on the critical fastener reaching its ultimate load capacity. If the connection failure load is multiplied by the ratio of fastener design strength to the fastener ultimate strength of 74 kips (Crawford and Kulak, 1971), we obtain the connection capacity.

From part (a), the design strength of one bolt (based on shear) is 21.6 kips.

$$\text{Maximum factored load} \;=\; 206(21.6/74) \;=\; 60.1 \text{ kips} > 53 \text{ kips} \quad \text{(OK)}$$

ANSWER   The connection is satisfactory by ultimate strength analysis. ■

Tables 8-18 through 8-25 in Part 8 of the *Manual* (Volume II) give coefficients for the design or analysis of common configurations of fastener groups subjected to eccentric loads. For each arrangement of fasteners considered, the tables give a value for C, the ratio of connection failure load to fastener ultimate strength. To obtain a safe connection load, this constant must be multiplied by the design strength of the particular fastener used. For eccentric connections not included in the tables, the elastic method, which is conservative, may be used. Of course, a computer program or spreadsheet software may also be used to perform an ultimate strength analysis.

■ EXAMPLE 8.3

Use the tables in Part 8 of the *Manual* to determine the factored load capacity P_u, based on bolt shear, for the connection shown in Figure 8.11. The bolts are ¾-inch A325 bearing-type with the threads in the plane of shear. The bolts are in single shear.

SOLUTION   This connection corresponds to the connections in Table 8-19, for Angle $= 0°$. The eccentricity is

$$e_x \;=\; 8 \;+\; 1.5 \;=\; 9.5 \text{ in.}$$

The number of bolts per vertical row is

$$n \;=\; 3$$

From Table 8-19,

$$C \;=\; 1.53 \text{ by interpolation}$$

■ **FIGURE 8.11**

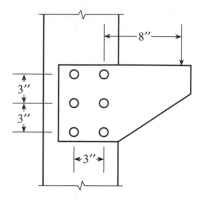

The design strength of a $\frac{3}{4}$-inch-diameter bolt in single shear is

$$\phi r_n = \phi(48)A_b = 0.75(48)(0.4418) = 15.90 \text{ kips}$$

Since $C = P_u/\phi r_n$,

$$P_u = C\phi r_n = 1.53(15.90) = 24.3 \text{ kips}$$

ANSWER   The maximum factored load capacity of the connection is 24.3 kips.   ■

8.3   ECCENTRIC BOLTED CONNECTIONS: SHEAR PLUS TENSION

In a connection such as the one for the tee stub bracket of Figure 8.12, an eccentric load creates a couple that will increase the tension in the upper row of fasteners and decrease it in the lower row. If the fasteners are bolts with no initial tension, the upper bolts will be put into tension, and the lower ones will not be affected. Regardless of the type of fastener, each one will receive an equal share of the *shear* load.

■ **FIGURE 8.12**

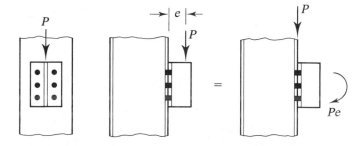

If the fasteners are pretensioned high-strength bolts, the contact surface between the column flange and the bracket flange will be uniformly compressed before the external load is applied. The bearing pressure will equal the total bolt tension divided by the area of contact. As the load P is gradually applied, the compression at the top will be relieved and the compression at the bottom will increase, as shown in Figure 8.13a. When the compression at the top has been completely overcome, the components will separate, and the couple Pe will be resisted by tensile bolt forces and compression on the remaining surface of contact, as shown in Figure 8.13b. As the ultimate load is approached, the forces in the bolts will approach their ultimate tensile strengths.

A conservative, simplified method will be used here. The neutral axis of the connection is assumed to pass through the centroid of the bolt areas. Bolts above this axis are subjected to tension, and bolts below the axis are assumed to be subjected to compressive forces, as shown in Figure 8.13c. Each bolt is assumed to have reached an ultimate value of r_{ut}. Since there are two bolts at each level (Figure 8.13c), each force is shown as $2r_{ut}$. The resultant of the tensile and compressive forces is a couple that equals the resisting moment of the connection. The moment of this couple can be found by summing moments of the bolt forces about any convenient axis, such as the neutral axis. When the resisting moment is equated to the applied moment, the resulting equation can be solved for the unknown bolt tensile force r_{ut}. (This method is the same as Case II in Part 8 of the *Manual,* Volume II.)

■ **FIGURE 8.13**

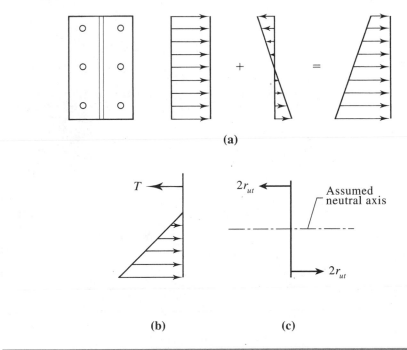

(a)

(b)   (c)

■ EXAMPLE 8.4

A beam-to-column connection is made with a structural tee as shown in Figure 8.14. Eight ¾-inch-diameter A325 fully tightened bearing-type bolts are used to attach the flange of the tee to the column flange. Investigate the adequacy of this connection (the tee-to-column connection) if it is subjected to a factored load of 88 kips at an eccentricity of 3 inches. Assume that the bolt threads are in the plane of shear. All structural steel is A36.

SOLUTION   The shear/bearing load per bolt is $^{88}\!/_8 = 11$ kips. For the design strength in bearing, use a hole diameter of

$$h = d + \frac{1}{16} = \frac{3}{4} + \frac{1}{16} = \frac{13}{16} \text{ in.}$$

For the hole nearest the edge, use $L_e = 1.5$ in. Then,

$$L_c = L_e - \frac{h}{2} = 1.5 - \frac{13/16}{2} = 1.094 \text{ in.}$$

$$2d = 2\left(\frac{3}{4}\right) = 1.5 \text{ in.}$$

Because $L_c < 2d$,

$$\phi R_n = \phi(1.2 L_c t F_u) = 0.75(1.2)(1.094)(0.560)(58) = 31.98 \text{ kips} > 11 \text{ kips} \quad \text{(OK)}$$

■ FIGURE 8.14

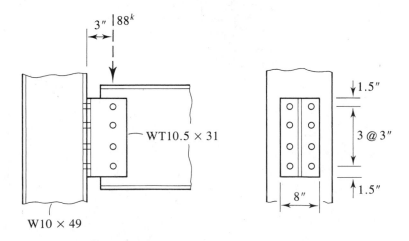

For the other holes, use $s = 3$ in. Then,

$$L_c = s - h = 3 - \frac{13}{16} = 2.188 \text{ in.} > 2d$$

$$\therefore \phi R_n = \phi(2.4dtF_u) = 0.75(2.4)\left(\frac{3}{4}\right)(0.560)(58) = 43.85 \text{ kips} > 11 \text{ kips (OK)}$$

The connection is satisfactory in bearing.

For the design strength in shear,

$$A_b = \frac{\pi(3/4)^2}{4} = 0.4418 \text{ in.}^2$$

$$\phi R_n = \phi F_v A_b = 0.75(48)(0.4418) = 15.90 \text{ kips}$$

Compute the tensile force per bolt and then check the tension–shear interaction. Because of symmetry, the centroid of the connection is at middepth. Figure 8.15 shows the bolt areas and the distribution of bolt tensile forces.

The moment of the resisting couple is found by summing moments about the neutral axis:

$$\Sigma M_{NA} = 2r_{ut}(4.5 + 1.5 + 1.5 + 4.5) = 24r_{ut}$$

The applied moment is

$$M_u = P_u e = 88(3) = 264 \text{ in.-kips}$$

Equating the resisting and applied moments, we get

$$24r_{ut} = 264, \quad \text{or} \quad r_{ut} = 11 \text{ kips}$$

The tensile design strength is

$$\phi R_n = \phi F_t A_b = 0.75(90)(0.4418) = 29.82 \text{ kips}$$

Check RCSC Equation LRFD 4.2 from the bolt specifications (RCSC, 1994) with P_u = r_{ut} = 11 kips, and V_u = bolt shear force = 11 kips:

$$\left[\frac{P_u}{(\phi R_n)_t}\right]^2 + \left[\frac{V_u}{(\phi R_n)_v}\right]^2 = \left(\frac{11}{29.82}\right)^2 + \left(\frac{11}{15.9}\right)^2 = 0.615 < 1.0 \qquad \text{(OK)}$$

ANSWER   The connection is satisfactory. ■

When bolts in slip-critical connections are subjected to tension, the slip-critical strength is ordinarily reduced by the factor given by AISC Equation A-J3-2 (see Section 7.9). The reason is that the clamping effect, and hence the friction force, is reduced. In a connection of the type just considered, however, there is additional compression on the lower part of the connection that increases the friction, thereby compensating for the reduction in the upper part of the connection. For this reason, the slip-critical strength should *not* be reduced in this type of connection.

8.4   ECCENTRIC WELDED CONNECTIONS: SHEAR ONLY

Eccentric welded connections are analyzed in much the same way as bolted connections, except that unit lengths of weld replace individual fasteners in the computations. As in the case of eccentric bolted connections loaded in shear, welded shear connections can be investigated by either elastic or ultimate strength methods.

Elastic Analysis

The load on the bracket shown in Figure 8.16a may be considered to act in the plane of the weld — that is, the plane of the throat. If this slight approximation is made, the load

---

■ **FIGURE 8.16**

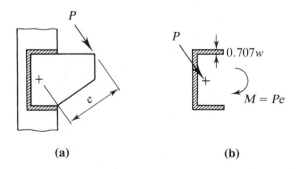

(a)   (b)

will be resisted by the area of weld shown in Figure 8.16b. Computations are simplified, however, if a unit throat dimension is used. The calculated load can then be multiplied by 0.707 times the weld size to obtain the actual load.

An eccentric load in the plane of the weld subjects the weld to both direct shear and torsional shear. Since all elements of the weld resist an equal portion of the direct shear, the direct shear stress is

$$f_1 = \frac{P}{L}$$

where L is the total length of the weld and numerically equals the shear area, because a unit throat size has been assumed. If rectangular components are used,

$$f_{1x} = \frac{P_x}{L} \text{ and } f_{1y} = \frac{P_y}{L}$$

where P_x and P_y are the x- and y-components of the applied load. The shearing stress caused by the couple is found with the torsion formula

$$f_2 = \frac{Md}{J}$$

where

d = distance from the centroid of the shear area to the point where the stress is being computed

J = polar moment of inertia of that area.

Figure 8.17 shows this stress at the upper right-hand corner of the given weld. In terms of rectangular components,

$$f_{2x} = \frac{My}{J} \text{ and } f_{2y} = \frac{Mx}{J}$$

Also,

$$J = \int_A r^2 dA = \int_A (x^2 + y^2) dA = \int_A x^2 dA + \int_A y^2 dA = I_y + I_x$$

■ **FIGURE 8.17**

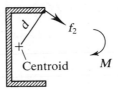

Centroid   M

where I_x and I_y are the rectangular moments of inertia of the shear area. Once all rectangular components have been found, they can be added vectorially to obtain the resultant shearing stress at the point of interest, or

$$f_v = \sqrt{(\Sigma\, f_x)^2 + (\Sigma\, f_y)^2}$$

As with bolted connections, the critical location for this resultant stress can usually be determined from an inspection of the relative magnitudes and directions of the direct and torsional shearing stress components.

Because a unit width of weld is used, the computations for centroid and moment of inertia are the same as for a line. In this book, we treat all weld segments as line segments, which we assume to be the same length as the edge of the connected part that they are adjacent to. Furthermore, we neglect the moment of inertia of a line segment about the axis coinciding with the line.

■ **EXAMPLE 8.5**

Determine the size of weld required for the bracket connection in Figure 8.18. The 60-kip load is a factored load. A36 steel is used for both the bracket and the column.

SOLUTION

The eccentric load may be replaced by a concentric load and a couple, as shown in Figure 8.18. The direct shearing stress, in kips per inch, is the same for all segments of the weld and is equal to

$$f_{1y} = \frac{60}{8 + 12 + 8} = \frac{60}{28} = 2.143 \text{ kips/in.}$$

Before computing the torsional component of shearing stress, the location of the centroid of the weld shear area must be determined. From the principle of moments with summation of moments about the y-axis,

$$\bar{x}(28) = 8(4)(2), \text{ or } \bar{x} = 2.286 \text{ in.}$$

The eccentricity e is $10 + 8 - 2.286 = 15.71$ in., and the torsional moment is

$$M = Pe = 60(15.71) = 942.6 \text{ in.-kips}$$

If the moment of inertia of each horizontal weld about its own centroidal axis is neglected, the moment of inertia of the total weld area about its horizontal centroidal axis is

$$I_x = \frac{1}{12}(1)(12)^3 + 2(8)(6)^2 = 720.0 \text{ in.}^4$$

Similarly,

$$I_y = 2\left[\frac{1}{12}(1)(8)^3 + 8(4 - 2.286)^2\right] + 12(2.286)^2 = 195.0 \text{ in.}^4$$

■ **FIGURE 8.18**

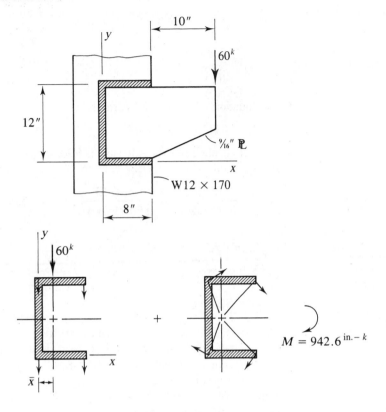

and

$$J = I_x + I_y = 720.0 + 195.0 = 915.0 \text{ in.}^4$$

Figure 8.18 shows the directions of both components of stress at each corner of the connection. By inspection, either the upper right-hand corner or the lower right-hand corner may be taken as the critical location. If the lower right-hand corner is selected,

$$f_{2x} = \frac{My}{J} = \frac{942.6(6)}{915.0} = 6.181 \text{ kips/in.}$$

$$f_{2y} = \frac{Mx}{J} = \frac{942.6(8 - 2.286)}{915.0} = 5.886 \text{ kips/in.}$$

$$f_v = \sqrt{(6.181)^2 + (2.143 + 5.886)^2} = 10.13 \text{ kips/in.}$$

Check the strength of the base metal. From Equation 7.21,

$$\phi R_n = \phi F_{BM} \times \text{area subject to shear}$$

$$= \phi F_{BM} \times t = 0.54 F_y t = 0.54(36)\left(\frac{9}{16}\right)$$

$$= 10.94 \text{ kips/in.} > 10.13 \text{ kips/in.} \quad \text{(OK)}$$

From Equation 7.20, the weld strength is

$$\phi R_n = 0.707 \times w \times L \times \phi F_W$$

The matching electrode for A36 steel is E70, with $\phi F_W = 31.5$ ksi. The required weld size is therefore

$$w = \frac{\phi R_n}{0.707 L \phi F_W} = \frac{10.13}{0.707(1.0)(31.5)} = 0.455 \text{ in.}$$

ANSWER   Use a $\frac{1}{2}$-inch fillet weld, E70 electrode. ■

Ultimate Strength Analysis

Eccentric welded shear connections may be safely designed by elastic methods, but the factor of safety will be larger than necessary and will vary from connection to connection (Butler, Pal, and Kulak, 1972). This type of analysis suffers from some of the same shortcomings as the elastic method for eccentric bolted connections, including the assumption of a linear load-deformation relationship for the weld. Another source of error is the assumption that the strength of the weld is independent of the direction of the applied load. An ultimate strength procedure is presented in Part 8 of the *Manual* (Volume II) and is summarized here. It is based on research by Butler *et al.* (1972) and Timler (1984) and closely parallels the method developed for eccentric bolted connections by Crawford and Kulak (1971).

Instead of considering individual fasteners, we treat the continuous weld as an assembly of discrete segments. At failure, the applied connection load is resisted by forces in each element, with each force acting perpendicular to the radius constructed from an instantaneous center of rotation to the centroid of the segment, as shown in Figure 8.19. This concept is essentially the same as that used for the fasteners. However, determining which element has the maximum deformation and computing the force in each element at failure is much more difficult. To determine the critical element, the ratio Δ_{max}/r is computed for each element, where

$$\Delta_{max} = 1.087 w (\theta + 6)^{-0.65} \le 0.17 w$$

θ = the angle the resisting force makes with the axis of the weld segment (see Figure 8.19)

w = weld leg size

r = the distance from the IC to the centroid of the segment

■ **FIGURE 8.19**

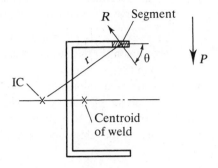

The element with the smallest ratio is the one that reaches its ultimate capacity first. The deformation of the other elements can then be computed as

$$\Delta = \frac{r}{r_{max}} \Delta_{max}$$

where

r = radius for the element

$\dfrac{\Delta_{max}}{r_{max}} = \dfrac{\Delta}{r}$ for the critical element.

The resisting force for each element can be found from

$$R = 0.60 F_{EXX} (1.0 + 0.50 \sin^{1.5} \theta)[p(1.9 - 0.9p)]^{0.3}$$

where

F_{EXX} = weld electrode tensile strength

$$p = \frac{\Delta}{\Delta_{max}}$$

(Unlike the bolted case, R is a function of θ.) The preceding computations are based on an assumed location of the instantaneous center of rotation. If it is the actual location, the equations of equilibrium will be satisfied. The remaining details are the same as for a bolted connection.

1. Solve for the load capacity from the equation

$$\Sigma \, M_{IC} = 0$$

where IC is the instantaneous center.

2. If the two force equilibrium equations are satisfied, the assumed location of the instantaneous center and the load found in Step 1 are correct; otherwise, assume a new location and repeat the entire process.

The absolute necessity for the use of a computer is obvious. Computer solutions for various common configurations of eccentric welded shear connections are given in tabular form in Part 8 of the *Manual*. Tables 8-38 through 8-45 give factored load capacities for various common combinations of horizontal and vertical weld segments based on an ultimate strength analysis. These tables may be used for either design or analysis and will cover almost any situation you are likely to encounter. For those connections not covered by the tables, the more conservative elastic method may be used.

■ EXAMPLE 8.6

Determine the weld size required for the connection in Example 8.5, based on ultimate strength considerations. Use the tables for eccentrically loaded weld groups given in Part 8 of the *Manual*.

SOLUTION   The weld of Example 8.5 is the same type as the one shown in Table 8-42 (angle $= 0°$), and the loading is similar. The following geometric constants are required for entry into the table:

$$a = \frac{a\ell}{\ell} = \frac{e}{\ell} = \frac{15.7}{12} = 1.3$$

$$k = \frac{k\ell}{\ell} = \frac{8}{12} = 0.67$$

By interpolation in Table 8-42 for $a = 1.3$,

$$C = 1.14 \quad \text{for} \quad k = 0.6 \quad \text{and} \quad C = 1.30 \quad \text{for} \quad k = 0.7$$

Interpolating between these two values for $k = 0.67$ gives

$$C = 1.25$$

For E70XX electrodes, $C_1 = 1.0$.
The required value of D is

$$D = \frac{P_u}{CC_1\ell} = \frac{60}{1.25(1.0)(12)} = 4.0$$

The required weld size is therefore

$$\frac{1}{16}(4.0) = 0.25 \quad \text{(versus 0.455 inch required in Example 8.5)}$$

ANSWER   Use a ¼-inch fillet weld, E70 electrode. ■

Special Provision for Axially Loaded Members

When a structural member is axially loaded, the stress is uniform over the cross section, and the resultant force may be considered to act along the *gravity axis,* which is a longitudinal axis through the centroid. For the member to be concentrically loaded at its ends, the resultant resisting force furnished by the connection must also act along this axis. If the member has a symmetrical cross section, this result can be accomplished by placing the welds or bolts symmetrically. If the member is one with an unsymmetrical cross section, such as the double-angle section in Figure 8.20, a symmetrical placement of welds or bolts will result in an eccentrically loaded connection, with a couple of *Te,* as shown in Figure 8.20b.

AISC J1.8 permits this eccentricity to be neglected in statically loaded members. When the member is subjected to fatigue caused by repeated loading or reversal of stress, the eccentricity must be either accounted for or eliminated by an appropriate placement of the welds or bolts. (Of course, this solution may be used even if the member is subjected to static loads only.) The correct placement can be determined by applying the force and moment equilibrium equations. For the welded connection shown in Figure 8.21, the first equation can be obtained by summing moments about the lower longitudinal weld:

$$\Sigma M_{L_2} = Tc - P_3 \frac{L_3}{2} - P_1 L_3 = 0$$

■ **FIGURE 8.20**

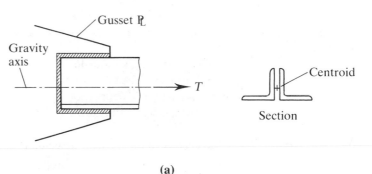

(a)

(b)

■ **FIGURE 8.21**

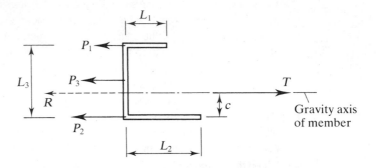

This equation can be solved for P_1, the required resisting force in the upper longitudinal weld. This value can then be substituted into the force equilibrium equation:

$$\Sigma F = T - P_1 - P_2 - P_3 = 0$$

This equation can be solved for P_2, the required resisting force in the lower longitudinal weld. For any size weld, the lengths L_1 and L_2 can then be determined. We illustrate this procedure, known as *balancing the welds,* in Example 8.7.

■ **EXAMPLE 8.7**

A tension member consists of a double-angle section, 2L5 × 3 × ½, with the long legs placed back-to-back. The angles are attached to a ⅜-inch-thick gusset plate. All steel is A36. Design a welded connection, balanced to eliminate eccentricity, that will resist the full tensile capacity of the member.

SOLUTION

The load capacity of the member based on the gross section is

$$\phi_t P_n = 0.90 F_y A_g = 0.90(36)(7.5) = 243.0 \text{ kips}$$

The load capacity based on the net section requires a value of U.

The length of the connection is not yet known, so U cannot be computed from AISC Equation B3-2. Using an average value of 0.85 gives

$$A_e = U A_g = 0.85(7.5) = 6.375 \text{ in.}^2$$
$$\phi_t P_n = 0.75 F_u A_e = 0.75(58)(6.375) = 277.3 \text{ kips} > 243.0 \text{ kips}$$

Yielding of the gross section is the controlling limit state, so $\phi_t P_n = 243.0$ kips. For one angle, the load to be resisted is

$$\frac{243.0}{2} = 121.5 \text{ kips}$$

For A36 steel, the appropriate electrode is E70XX, and

$$\text{Minimum weld size } = \frac{3}{16} \text{ in.} \quad \text{(AISC Table J2.4)}$$

$$\text{Maximum size } = \frac{1}{2} - \frac{1}{16} = \frac{7}{16} \text{ in.} \quad \text{(AISC J2.2b)}$$

Try a 5/16-inch fillet weld, E70 electrode:

$$\text{Capacity per inch of length } = 0.707w(\phi F_W)$$

$$= 0.707\left(\frac{5}{16}\right)(31.5)$$

$$= 6.960 \text{ kips/in.}$$

$$\text{Capacity of base metal in shear } = t(\phi F_{BM}) = t(0.54F_y)$$

$$= \left(\frac{3}{8}\right)(0.54)(36)$$

$$= 7.29 \text{ kips/in.}$$

The weld strength controls, so use 6.960 kips/in.

Refer to Figure 8.22. The capacity of the weld across the end of the angle is

$$P_3 = 6.960(5) = 34.80 \text{ kips.}$$

$$\Sigma M_{L_2} = 121.5(3.25) - 34.80\left(\frac{5}{2}\right) - P_1(5) = 0$$

$$P_1 = 61.58 \text{ kips}$$

$$\Sigma F = 121.5 - 61.58 - 34.80 - P_2 = 0, \quad P_2 = 25.12 \text{ kips}$$

$$L_1 = \frac{P_1}{6.960} = \frac{61.58}{6.960} = 8.85 \text{ in.} \quad \text{use 9 in.}$$

$$L_2 = \frac{25.12}{6.960} = 3.61 \text{ in.} \quad \text{use 4 in.}$$

■ **FIGURE 8.22**

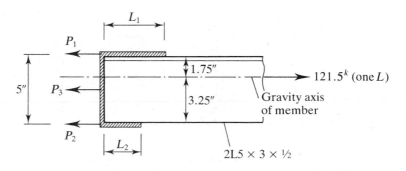

■ **FIGURE 8.23**

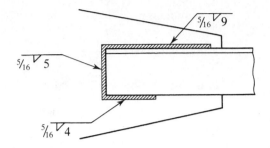

ANSWER   Use the weld shown in Figure 8.23.   ■

| 8.5 | **ECCENTRIC WELDED CONNECTIONS: SHEAR PLUS TENSION** |

Many eccentric connections, particularly beam-to-column connections, place the welds in tension as well as shear. Two such connections are illustrated in Figure 8.24.

■ **FIGURE 8.24**

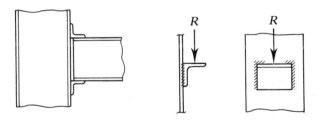

(a) Seated Beam Connection

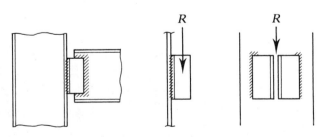

(b) Framed Beam Connection

The seated beam connection consists primarily of a short length of angle that serves as a "shelf" to support the beam. The welds attaching this angle to the column must resist the moment caused by the eccentricity of the reaction as well as the beam reaction in direct shear. The angle connecting the top flange provides torsional stability to the beam at its end and does not assist in supporting the reaction. It may be attached to the beam web instead of the top flange. The beam-to-angle connections can be made with either welds or bolts and will not carry any calculated load.

The framed beam connection, which is very common, subjects the vertical angle-to-column welds to the same type of load as the seated beam connection. The beam-to-angle part of the connection is also eccentric, but the load is in the plane of shear, so there is no tension. Both the seated and the framed connections have their bolted counterparts.

In each of the connections discussed, the vertical welds on the column flange are loaded as shown in Figure 8.25. As with the bolted connection in Section 8.3, the eccentric load P can be replaced by a concentric load P and a couple $M = Pe$. The shearing stress is

$$f_v = \frac{P}{A}$$

where A is the total throat area of the weld. The maximum tensile stress can be computed from the flexure formula

$$f_t = \frac{Mc}{I}$$

where I is the moment of inertia about the centroidal axis of the area consisting of the total throat area of the weld, and c is the distance from the centroidal axis to the farthest point on the tension side. The maximum resultant stress can be found by adding these two components vectorially:

$$f_r = \sqrt{f_v^2 + f_t^2}$$

■ **FIGURE 8.25**

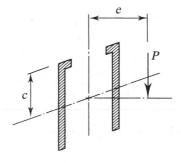

For units of kips and inches, this stress will be in kips per square inch. If a unit throat size is used in the computations, the same numerical value can also be expressed as kips per linear inch. If f_r is derived from factored loads, it may be compared with the design strength of a unit length of weld. Although this procedure assumes elastic behavior, it will be conservative when used in an LRFD context.

■ EXAMPLE 8.8

An L6 × 4 × ½ is used in a seated beam connection, as shown in Figure 8.26. It must support a factored load reaction of 22 kips. All structural steel is A36, and E70XX electrodes are to be used. What size fillet welds are required for the connection to the column flange?

SOLUTION
As in previous design examples, a unit throat size will be used in the calculations. Although an end return is needed for a weld of this type, conservatively for simplicity it will be neglected in the following calculations. In any event, its length could only be estimated at this point because the weld size has not yet been determined.

For the beam setback of ¾ inch, the beam is supported by 3.25 inches of the 4-inch outstanding leg of the angle. If the reaction is assumed to act through the center of this contact length, the eccentricity of the reaction with respect to the weld is

$$e = 0.75 + \frac{3.25}{2} = 2.375 \text{ in.}$$

and the moment is

$$M = Pe = 22(2.375) = 52.25 \text{ in.-kips}$$

■ FIGURE 8.26

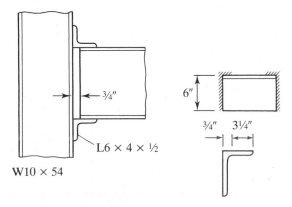

W10 × 54

For the assumed weld configuration in Figure 8.27,

$$I = \frac{2(1)(6)^3}{12} = 36 \text{ in.}^4, \qquad c = \frac{6}{2} = 3 \text{ in.}$$

$$f_t = \frac{Mc}{I} = \frac{52.25(3)}{36} = 4.354 \text{ kips/in.}$$

$$f_v = \frac{P}{A} = \frac{22}{2(1)(6)} = 1.833 \text{ kips/in.}$$

$$f_r = \sqrt{f_t^2 + f_v^2} = \sqrt{(4.354)^2 + (1.833)^2} = 4.724 \text{ kips/in.}$$

The required weld size w can be found by equating f_r to the weld capacity per inch of length:

$$f_r = 0.707w(\phi F_W)$$
$$4.724 = 0.707w(31.5), \qquad w = 0.212 \text{ in.}$$

From AISC Table J2.4,

$$\text{Minimum weld size} = \frac{1}{4} \text{ in. (based on column flange thickness of } \frac{5}{8} \text{ inch)}$$

From AISC J2.2b,

$$\text{Maximum size} = \frac{1}{2} - \frac{1}{16} = \frac{7}{16} \text{ in.}$$

Check the shear capacity of the base metal:

$$\text{Applied direct shear} = f_v = 1.833 \text{ kips/in.}$$

$$\text{Shear capacity of angle leg} = t(\phi F_{BM}) = t(0.54 F_y) = \frac{1}{2}(0.54)(36)$$

$$= 9.72 \text{ kips/in.} > 1.833 \text{ kips/in.} \qquad \text{(OK)}$$

■ **FIGURE 8.27**

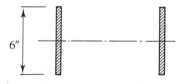

ANSWER   Use a ¼-inch fillet weld, E70XX electrodes.

We neglected the end returns in Example 8.8 but could have included them by performing a second iteration with an end return length of twice the weld size found in the first iteration. (This additional step was not warranted here because the minimum weld size governed the design.) End returns are accounted for in Example 8.9.

■ EXAMPLE 8.9

A welded framed beam connection is shown in Figure 8.28. The framing angles are 4 × 3 × ½, and the column is a W12 × 72. A36 steel is used throughout, and the welds are ⅜-inch fillet welds made with E70XX electrodes. Determine the maximum factored load beam reaction as limited by the welds at the column flange.

SOLUTION

The beam reaction will be assumed to act through the center of gravity of the connection to the framing angles. The eccentricity of the load with respect to the welds at the column flange will therefore be the distance from this center of gravity to the column flange. For a unit throat size and the weld shown in Figure 8.29a,

$$\bar{x} = \frac{2(2.5)(1.25)}{32 + 2(2.5)} = 0.1689 \text{ in. and } e = 3 - 0.1689 = 2.831 \text{ in.}$$

The moment on the column flange welds is

$$M = Re = 2.831R \text{ in.-kips}$$

where R is the beam reaction in kips.

From the dimensions given in Figure 8.29b, the properties of the column flange welds can be computed as

$$\bar{y} = \frac{32(16)}{32 + 0.75} = 15.63 \text{ in.}$$

$$I = \frac{1(32)^3}{12} + 32(16 - 15.63)^2 + 0.75(15.63)^2 = 2918 \text{ in.}^4$$

■ FIGURE 8.28

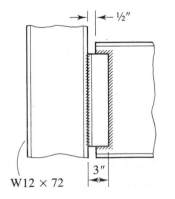

W12 × 72

½"

3"

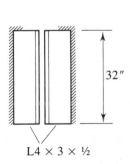

32"

L4 × 3 × ½

■ **FIGURE 8.29**

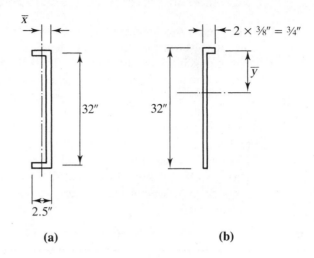

(a)   (b)

For the two welds,

$$I = 2(2918) = 5836 \text{ in.}^4$$

$$f_t = \frac{Mc}{I} = \frac{2.831R(15.63)}{5836} = 0.007582R \text{ kips/in.}$$

$$f_v = \frac{R}{A} = \frac{R}{2(32 + 0.75)} = 0.01527R \text{ kips/in.}$$

$$f_r = \sqrt{(0.007582R)^2 + (0.01527R)^2} = 0.01705R \text{ kips/in.}$$

Let $0.01705R = 0.707w(\phi F_w)$ and solve for R:

$$0.01705R = 0.707\left(\frac{3}{8}\right)(31.5), \quad R = 489.8 \text{ kips}$$

Check the shear capacity of the base metal (thickness of angle controls):

$$t(\phi F_{BM}) = t(0.54F_y) = 0.5(0.54)(36) = 9.72 \text{ kips/in.}$$

The direct shear to be resisted is

$$\frac{489.8}{2(32.75)} = 7.48 \text{ kips/in.} < 9.72 \text{ kips/in.} \quad (\text{OK})$$

ANSWER   The maximum factored load beam reaction = 490 kips.   ■

8.6   MOMENT-RESISTING CONNECTIONS

In all beam-to-column or beam-to-beam connections, there is some degree of moment restraint, even in connections designed to be simple, or moment-free. On the one hand, it is impossible to construct a perfectly frictionless pin or hinge, and most connections designed to be moment-free fail by far to achieve that condition. On the other hand, it is also very difficult to fabricate a perfectly rigid joint that is capable of transferring 100% of the moment capacity of one member to another. Thus, although the framed and seated beam connections shown in Figure 8.24 might seem to be somewhat rigid, they will actually transmit very little moment if the connecting angles are sufficiently flexible. As noted earlier, the eccentricity of the load with respect to the bolts or welds is very small and is usually neglected.

The AISC Specification defines two types of construction in Section A2.2, "Types of Construction."

> **Type FR — Fully Restrained (Rigid, or Continuous, Framing).** This type of structure has moment-resisting joints capable of transmitting any moment the member can resist, with no relative rotation of the members that meet at the joint. This type of construction is authorized without any restrictions. If the frame is designed as a rigid frame, the connections must be designed accordingly — that is, as moment connections.

> **Type PR — Partially Restrained (Semirigid Framing).** In this type of construction, the frame is designed on the basis of a known amount of restraint, intermediate between simple and rigid, at each joint. In general, moment restraint will be on the order of 20% to 90% of the member moment capacity. The chief drawback to this type of framing is the requirement for a rigorous structural analysis that incorporates this partial joint restraint. Implicit in this requirement is the need for moment-rotation curves for the connections.

If the partial restraint is neglected, beams can be treated as simply supported with no moment resistance at the joints. The previously mentioned framed and seated beam connections fall into this category. In general, connections that transfer less than 20% of the member capacity are considered simple. Beam supports designed in this manner are sometimes called *shear connections,* as only the reaction, or end shear, is transmitted.

Frames with shear connections must be braced in the plane of the frame because there is no "frame action" to provide lateral stability. This bracing can take several forms: diagonal bracing members, shear walls, or lateral support from an adjacent structure. The moments resulting from lateral loads (usually wind or seismic) can also be provided for in the design of selected beam-to-column connections. In this approach, the connection is assumed to act as a simple connection in resisting the dead and live loads (gravity loads) and as a moment connection with the limited capability of resisting the wind moments. The connection will actually be a partially restrained connection regardless of the loading. If the beam is designed as if it were simply supported, the maximum gravity load moment will be overestimated, and the beam will be somewhat overdesigned. In many cases, however, the wind moments will be small and the overconservatism will be slight. If this concept of simple framing is used, the Specification requires that the following conditions be met.

1. Although the beams (girders) are not simply supported, they must be capable of supporting the gravity loads as if they were.

2. The connections and the connected members (beams and columns) must be capable of resisting the wind moments.

3. The connections must have enough inelastic rotational capacity that the fasteners or welds will not be overloaded under the combined action of gravity loads and wind.

In this book, we consider only two types of connections: *simple connections* designed for gravity loads (with lateral frame stability provided by a positive bracing system) and *rigid connections* designed for more than 90% of the moment capacity of the beam. We have already considered simple connections in the discussion of framed and seated beam connections and now turn our attention to rigid connections.

Several examples of commonly used moment connections are illustrated in Figure 8.30. As a general rule, most of the moment transfer is through the beam flanges, and most of the moment capacity is developed there. The connection in Figure 8.30a typifies this concept. The plate connecting the beam web to the column is shop welded to the column and field bolted to the beam. With this arrangement, the beam is conveniently held in position so that the flanges can be field welded to the column. The plate connection is designed to resist only shear and takes care of the beam reaction. Complete penetration groove welds connect the beam flanges to the column and can transfer a moment equal to the moment capacity of the beam flanges. This will constitute most of the moment capacity of the beam, but a small amount of restraint will also be provided by the plate connection. (Because of strain hardening, the full plastic moment capacity of the beam can actually be developed through the flanges.) Making the flange connection requires that a small portion of the beam web be removed and a "backing bar" used at each flange to permit all welding to be done from the top. When the flange welds cool, they will shrink, typically about $1/8$ inch. The resulting longitudinal displacement can be accounted for by using slotted bolt holes and tightening the bolts after the welds have cooled. This type of connection also uses column stiffeners, which are not always required (see Section 8.7).

The moment connection of Figure 8.30a also illustrates a recommended connection design practice: Whenever possible, welding should be done in the fabricating shop, and bolting should be done in the field. Shop welding is less expensive and can be more closely controlled.

In most beam-to-column moment connections, the members are part of a plane frame and are oriented as shown in Figure 8.30a — that is, with the webs in the plane of the frame so that bending of each member is about its major axis. When a beam must frame into the web of a column rather than its flange (for example, in a space frame), the connection shown in Figure 8.30b can be used. This connection is similar to the one shown in Figure 8.30a but requires the use of column stiffeners to make the connections to the beam flanges.

■ **FIGURE 8.30**

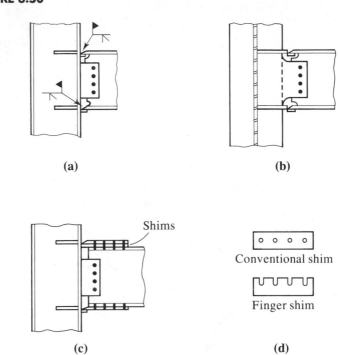

(a)

(b)

(c)

(d)

Although the connection shown in Figure 8.30a is simple in concept, its execution requires close tolerances. If the beam is shorter than anticipated, the gap between the column and the beam flange may cause difficulties in welding, even when a backing bar is used. The three-plate connection shown in Figure 8.30c does not have this handicap, and it has the additional advantage of being completely field bolted. The flange plates and the web plate are all shop welded to the column flange and field bolted to the beam. To provide for variation in the beam depth, the distance between flange plates is made larger than the nominal depth of the beam, usually by about ⅜ inch. This gap is filled at the top flange during erection with *shims,* which are thin strips of steel used for adjusting the fit at joints. Shims may be one of two types: either conventional or "finger" shims, which can be inserted after the bolts are in place, as shown in Figure 8.30d. In regions where seismic forces are large, the connection shown in Figure 8.30a requires special design procedures (FEMA, 1995).

Example 8.10 illustrates the design of a three-plate moment connection, including the requirements for connecting elements, which are covered by AISC J5.

■ EXAMPLE 8.10

Design a three-plate moment connection of the type shown in Figure 8.31 for the connection of a W21 × 50 beam to the flange of a W14 × 99 column. Assume a beam setback of ½ inch. A frame analysis shows that the connection must transfer a factored load moment of 210 ft-kips and a factored load shear of 33 kips. All plates are to be shop welded to the column with E70XX electrodes and field bolted to the beam with A325 bearing-type bolts. A36 steel is used throughout.

SOLUTION   For the web plate (neglect eccentricity), **try ¾-inch diameter bolts.** Assume that the threads are in the plane of shear. The shear capacity of one bolt is

$$\phi F_v A_b = 0.75(48)(0.4418) = 15.90 \text{ kips}$$

$$\text{Number of bolts required} = \frac{33}{15.90} = 2.08$$

Try 3 bolts and determine the plate thickness required for bearing. Use the spacing and edge distance shown in Figure 8.32a and a hole diameter of

$$h = d + \frac{1}{16} = \frac{3}{4} + \frac{1}{16} = \frac{13}{16} \text{ in.}$$

For the hole nearest the edge,

$$L_c = L_e - \frac{h}{2} = 1.5 - \frac{13/16}{2} = 1.094 \text{ in.}$$

$$2d = 2\left(\frac{3}{4}\right) = 1.5 \text{ in.}$$

Because $L_c < 2d$, the bearing strength is

$$\phi R_n = \phi(1.2 L_c t F_u) = 0.75(1.2)(1.094)t(58) = 57.11t \text{ kips/bolt}$$

■ FIGURE 8.31

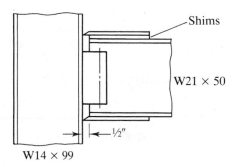

Shims

W21 × 50

½"

W14 × 99

■ **FIGURE 8.32**

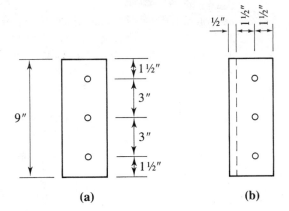

(a)   (b)

For the other holes,

$$L_c = s - h = 3 - \frac{13}{16} = 2.188 \text{ in.} > 2d$$

$$\therefore \phi R_n = \phi(2.4dtF_u) = 0.75(2.4)\left(\frac{3}{4}\right)t(58) = 78.30t \text{ kips/bolt}$$

To find the required thickness, equate the total bearing strength to the applied load:

$$57.11t + 2(78.30t) = 33, \quad \text{or} \quad t = 0.154 \text{ in.}$$

For the beam web, $t_w = 0.380$ in. > 0.154 in.　　(OK)

To determine the plate thickness required for shear, consider a vertical section through the plate. From AISC J5, "Connecting Elements,"

$$\phi R_n = 0.90[0.60A_gF_y] \hspace{3cm} \text{(AISC Equation J5-3)}$$
$$33 = 0.90[0.60(9t)(36)]$$
$$t = 0.189 \text{ in.} \quad \text{(controls)} \quad \text{use } t = \frac{1}{4} \text{ in.}$$

For the connection of the shear plate to the column flange, the minimum fillet weld size is ¼ inch. (Based on the thicker connected part, the minimum is ⁵⁄₁₆ inch, but it need not be larger than the thickness of the thinner part.) Then, the

$$\text{Capacity per inch of length} = 0.707w(\phi F_W) = 0.707\left(\frac{1}{4}\right)(31.5)$$

$$= 5.568 \text{ kips/in.}$$

The shear capacity of the base metal is

$$t\phi F_{BM} = t(0.54F_y) = \frac{1}{4}(0.54)(36) = 4.86 \text{ kips/in.} \quad \text{(controls)}$$

The required length of the ¼-inch fillet weld is therefore

$$\frac{33}{4.86} = 6.79 \text{ in.}$$

A continuous weld on one side of the plate would suffice, but both sides are usually welded, and that will be done here.

The minimum width of the plate can be determined from a consideration of edge distances. The load being resisted (the beam reaction) is vertical, so the horizontal edge distance need only conform to the clearance requirements of AISC Table J3.4. If we assume a sheared edge, the minimum edge distance is 1¼ inch.

With a beam setback of ½ inch and edge distances of 1½ inches, as shown in Figure 8.32b, the width of the plate is

$$0.5 + 2(1.5) = 3.5 \text{ in.} \qquad \text{use a bar } 3\tfrac{1}{2} \times \tfrac{1}{4}$$

For the flange plates, find the force at interface between beam flange and plate. From Figure 8.33.

$$M = Hd \quad \text{and} \quad H = \frac{M}{d} = \frac{210(12)}{20.83} = 121.0 \text{ kips}$$

Try ¾-inch A325 bolts. (Since ¾-inch-diameter bolts were selected for the shear connection, try the same size here.) If bolt shear controls, the number of bolts required is

$$\frac{121.0}{15.90} = 7.61 \qquad \text{use 8 (4 pairs)}$$

Use edge distances of 1½ inches, spacings of 3 inches, and determine the minimum plate thickness required for bearing. Use a hole diameter of

$$h = d + \frac{1}{16} = \frac{3}{4} + \frac{1}{16} = \frac{13}{16} \text{ in.}$$

For the hole nearest the edge,

$$L_c = L_e - \frac{h}{2} = 1.5 - \frac{13/16}{2} = 1.094 \text{ in.}$$
$$2d = 2(3/4) = 1.5 \text{ in.}$$

■ **FIGURE 8.33**

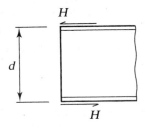

Because $L_c < 2d$, the bearing strength is

$$\phi R_n = \phi(1.2L_c t F_u) = 0.75(1.2)(1.094)t(58) = 57.11t \text{ kips/bolt}$$

For the other holes,

$$L_c = s - h = 3 - \frac{13}{16} = 2.188 \text{ in.} > 2d$$

$$\therefore \phi R_n = \phi(2.4dtF_u) = 0.75(2.4)\left(\frac{3}{4}\right)t(58) = 78.30t \text{ kips/bolt}$$

To find the required thickness, equate the total bearing strength to the applied load:

$$2(57.11t) + 6(78.30t) = 121.0 \quad \text{or} \quad t = 0.207 \text{ in.}$$

Both flange plates will be designed as tension connecting elements. (Although one of the plates will be in compression, the connection details preclude any stability problems.) The minimum cross section required for tension on the gross and net areas will now be determined. From AISC Equation J5-1,

$$\phi R_n = 0.90(A_g F_y)$$

$$\text{Required } A_g = \frac{\phi R_n}{0.90 F_y} = \frac{H}{0.90 F_y} = \frac{121.0}{0.90(36)} = 3.735 \text{ in.}^2$$

From AISC Equation J5-2,

$$\phi R_n = 0.75 A_n F_u$$

$$\text{Required } A_n = \frac{\phi R_n}{0.75 F_u} = \frac{H}{0.75 F_u} = \frac{121.0}{0.75(58)} = 2.782 \text{ in.}^2$$

Try a plate width of w_g = 6.5 in. (equal to the beam flange width). Compute the thickness needed to satisfy the gross area requirement.

$$A_g = 6.5t = 3.735 \text{ in.}^2 \quad \text{or} \quad t = 0.575 \text{ in.}$$

Compute the thickness needed to satisfy the net area requirement.

$$A_n = tw_n = t(w_g - \Sigma d_{\text{hole}}) = t\left[6.5 - 2\left(\frac{7}{8}\right)\right] = 4.750t$$

Let

$$4.750t = 2.782 \text{ in.}^2 \quad \text{or} \quad t = 0.586 \text{ in.} \quad \text{(controls)}$$

This thickness is also greater than that required for bearing, so it will be the minimum acceptable thickness. **Try a bar $6\frac{1}{2} \times \frac{5}{8}$.** This plate is a tension connecting element, so its net area cannot exceed $0.85A_g$ in the computations (AISC J5.2):

$$A_n = \frac{5}{8}\left[6.5 - 2\left(\frac{7}{8}\right)\right] = 2.969 \text{ in.}^2$$

$$0.85A_g = 0.85(0.625)(6.5) = 3.453 \text{ in.}^2 > 2.969 \text{ in.}^2 \quad \text{(OK) use a bar } 6\,\tfrac{1}{2} \times \tfrac{5}{}$$

Part of the flange area of the beam will be lost because of the bolt holes, and the moment capacity will be reduced. AISC B10 permits this reduction to be neglected when

$$0.75F_u A_{fn} \geq 0.9F_y A_{fg}$$ (AISC Equation B10-1)

where

A_{fg} = gross flange area

   = $b_f t_f$ = 6.530(0.535) = 3.494 in.^{2}

A_{fn} = net flange area

   = $t_f(b_f - \Sigma d_h)$ = $0.535\left[6.530 - 2\left(\dfrac{7}{8}\right)\right]$ = 2.557 in.^{2}

Evaluating AISC Equation B10-1 gives

$0.75F_u A_{fn}$ = 0.75(58)(2.557) = 111.2 kips

$0.9F_y A_{fg}$ = 0.9(36)(3.494) = 113.2 kips > 111.2 kips

Since AISC Equation B10-1 is not satisfied, the flexural strength should be based on an effective flange area of

$$A_{fe} = \frac{5}{6}\frac{F_u}{F_y}A_{fn}$$

(AISC Equation B10-3)

$$= \frac{5}{6}\left(\frac{58}{36}\right)(2.557) = 3.433 \text{ in.}^2$$

■ **FIGURE 8.34**

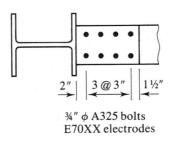

¾" φ A325 bolts
E70XX electrodes

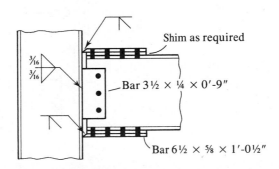

This area does not differ significantly from the actual gross flange area of 3.494 in.^{2}, so the flexural strength will not be modified.

ANSWER   Use the connection shown in Figure 8.34 (column stiffener requirements will be considered in Section 8.7).*

———————————

*Figure 8.34 also shows the symbol for a bevel groove weld, used here for the beam flange plate-to-column connection

8.7 COLUMN STIFFENERS AND OTHER REINFORCEMENT

Most of the moment transferred from the beam to the column in a rigid connection takes the form of a couple consisting of the tensile and compressive forces in the beam flanges. The application of these relatively large concentrated forces may require reinforcement of the column. For negative moment, as would be the case with gravity loading, these forces are directed as shown in Figure 8.35, with the top flange of the beam delivering a tensile force to the column, and the bottom flange delivering a compressive force.

Both forces are transmitted to the column web, with compression being more critical because of the stability problem. The tensile load at the top can distort the column flange (exaggerated in Figure 8.35c), creating additional load on the weld connecting the beam flange to the column flange. A stiffener of the type shown can provide anchorage for the column flange. Obviously, this stiffener must be welded to both web and flange. If the applied moment never changes direction, the stiffener resisting the com-

■ **FIGURE 8.35**

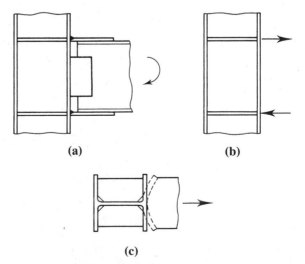

(a)   (b)

(c)

pressive load (the bottom stiffener in this illustration) can be fitted to bear on the flange and need not be welded to it.

AISC Specification Requirements

The AISC requirements for column web reinforcement are covered in Chapter K, "Strength Design Considerations." For the most part, these provisions are based on theoretical analyses that have been modified to fit test results. If the applied factored load transmitted by the beam flange or flange plate exceeds the design strength, ϕR_n, for any of the limit states considered, stiffeners must be used.

To avoid a local bending failure of the column flange, the tensile load from the beam flange must not exceed

$$\phi R_n = \phi(6.25t_f^2 F_{yf}) \qquad \text{(AISC Equation K1-1)}$$

where

$\psi = 0.90$

$t_f = $ thickness of the column flange

$F_{yf} = $ yield stress of the column flange

For the limit state of local web yielding in compression,

$$\phi R_n = \phi[(5k + N)F_{yw}t_w] \qquad \text{(AISC Equation K1-2)}$$

or, when the load is applied within a distance from the end of the member that equals the depth of the member,

$$\phi R_n = \phi[(2.5k + N)F_{yw}t_w] \qquad \text{(AISC Equation K1-3)}$$

where

$\phi = 1.0$

$k = $ distance from the outer flange surface of the column to the toe of the fillet in the web

$N = $ length of applied load $= $ thickness of beam flange or flange plate

$F_{yw} = $ yield stress of the column web

$t_w = $ thickness of the column web

We also used AISC Equations K1-2 and K1-3 in Section 5.13 to investigate web yielding in beams subjected to concentrated loads.

To prevent web crippling when the compressive load is delivered to one flange only, as in the case of an exterior column with a beam connected to one side, the applied load must not exceed the design strength given by one of the following equations. (We also addressed web crippling in Section 5.13.) When the load is applied at a distance of at least $d/2$ from the end of the column,

$$\phi R_n = \phi 135t_w^2\left[1 + 3\left(\frac{N}{d}\right)\left(\frac{t_w}{t_f}\right)^{1.5}\right]\sqrt{\frac{F_{yw}t_f}{t_w}} \qquad \text{(AISC Equation K1-4)}$$

where

$\phi = 0.75$

d = total column depth

If the load is applied at the end of the column,

$$\phi R_n = \phi 68t_w^2 \left[1 + 3\left(\frac{N}{d}\right)\left(\frac{t_w}{t_f}\right)^{1.5} \right] \sqrt{\frac{F_{yw}t_f}{t_w}}, \qquad \text{for } \frac{N}{d} \leq 0.2$$

(AISC Equation K1-5a)

or

$$\phi R_n = \phi 68t_w^2 \left[1 + \left(4\frac{N}{d} - 0.2\right)\left(\frac{t_w}{t_f}\right)^{1.5} \right] \sqrt{\frac{F_{yw}t_f}{t_w}}, \qquad \text{for } \frac{N}{d} > 0.2$$

(AISC Equation K1-5b)

Compression buckling of the web must be investigated when loads are delivered to *both* column flanges. Such loading would occur at an interior column with beams connected to both sides. The design strength for this limit state is

$$\phi R_n = \phi \left[\frac{4100t_w^3 \sqrt{F_{yw}}}{h} \right]$$

(AISC Equation K1-8)

where

$\phi = 0.90$

h = column web depth from toe of fillet to toe of fillet (Figure 8.36)

■ **FIGURE 8.36**

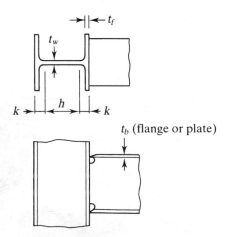

If the connection is near the end of the column (i.e., if the load is applied within a distance $d/2$ from the end), the strength given by AISC Equation K1-8 should be reduced by half.

In summary, to investigate the need for column stiffeners, three limit states should be checked:

1. Local flange bending (AISC Equation K1-1)
2. Local web yielding (AISC Equation K1-2 or K1-3)
3. Web crippling *or* compression buckling of the web. (If the compressive load is applied to *one* flange only, check web crippling [AISC Equation K1-4 or K1-5]. If the compressive load is applied to *both* flanges, check compressive buckling of the web [AISC Equation K1-8].)

If stiffeners are required by AISC Equation K1-2 for local web yielding, the required cross-sectional area of the stiffeners can be found as follows. Assume that the additional design strength can be obtained from an area of stiffener A_{st} that has yielded. Thus, from AISC Equation K1-2.

$$\phi R_n = \phi[(5k + N)F_{yw}t_w + A_{st}F_{yst}]$$

where F_{yst} is the yield stress of the stiffener. Equating the right side of this equation to the applied load, denoted P_{bf}, and solving for A_{st} gives

$$A_{st} = \frac{P_{bf}/\phi - (5k + N)F_{yw}t_w}{F_{yst}}$$

$$= \frac{P_{bf} - (5k + t_b)F_{yw}t_w}{F_{yst}} \tag{8.6}$$

where $\phi = 1.0$ and t_b is the thickness of the beam flange or flange plate. Equation 8.6 can also be used to check the local web yielding strength of the column. Solve for A_{st}; if a negative result is obtained, no stiffeners are needed for this limit state.

If stiffeners are required because of any of these provisions, AISC K1.9 gives the following guidelines for their proportioning.

- The width of the stiffener plus half the column web thickness must be equal to at least one third of the width of the beam flange or plate delivering the force to the column or, from Figure 8.37,

$$b + \frac{t_w}{2} \geq \frac{b_b}{3} \qquad \therefore\ b \geq \frac{b_b}{3} - \frac{t_w}{2}$$

- The stiffener thickness must be at least half the thickness of the beam flange or plate, or

$$t_{st} \geq \frac{t_b}{2}$$

■ **FIGURE 8.37**

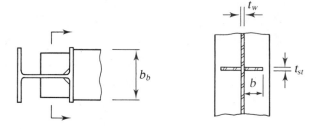

- The following width-thickness requirement should be met:

$$\frac{b}{t_{st}} \leq \frac{95}{\sqrt{F_y}}$$

Full-depth stiffeners are required for the compression buckling case, but half-depth stiffeners are permitted for the other limit states. Thus full-depth stiffeners are required only when beams are connected to both sides of the column.

For any of the limit states, the decision on whether to weld the stiffener to the flange should be based on the following criteria.

- On the tension side, the stiffeners should be welded to both the web and flange.
- On the compression side, the stiffeners need only bear on the flange but may be welded.

Part 3 of the *Manual,* "Column Design," contains tabulated constants that expedite the evaluation of the need for stiffeners. Their use is illustrated in an example in the "General Notes" preceding the column load tables and is not covered here.

Shear in the Column Web

The transfer of a large moment to a column can produce large shearing stresses in the column web within the boundaries of the connection; for example, region ABCD in Figure 8.38. This region is sometimes called the *panel zone*. The *net* moment is of concern, so if beams are connected to both sides of the column, the algebraic sum of the moments induces this web shear.

If the beam flange forces are assumed to act at a distance of $0.95d_b$ apart, where d_b is the beam depth, each flange force can be taken as

$$H = \frac{M_1 + M_2}{0.95d_b}$$

■ **FIGURE 8.38**

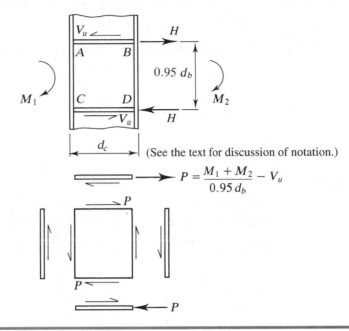

(See the text for discussion of notation.)

If the column shear adjacent to the panel is V_u and is directed as shown, the total shear force in the panel is

$$P = H - V_u = \frac{M_1 + M_2}{0.95d_b} - V_u \tag{8.7}$$

The web shear strength is given in AISC K1.7 as ϕR_v, where $\phi = 0.90$ and R_v is a function of the axial load in the column. When $P_u \leq 0.4P_y$,

$$R_v = 0.60F_y d_c t_w \qquad \text{(AISC Equation K1-9)}$$

When $P_u > 0.4P_y$,

$$R_v = 0.60F_y d_c t_w \left[1.4 - \left(\frac{P_u}{P_y}\right)\right] \qquad \text{(AISC Equation K1-10)}$$

where

P_u = factored axial load in column

P_y = axial yield strength of column = AF_y

A = cross-sectional area of column, including any reinforcement (i.e., doubler plates)

d_c = total column depth

t_w = column web thickness, including any reinforcing plates

F_y = yield stress of column web

If the column web has insufficient shear strength, it must be reinforced. A *doubler plate* with sufficient thickness to make up the deficiency can be welded to the web or a pair of diagonal stiffeners can be used. Stiffeners are usually more practical.

AISC K1.7 also provides equations to be used when frame stability, including plastic deformation of the panel zone, is considered in the analysis. They are not covered here.

■ EXAMPLE 8.11

Determine whether stiffeners or other column web reinforcement is required for the connection of Example 8.10. Assume that $V_u = 0$ and $P_u/P_y = 0.4$

SOLUTION From Example 8.10, the flange force can be conservatively taken as

$$P_{bf} = H = 121.0 \text{ kips}$$

Check local flange bending with AISC Equation K1-1:

$$\phi R_n = \phi(6.25t_f^2 F_{yf})$$

$$= 0.90[6.25(0.780)^2(36)] = 123 \text{ kips} > 121 \text{ kips} \quad \text{(OK)}$$

Check local web yielding with AISC Equation K1-2:

$$\phi R_n = \phi[(5k + N)F_{yw}t_w]$$

$$= 1.0\left[5(1.438) + \frac{5}{8}\right](36)(0.485) = 136 \text{ kips} > 121 \text{ kips} \quad \text{(OK)}$$

Check web crippling with AISC Equation K1-4:

$$\phi R_n = \phi 135t_w^2\left[1 + 3\left(\frac{N}{d}\right)\left(\frac{t_w}{t_f}\right)^{1.5}\right]\sqrt{\frac{F_{yw}t_f}{t_w}}$$

$$= 0.75(135)(0.485)^2\left[1 + 3\left(\frac{5/8}{14.16}\right)\left(\frac{0.485}{0.780}\right)^{1.5}\right]\sqrt{\frac{36(0.780)}{0.485}}$$

$$= 193 \text{ kips} > 121 \text{ kips} \quad \text{(OK)}$$

ANSWER Column stiffeners are not required.

For shear in the column web, from Equation 8.7 and neglecting the thickness of shims in the computation of d_b, the factored load shear force in the column web panel zone is

$$P = \frac{(M_1 + M_2)}{0.95d_b} - V_u$$

$$= \frac{210(12)}{0.95[20.83 + 2(5/8)]} - 0 = 120 \text{ kips}$$

Because $P_u = 0.4P_y$, use AISC Equation K1-9:

$$R_v = 0.60F_y d_c t_w = 0.60(36)(14.16)(0.485) = 148.3 \text{ kips}$$

The design strength is

$$\phi R_v = 0.90(148.3) = 134 \text{ kips} > 120 \text{ kips} \qquad \text{(OK)}$$

ANSWER Column web reinforcement is not required.   ■

■ EXAMPLE 8.12

The beam-to-column connection shown in Figure 8.39 must transfer a factored moment of 142 ft-kips. This moment is caused by dead and live gravity loads. All steel is A36, and E70 electrodes are used. Investigate column stiffener and web panel–zone reinforcement requirements. Assume that $V_u = 0$ and $P_u < 0.4P_y$.

SOLUTION The flange force is

$$P_{bf} = \frac{M}{d_b - t_b} = \frac{142(12)}{17.90 - 0.525} = 98.07 \text{ kips}$$

To check local flange bending, use AISC Equation K1-1:

$$\phi R_n = \phi(6.25t_f^2 F_{yf})$$

$$= 0.90[6.25(0.560)^2(36)] = 63.50 \text{ kips} < 98.07 \text{ kips} \qquad \text{(N.G.)}$$

∴ stiffeners are required to prevent local flange bending.

To check for local web yielding, use Equation 8.6 in lieu of AISC Equation K1-2:

$$A_{st} = \frac{P_{bf} - (5k + t_b)F_{yw}t_w}{F_{yst}}$$

$$= \frac{98.07 - [5(1.062) + 0.525](36)(0.360)}{36} = 0.6236 \text{ in.}^2$$

Because A_{st} is positive, a pair of stiffeners with a combined cross-sectional area of at least 0.6236 in.² will be required.

■ **FIGURE 8.39**

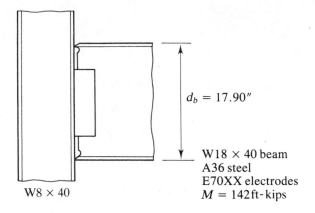

$d_b = 17.90''$

W18 × 40 beam
A36 steel
E70XX electrodes
$M = 142$ ft-kips

W8 × 40

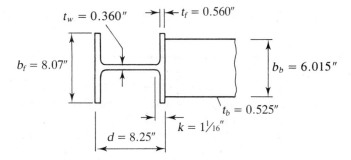

$t_w = 0.360''$ $t_f = 0.560''$

$b_f = 8.07''$ $b_b = 6.015''$

$t_b = 0.525''$

$k = 1\frac{1}{16}''$

$d = 8.25''$

To check web crippling strength, use AISC Equation K1-4:

$$\phi R_n = \phi 135 t_w^2 \left[1 + 3 \left(\frac{N}{d} \right) \left(\frac{t_w}{t_f} \right)^{1.5} \right] \sqrt{\frac{F_{yw} t_f}{t_w}}$$

$$= 0.75(135)(0.360)^2 \left[1 + 3 \left(\frac{0.525}{8.25} \right) \left(\frac{0.360}{0.560} \right)^{1.5} \right] \sqrt{\frac{36(0.560)}{0.360}}$$

$$= 107.9 \text{ kips} > 98.07 \text{ kips} \quad \text{(OK)}$$

The stiffener dimensions will be selected on the basis of the criteria given in AISC Section K1.9, and the area of the resulting cross section will then be checked.

The minimum width is

$$b \geq \frac{b_f}{3} - \frac{t_w}{2} = \frac{6.015}{3} - \frac{0.360}{2} = 1.825 \text{ in.}$$

If the stiffeners are not permitted to extend beyond the edges of the column flange, the maximum width is

$$b \leq \frac{8.07 - 0.360}{2} = 3.855 \text{ in.}$$

The minimum thickness is

$$\frac{t_b}{2} = \frac{0.525}{2} = 0.2625 \text{ in.}$$

Try a bar 3 × ⁵⁄₁₆:

$$A_{st} = 3\left(\frac{5}{16}\right) \times 2 \text{ stiffeners} = 1.875 \text{ in.}^2 > 0.6236 \text{ in.}^2 \qquad \text{(OK)}$$

Check the width–thickness ratio:

$$\frac{b}{t_{st}} = \frac{3}{5/16} = 9.6$$

$$\frac{95}{\sqrt{F_y}} = \frac{95}{\sqrt{36}} = 15.8 > 9.6 \qquad \text{(OK)}$$

This connection is one-sided, so full-depth stiffeners are not required. Hence

$$\frac{d}{2} = \frac{8.25}{2} = 4.125 \text{ in.} \qquad \text{use } 4\frac{1}{2} \text{ inches}$$

ANSWER   Use 2 bars 3 × ⁵⁄₁₆ × 4½. (Clip the inside corners to avoid the fillets at the column flange-to-web intersection. Clip at a 45° angle for ⁵⁄₈ inch.)

For the stiffener to column web welds,

$$\text{Minimum size} = \frac{3}{16} \text{ in.} \qquad \text{(AISC Table J2.4, based on web thickness).}$$

The size required for strength is

$$w = \frac{\text{force resisted by stiffener}}{0.707L(\phi F_W)}$$

From Equation 8.6, the force to be resisted by the stiffener is

$$A_{st}F_{yst} = P_{bf} - (5k + t_b)F_{yw}t_w$$
$$= 98.07 - [5(1.062) + 0.525](36)(0.360) = 22.45 \text{ kips}$$

The length available for welding the stiffener to the web is

$$L = \left(4.5 - \frac{5}{8}\right) \times 2 \text{ sides} \times 2 \text{ stiffeners} = 15.5 \text{ in.} \qquad \text{(see Figure 8.40)}$$

$$w = \frac{22.45}{0.707(15.5)(31.5)} = 0.0650 \text{ in.} < \frac{3}{16} \text{ in. minimum}$$

The shear strength of the base metal is

$$\phi R_n = \phi F_{BM}t = 0.54F_y t_{st} = 0.54(36)\left(\frac{5}{16}\right) = 6.075 \text{ kips/in.}$$

■ **FIGURE 8.40**

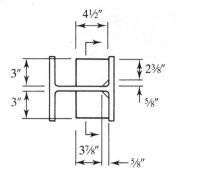

Section

and the

$$\text{Required weld capacity (for one stiffener)} = 0.0650(0.707)(31.5)(2)$$
$$= 2.90 \text{ kips/in.} < 6.075 \text{ kips/in.}$$
(OK)

ANSWER   Use a $^3\!/_{16}$-inch fillet weld.

For the stiffener to column flange welds,

$$\text{Minimum size} = \frac{1}{4} \text{ in.} \quad \text{(AISC Table J2.4, based on flange thickness)}$$

$$\text{Capacity per inch} = 0.707\!\left(\frac{1}{4}\right)\!(31.5) = 5.568 \text{ kips/in.}$$

$$< 0.54 F_y t_{st} = 6.075 \text{ kips/in.} \quad \text{(OK)}$$

$$\text{Length available for welding} = \left(3 - \frac{5}{8}\right)\!(2)(2) = 9.5 \text{ in.}$$

The size required for strength is

$$w = \frac{\text{force resisted by stiffener}}{0.707 L(\phi F_W)} = \frac{22.45}{0.707(9.5)(31.5)} = 0.106 \text{ in.} < \frac{1}{4} \text{ in.}$$

ANSWER   Use a $^1\!/_4$-inch fillet weld. (The applied moment is caused by gravity loads and is not reversible, so the stiffeners opposite the compression flange of the beam can be fitted to bear on the column flange and need not be welded, but this option is not exercised here.)
Check the column web for shear. From Equation 8.7,

$$P = \frac{(M_1 + M_2)}{0.95 d_b} - V_u = \frac{142(12)}{0.95(17.90)} - 0 = 100.2 \text{ kips}$$

From AISC Equation K1-9,

$$R_v = 0.60F_y d_c t_w = 0.60(36)(8.25)(0.360) = 64.15 \text{ kips}$$

The design strength is

$$\phi R_v = 0.90(64.15) = 57.74 \text{ kips} < 100.2 \text{ kips} \qquad (\text{N.G.})$$

Use AISC Equation K1-9 to find the required web thickness. Multiplying both sides by ϕ and solving for t_w gives

$$t_w = \frac{\phi R_v}{\phi(0.60F_y d_c)} = \frac{100.2}{0.90(0.60)(36)(8.25)} = 0.625 \text{ in.}$$

Use a web doubler plate. Let

t_d = required doubler plate thickness

$= 0.625 - 0.360 = 0.265 \text{ in.}$

Try $t_d = {}^5\!/_{16}$ in. Provide a weld to match the shear strength of the *required* thickness of doubler plate. Let

$$\phi F_{BM} t_d = 0.707w(\phi F_W)$$

or

$$w = \frac{\phi F_{BM} t_d}{0.707(\phi F_W)} = \frac{0.54(36)(0.265)}{0.707(31.5)} = 0.231 \text{ in.}$$

To the nearest $\frac{1}{16}$ inch, the required weld size is $\frac{1}{4}$ inch.
From AISC J2.2b, the maximum weld size is

$$t_d - \frac{1}{16} = \frac{5}{16} - \frac{1}{16} = \frac{1}{4} \text{ in.} \qquad (\text{OK})$$

ANSWER Use a $\frac{5}{16}$-inch doubler plate and a $\frac{1}{4}$-inch fillet weld.

Use a diagonal stiffener. Use full-depth horizontal stiffeners, as shown in Figure 8.41, with this alternative.

The shear force to be resisted by the web reinforcement is $100.2 - 57.74 = 42.46$ kips. If this force is taken as the horizontal component of an axial force P in the stiffener,

$$P \cos \theta = 42.46 \text{ kips}$$

where

$$\theta = \tan^{-1}\!\left(\frac{d_b}{d_c}\right) = \tan^{-1}\!\left(\frac{17.90}{8.25}\right) = 65.26°$$

$$P = \frac{42.46}{\cos(65.26°)} = 101.5 \text{ kips}$$

■ **FIGURE 8.41**

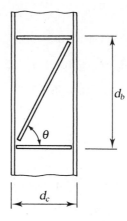

Let

$$\phi R_n = \phi A_{st} F_y = 0.9 A_{st}(36) = 101.5 \text{ kips}$$

Then,

$$A_{st} = \frac{101.5}{0.9(36)} = 3.13 \text{ in.}^2$$

Use two stiffeners, $3 \times \frac{9}{16}$, one on each side of the web.

$$A_{st} \text{ provided} = 2(3)\left(\frac{9}{16}\right) = 3.38 \text{ in.}^2 > 3.13 \text{ in.}^2 \text{ required} \qquad \text{(OK)}$$

Check the width–thickness ratio:

$$\frac{b}{t_{st}} = \frac{3}{9/16} = 5.3 < \frac{95}{\sqrt{36}} = 15.8 \qquad \text{(OK)}$$

Design the welds. The approximate length of each diagonal stiffener is

$$\frac{d_c}{\cos \theta} = \frac{8.25}{\cos (65.26°)} = 19.7 \text{ in.}$$

If welds are used on both sides of the stiffeners, the available length for welding is

$$L = 19.7(4) = 78.8 \text{ in.}$$

The weld size required for strength is

$$w = \frac{P}{0.707L(\phi F_W)} = \frac{101.5}{0.707(78.8)(31.5)} = 0.058 \text{ in.}$$

Use the minimum size of ¼ inch (AISC Table J2.4).

Because of the small size required for strength, investigate the possibility of using intermittent welds. From AISC J2.2b, the

$$\text{Minimum length} = 4w = 4\left(\frac{1}{4}\right)$$

$$= 1.0 \text{ in. but not less than 1.5 in.} \qquad (1.5 \text{ in. controls})$$

The capacities and spacing of a group of four welds are

$$4(0.707)wL(\phi F_W) = 4(0.707)\left(\frac{1}{4}\right)(1.5)(31.5) = 33.41 \text{ kips}$$

$$\text{Required capacity per inch} = \frac{101.5}{19.7} = 5.152 \text{ kips/in.}$$

$$\text{Required spacing of welds} = \frac{33.41}{5.152} = 6.48 \text{ in.}$$

$$\text{Shear capacity of base metal} = 0.54 F_y t_w = 0.54(36)(0.360) = 7.00 \text{ kips/in.}$$

$$\text{Required capacity of weld} = 0.707w(\phi F_W) = 0.707\left(\frac{1}{4}\right)(31.5)$$

$$= 5.57 \text{ kips/in.} < 7.00 \text{ kips/in.} \qquad (\text{OK})$$

ANSWER   Use ¼-inch × 1½-inch intermittent fillet welds spaced at 6 inches on centers, on each side of each diagonal stiffener. ■

As previously mentioned, diagonal stiffeners are usually more practical than doubler plates, but the most economical alternative may be simply to use a larger column section. The labor costs associated with doubler plates and stiffeners (including those opposite the beam flanges) may outweigh the additional cost of material for a larger column.

8.8 END PLATE CONNECTIONS

The end plate connection is a popular beam-to-column and beam-to-beam connection that has been in use since the mid-1950s. Figure 8.42 illustrates two types: the simple, or shear only, connection (Type PR construction) and the rigid (moment resisting) connection (Type FR construction). The rigid version is also called an extended end plate connection. The basis of both types is a plate that is shop welded to the end of a beam and field bolted to a column or another beam. This feature is one of the chief advantages of this type of connection, the other being that ordinarily fewer bolts are required than with other types of connections, thus making erection faster. However, there is little room for error in beam length, and the end must be square. Cambering will make the fit even more critical. Some latitude can be provided in beam length by fabricating it short and achieving the final fit with shims.

■ **FIGURE 8.42**

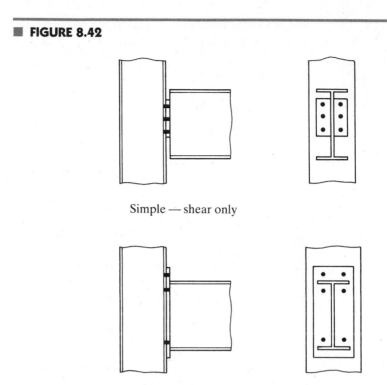

Simple — shear only

Rigid — moment resisting

In a simple connection, care must be taken to make the connection sufficiently flexible that enough end rotation of the beam is possible. This flexibility can be achieved if the plate is relatively short and thin, compared to the fully restrained version of this connection. The *Manual of Steel Construction,* in Part 9, "Simple Shear Connections," recommends that the thickness be between ¼ inch and ⅜ inch to achieve this flexibility. This part of the *Manual* also presents other guidelines and a design example as an introduction to Table 9-5, which contains reaction capacities for various combinations of plates and bolts.

The design of moment-resisting end plate connections requires determination of the plate thickness, the weld size or sizes, and the bolt details. The design of the welds and the bolts is a fairly straightforward application of traditional analysis procedures. Determination of the plate thickness, however, relies on the results of experimental and statistical research (Krishnamurthy, 1978). The tension side of the connection is critical; the bolts on the compression side serve mainly to keep the connection in alignment. If the moment is reversible, the design for the tension side is used on both sides. The general approach is as follows.

1. Determine the force in the tension flange of the beam.

2. Select the bolts needed to resist this force and arrange them symmetrically about the tension flange. If the moment is reversible, use this same arrangement at the compression side. Otherwise, use a nominal number of bolts for alignment purposes. The total number of bolts must be adequate to resist the shear from the beam reaction.

3. Consider a portion of the beam flange and adjacent plate to act as a tee-shape subjected to a tensile load applied to its stem, as shown in Figure 8.43.

4. Select the width and thickness of the "flange" of this tee to satisfy flexural requirements, in much the same way as the design of a tee hanger is approached (see Section 7.8).

■ **FIGURE 8.43**

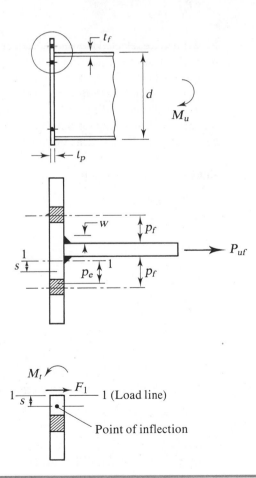

5. Check the shear in the plate.

6. Design the welds.

The *Manual of Steel Construction* (Volume II), presents a detailed design procedure with examples in Part 10, "Fully Restrained (FR) Moment Connections." It is essentially the same as the procedure just outlined, sometimes called the *split-tee method,* with modifications (Krishnamurthy, 1978). The main difference is in Step 4, in the computation of the bending moment in the plate. Traditional analysis accounts for prying forces in the manner discussed in Section 7.8. In the current approach, the selection of the bolts and the plate thickness is independent of any consideration of prying action. The moment computation is based on the statistical analysis of finite element studies that have been confirmed experimentally. The first step in the procedure is to compute the force in the beam tension flange:

$$P_{uf} = \frac{M_u}{d - t_f}$$

Next, bolts are selected to resist this tensile force and are arranged symmetrically in two rows about the beam tension flange. Additional bolts are placed at the compression flange as required for the beam reaction, with a minimum of two. The number needed to resist the beam reaction will be based on the shear capacity or the slip-critical capacity of the bolts, depending on the type of connection. If the connection is a bearing-type, interaction of shear and tension in the bolts should be checked. This investigation may be omitted in a slip-critical connection.

The maximum moment in the split-tee will occur at the "load line," Section 1.1 shown in Figure 8.43, and is

$$M_t = F_1 s$$

where

$$F_1 = \text{shear force} = \frac{P_{uf}}{2}$$

$$s = \text{distance from load line to point of inflection} = \frac{p_e}{2}$$

$$p_e = p_f - 0.25d_b - 0.707w$$

From Figure 8.43, p_f is the distance from the bolt line to the beam flange, which usually equals the bolt diameter d_b plus ½ inch, and w is the weld size; p_f is called the bolt distance, and p_e is the effective bolt distance or effective span. The moment M_t is modified by the factor α_m to obtain the effective moment M_{eu}, or

$$M_{eu} = \alpha_m M_t$$

where

$$\alpha_m = C_a C_b (A_f/A_w)^{1/3} (p_e/d_b)^{1/4}$$

$$C_a = \text{a constant relating the material properties of the bolts and the plate}$$

$$C_b = \sqrt{b_f / b_p}$$

b_f = width of beam flange

b_p = width of end plate [Krishnamurthy (1978) recommends a maximum *effective* width of $b_f + 2w + t_p$, where t_p is the thickness of the end plate. The *Manual* recommends a maximum *actual* width of $b_f + 1$ in.]

A_f = area of beam flange

A_w = area of beam web between fillets

The constant C_a is a function of only the properties of the materials and can be tabulated for the usual grades of structural steel and high-strength bolts. This tabulation is presented in Table 10-1 in Part 10 of the *Manual*. Table 10-2 gives values of A_f / A_w for commonly used beam shapes. Once the moment M_{eu} has been computed, it can be equated to the design strength, and the minimum required plate thickness, $t_{p_{req}}$, can be found. For a rectangular cross section bent about its minor axis, the design strength is

$$\phi_b M_n = \phi_b M_p = \phi_b Z F_y = 0.90 \left(\frac{b_p t_{p_{req}}^2}{4} \right) F_y$$

Equating this expression to the factored load moment and solving for the plate thickness gives

$$0.90 \left(\frac{b_p t_{p_{req}}^2}{4} \right) F_y = M_{eu} \quad \text{or} \quad t_{p_{req}} = \sqrt{\frac{4 M_{eu}}{0.90 b_p F_y}}$$

The tension flange of the beam can be attached to the plate with either a full penetration groove weld or a fillet weld completely surrounding the flange. The full flange force must be developed on the tension side. The web should be welded on both sides with fillet welds capable of resisting the beam reaction. The following additional guidelines must be met in order to satisfy assumptions that underlie the method.

1. Both the plate and the beam must have the same yield stress F_y.
2. The bolt diameter, d_b, should not exceed $1\frac{1}{2}$ inches.
3. The bolts must be tensioned according to AISC Table J3.1.
4. The vertical edge distance should be approximately $1\frac{3}{4} d_b$, but no less than $1\frac{1}{2} d_b$.

■ EXAMPLE 8.13

Design an end plate connection for a W18 × 35 beam. This connection must be capable of transferring a factored load moment of 173 ft-kips and a factored load shear of 34 kips. Use A36 steel, E70XX electrodes, and A325 slip-critical bolts.

SOLUTION   The flange force is

$$P_{uf} = \frac{M_u}{d - t_f} = \frac{173(12)}{17.7 - 0.425} = 120.2 \text{ kips}$$

Try two rows of two bolts at the top flange and two bolts at the bottom flange for a total of six bolts. The design strength in tension for one bolt is

$$\phi R_n = 0.75(90)A_b$$

and the required area of one bolt is

$$A_b = \frac{\text{Required } \phi R_n}{0.75(90)} = \frac{120.2/4}{0.75(90)} = 0.445 \text{ in.}^2$$

ANSWER   Use ⅞-inch-diameter A325 bolts ($A_b = 0.6013 \text{ in.}^2$)

The maximum shear that can be supported by this connection can be determined from a consideration of the slip-critical strength of the bolts (which will be smaller than the shear strength). For six bolts,

$$\phi R_{str} = \phi(1.13\mu T_m N_b N_s) = 1.0(1.13)(0.33)(39)(6)(1) = 87.3 \text{ kips} > 34 \text{ kips}$$
$$\text{(OK)}$$

(The thickness of the column flange is not given and the end plate thickness is not yet known, so the bearing strength cannot be investigated at this time. Once all connected parts have been designed, however, the bearing strength should be checked.) Since this is a slip-critical connection, interaction of shear and tension does not have to be checked.

ANSWER   Use six bolts, with four arranged symmetrically about the tension flange and two located at the compression flange.

For the flange weld, the available length is

$$L = 2b_f + 2t_f - t_w = 2(6.00) + 2(0.425) - 0.300 = 12.55 \text{ in.}$$

The required weld size is

$$w = \frac{P_{uf}}{0.707L(\phi F_w)} = \frac{120.2}{0.707(12.55)(31.5)} = 0.4301 \text{ in.}$$

Although the thickness of the end plate is not yet known, the minimum weld size from AISC Table J2.4 will never exceed 5⁄16 inch, so the 0.430 inch required for strength will control.

ANSWER   Use a 7⁄16-inch fillet weld.

For the end plate, use

$$p_f = d_b + \frac{1}{2} = 0.875 + 0.500 = 1.375 \text{ in.}$$

$$p_e = p_f - 0.25d_b - 0.707w$$

$$= 1.375 - 0.25(0.875) - 0.707\left(\frac{7}{16}\right) = 0.8470 \text{ in.}$$

For the plate width, use

$$b_p = b_f + 1 = 6.00 + 1 = 7.00 \text{ in.}$$

Then,

$$M_t = F_1 s = \left(\frac{P_{uf}}{2}\right)\left(\frac{p_e}{2}\right) = \left(\frac{120.2}{2}\right)\left(\frac{0.8470}{2}\right) = 25.45 \text{ in.-kips}$$

$$C_a = 1.36 \qquad \text{(Table 10-1, Part 10 of the \textit{Manual})}$$

$$C_b = \sqrt{\frac{b_f}{b_p}} = \sqrt{\frac{6.00}{7.00}} = 0.9258$$

$$\frac{A_f}{A_w} = 0.504 \qquad \text{(Table 10-2, Part 10 of the \textit{Manual})}$$

$$\alpha_m = C_a C_b \left(\frac{A_f}{A_w}\right)^{1/3} \left(\frac{p_e}{d_b}\right)^{1/4}$$

$$= 1.36(0.9258)(0.504)^{1/3}(0.8470/0.875)^{1/4} = 0.9939$$

$$M_{eu} = \alpha_m M_t = 0.9939(25.45) = 25.29 \text{ in.-kips}$$

$$t_{P_{req}} = \sqrt{\frac{4M_{eu}}{0.90 b_p F_y}} = \sqrt{\frac{4(25.29)}{0.90(7.00)(36)}} = 0.668 \text{ in.}$$

ANSWER   Use a plate thickness of ¾ inch.

The maximum effective plate width recommended by Krishnamurthy (1978) is

$$b_f + 2w + t_p = 6.00 + 2\left(\frac{7}{16}\right) + \frac{3}{4} = 7.62 \text{ in.} > 7.00 \qquad \text{(OK)}$$

Check shear. The shear force in the plate is

$$F_1 = \frac{P_{uf}}{2} = \frac{120.2}{2} = 60.1 \text{ kips}$$

From AISC J5, the shear strength is

$$\phi R_n = 0.90(0.60 A_g F_y) = 0.90(0.60)\left(7 \times \frac{3}{4}\right)(36)$$

$$= 102 \text{ kips} > 60.1 \text{ kips} \qquad \text{(OK)}$$

To match the shear strength of the web, the required weld strength (two welds, one each side of web) is

$$\frac{\phi_v V_n}{d} = \frac{103}{17.7} = 5.819 \text{ kips/in.}$$

The required weld size is

$$w = \frac{5.819/2}{0.707(31.5)} = 0.131 \text{ in.}$$

Determine the size required to resist bending in the web. When the plastic moment has been reached, the stress in the web equals the yield stress F_y, and the load per unit length of web is

$$\phi_b(F_y \times t_w \times 1) = 0.90(36)(0.300) = 9.720 \text{ kips/in.}$$

The load per weld is $^{9.720}\!/_2 = 4.860$ kips/in. and the required weld size is

$$w = \frac{4.860}{0.707(31.5)} = 0.2182 \text{ in.} > 0.131 \text{ in.}$$

The minimum size is ¼ inch (AISC Table J2.4, based on plate thickness).

ANSWER Use a ¼-inch fillet weld. (The design is summarized in Figure 8.44.) ■

■ **FIGURE 8.44**

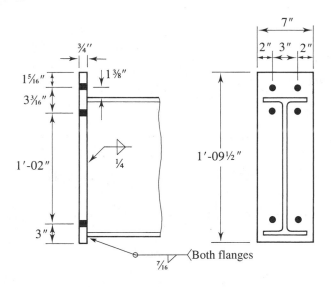

Column Web Stiffener Considerations

A minor liberalization of AISC Equation K1-2, which prevents column web yielding in beam-to-column connections, can be made when end plates are used. This equation is based on limiting the stress on a cross section of the web formed by its thickness and a length of $t_b + 5k$, as shown in Figure 8.45a. As shown in Figure 8.45b, a larger area

■ **FIGURE 8.45**

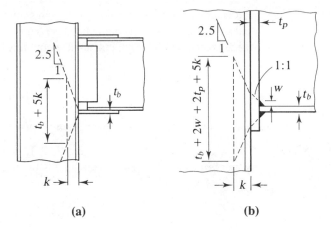

(a)   (b)

will be available when the load is transmitted through the additional thickness of the end plate. If the beam flange-to-plate welds are taken into account and the load is assumed to disperse at a slope of 1 : 1 through the plate, the loaded length of web will be $t_b + 2w + 2t_p + 5k$. Based on experimental studies (Hendrick and Murray, 1984), the term $5k$ can be replaced with $6k$, resulting in the following equation for web yielding strength:

$$\phi R_n = \phi[(6k + t_b + 2w + 2t_p)F_{yw}t_w]$$

where w = weld size.

In addition, local flange bending and web stability (either web crippling or compression buckling) should be investigated. Part 10 of the *Manual* contains guidelines that may be used to account for local flange bending.

8.9   CONCLUDING REMARKS

In this chapter we emphasized the design and analysis of bolts and welds rather than connection fittings, such as framing angles and beam seats. In most cases, the provisions for bearing in bolted connections and base metal shear in welded connections will ensure adequate strength of these parts. Sometimes, however, additional shear investigation is needed. At other times, direct tension or bending must be considered.

Flexibility of the connection is another important consideration. In a shear connection (simple framing), the connecting parts must be flexible enough to permit the connection to rotate under load. Type FR connections (rigid connections), however, should be stiff enough so that relative rotation of the connected members is kept to a minimum.

This chapter is intended to be introductory only and is by no means a complete guide to the design of building connections. Blodgett (1966) is a useful source of detailed welded connection information. Although somewhat dated, it contains numerous practical suggestions. Also recommended is *Detailing for Steel Construction* (AISC, 1983), which is intended for detailers but also contains information useful for designers.

■ PROBLEMS

Eccentric Bolted Connections: Shear Only

8.2-1   For the connection shown in Figure P8.2-1, determine the maximum bolt shear force by an elastic analysis. The 50-kip load is factored.

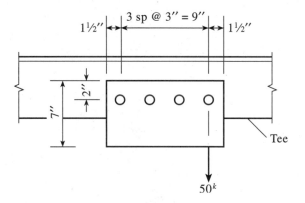

FIGURE P8.2-1

8.2-2   For the connection shown in Figure P8.2-2, determine the maximum bolt shear force by an elastic analysis. The 60-kip load is factored.

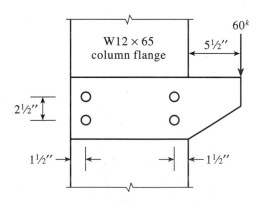

FIGURE P8.2-2

8.2-3 A 100-kip factored load acts on a bolt group, as shown in Figure P8.2-3. The bolts are in single shear. Use an elastic analysis to determine the required diameter for A325 slip-critical bolts. Assume that the bearing strength is adequate.

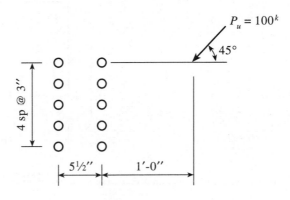

FIGURE P8.2-3

8.2-4 The bolt group shown in Figure P8.2-4 consists of ¾-inch-diameter A325 bearing-type bolts in single shear. Use an elastic analysis to determine the maximum permissible value of the factored load P_u. Assume that the bearing strength is adequate.

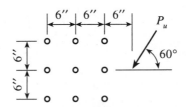

FIGURE P8.2-4

8.2-5 What size A325 slip-critical bolts are needed for the connection shown in Figure P8.2-5? The loads are service loads, and all steel is A36.

8.2-6 A portion of a beam-to-column connection is shown in Figure P8.2-6. The force to be resisted by the bolts in the beam web is the beam reaction V_u. Use an elastic analysis to determine the maximum bolt force in terms of V_u.

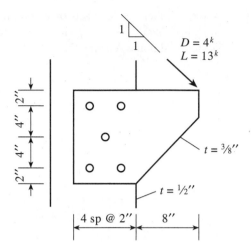

FIGURE P8.2-5

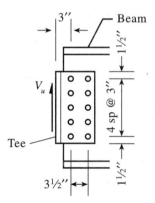

FIGURE P8.2-6

8.2-7 For the connection shown in Figure P8.2-7, determine the maximum permissible value of the factored load P_u. The bolts are $3/4$-inch-diameter A325 slip-critical, and the steel is A572 Grade 50.

8.2-8 For the bolt group shown in Figure P8.2-8,

a. determine the maximum bolt force in terms of P_u.

b. If P_u is a factored load of 65 kips, what size A325 bearing-type bolts are required? The bolts are in single shear. Assume that the bearing strength of the connected parts is adequate.

8.2-9 Determine the required A490 bolt size for the service loads shown in Figure P8.2-9. Slip is not permitted. Assume that the bearing strength of the connected parts is adequate.

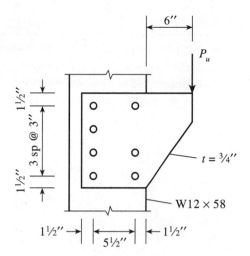

FIGURE P8.2-7

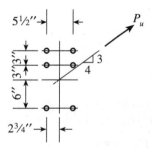

FIGURE P8.2-8

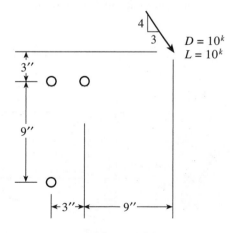

FIGURE P8.2-9

8.2-10 For the bolt group shown in Figure P8.2-10,

    a. compute the maximum bolt force caused by the 10-kip load.

    b. If the 10-kip load is a service load with a live load to dead load ratio of 3 : 1, what diameter A325 bolt is required? Assume that slip is permissible and that the bearing strength of the connected parts is adequate.

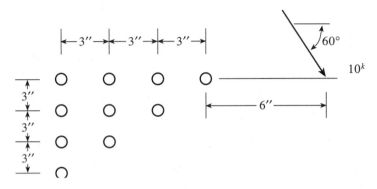

FIGURE P8.2-10

8.2-11 An L7 × 4 × ⅝ tension member is connected with five bolts as shown in Figure P8.2-11. The bolts are in a single line at the usual gage distance (see Section 3.6), which does not coincide with the centroidal axis of the angle.

    a. Use elastic analysis to compute the maximum bolt shear force without considering the eccentricity.

    b. Account for the eccentricity and compute the maximum bolt shear force. What is the percentage increase over the value from part (a)?

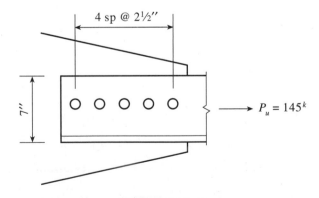

FIGURE P8.2-11

8.2-12 The bracket shown in Figure P8.2-12 must support a 66-kip service load, which is half dead load and half live load.

   a. Use an elastic analysis and compute the maximum bolt force.

   b. Find the maximum bolt force by the ultimate strength method (use the tables in Part 8 of the *Manual,* Volume II). What is the percent difference from the value obtained by elastic analysis?

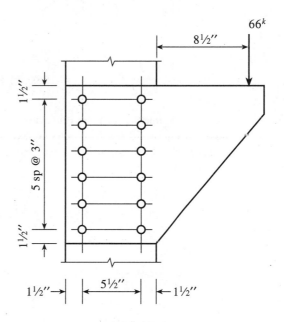

FIGURE P8.2-12

8.2-13 Solve Problem 8.2-3 by the ultimate strength method (use the tables in Part 8 of the *Manual,* Volume II).

8.2-14 **a.** Solve Problem 8.2-4 by the ultimate strength method (use the tables in Part 8 of the *Manual,* Volume II).

   b. If the factored load P_u is 175 kips, how many bolts per vertical row are needed (instead of three, as shown)?

Eccentric Bolted Connections: Shear Plus Tension

8.3-1 A bracket must support the service loads shown in Figure P8.3-1. The connection to the column flange is with ⅞-inch-diameter A325 bearing-type bolts. A36 steel is used. Is the connection adequate?

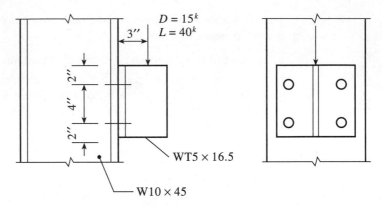

FIGURE P8.3-1

8.3-2 A beam is connected to a column with ¾-inch-diameter A325 bearing-type bolts, as shown in Figure P8.3-2. Ten bolts connect the angles to the column. A572 Grade 50 steel is used. Is the angle-to-column connection adequate? The 150-kip load is a factored beam reaction.

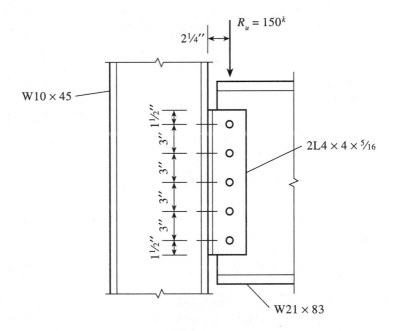

FIGURE P8.3-2

8.3-3 A beam is connected to a column with ⅞-inch-diameter A325 slip-critical bolts, as shown in Figure P8.3-3. A572 Grade 50 steel is used. Based on the eight tee-to-column bolts, what is the maximum permissible value of the factored load, R_u?

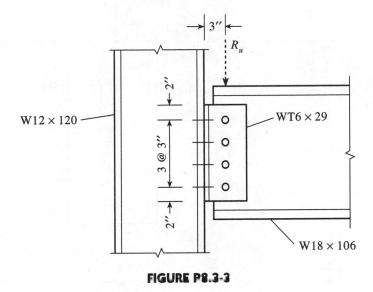

FIGURE P8.3-3

8.3-4 The connection shown in Figure P8.3-4 is with six ⅝-inch-diameter A325 bearing-type bolts. All steel is A36. How much factored load R_u can be supported by the bolts?

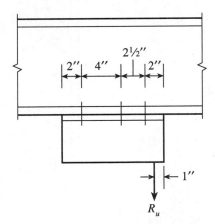

FIGURE P8.3-4

8.3-5 Two angles, $4 \times 4 \times \frac{3}{8}$, are used as a bracket, as shown in Figure P8.3-5. There is a ⅜-inch separation between the angle legs. A325 bearing-type bolts, 1 inch in diameter, are placed at the usual gage distance (see Section 3.6). A572 Grade 50 steel is used. Investigate the adequacy of the ten bolts in the column flange for the factored load of 265 kips.

8.3-6 A bracket is connected to a column flange with eight ⅞-inch-diameter A325 slip-critical bolts, as shown in Figure P8.3-6. A572 Grade 50 steel is used. Are these bolts adequate?

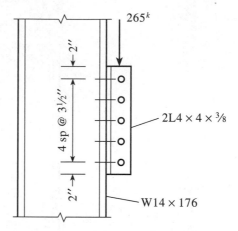

FIGURE P8.3-5

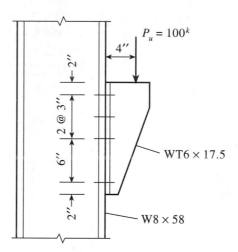

FIGURE P8.3-6

8.3-7 A bracket cut from a W12 × 120 is connected to a W12 × 120 column flange with twelve A325 bearing-type bolts, as shown in Figure P8.3-7. A36 steel is used. What size bolt is required?

8.3-8 A bracket cut from a WT-shape is connected to a column flange with ten A325 slip-critical bolts, as shown in Figure P8.3-8. A572 Grade 50 steel is used. What size bolt is required for the factored loads shown?

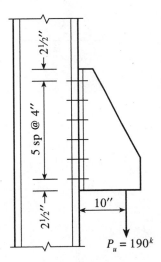

FIGURE P8.3-7

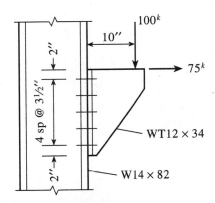

FIGURE P8.3-8

8.3-9   Use A572 Grade 50 steel and A325 slip-critical bolts for the following designs.

a. Design a simply supported beam for the conditions shown in Figure P8.3-9. In addition to its own weight, the beam must support a service live load of 12 kips/ft. Assume continuous lateral support of the compression flange.

b. Design an all-bolted, double-angle connection. Do not consider eccentricity.

c. Consider eccentricity and check the connection designed in part (b). Revise the design if necessary.

d. Prepare a detailed sketch of your recommended connection.

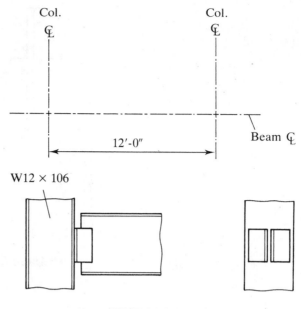

Col.
℄

Col.
℄

Beam ℄

12'-0"

W12 × 106

FIGURE P8.3-9

Eccentric Welded Connections: Shear Only

8.4-1   Use an elastic analysis to determine the maximum load on the weld in kips/inch for the connection shown in Figure P8.4-1.

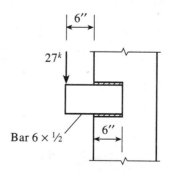

6"

27^k

Bar 6 × ½

6"

FIGURE P8.4-1

8.4-2   Use an elastic analysis to determine the maximum load on the weld in kips/inch for the connection shown in Figure P8.4-2.

8.4-3   Use an elastic analysis to determine the maximum load on the weld in kips/inch in terms of P_u for the connection shown in Figure P8.4-3.

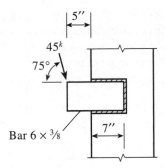

FIGURE P8.4-2

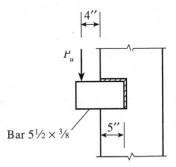

FIGURE P8.4-3

8.4-4 Use the elastic method of analysis and determine the required weld size for E70 electrodes for the connection shown in Figure P8.4-4. A36 steel is used.

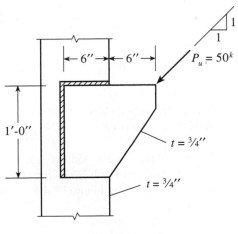

FIGURE P8.4-4

8.4-5 A double-angle shape is used as a bracket, as shown in Figure P8.4-5. The connection to the column flange is with ³⁄₁₆-inch fillet welds. E70 electrodes and A36 steel are used. Is the connection adequate? Use the elastic method and assume that the shear strength of the base metal is adequate.

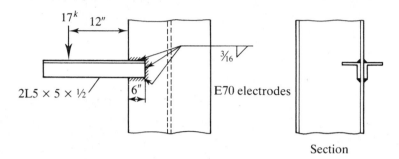

FIGURE P8.4-5

8.4-6 A bar 7 × ⁵⁄₈ is used as a bracket connected to a column flange as shown in Figure P8.4-6. The vertical 3-inch weld segment is on the back side of the bar. Determine the maximum load on the weld in kips/in.

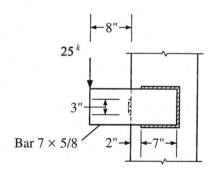

FIGURE P8.4-6

8.4-7 For the connection shown in Figure P8.4-7, determine the required weld size in terms of the factored load P_u. Use E70 electrodes.

8.4-8 The load P is a service load with a live load to dead load ratio of 3 : 1. If the weld is a ³⁄₁₆-inch fillet weld with E70 electrodes, what is the maximum service load P that the connection shown in Figure P8.4-8 can support?

8.4-9 Solve Problem 8.4-1 by the ultimate strength method (use the tables in Part 8 of the *Manual,* Volume II).

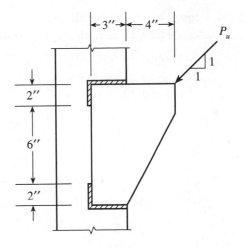

FIGURE P8.4-7

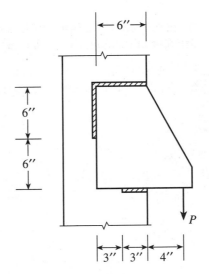

FIGURE P8.4-8

8.4-10 Solve Problem 8.4-2 by the ultimate strength method (use the tables in Part 8 of the *Manual,* Volume II).

8.4-11 Solve Problem 8.4-3 by the ultimate strength method (use the tables in Part 8 of the *Manual,* Volume II).

8.4-12 A connection is to be made with the weld shown in Figure P8.4-12. The applied load is a service load.

  a. Determine the required weld size. Use an elastic analysis.

  b. Determine the required weld size by the ultimate strength method (use the tables in Part 8 of the *Manual*, Volume II).

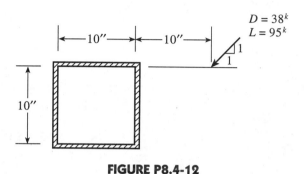

FIGURE P8.4-12

8.4-13 A bracket is welded to a member, as shown in Figure P8.4-13. Use E70 electrodes and determine the required weld size, in terms of the factored load P_u, as follows.

  a. Use an elastic analysis.

  b. Use the ultimate strength method (use the tables in Part 8 of the *Manual*, Volume II).

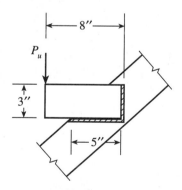

FIGURE P8.4-13

8.4-14 A single-angle tension member is welded to a gusset plate, as shown in Figure P8.4-14.

  a. Determine the maximum load on the weld in kips/in. if the eccentricity is neglected.

  b. Determine the maximum load if the eccentricity is accounted for. What is the percentage difference between this load and that in part (a)?

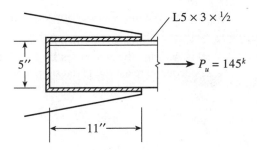

FIGURE P8.4-14

8.4-15 A single-angle tension member is connected to a gusset plate, as shown in Figure P8.4-15. A36 steel is used. Use the minimum size fillet weld and design a welded connection in the following ways.

a. Do not balance the welds. Provide a sketch of your design.

b. Balance the welds. Provide a sketch of your design.

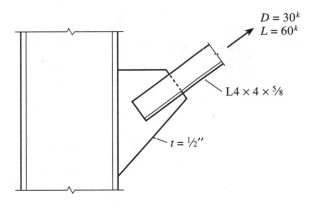

FIGURE P8.4-15

8.4-16 Rework Problem 7.11-9, using balanced welds. Provide a sketch of your design.

8.4-17 Design a welded connection for the $2L5 \times 3 \times \frac{1}{2}$ tension member in Problem 7.9-5. Balance the welds to eliminate eccentricity. Provide a sketch of your design.

8.4-18 Rework Problem 7.11-3, assuming that the 8 inches of weld are distributed so as to eliminate eccentricity. Use weld lengths to the nearest $\frac{1}{4}$ inch, and show the weld placement on a sketch.

Eccentric Welded Connections: Shear Plus Tension

8.5-1   A WT5 × 15 bracket is welded to a W10 × 33 column, as shown in Figure P8.5-1. The steel is A36 and the electrodes are E70. Neglect the end returns and compute the weld size needed to resist the given service loads.

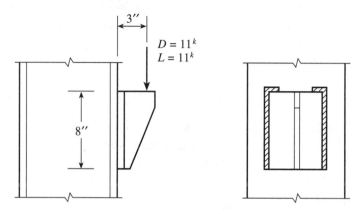

FIGURE P8.5-1

8.5-2   Compute the maximum load on the weld shown in Figure P8.5-2 in kips/in. in terms of the load P_u.

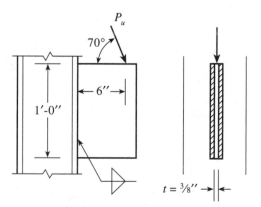

FIGURE P8.5-2

8.5-3   A W24 × 55 beam is connected to a W10 × 60 column with two angles, 4 × 3 × ⅜, as shown in Figure P8.5-3. All steel is A572 Grade 50. The angle-to-column welds are ¼-inch E70 fillet welds. Check the adequacy of the welds.

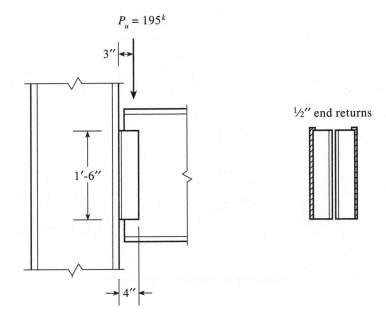

$P_u = 195^k$

3″

1′-6″

4″

½″ end returns

FIGURE P8.5-3

8.5-4 A WT5 × 15 bracket is connected to a W10 × 30 column with ³⁄₁₆-inch E70 fillet welds, as shown in Figure P8.5-4. All steel is A572 Grade 50. What is the maximum factored load P_u that can be supported?

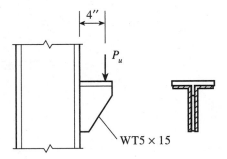

4″

P_u

WT5 × 15

FIGURE P8.5-4

8.5-5 An L6 × 4 × ½ of A36 steel is used to support a factored beam reaction of 35 kips. This seat angle is 6 inches long and is welded to a W12 × 40 column, as shown in Figure P8.5-5.

a. Neglect the end returns and determine the required fillet weld size. Use E70 electrodes.

b. Use the size obtained in part (a) to determine the end return length to the nearest ½ inch, then determine the required weld size accounting for these end returns.

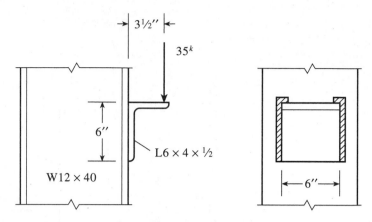

FIGURE P8.5-5

Moment-Resisting Connections

8.6-1 What is the moment strength of the connection in Figure P8.6-1? The fasteners are ⅝-inch diameter A325 bolts, and slip is permitted. Structural steel is A36.

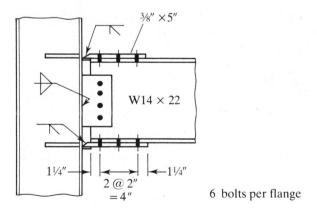

FIGURE P8.6-1

8.6-2 Design a connection of the type shown in Figure P8.6-1 for a W10 × 45 column and a W16 × 45 beam. The maximum factored load moment and shear, which are caused by gravity loads, are M_u = 200 ft-kips and V_u = 84 kips. Use A572 Grade 50 steel, E70XX electrodes, and A325 bolts. Assume that slip of the connection is permissible.

Column Stiffeners and Other Reinforcement

8.7-1 Determine whether column stiffeners are required for the maximum force that can be developed in the beam flange plates shown in Figure P8.7-1. If they are, specify the required dimensions. The plates are welded to the beam flanges, and A572 Grade 50 steel is used.

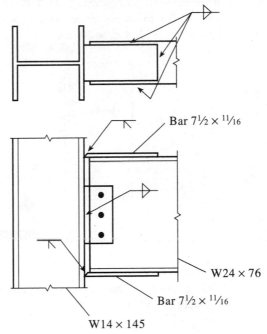

Bar $7\frac{1}{2} \times \frac{11}{16}$

W24 × 76

Bar $7\frac{1}{2} \times \frac{11}{16}$

W14 × 145

FIGURE P8.7-1

8.7-2 Determine whether column stiffeners are required for the maximum force to be developed in the flange of the beam shown in Figure P8.7-2. If they are, specify the required dimensions. Use A36 steel.

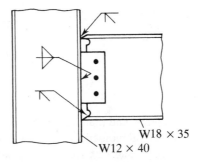

W18 × 35
W12 × 40

FIGURE P8.7-2

8.7-3 **a.** Design a connection similar to the one shown in Figure 8.29a for the conditions of Problem 8.6-2. Provide stiffeners if needed.
b. If panel zone reinforcement is required, provide two alternatives: (1) diagonal stiffeners and (2) a larger column.

End Plate Connections

8.8-1   Investigate the adequacy of the bolts in the end plate connection shown in Figure P8.8-1. The loads are service loads, consisting of 25% dead load and 75% live load. All steel is A572 Grade 50 steel.

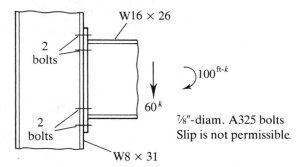

FIGURE P8.8-1

8.8-2   Investigate the adequacy of the bolts in the end plate connection shown in Figure P8.8-2. The loads are service loads, consisting of 25% dead load and 75% live load. All steel is A572 Grade 50 steel.

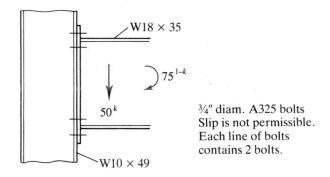

FIGURE P8.8-2

8.8-3   Design an end plate connection for a W21 × 50 beam to a W10 × 49 column. Design for the full moment and shear capacities of the beam. Use A572 Grade 50 steel for the members and A36 for the end plate. Use A325 bearing-type bolts.

8.8-4   Design an end plate connection for a W16 × 31 beam to a W10 × 60 column. The shear consists of a 14-kip service dead load and a 50-kip service live load. The service dead load moment is 10 ft-kips, and the service live load moment is 20 ft-kips. Use A572 Grade 50 steel for all components and A325 bolts. Slip of the connection is not permitted.

9 Composite Construction

9.1 ## INTRODUCTION

Composite construction employs structural members that are composed of two materials: structural steel and reinforced concrete. Strictly speaking, any structural member formed with two or more materials is composite. In buildings and bridges, however, that usually means structural steel and reinforced concrete, and that usually means composite beams or columns. Composite columns are being used again in some structures after a period of disuse; we cover them later in this chapter. Our coverage of beams is restricted to those that are part of a floor or roof system. Composite construction is covered in AISC Chapter I, "Composite Members."

Composite beams can take several forms. The earliest versions consisted of beams encased in concrete (Figure 9.1a). This was a practical alternative when the primary means of fireproofing structural steel was to encase it in concrete; the rationale was that if the concrete was there, we might as well account for its contribution to the strength of the beam. Currently, lighter and more economical methods of fireproofing are available, and encased composite beams are rarely used. Instead, composite behavior is achieved by connecting the steel beam to the reinforced concrete slab it supports, causing the two parts to act as a unit. In a floor or roof system, a portion of the slab acts with each steel beam to form a composite beam consisting of the rolled steel shape augmented by a concrete flange at the top (Figure 9.1b).

This unified behavior is possible only if horizontal slippage between the two components is prevented. That can be accomplished if the horizontal shear at the interface is resisted by connecting devices known as *shear connectors*. These devices — which can be headed studs, spiral reinforcing steel, or short lengths of small channel shapes — are welded to the top flange of the steel beam at prescribed intervals and provide the connection mechanically through anchorage in the hardened concrete (Figure 9.1c). Studs are the most commonly used type of shear connectors, and more than one can be used at each location if the flange is wide enough to accommodate them (which depends on the allowable spacing, which we consider in Section 9.4). One reason for the popularity of shear studs is their ease of installation. It is essentially a one-worker job made possible by an automatic tool that allows the operator to position the stud and weld it to the beam in one operation.

A certain number of shear connectors will be required to make a beam fully composite. Any fewer than this number will permit some slippage to occur between the steel and concrete; such a beam is said to be *partially composite*. Partially composite beams (which actually are more efficient than fully composite beams) are covered in Section 9.7.

■ **FIGURE 9.1**

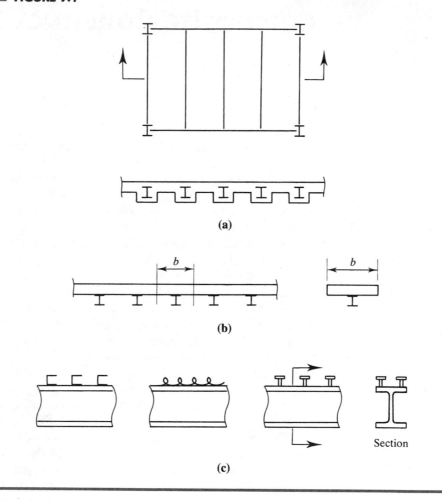

(a)

(b)

(c)

Section

Most composite construction in buildings utilizes formed steel deck, which serves as formwork for the concrete slab and is left in place after the concrete cures. This metal deck also contributes to the strength of the slab, the design of which we do not consider here. The deck can be used with its ribs oriented either transverse or parallel to the beams. In the usual floor system, the ribs will be perpendicular to the floor beams and parallel to the supporting girders. The shear studs are welded to the beams from above, through the deck. Since they can be placed only in the ribs, the spacing of the studs along the length of the beam is limited to multiples of the rib spacing. Figure 9.2 shows a slab with formed steel deck and the ribs perpendicular to the beam axis.

Almost all highway bridges that use steel beams are of composite construction, and composite beams are frequently the most economical alternative in buildings. Although

■ **FIGURE 9.2**

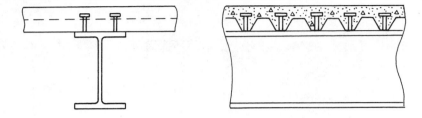

smaller, lighter rolled steel beams can be used with composite construction, this advantage will sometimes be offset by the additional cost of the shear connectors. Even so, other advantages may make composite construction attractive. Shallower beams can be used, and deflections will be smaller than with conventional noncomposite construction.

Elastic Stresses in Composite Beams

Although the design strength of composite beams is usually based on conditions at failure, an understanding of the behavior at service loads is important for several reasons. Deflections are always investigated at service loads, and in some cases, the design strength is based on the limit state of first yield.

Flexural and shearing stresses in beams of homogeneous materials can be computed from the formulas

$$f_b = \frac{Mc}{I} \quad \text{and} \quad f_v = \frac{VQ}{It}$$

A composite beam is not homogeneous, however, and these formulas are not valid. To be able to use them, an artifice known as the *transformed section* is employed to "convert" the concrete into an amount of steel that has the same effect as the concrete. This procedure requires the strains in the fictitious steel to be the same as those in the concrete it replaces. Figure 9.3 shows a segment of a composite beam with stress and strain diagrams superimposed. If the slab is properly attached to the rolled steel shape, the strains must be as shown, consistent with small displacement theory, which states that cross sections that are plane before bending remain plane after bending. However, the linear stress distribution shown is valid only if the beam is assumed to be homogeneous. We first require that the strain in the concrete at any point be equal to the strain in any replacement steel at that point:

$$\varepsilon_c = \varepsilon_s \quad \text{or} \quad \frac{f_c}{E_c} = \frac{f_s}{E_s}$$

■ **FIGURE 9.3**

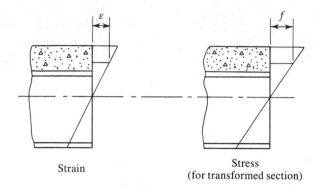

Strain	Stress (for transformed section)

---

and

$$f_s = \frac{E_s}{E_c} f_c = nf_c \qquad\qquad (9.1)$$

where

E_c = modulus of elasticity of concrete

$n = \dfrac{E_s}{E_c}$ = modular ratio

AISC I2.2 gives the modulus of elasticity of concrete as*

$$E_c = w_c^{1.5} \sqrt{f_c'}$$

where

w_c = unit weight of concrete (lb/ft³)

f_c' = 28-day compressive strength of concrete (kips/in.²)

Normal-weight concrete weighs approximately 145 lb/ft³.

   Equation 9.1 can be interpreted as follows: n square inches of concrete are required to resist the same force as one square inch of steel. To determine the area of steel that will resist the same force as the concrete, divide the concrete area by n. That is, replace A_c by A_c/n. The result is the *transformed area*.

   Consider the composite section shown in Figure 9.4a (determination of the effective flange width b when the beam is part of a floor system is discussed presently). To

---

*The ACI Building Code (ACI, 1995) gives the value of E_c as $w_c^{1.5}(33)\sqrt{f_c'}$, where f_c' is in pounds per square inch.

■ **FIGURE 9.4**

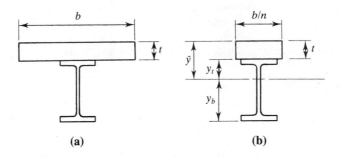

(a)   (b)

transform the concrete area, A_c, we must divide by n. The most convenient way to do this is to divide the width by n and leave the thickness unchanged. Doing so results in the homogeneous steel section of Figure 9.4b. To compute stresses, we locate the neutral axis of this composite shape and compute the corresponding moment of inertia. We can then compute bending stresses with the flexure formula. At the top of the steel,

$$f_{st} = \frac{My_t}{I_{tr}}$$

At the bottom of the steel,

$$f_{sb} = \frac{My_b}{I_{tr}}$$

where

> M = applied bending moment
>
> I_{tr} = moment of inertia about the neutral axis (same as the centroidal axis for this homogeneous section).
>
> y_t = distance from the neutral axis to the top of the steel
>
> y_b = distance from the neutral axis to the bottom of the steel

The stress in the concrete may be computed in the same way, but because the material under consideration is steel, the result must be divided by n (see Equation 9.1) so that

$$\text{Maximum } f_c = \frac{M\bar{y}}{nI_{tr}}$$

where $\bar{y}$ is the distance from the neutral axis to the top of the concrete.

This procedure is valid only for a positive bending moment, with compression at the top, because concrete has negligible tensile strength.

■ EXAMPLE 9.1

A composite beam consists of a W16 × 36 of A36 steel with a 5-inch thick × 87-inch wide reinforced concrete slab at the top. The strength of the concrete is $f_c' = 4000$ psi. Determine the maximum stresses in the steel and concrete resulting from a positive bending moment of 160 ft-kips.

SOLUTION

$$E_c = w_c^{1.5}\sqrt{f_c'} = (145)^{1.5}\sqrt{4} = 3492 \text{ ksi}$$

$$n = \frac{E_s}{E_c} = \frac{29{,}000}{3492} = 8.3 \quad \text{use } n = 8$$

Since the modulus of elasticity of concrete can only be approximated, the usual practice of rounding n to the nearest whole number is sufficiently accurate. Thus

$$\frac{b}{n} = \frac{87}{8} = 10.88 \text{ in.}$$

The transformed section is shown in Figure 9.5.

The location of the neutral axis can be found by applying the principle of moments with the axis of moments at the top of the slab. The computations are summarized in Table 9.1, and the distance from the top of the slab to the centroid is

$$\bar{y} = \frac{\Sigma Ay}{\Sigma A} = \frac{273.1}{65.00} = 4.202 \text{ in.}$$

Applying the parallel axis theorem and tabulating the computations in Table 9.2, we obtain the moment of inertia of the transformed section as

$$I_{tr} = 1526 \text{ in.}^4$$

The stress at the top of the steel is

$$y_t = \bar{y} - t = 4.202 - 5.000 = -0.7980 \text{ in.}$$

where t is the thickness of the slab.

$$f_{st} = \frac{My_t}{I_{tr}} = \frac{(160 \times 12)(0.7980)}{1526} = 1.00 \text{ ksi} \quad \text{(tension)}$$

■ **FIGURE 9.5**

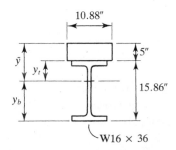

■ **TABLE 9.1**

Component	A	y	Ay
Concrete	54.40	2.50	136.0
W16 × 36	10.6	12.93	137.1
	65.00		273.1

■ **TABLE 9.2**

Component	A	y	$\bar{I}$	d	$\bar{I} + Ad^2$
Concrete	54.40	2.50	113.3	1.702	270.9
W16 × 36	10.6	12.93	448	8.728	1255
					1525.9

(The top of the steel is below the neutral axis, so f_{st} is a *tensile* stress.)
Stress at the bottom of the steel:

$$y_b = t + d - \bar{y} = 5 + 15.86 - 4.202 = 16.66 \text{ in.}$$

$$f_{sb} = \frac{My_b}{I_{tr}} = \frac{(160 \times 12)(16.66)}{1526} = 21.0 \text{ ksi} \quad \text{(tension)}$$

The stress at the top of the concrete is

$$f_c = \frac{M\bar{y}}{nI_{tr}} = \frac{(160 \times 12)(4.202)}{8(1526)} = 0.661 \text{ ksi}$$

If the concrete is assumed to have no tensile strength, the concrete below the neutral axis should be discounted. The geometry of the transformed section will then be different from what was originally assumed; to obtain an accurate result, the location of the neutral axis should be recomputed on the basis of this new geometry. Referring to Figure 9.6 and Table 9.3, we can compute the new location of the neutral axis as follows:

$$\bar{y} = \frac{\Sigma Ay}{\Sigma A} = \frac{5.44\bar{y}^2 + 137.1}{10.88\bar{y} + 10.6}$$

$$\bar{y}(10.88\bar{y} + 10.6) = 5.44\bar{y}^2 + 137.1$$

$$5.44\bar{y}^2 + 10.6\bar{y} - 137.1 = 0$$

$$\bar{y} = 4.140 \text{ in.}$$

The moment of inertia of this revised composite area is

$$I_{tr} = \frac{1}{3}(10.88)(4.140)^3 + 448 + 10.6(12.93 - 4.140)^2 = 1524 \text{ in.}^4$$

■ **FIGURE 9.6**

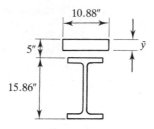

■ **TABLE 9.3**

Component	A	y	Ay
Concrete	$10.88\bar{y}$	$\bar{y}/2$	$5.44\bar{y}^2$
W16 × 36	10.6	12.93	137.1

and the stresses are

$$f_{st} = \frac{(160 \times 12)(5 - 4.140)}{1524} = 1.08 \text{ ksi} \qquad \text{(tension)}$$

$$f_{sb} = \frac{(160 \times 12)(5 + 15.86 - 4.140)}{1524} = 21.1 \text{ ksi} \qquad \text{(tension)}$$

$$f_c = \frac{(160 \times 12)(4.140)}{8(1524)} = 0.652 \text{ ksi}$$

The difference between the two analyses is negligible, and the refinement in locating the neutral axis is not necessary.

ANSWER   The maximum stress in the steel is 21.1 ksi tension, and the maximum stress in the concrete is 0.652 ksi compression. ■

Flexural Strength

In most cases, the nominal flexural strength will be reached when the entire steel cross section yields and the concrete crushes in compression. The corresponding stress distribution on the composite section is called a *plastic stress distribution*. The AISC Specification gives the design strength for *positive* bending as $\phi_b M_n$, which is defined as follows.

For shapes with compact webs — that is, $h/t_w \leq 640/\sqrt{F_y}$ — the resistance factor ϕ_b is 0.85, and M_n is obtained from the plastic stress distribution.

For shapes with noncompact webs $(h/t_w > 640/\sqrt{F_y})$, ϕ_b is 0.9, and M_n is obtained from the elastic stress distribution corresponding to first yielding of the steel.

All shapes tabulated in the *Manual* have compact webs, so the first condition will govern for all composite beams except those with built-up steel shapes. We consider only compact shapes in this chapter.

When a composite beam has reached the plastic limit state, the stresses will be distributed in one of the three ways shown in Figure 9.7. The concrete stress is shown as a uniform compressive stress of $0.85f_c'$, extending from the top of the slab to a depth that may be equal to or less than the total slab thickness. This distribution is the *Whitney equivalent stress distribution,* which has a resultant that matches that of the actual stress distribution (ACI, 1995). Figure 9.7a shows the distribution corresponding to full tensile yielding of the steel and partial compression of the concrete, with the plastic neutral axis (PNA) in the slab. The tensile strength of concrete is small and is discounted, so no stress is shown where tension is applied to the concrete. This condition will usually prevail when there are enough shear connectors provided to prevent slip completely — that is, to ensure full composite behavior. In Figure 9.7b, the concrete stress block extends the full depth of the slab, and the PNA is in the flange of the steel shape. Part of the flange will therefore be in compression to augment the compressive force in the slab. The third possibility, the PNA in the web, is shown in Figure 9.7c. Note that the concrete stress block need not extend the full depth of the slab for any of these three cases.

In each case shown in Figure 9.7, we can find the nominal moment capacity by computing the moment of the couple formed by the compressive and tensile resultants. This can be accomplished by summing the moments of the resultants about any convenient point. Because of the connection of the steel shape to the concrete slab, lateral-torsional buckling is no problem once the concrete has cured and composite action has been achieved.

To determine which of the three cases governs, compute the compressive resultant as the smallest of

1. $A_s F_y$
2. $0.85f_c' A_c$
3. $\sum Q_n$

where

A_s = cross-sectional area of steel shape

A_c = area of concrete

= tb (see Figure 9.7)

$\sum Q_n$ = total shear strength of the shear connectors

Each possibility represents a horizontal shear force at the interface between the steel and the concrete. When the first possibility controls, the steel is being fully utilized, and the stress distribution of Figure 9.7a applies. The second possibility corresponds to the

■ **FIGURE 9.7**

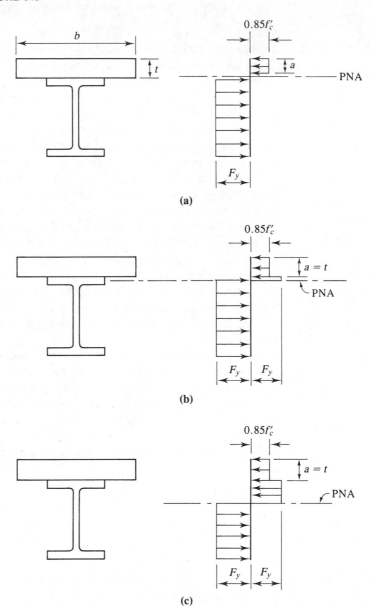

concrete controlling, and the PNA will be in the steel (Figure 9.7b or c). The third case governs only when there are fewer shear connectors than required for full composite behavior, resulting in partial composite behavior. Although partial composite action can exist with either solid slabs or slabs with formed steel deck, it will be covered in Section 9.7, "Composite Beams with Formed Steel Deck."

■ EXAMPLE 9.2

Compute the design strength of the composite beam of Example 9.1. Assume that sufficient shear connectors are provided for full composite behavior.

SOLUTION

Determine the compressive force C in the concrete (horizontal shear force at the interface between the concrete and steel). Because there will be full composite action, this force will be the smaller of A_sF_y and $0.85f_c'A_c$:

$$A_sF_y = 10.6(36) = 381.6 \text{ kips}$$
$$0.85f_c'A_c = 0.85(4)(5 \times 87) = 1479 \text{ kips}$$

The steel controls; $C = 381.6$ kips. This means that the full depth of the slab is not needed to develop the required compression force. The stress distribution shown in Figure 9.8 will result.

The resultant compressive force can also be expressed as

$$C = 0.85f_c'ab$$

from which we obtain

$$a = \frac{C}{0.85f_c'b} = \frac{381.6}{0.85(4)(87)} = 1.290 \text{ in.}$$

■ FIGURE 9.8

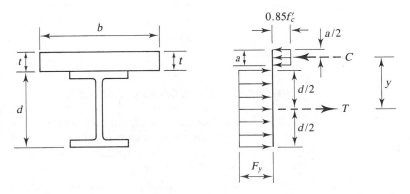

The force C will be located at the centroid of the compressive area at a depth of $a/2$ from the top of the slab. The resultant tensile force T (equal to C) will be located at the centroid of the steel area. The moment arm of the couple formed by C and T is

$$y = \frac{d}{2} + t - \frac{a}{2} = \frac{15.86}{2} + 5 - \frac{1.290}{2} = 12.28 \text{ in.}$$

The nominal strength is the moment of the couple, or

$$M_n = Cy = Ty = 381.6(12.28) = 4686 \text{ in.-kips} = 390.5 \text{ ft-kips}$$

and the design strength is

$$\phi_b M_n = 0.85(390.5) = 332 \text{ ft-kips}$$

ANSWER   Design strength = 332 ft-kips ■

When full composite behavior exists, the conditions of Example 9.2 will be the norm. Analysis for the case of the plastic neutral axis located within the steel section will be deferred until partial composite action has been covered.

9.2   SHORED VERSUS UNSHORED CONSTRUCTION

Until the concrete has cured and attained its design strength (at least 75% of its 28-day compressive strength f_c'), there can be no composite behavior, and the weight of the slab must be supported by some other means. Once the concrete has cured, composite action is possible, and all subsequently applied loads will be resisted by the composite beam. If the steel shape is supported at a sufficient number of points along its length before the slab is placed, the weight of the wet concrete will be supported by these temporary *shores* rather than by the steel. Once the concrete has cured, the temporary shoring can be removed, and the weight of the slab, as well as any additional loads, will be carried by the composite beam. If shoring is not used, however, the rolled steel shape must support not only its own weight, but also the weight of the slab and its formwork during the curing period. Once composite behavior is attained, additional loads, both dead and live, will be supported by the composite beam. We now consider these different conditions in more detail.

Unshored: Before Concrete Cures

AISC I3.4 requires that when temporary shoring is not provided, the steel shape alone must have sufficient strength to resist all loads applied before the concrete attains 75% of its strength, f_c'. The flexural strength is computed in the usual way, based on Chapter F of the Specification (Chapter 5 of this book). Depending on its design, the formwork for the concrete slab may or may not provide lateral support for the steel beam.

If not, the unbraced length L_b must be taken into account, and lateral-torsional buckling may control the flexural strength. If temporary shoring is not used, the steel beam may also be called on to resist incidental construction loads. To account for these loads, an additional 20 pounds per square foot is recommended (Hansell et al., 1978).

Unshored: After Concrete Cures

After composite behavior is achieved, all loads subsequently applied will be supported by the composite beam. At failure, however, all loads will be resisted by the internal couple corresponding to the stress distribution at failure. Thus the composite section must have adequate strength to resist all loads, including those applied to the steel beam before the concrete cures.

Shored Construction

In shored construction, only the composite beam need be considered, because the steel shape will not be required to support anything other than its own weight.

Shear Strength

AISC I3.6 conservatively requires that all shear be resisted by the web of the steel shape, as provided for in Chapter F of the Specification.

■ EXAMPLE 9.3

A W12 × 50 acts compositely with a 4-inch-thick concrete slab. The effective slab width is 72 inches. Shoring is not used, and the applied bending moments are as follows: from the beam weight, $M_{\text{beam}} = 13$ ft-kips; from the slab weight, $M_{\text{slab}} = 77$ ft-kips; and from the live load, $M_L = 38$ ft-kips. (Do not consider any additional construction loads in this example.) Steel is A36, and $f_c' = 4000$ psi. Determine whether the flexural strength of this beam is adequate. Assume full composite action and assume that the formwork provides lateral support of the steel section before curing of the concrete.

SOLUTION Before the concrete cures, there is only dead load (no construction load in this example). Hence load combination A4-1 controls, and the factored load moment is

$$M_u = 1.4(M_D) = 1.4(13 + 77) = 126 \text{ ft-kips}$$

From the beam design charts in Part 4 of the *Manual,* for A36 steel,

$$\phi_b M_n = 195 \text{ ft-kips} > 126 \text{ ft-kips} \qquad \text{(OK)}$$

After the concrete has cured, the factored load moment that must be resisted by the composite beam is

$$M_u = 1.2 M_D + 1.6 M_L = 1.2(13 + 77) + 1.6(38) = 168.8 \text{ ft-kips}$$

The compressive force C is the smaller of

$$A_s F_y = 14.7(36) = 529.2 \text{ kips}$$

or

$$0.85 f_c' A_c = 0.85(4)(4 \times 72) = 979.2 \text{ kips}$$

The PNA is in the concrete, and $C = 529.2$ kips. From Figure 9.8, the depth of the compressive stress block is

$$a = \frac{C}{0.85 f_c' b} = \frac{529.2}{0.85(4)(72)} = 2.162 \text{ in.}$$

The moment arm is

$$y = \frac{d}{2} + t - \frac{a}{2} = \frac{12.19}{2} + 4 - \frac{2.162}{2} = 9.014 \text{ in.}$$

The design moment is

$$\phi_b M_n = \phi_b C y = 0.85(529.2)(9.014)$$
$$= 4055 \text{ in.-kips} = 338 \text{ ft-kips} > 168.8 \text{ ft-kips} \quad \text{(OK)}$$

ANSWER   The beam has sufficient flexural strength.   ■

Obviously, shored construction is more efficient than unshored construction because the steel section is not called on to support anything other than its own weight. In some situations, the use of shoring will enable a smaller steel shape to be used. Most composite construction is unshored, however, because the additional cost of the shores, especially the labor cost, outweighs the small savings in steel weight that may result. Consequently, we devote the remainder of this chapter to unshored composite construction.

9.3   EFFECTIVE FLANGE WIDTH

The portion of the floor slab that acts compositely with the steel beam is a function of several factors, including the span length and beam spacing. AISC I3.1 requires that the effective width of floor slab on *each side* of the beam centerline be taken as the smallest of

(a) one eighth of the span length,

(b) one half of the beam center-to-center spacing, or

(c) the distance from the beam centerline to the edge of the slab.

The third criterion will apply to edge beams only, so for *interior* beams, the full effective width will be the smaller of one fourth the span length or the center-to-center spacing of the beams (assuming that the beams are evenly spaced).

■ **EXAMPLE 9.4**

A composite floor system consists of W21 × 44 steel beams spaced at 9 feet and supporting a 4.5-inch-thick reinforced concrete slab. The span length is 30 feet. In addition to the weight of the slab, there is a 20 psf partition load and a live load of 125 psf (light manufacturing). The steel is A36, and the concrete strength is $f_c' = 4000$ psi. Investigate a typical interior beam for compliance with the AISC Specification if no temporary shores are used. Assume full lateral support during construction and an additional construction load of 20 psf. Sufficient shear connectors are provided for full composite action.

SOLUTION

Loads applied before the concrete cures include the weight of the slab, or (4.5/12)(150) = 56.25 psf. (Although normal-weight concrete weights 145 pcf, recall that *reinforced concrete* is assumed to weigh 150 pcf.) For a beam spacing of 9 ft, the dead load is

$$56.25(9) = 506 \text{ lb/ft}$$
$$+ \text{ Beam weight} = \underline{\quad 44 \text{ lb/ft}}$$
$$550 \text{ lb/ft}$$

The construction load is 20(9) = 180 lb/ft, which is treated as a live load. The factored load and moment are

$$w_u = 1.2w_D + 1.6w_L = 1.2(550) + 1.6(180) = 948 \text{ lb/ft}$$
$$M_u = \frac{1}{8}(0.948)(30)^2 = 106.6 \text{ ft-kips}$$

From the Load Factor Design Selection Table,

$$\phi_b M_n = \phi_b M_p = 258 \text{ ft-kips} > 106.6 \text{ ft-kips} \qquad \text{(OK)}$$

After the concrete cures, the construction loads do not act, but the partition load does, and it will be treated as a dead load (see Example 5.13):

$$w_{\text{part}} = 20(9) = 180 \text{ lb/ft}$$
$$w_D = 506 + 44 + 180 = 730 \text{ lb/ft}$$

The live load is

$$w_L = 125(9) = 1125 \text{ lb/ft}$$

The factored load and moment are

$$w_u = 1.2w_D + 1.6w_L = 1.2(730) + 1.6(1125) = 2676 \text{ lb/ft}$$
$$M_u = \frac{1}{8}(2.676)(30)^2 = 301 \text{ ft-kips}$$

To obtain the effective flange width, use either

$$\frac{\text{Span}}{4} = \frac{30(12)}{4} = 90 \text{ in. or Beam spacing} = 9(12) = 108 \text{ in.}$$

Since this member is an interior beam, the third criterion is not applicable. Use $b = 90$ inches as the effective flange width. Then, as shown in Figure 9.9, the compressive force is the smaller of

$$A_s F_y = 13(36) = 468 \text{ kips}$$

or

$$0.85 f_c' A_c = 0.85(4)(4.5)(90) = 1377 \text{ kips}$$

Use $C = 468$ kips. From Figure 9.9,

$$a = \frac{C}{0.85 f_c' b} = \frac{468}{0.85(4)(90)} = 1.529 \text{ in.}$$

$$y = \frac{d}{2} + t - \frac{a}{2} = 10.33 + 4.5 - \frac{1.529}{2} = 14.07 \text{ in.}$$

$$\phi_b M_n = \phi_b C y = 0.85(468)(14.07)$$
$$= 5595 \text{ in.-kips} = 466 \text{ ft-kips} > 301 \text{ ft-kips} \quad \text{(OK)}$$

Check the shear:

$$V_u = \frac{w_u L}{2} = \frac{2.676(30)}{2} = 40.1 \text{ kips}$$

From the factored uniform load tables,

$$\phi_v V_n = 141 \text{ kips} > 40.1 \text{ kips} \quad \text{(OK)}$$

■ **FIGURE 9.9**

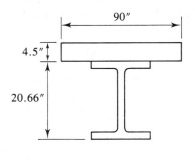
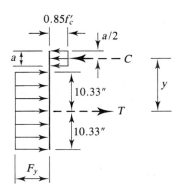

ANSWER   The beam complies with the AISC Specification.

SHEAR CONNECTORS

As we have shown, the horizontal shear force to be transferred between the concrete and the steel is equal to the compressive force in the concrete, C. We denote this horizontal shear force V_h. Thus V_h is given by the smallest of $A_s F_y$, $0.85 f_c' A_c$, or ΣQ_n. If $A_s F_y$ or $0.85 f_c' A_c$ controls, full composite action will exist and the number of shear connectors required between the points of zero moment and maximum moment is

$$N_1 = \frac{V_h}{Q_n} \tag{9.2}$$

where Q_n is the nominal shear strength of one connector. The N_1 connectors should be uniformly spaced within the length where they are required. The AISC Specification gives equations for the strength of both stud and channel shear connectors. As indicated at the beginning of this chapter, stud connectors are the most common, and we consider only this type. For one stud shear connector,

$$Q_n = 0.5 A_{sc} \sqrt{f_c' E_c} \leq A_{sc} F_u \qquad \text{(AISC Equation I5-1)}$$

where

A_{sc} = cross-sectional area of stud (in.^{2})

f_c' = 28-day compressive strength of the concrete (ksi)

E_c = modulus of elasticity of the concrete (ksi)

F_u = minimum tensile strength of stud (ksi)

For studs used as shear connectors in composite beams, the tensile strength F_u is 60 ksi (AWS, 1996). Values given by AISC Equation I5-1 are based on experimental studies (Ollgaard, Stutter, and Fisher, 1971). No resistance factor is applied to Q_n; the overall flexural resistance factor ϕ_b accounts for all strength uncertainties.

Equation 9.2 gives the number of shear connectors required between the point of zero moment and the point of maximum moment. Consequently, for a simply supported, uniformly loaded beam, $2N_1$ connectors will be required, and they should be equally spaced. When concentrated loads are present, AISC I5.6 requires that enough of the N_1 connectors be placed between the concentrated load and the adjacent point of zero moment to develop the moment required at the load. This portion is denoted N_2, and this requirement is illustrated in Figure 9.10. Note that the total number of shear connectors is not affected by this requirement.

Miscellaneous Requirements for Headed Studs (AISC I5)

- Maximum diameter = 2.5 × flange thickness of steel shape
- Minimum length = 4 × stud diameter
- Minimum longitudinal spacing (center-to-center) = 6 × stud diameter
- Maximum longitudinal spacing (center-to-center) = 8 × slab thickness

■ **FIGURE 9.10**

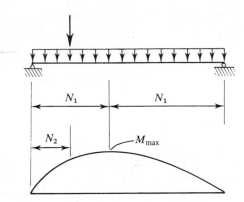

- Minimum transverse spacing (center-to-center) = 4 × stud diameter
- Minimum lateral cover = 1 inch (there is no minimum *vertical* cover.)

The AWS Structural Code (AWS 1996) lists standard stud diameters of ½, ⅝, ¾, ⅞, and 1 inch. Matching these diameters with the minimum lengths prescribed by AISC, we get the common stud sizes of ½ × 2, ⅝ × 2½, ¾ × 3, ⅞ × 3½, and 1 × 4 (but longer studs may be used).

■ **EXAMPLE 9.5**

Design shear connectors for the floor system in Example 9.4.

SOLUTION Summary of data from Example 9.4:

> W21 × 44, A36 steel
> $f_c' = 4000$ psi
> Slab thickness $t = 4.5$ in.
> Span length = 30 ft

From Example 9.4, the horizontal shear force V_h corresponding to full composite action is

$$V_h = C = 468 \text{ kips}$$

Try ½ in. × 2-inch studs. The maximum permissible diameter is

$$2.5t_f = 2.5(0.450) = 1.125 \text{ in.} > 0.5 \text{ in.} \quad \text{(OK)}$$

The cross-sectional area of one shear connector is

$$A_{sc} = \frac{\pi(0.5)^2}{4} = 0.1963 \text{ in.}^2$$

If we assume normal-weight concrete, the modulus of elasticity of the concrete is

$$E_c = w_c^{1.5}\sqrt{f_c'} = (145)^{1.5}\sqrt{4} = 3492 \text{ ksi}$$

From AISC Equation I5-1, the shear strength of one connector is

$$Q_n = 0.5A_{sc}\sqrt{f_c'E_c} \le A_{sc}F_u$$
$$= 0.5(0.1963)\sqrt{4(3492)} = 11.60 \text{ kips}$$
$$A_{sc}F_u = 0.1963(60) = 11.78 \text{ kips} > 11.60 \text{ kips} \qquad \therefore \text{ use } Q_n = 11.60 \text{ kips}$$

and

Minimum longitudinal spacing is $6d = 6(0.5) = 3$ in.

Minimum transverse spacing is $4d = 4(0.5) = 2$ in.

Maximum longitudinal spacing is $8t = 8(4.5) = 36$ in.

The number of studs required between the end of the beam and midspan is

$$N_1 = \frac{V_h}{Q_n} = \frac{468}{11.60} = 40.3$$

Use a minimum of 41 for half the beam, or 82 total. If one stud is used at each section, the required spacing will be

$$s = \frac{30(12)}{82} = 4.4 \text{ in.} \qquad \text{say 4 in.}$$

For two studs per section,

$$s = \frac{30(12)}{82/2} = 8.8 \text{ in.} \qquad \text{say } 8\tfrac{1}{2} \text{ in.}$$

Either arrangement is satisfactory; either spacing will be between the lower and upper limits. The layout shown in Figure 9.11 will be recommended. Although this layout requires more shear connectors than needed, the spacings are convenient.

■ **FIGURE 9.11**

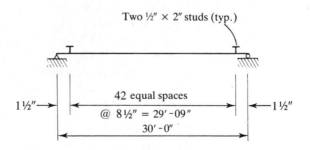

ANSWER Use 86 studs, ½ in. × 2 in., placed as shown in Figure 9.11.

| 9.5 | **DESIGN** |

The first step in the design of a floor system is to select the thickness of the floor slab, whether it is solid or ribbed (formed with steel deck). The thickness will be a function of the beam spacing, and several combinations of slab thickness and beam spacing may need to be investigated so that the most economical system can be found. The design of the slab is beyond the scope of this text, however, and we will assume that the slab thickness and beam spacing are known. Having made this assumption, we can take the following steps to complete the design of an unshored floor system.

1. Compute the factored load moments acting before and after the concrete cures.

2. Select a steel shape for trial.

3. Compute the design strength of the steel shape and compare it to the factored moment acting before the concrete cures. Account for the unbraced length if the formwork does not provide adequate lateral support. If this shape is not satisfactory, try a larger one.

4. Compute the design strength of the composite section and compare it to the *total* factored load moment. If the composite section is inadequate, select another steel shape for trial.

5. Check the shear strength of the steel shape.

6. Design the shear connectors:

   a. Compute V_h, the horizontal shear force at the interface between the concrete and the steel.

   b. Divide this force by Q_n, the shear capacity of a single connector, to obtain the total number of shear connectors required. This number of connectors will provide full composite action. If partial composite behavior is desired, the number of connectors can be reduced (we cover this in Section 9.7).

7. Check deflections (we cover this in Section 9.6).

The major task in the trial-and-error procedure just outlined is the selection of a trial steel shape. A formula that will give the required area (or, alternatively, the required weight per foot of length) can be developed if a beam depth is assumed. Assuming full composite action and the PNA in the slab (i.e., steel controlling, the most common case), we can write the design strength (refer to Figure 9.12) as

$$\phi_b M_n = \phi_b(Ty) = \phi_b(A_s F_y y)$$

Equating the design strength to the factored load moment and solving for A_s, we obtain

$$\phi_b A_s F_y y = M_u \quad \text{and} \quad A_s = \frac{M_u}{\phi_b F_y y}$$

■ **FIGURE 9.12**

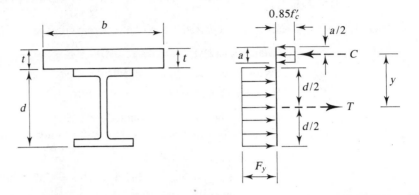

or

$$A_s = \frac{M_u}{\phi_b F_y(d/2 + t - a/2)}$$ (9.3)

Equation 9.3 can also be written in terms of weight rather than area. Since 1 foot of length has a volume of $A_s/144$ cubic feet and structural steel weighs 490 pounds per cubic foot,

$$w = \frac{A_s}{144}(490) = 3.4A_s \text{ lb/ft} \qquad \text{(for } A_s \text{ in square inches)}$$

From Equation 9.3, the estimated weight per foot is therefore

$$w = \frac{3.4M_u}{\phi_b F_y(d/2 + t - a/2)} \text{ lb/ft}$$ (9.4)

where M_u is in in.-kips; F_y is in ksi; and d, t, and a are in inches. Either Equation 9.3 or Equation 9.4 can be used to select a trial shape. Both require an assumed depth and an estimate of $a/2$. The stress block depth will generally be very small; consequently, an error in the estimate of $a/2$ will have only a slight effect on the estimated value of A_s. An assumed value of $a/2 = 1.0$ is suggested.

If Equation 9.4 is used, and a nominal depth is assumed for d, the selection of a trial shape can be quickly made. The use of this equation also provides an estimate of the beam weight directly.

■ **EXAMPLE 9.6**

The span length of a certain floor system is 30 feet, and the beam spacing is 10 feet center-to-center. Select a rolled steel shape and the shear connectors needed to achieve full composite behavior with a 3.5-inch-thick reinforced concrete floor slab. Superimposed loading consists of a 10 psf partition load and a 55 psf live load. Concrete strength is $f_c' = 4000$ psi, and A36 steel is to be used. Assume that the beam has full lateral support during construction and that there is a 20 psf construction load.

SOLUTION   Loads to be supported before the concrete cures are

Slab: $(3.5/12)(150) = 43.75$ psf

Weight per linear foot: $43.75(10) = 437.5$ lb/ft

Construction load: $20(10) = 200$ lb/ft

(The beam weight will be accounted for later.)
Loads to be supported after the concrete cures are

$w_{part} = 10(10) = 100$ lb/ft

$w_D = w_{slab} + w_{part} = 437.5 + 100 = 537.5$ lb/ft

$w_L = 55(10) = 550$ lb/ft

$w_u = 1.2w_D + 1.6w_L = 1.2(0.5375) + 1.6(0.550) = 1.525$ kips/ft

$M_u = \dfrac{1}{8}(1.525)(30)^2 = 171.6$ ft-kips

Try a nominal depth of $d = 16$ inches. From Equation 9.4, the estimated beam weight is

$$w = \frac{3.4M_u}{\phi_b F_y(d/2 + t - a/2)} = \frac{3.4(171.6 \times 12)}{0.85(36)(16/2 + 3.5 - 1)} = 21.8 \text{ lb/ft}$$

Try a W16 × 26. Check the unshored steel shape for loads applied before the concrete cures (the weight of the slab, the weight of the beam, and the construction load.)

$w_u = 1.2(0.4375 + 0.026) + 1.6(0.200) = 0.8762$ kips/ft

$M_u = \dfrac{1}{8}(0.8762)(30)^2 = 98.6$ ft-kips

From the Load Factor Design Selection Table,

$\phi_b M_n = \phi_b M_p = 119$ ft-kips > 98.6 ft-kips   (OK)

After the concrete cures and composite behavior has been achieved,

$w_D = w_{slab} + w_{part} + w_{beam} = 0.4375 + 0.100 + 0.016 = 0.5535$ kips/ft

$w_u = 1.2w_D + 1.6w_L = 1.2(0.5535) + 1.6(0.550) = 1.544$ kips/ft

$M_u = \dfrac{1}{8}(1.544)(30)^2 = 174$ ft-kips

Before computing the design strength of the composite section, we must first determine the effective slab width. For an interior beam, the effective width is the smaller of

$\dfrac{\text{Span}}{4} = \dfrac{30(12)}{4} = 90$ in. or Beam spacing $= 10(12) = 120$ in.

Use b = 90 in. For full composite behavior, the compressive force in the concrete at ultimate (equal to the horizontal shear at the interface between the concrete and steel) will be the smaller of

$$A_s F_y = 7.68(36) = 276.5 \text{ kips}$$

or

$$0.85 f_c' A_c = 0.85(4)(90)(3.5) = 1071 \text{ kips}$$

Use $C = V_h = 276.5$ kips. The depth of the compressive stress block in the slab is

$$a = \frac{C}{0.85 f_c' b} = \frac{276.5}{0.85(4)(90)} = 0.9036 \text{ in.}$$

and the moment arm of the internal resisting couple is

$$y = \frac{d}{2} + t - \frac{a}{2} = \frac{15.69}{2} + 3.5 - \frac{0.9036}{2} = 10.89 \text{ in.}$$

The design flexural strength is

$$\phi_b M_n = \phi_b(Cy) = 0.85(276.5)(10.89)$$
$$= 2550 \text{ in.-kips} = 213 \text{ ft-kips} > 174 \text{ ft-kips} \qquad \text{(OK)}$$

Check the shear:

$$V_u = \frac{w_u L}{2} = \frac{1.544(30)}{2} = 23.2 \text{ kips}$$

From the factored uniform load tables,

$$\phi_v V_n = 76.3 \text{ kips} > 23.2 \text{ kips} \qquad \text{(OK)}$$

ANSWER Use a W16 × 26.

The modulus of elasticity of the concrete will be needed for the shear connector design. From Example 9.5, E_c = 3492 ksi for normal weight concrete with f_c' = 4000 psi. **Try ½-in. × 2-in. studs** (A_{sc} = 0.1963 in.²):

$$\text{Maximum diameter} = 2.5 t_f = 2.5(0.345) = 0.8625 \text{ in.} > 0.5 \text{ in.} \qquad \text{(OK)}$$

From AISC Equation I5-1, the shear strength of one connector is

$$Q_n = 0.5 A_{sc} \sqrt{f_c' E_c} \le A_{sc} F_u$$
$$= 0.5(0.1963)\sqrt{4(3492)}$$
$$= 11.60 \text{ kips}$$
$$A_{sc} F_u = 0.1963(60) = 11.78 \text{ kips} > 11.60 \text{ kips} \qquad \therefore \text{ use } Q_n = 11.60 \text{ kips}$$

The number of studs required between the end of the beam and midspan is

$$N_1 = \frac{V_h}{Q_n} = \frac{276.5}{11.60} = 23.8 \qquad \text{use 24 for half the beam, or 48 total}$$

and

Minimum longitudinal spacing is $6d = 6(0.5) = 3$ in.

Minimum transverse spacing is $4d = 4(0.5) = 2$ in.

Maximum longitudinal spacing is $8t = 8(3.5) = 28$ in.

If one stud is used at each section, the approximate spacing will be

$$s = \frac{30(12)}{48} = 7.5 \text{ in.}$$

This spacing is between the upper and lower limits and is therefore satisfactory.

ANSWER   Use the design shown in Figure 9.13.   ■

■ **FIGURE 9.13**

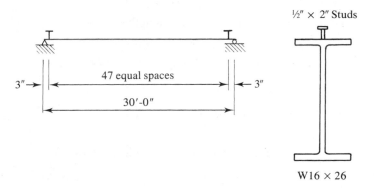

½″ × 2″ Studs

3″ → |← 47 equal spaces →| ← 3″

30′-0″

W16 × 26

<table>
<tr><td>**9.6**</td><td></td></tr>
</table>

DEFLECTIONS

Because of the large moment of inertia of the transformed section, deflections in composite beams are smaller than in noncomposite beams. This larger moment of inertia is available only after the concrete slab has cured, however. Deflections caused by loads applied before the concrete cures must be computed with the moment of inertia of the steel shape. An additional complication arises if the beam is subject to sustained loading, such as the weight of partitions, after the concrete cures. In positive moment regions, the concrete will be in compression continuously and is subject to a phenomenon known as *creep*. Creep is a deformation that takes place under sustained compressive load. After the initial deformation, additional deformation will take place at a very slow rate over a long period of time. The effect on a composite beam is to increase the curvature and hence the vertical deflection. Long-term deflection can only be estimated; the usual technique is to use a reduced area of concrete in the transformed section so as to obtain a smaller moment of inertia and a larger computed deflection. The reduced area is computed by using $2n$ or $3n$ instead of the actual modular ratio n. We use a value of $2n$ in this book. The increased deflection resulting from creep is not covered by the AISC Specification.

For unshored construction, as many as three different moments of inertia can be required for computing the total long-term deflection.

1. Use I_s, the moment of inertia of the rolled steel shape, for deflection caused by loads applied before the concrete cures.

2. Use I_{tr}, the moment of inertia of the transformed section, computed with b/n for deflection caused by *live* loads and for the initial deflection caused by dead loads applied after the concrete cures.

3. Use I_{tr} computed with $b/2n$ for long-term deflections caused by dead loads applied after the concrete cures.

■ EXAMPLE 9.7

Compute the immediate and long-term deflections for the beam in Example 9.4.

SOLUTION   Summary of data from Example 9.4:

W21 × 44, A36 steel.

Slab thickness $t = 4.5$ in. and effective width $b = 90$ in.

$f_c' = 4000$ psi

Dead load applied before concrete cures is $w_D = 550$ lb/ft (slab plus beam)

Construction load is $w_{const} = 180$ lb/ft

Live load is $w_L = 125(9) = 1125$ lb/ft

Partition load is $w_{part} = 20(9) = 180$ lb/ft

Immediate deflection. For the beam plus the slab, $w = 550$ lb/ft and

$$\Delta_1 = \frac{5wL^4}{384EI_s} = \frac{5(0.550/12)(30 \times 12)^4}{384(29,000)(843)} = 0.4100 \text{ in.}$$

The construction load is $w = 180$ lb/ft and

$$\Delta_2 = \frac{5wL^4}{384EI_s} = \frac{5(0.180/12)(30 \times 12)^4}{384(29,000)(843)} = 0.1342 \text{ in.}$$

The total immediate deflection is $\Delta_1 + \Delta_2 = 0.4100 + 0.1342 = 0.544$ in.

For the remaining deflections, the moments of inertia of two transformed sections will be needed: I_{tr}, with a transformed slab width of b/n, and I_{tr}, with a transformed slab width of $b/2n$. For a normal-weight concrete and $f_c' = 4000$ psi, $E_c = 3492$ ksi, and the modular ratio is

$$n = \frac{E_s}{E_c} = \frac{29,000}{3492} = 8.3 \qquad \text{use } n = 8$$

For deflections of the composite section that do not involve creep, the effective width is

$$\frac{b}{n} = \frac{90}{8} = 11.25 \text{ in.}$$

Figure 9.14 shows the corresponding transformed section. The computations for the neutral axis location and the moment of inertia are summarized in Table 9.4.

The initial deflection caused by the partition weight is

$$\Delta_3 = \frac{5w_{\text{part}}L^4}{384EI_{tr}} = \frac{5(0.180/12)(30 \times 12)^4}{384(29,000)(2566)} = 0.0441 \text{ in.}$$

The deflection caused by the live load is

$$\Delta_4 = \frac{5w_L L^4}{384EI_{tr}} = \frac{5(1.125/12)(30 \times 12)^4}{384(29,000)(2566)} = 0.2755 \text{ in.}$$

Long-term deflection caused by creep. Use a transformed slab width of

$$\frac{b}{2n} = \frac{90}{2(8)} = 5.625 \text{ in.}$$

■ **FIGURE 9.14**

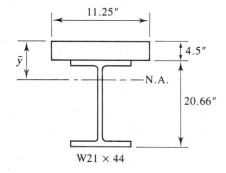

W21 × 44

■ **TABLE 9.4**

Component	A	y	Ay	$\bar{I}$	d	$I + Ad^2$
Concrete	50.62	2.25	113.9	85.43	2.571	420
W21 × 44	13.00	14.83	192.8	843	10.01	2146
	63.62		306.7			2566 in.⁴

$$\bar{y} = \frac{\Sigma Ay}{\Sigma A} = \frac{306.7}{63.62} = 4.821 \text{ in.}$$

■ **FIGURE 9.15**

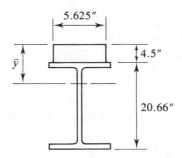

The transformed section is shown in Figure 9.15. The calculations for the centroid and moment of inertia are summarized in Table 9.5. Calling this transformed moment of inertia I_{tr}', we can compute the long-term deflection caused by creep as

$$\Delta_5 = \frac{5 w_{part} L^4}{384 E I_{tr}'} = \frac{5(0.180/12)(30 \times 12)^4}{384(29,000)(2245)} = 0.0504 \text{ in.}$$

ANSWER The following is a summary of the deflections:

Immediate deflection, before composite behavior is attained:

$$\Delta_1 + \Delta_2 = 0.4100 + 0.1342 = 0.544 \text{ in.}$$

Short-term deflection with partitions but no live load:

$$\Delta_1 + \Delta_3 = 0.4100 + 0.0441 = 0.454 \text{ in.}$$

Short-term deflection, live load added:

$$\Delta_1 + \Delta_3 + \Delta_4 = 0.4100 + 0.0441 + 0.2755 = 0.730 \text{ in.}$$

Long-term deflection, without live load:

$$\Delta_1 + \Delta_5 = 0.4100 + 0.0504 = 0.460 \text{ in.}$$

■ **TABLE 9.5**

Component	A	y	Ay	$\bar{I}$	d	$\bar{I} + Ad^2$
Concrete	25.31	2.25	56.95	42.71	4.269	504
W21 × 44	13.00	14.83	192.8	843	8.311	1741
	38.31		249.8			2245 in.⁴

$$\bar{y} = \frac{\sum Ay}{\sum A} = \frac{249.8}{38.31} = 6.519 \text{ in.}$$

Long-term deflection, with live load:

$$\Delta_1 + \Delta_4 + \Delta_5 = 0.4100 + 0.2755 + 0.0504 = 0.736 \text{ in.}$$

Because of the small dead load applied after the concrete has cured, the deflection resulting from creep is insignificant in this example. ■

9.7 COMPOSITE BEAMS WITH FORMED STEEL DECK

The floor slab in many steel-framed buildings is formed on ribbed steel deck, which is left in place to become an integral part of the structure. Although there are exceptions, the ribs of the deck are usually oriented perpendicular to floor beams and parallel to supporting girders. In Figure 9.16 the ribs are shown perpendicular to the beam. The installation of shear studs is done in the same way as without the deck; the studs are welded to the beam flange directly through the deck. The attachment of the deck to the beam can be considered to provide lateral support for the beam before the concrete has cured. The design or analysis of composite beams with formed steel deck is essentially the same as with slabs of uniform thickness, with the following exceptions.

1. The concrete in the ribs — that is, below the top of the deck — is neglected when the ribs are perpendicular to the beam (AISC I3.5b). When the ribs are parallel to the beam, the concrete may be included in determining section properties, and it *must* be included in computing A_c (AISC I3.5c).

2. The capacity of the shear connectors will possibly be reduced.

3. Full composite behavior will not usually be possible. The reason is that the spacing of the shear connectors is limited by the spacing of the ribs, and the exact number of required connectors cannot always be used. Although partial composite design can be used without formed steel deck, it is covered here because it is almost a necessity with formed steel deck. This is not a disadvantage; in fact, it will be the most economical alternative.

■ **FIGURE 9.16**

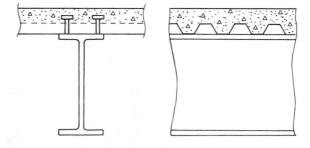

Most composite beams with formed steel deck are floor beams with the deck ribs oriented perpendicular to the beam, and we limit our coverage to this case. Special requirements that apply when the ribs are parallel to the beam are presented in AISC I3.5c.

Reduced Capacity of Shear Connectors

Based on tests, AISC I3.5b requires that Q_n, the shear strength of the shear connectors, be multiplied by the following reduction factor when the ribs are perpendicular to the beam:

$$\frac{0.85}{\sqrt{N_r}}\left(\frac{w_r}{h_r}\right)\left[\left(\frac{H_s}{h_r}\right) - 1.0\right] \leq 1.0 \qquad \text{(AISC Equation I3-1)}$$

where

N_r = number of studs per rib at a beam intersection (limited to three in the computations)

w_r = average width of rib in inches

h_r = height of rib in inches

H_s = length of stud in inches, not to exceed (h_r + 3) in the computations.

These dimensions are illustrated in Figure 9.17.

■ **FIGURE 9.17**

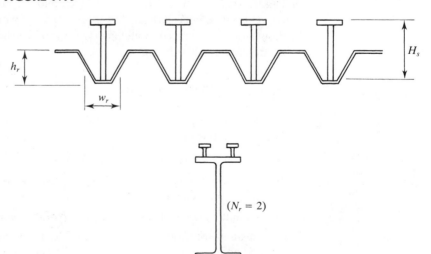

(N_r = 2)

Partial Composite Action

Partial composite action exists when there are not enough shear connectors to completely prevent slip between the concrete and steel. Neither the full strength of the concrete nor that of the steel can be developed, and the compressive force is limited to the maximum force that can be transferred across the interface between the steel and the concrete — that is, the strength of the shear connectors, ΣQ_n. Recall that C is the smallest of $A_s F_y$, $0.85 f_c' A_c$, and ΣQ_n.

With partial composite action, the plastic neutral axis (PNA) will usually fall within the steel cross section. This location will make the strength analysis somewhat more difficult than if the PNA were in the slab, but the basic principles are the same.

When an elastic analysis must be made, as when deflections are being computed, an estimate of the moment of inertia of the partially composite section must be made. A parabolic transition from I_s (for the steel shape alone) to I_{tr} (for the fully composite section) works well (Hansell et al., 1978). This approximation results in the following equation for an effective moment of inertia, presented in the Commentary to the AISC Specification:

$$I_{\text{eff}} = I_s + \sqrt{\Sigma Q_n / C_f} \, (I_{tr} - I_s)$$ (AISC Equation C-I3-6)

where C_f is the compressive force in the concrete for the fully composite condition — the smaller of $A_s F_y$ and $0.85 f_c' A_c$. Since ΣQ_n is the actual compressive force for the partially composite case, the ratio $\Sigma Q_n / C_f$ is the fraction of "compositeness" that exists. If the ratio is less than 0.25, AISC Equation C-I3-6 should not be used (Hansell et al., 1978).

The steel strength will not be fully developed in a partially composite beam, so a larger shape will be required than with full composite behavior. However, fewer shear connectors will be required, and the costs of both the steel and the shear connectors (including the cost of installation) must be taken into account in any economic analysis. Whenever a fully composite beam has excess capacity, which almost always is the case, the design can be fine-tuned by eliminating some of the shear connectors, thereby creating a partially composite beam.

Miscellaneous Requirements

The following requirements are from AISC Sections I3.5a and b. Only those provisions not already discussed are listed.

- Maximum rib height $h_r = 3$ inches.
- Minimum average width of rib $w_r = 2$ inches, but the value of w_r used in the calculations shall not exceed the clear width at the top of the deck.
- Minimum slab thickness above the top of the deck = 2 inches.
- Maximum stud diameter = ¾ inch. This requirement for formed steel deck is in addition to the usual maximum diameter of $2.5 t_f$.
- Minimum height of stud above the top of the deck = 1½ inches.

- Maximum longitudinal spacing of shear studs = 36 inches.
- The deck must be attached to the beam flange at intervals of no more than 18 inches, either by the studs or spot welds. This is for the purpose of resisting uplift.

Slab and Deck Weight

To simplify computation of the slab weight, we use the full depth of the slab, from bottom of deck to top of slab. Although this approach overestimates the volume of concrete, it is conservative. For the unit weight of reinforced concrete, we use the weight of plain concrete plus 5 pcf. Because slabs on formed metal deck are usually lightly reinforced (sometimes welded wire mesh, rather than reinforcing bars, is used), adding 5 pcf for reinforcement may seem excessive, but the deck itself can weigh between 2 and 3 psf.

An alternative approach is to use the thickness of the slab above the deck plus half the height of the rib as the thickness of concrete in computing the weight of the slab. In practice, the combined weight of the slab and deck can usually be found in tables furnished by the deck manufacturer.

■ **EXAMPLE 9.8**

Floor beams are to be used with the formed steel deck shown in Figure 9.18 and a reinforced concrete slab whose total thickness is 4.75 inches. The deck ribs are perpendicular to the beams. The span length is 30 feet, and the beams are spaced at 10 feet center-to-center. The structural steel is A36, and the concrete strength is $f_c' = 3000$ psi. The slab and deck combination weighs 50 psf. The live load is 40 psf, and there is a partition load of 10 psf. No shoring is used, and there is a construction load of 20 psf.

   a. Select a W-shape.
   b. Design the shear connectors.
   c. Check deflections. The maximum permissible total long-term deflection is $\frac{1}{240}$ of the span length.

■ **FIGURE 9.18**

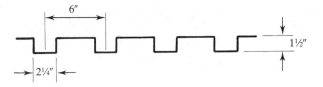

SOLUTION   **a. Beam design.** Select a trial shape based on full composite behavior.

Slab: $50(10) = 500$ lb/ft

Partitions: $10(10) = 100$ lb/ft

Live load: $40(10) = 400$ lb/ft

$w_u = 1.2w_D + 1.6w_L = 1.2(0.500 + 0.100) + 1.6(0.400) = 1.360$ kips/ft

$M_u = (1/8)(1.360)(30)^2 = 153$ ft-kips

Assume that $d = 16$ in., $a/2 = 1.0$ in., and estimate the beam weight from Equation 9.4:

$$ w = \frac{3.4M_u}{\phi_b F_y(d/2 + t - a/2)} = \frac{3.4(153 \times 12)}{0.85(36)(16/2 + 4.75 - 1)} = 17.4 \text{ lb/ft} $$

Try a W16 × 26. Check the flexural strength before the concrete has cured.

Construction load: $20(10) = 200$ lb/ft

$w_u = 1.2w_D + 1.6w_L = 1.2(0.500 + 0.026) + 1.6(0.200) = 0.9512$ kips/ft

$M_u = (1/8)(0.9512)(30)^2 = 107$ ft-kips

A W16 × 26 is compact for A36 steel, and since the steel deck will provide adequate lateral support, the nominal strength M_n is equal to the plastic moment strength M_p. From the Load Factor Design Selection Table,

$\phi_b M_p = 119$ ft-kips > 107 ft-kips   (OK)

After the concrete has cured, the total factored load to be resisted by the composite beam, adjusted for the weight of the steel shape, is

$w_u = 1.2(0.500 + 0.026 + 0.100) + 1.6(0.400)$

$\quad = 1.391$ kips/ft

and the factored moment is

$$ M_u = \frac{1}{8}(1.391)(30)^2 = 156 \text{ ft-kips} $$

The effective slab width of the composite section will be the smaller of

$$ \frac{\text{Span}}{4} = \frac{30(12)}{4} = 90 \text{ in. or Beam spacing} = 10(12) = 120 \text{ in.} $$

Use $b = 90$ in. For full composite action, the compressive force C in the concrete is the smaller of

$A_s F_y = 7.68(36) = 276.5$ kips

or

$0.85f_c'A_c = 0.85(3)[90(4.75 - 1.5)] = 745.9$ kips

■ **FIGURE 9.19**

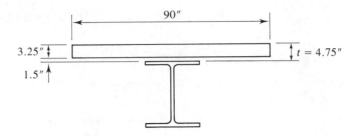

where only the concrete above the top of the deck has been accounted for in the second equation, as illustrated in Figure 9.19. With $C = 276.5$ kips, the depth of the compressive stress distribution in the concrete is

$$a = \frac{C}{0.85 f_c' b} = \frac{276.5}{0.85(3)(90)} = 1.205 \text{ in.}$$

The moment arm of the internal resisting couple is

$$y = \frac{d}{2} + t - \frac{a}{2} = \frac{15.69}{2} + 4.75 - \frac{1.209}{2} = 11.99 \text{ in.}$$

and the design strength is

$$\phi_b M_n = \frac{0.85(276.5)(11.99)}{12}$$
$$= 235 \text{ ft-kips} > 156 \text{ ft-kips} \quad \text{(OK)}$$

Check the shear:

$$V_u = \frac{w_u L}{2} = \frac{1.391(30)}{2} = 20.9 \text{ kips}$$

From the factored uniform load tables,

$$\phi_v V_n = 76.3 \text{ kips} > 20.9 \text{ kips} \quad \text{(OK)}$$

ANSWER

a. Use a W16 × 26.

b. **Shear connectors.** Because this beam has a substantial excess of moment strength, it will benefit from partial composite behavior. We must first find the shear connector requirements for full composite behavior and then reduce the number of connectors. For the fully composite beam, $C = V_h = 276.5$ kips.

Try ¾-in. × 3-in. studs ($A_{sc} = 0.4418$ in.^{2}), one at each section:

Maximum diameter $= 2.5t_f = 2.5(0.345) = 0.8625$ in.

or $\dfrac{3}{4}$ in. (controls)

Actual diameter $= \dfrac{3}{4}$ in. (OK)

Compute the stud strength-reduction factor:

$N_r = 1$

Height of stud above top of deck $= 3 - 1.5 = 1.5$ in. $=$ permissible value

(OK)

From AISC Equation I3-1,

$$\text{Reduction factor} = \frac{0.85}{\sqrt{N_r}} \left(\frac{w_r}{h_r}\right)\left[\left(\frac{H_s}{h_r}\right) - 1.0\right] \leq 1.0$$

$$= \frac{0.85}{1.0}\left(\frac{2.25}{1.5}\right)\left(\frac{3}{1.5} - 1.0\right) = 1.275 > 1.0$$

No reduction of stud strength is required. For $f_c' = 3000$ psi, the modulus of elasticity of the concrete is

$$E_c = w_c^{1.5}\sqrt{f_c'} = 145^{1.5}\sqrt{3} = 3024 \text{ ksi}$$

From AISC Equation I5-1, the shear strength of one connector is

$$Q_n = 0.5A_{sc}\sqrt{f_c'E_c} \leq A_{sc}F_u \quad \text{— leo}$$

$$= 0.5(0.4418)\sqrt{3(3024)}$$

$$= 21.04 \text{ kips}$$

$$A_{sc}F_u = 0.4418(60) = 26.51 \text{ kips} > 21.04 \text{ kips} \quad \therefore \text{ use } Q_n = 21.04 \text{ kips}$$

The number of studs required between the end of the beam and midspan is

$$N_1 = \frac{V_h}{Q_n} = \frac{276.5}{21.04} = 13.1$$

Use 14 for half the beam, or 28 total.

With one stud in each rib, the spacing is 6 inches, and the maximum number that can be accommodated is

$$\frac{30(12)}{6} = 60 > 28 \text{ required}$$

With one stud in every other rib, 30 will be furnished, which is still too many. With one stud in every third rib, the spacing will be $3(6) = 18$ inches, and the number of studs will be $30(12)/18 = 20$, which is fewer than needed for full composite action. However, there is an excess of flexural strength, so partial composite action may be adequate.

Try 20 studs per beam so that N_1 provided $= {}^{20}\!/_2 = 10$.

$$\Sigma Q_n = 10(21.04)$$
$$= 210.4 \text{ kips} < 276.5 \text{ kips} \qquad \therefore C = V_h = 210.4 \text{ kips}$$

Because C is smaller than $A_s F_y$, part of the steel section must be in compression, and the plastic neutral axis is in the steel section.

To analyze this case, we must first determine whether the PNA is in the top flange or in the web. This can be done as follows. If the PNA were at the bottom of the flange, the entire flange would be in compression, and the resultant compressive force, as illustrated in Figure 9.20, would be

$$P_{yf} = b_f t_f F_y = 5.500(0.345)(36) = 68.31 \text{ kips}$$

The net force to be transferred at the interface between the steel and the concrete would be

$$T - C_s = T - P_{yf} = (A_s F_y - P_{yf}) - P_{yf} = 276.5 - 2(68.31)$$
$$= 139.9 \text{ kips}$$

This is less than the actual net tension force of 210.4 kips, so the top flange does not need to be in compression for its full thickness. This means that the PNA is in the flange. From Figure 9.21, the horizontal shear force to be transferred is

$$T - C_s = (A_s F_y - b_f t' F_y) - b_f t' F_y = V_h$$
$$276.5 - 2[5.500 t'(36)] = 210.4$$

Solving for the depth of compression in the flange, we obtain

$$t' = 0.1669 \text{ in.}$$

■ **FIGURE 9.20**

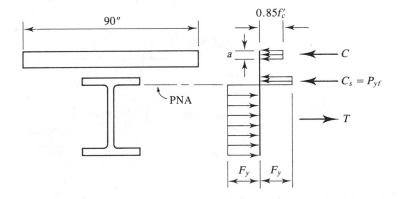

■ **FIGURE 9.21**

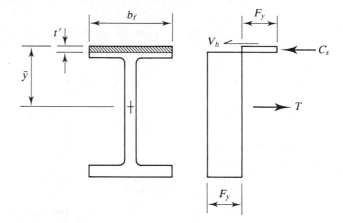

The tensile resultant force will act at the centroid of the area below the PNA. Before the moment strength can be computed, the location of this centroid must be determined. The calculations for $\bar{y}$, the distance from the top of the steel shape, are summarized in Table 9.6.

The depth of the compressive stress block in the concrete is

$$a = \frac{C}{0.85 f_c' b} = \frac{210.4}{0.85(3)(90)} = 0.9168 \text{ in.}$$

Moment arm for the concrete compressive force is

$$\bar{y} + t - \frac{a}{2} = 8.819 + 4.75 - \frac{0.9168}{2} = 13.11 \text{ in.}$$

Moment arm for the compressive force in the steel is

$$\bar{y} - \frac{t'}{2} = 8.819 - \frac{0.1669}{2} = 8.736 \text{ in.}$$

■ **TABLE 9.6**

Component	A		y	Ay
W16 × 26		7.68	15.69/2 = 7.845	60.25
Flange segment	−0.1669(5.500) =	−0.918	0.1669/2 = 0.0834	−0.08
Sum		6.762		60.17

$$\bar{y} = \frac{\Sigma Ay}{\Sigma A} = \frac{60.17}{6.762} = 8.898 \text{ in.}$$

Taking moments about the tensile force and using the notation of Figure 9.20, we obtain the nominal strength:

$$M_n = C(13.11) + C_s(8.736)$$
$$= 210.4(13.11) + 0.1669(5.500)(36)(8.736) = 3047 \text{ in.-kips} = 253.9 \text{ ft-kips}$$

The design strength is

$$\phi_b M_n = 0.85(253.9) = 216 \text{ ft-kips} > 156 \text{ ft-kips} \qquad \text{(OK)}$$

The deck will be attached to the beam flange at intervals of 18 inches, so no spot welds will be needed to resist uplift.

ANSWER   **b.** Use the shear connectors shown in Figure 9.22.

c. Deflections. Before the concrete has cured,

$$w_D = w_{\text{slab}} + w_{\text{beam}} = 0.500 + 0.026 = 0.526 \text{ kips/ft}$$
$$\Delta_1 = \frac{5w_D L^4}{384 EI_s} = \frac{5(0.526/12)(30 \times 12)^4}{384(29,000)(301)} = 1.098 \text{ in.}$$

The deflection caused by the construction load is

$$\Delta_2 = \frac{5w_{\text{const}} L^4}{384 EI_s} = \frac{5(0.200/12)(30 \times 12)^4}{384(29,000)(301)} = 0.418 \text{ in.}$$

The total deflection before the concrete has cured is

$$\Delta_1 + \Delta_2 = 1.098 + 0.418 = 1.52 \text{ in.}$$

For deflections that occur after the concrete has cured, the moments of inertia of two transformed sections will be needed: I_{tr} with a transformed slab width of b/n and I_{tr} with a transformed slab width of $b/2n$. The modular ratio is

$$n = \frac{E_s}{E_c} = \frac{29,000}{3024} = 9.6 \qquad \text{use } n = 10$$

■ **FIGURE 9.22**

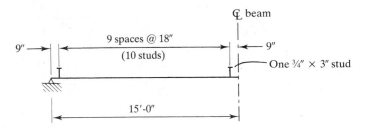

■ **FIGURE 9.23**

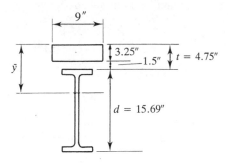

For deflections of the composite section that do not involve creep, the effective width is

$$\frac{b}{n} = \frac{90}{10} = 9 \text{ in.}$$

Figure 9.23 shows the corresponding transformed section. The computations for the neutral axis location and the moment of inertia are summarized in Table 9.7.

Since partial composite action is being used, a reduced transformed moment of inertia must be used. From AISC Equation C-I3-6, this effective moment of inertia is

$$I_{\text{eff}} = I_s + \left(\sqrt{\Sigma Q_n / C_f} \right)(I_{tr} - I_s)$$
$$= 301 + \sqrt{210.4/276.5}(1059 - 301) = 962.2 \text{ in.}^4$$

The deflection caused by the live load is

$$\Delta_3 = \frac{5 w_L L^4}{384 E I_{\text{eff}}} = \frac{5(0.400/12)(30 \times 12)^4}{384(29,000)(962.2)} = 0.2613 \text{ in.}$$

■ **TABLE 9.7**

Component	A	y	Ay	$\bar{I}$	d	$\bar{I} + Ad^2$
Concrete	29.25	1.625	47.53	25.75	2.282	178
W16 × 26	7.68	12.60	96.77	301	8.693	881
Sum	36.93		144.30			1059 in.⁴

$$\bar{y} = \frac{\Sigma Ay}{\Sigma A} = \frac{144.3}{36.93} = 3.907 \text{ in.}$$

■ **FIGURE 9.24**

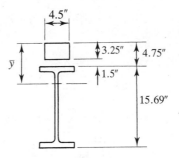

Deflection caused by dead load applied after the concrete has cured should be based on a transformed moment of inertia obtained with $2n$ rather than n. Therefore use a transformed slab width of

$$\frac{b}{2n} = \frac{90}{2(10)} = 4.5 \text{ in.}$$

From Figure 9.24 and Table 9.8, the transformed moment of inertia is

$$I'_{tr} = 920.4 \text{ in.}^4$$

The effective moment of inertia, which we will call I_{eff}', is

$$\begin{aligned} I'_{\text{eff}} &= I_s + \sqrt{\Sigma \, Q_n / C_f} \, (I'_{tr} - I_s) \\ &= 301 + \sqrt{210.4/276.5}(920.4 - 301) = 841.3 \text{ in.}^4 \end{aligned}$$

The long-term deflection caused by the dead load applied after the concrete has cured is

$$\Delta_4 = \frac{5(0.100/12)(30 \times 12)^4}{384(29,000)(841.3)} = 0.0747 \text{ in.}$$

TABLE 9.8

Component	A	y	Ay	$\bar{I}$	d	$\bar{I} + Ad^2$
Concrete	14.62	1.625	23.76	12.87	3.780	221.8
W16 × 26	7.68	12.60	96.77	301	7.195	698.6
Sum	22.30		120.53			920.4 in.^{4}

$$\bar{y} = \frac{\Sigma \, Ay}{\Sigma \, A} = \frac{120.5}{22.30} = 5.405 \text{ in.}$$

The total deflection is

$$\Delta_1 + \Delta_3 + \Delta_4 = 1.098 + 0.2613 + 0.0747 = 1.43 \text{ in.}$$

and

$$\frac{L}{240} = \frac{30(12)}{240} = 1.50 \text{ in.} > 1.43 \text{ in.} \qquad \text{(OK)}$$

ANSWER   **c.** The deflection is satisfactory.   ■

9.8   TABLES FOR COMPOSITE BEAM ANALYSIS AND DESIGN

When the plastic neutral axis is within the steel section, computation of the flexural strength can be laborious. Formulas to expedite this computation have been developed (Hansell et al., 1978), but the tables presented in Part 5 of the *Manual* are more convenient. Two sets of tables are presented: design strengths of various combinations of shapes and slabs, for $F_y = 36$ ksi and $F_y = 50$ ksi; and tables of "lower bound" moments of inertia for these same combinations.

The design strength table, called "Composite Beam Selection Table," is limited to shapes with compact webs and a total shear connector strength of $\Sigma Q_n \geq 0.25 A_s F_y$ (the recommended lower limit for partially composite beams).

The design strength $\phi_b M_n$ is given for seven specific locations of the plastic neutral axis, as shown in Figure 9.25: top of the top flange, bottom of the top flange, three

---

■ **FIGURE 9.25**

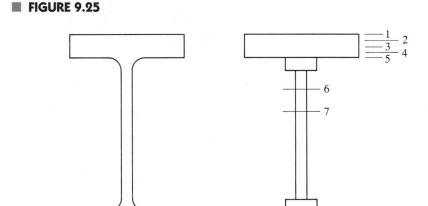

■ **FIGURE 9.26**

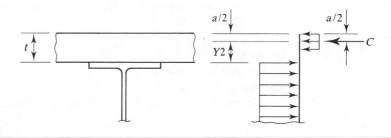

equally spaced levels within the top flange, and two locations in the web.* The lowest PNA location, level 7, corresponds to the recommended lower limit of $\Sigma Q_n = 0.25 A_s F_y$. PNA location 6 corresponds to a ΣQ_n midway between ΣQ_n for level 5 and ΣQ_n for level 7.

To use the tables for analysis of a composite beam, first find the portion of the table corresponding to the steel shape and proceed as follows.

1. Select ΣQ_n. This is the *Manual*'s notation for the compressive force C, which is the smallest of $A_s F_y$, $0.85 f_c' A_c$, and the total shear connector strength (which we have been calling ΣQ_n).

2. Select $Y2$, the distance from the top of the steel shape to the resultant compressive force in the concrete, computed as

$$Y2 = t - \frac{a}{2}$$

This dimension is illustrated in Figure 9.26.

3. Read ϕM_n, interpolating if necessary.

For design, the tables can be entered with the required ϕM_n, and a combination of steel shape and ΣQ_n can be selected. A value of $Y2$ will be needed, so the depth of the concrete compressive stress distribution will need to be assumed and then revised after an iteration. The discussion in the *Manual* preceding the tables gives the equivalent of Equation 9.4 for estimating the beam weight, but if the tables are used, it is not really needed.

The tables also give values of $\phi_b M_p$, which may be needed for checking unshored beams during the curing of the concrete; and $Y1$, the distance from the top of the steel to the plastic neutral axis.

---

*The notation ϕM_n is used in the table for the design strength of composite shapes, and $\phi_b M_p$ is used for the design strength of the steel shape alone. The resistance factor is denoted in two different ways because two different numerical values are used.

■ EXAMPLE 9.9

Compute the design strength of the composite beam in Examples 9.1 and 9.2. Use the tables in Part 5 of the *Manual*.

SOLUTION   From Example 9.1, the composite beam consists of a $W16 \times 36$ of A36 steel, a slab with a thickness $t = 5$ inches, and an effective width $b = 87$ inches. The 28-day compressive strength of the concrete $f_c' = 4000$ psi.

The compressive force in the concrete is the smaller of

$$A_s F_y = 10.6(36) = 381.6 \text{ kips}$$

or

$$0.85 f_c' A_c = 0.85(4)(5 \times 87) = 1487 \text{ kips}$$

Use $C = 381.6$ kips. The depth of the compressive stress block is

$$a = \frac{C}{0.85 f_c' b} = \frac{381.6}{0.85(4)(87)} = 1.290 \text{ in.}$$

The distance from the top of the steel to the compressive force C is

$$Y2 = t - \frac{a}{2} = 5 - \frac{1.290}{2} = 4.36 \text{ in.}$$

Enter the tables with $\Sigma Q_n = 382$ kips and $Y2 = 4.36$. By interpolation,

$$\phi M_n = 332 \text{ ft-kips}$$

which checks with the results of Example 9.2 but involves about the same amount of effort. The value of the tables becomes obvious when the plastic neutral axis is within the steel shape.

ANSWER   Design strength = 332 ft-kips.   ■

The tables for lower-bound moment of inertia, denoted I_{LB}, give a conservative estimate of the moment of inertia of the transformed section for the same beams that are in the design strength tables. The chief simplifying assumption made in the construction of these tables is that only the area of concrete used in resisting the moment is effective in computing the moment of inertia. The force in the concrete is $C = \Sigma Q_n$ and the corresponding area in the transformed section is

$$A_c = \frac{\Sigma Q_n}{\text{stress in transformed area}} = \frac{\Sigma Q_n}{F_y}$$

As a further simplification, the moment of inertia of the concrete about its own centroidal axis is neglected. To illustrate the procedure, one of the tabulated values will be derived in Example 9.10.

■ **EXAMPLE 9.10**

A design has resulted in a $W16 \times 31$ with $\Sigma Q_n = 241$ kips (PNA location 3), $Y2 = 4$ inches, and $F_y = 36$ ksi. Compute the lower-bound moment of inertia.

SOLUTION

The area of concrete to be used is

$$A_c = \frac{\Sigma Q_n}{F_y} = \frac{241}{36} = 6.694 \text{ in.}^2$$

The corresponding transformed section is shown in Figure 9.27, and the computations are summarized in Table 9.9. To locate the centroid, take moments about an axis at the bottom of the steel shape.

■ **FIGURE 9.27**

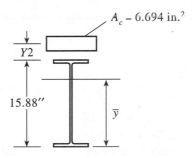

■ **TABLE 9.9**

Component	A	y	Ay	$\bar{I}$	d	$\bar{I} + Ad^2$
$W16 \times 31$	9.12	7.94	72.4	375	5.06	608.5
Concrete	6.694	19.88	133.1	—	6.88	316.9
Sum	15.81		205.5			925.4 in.⁴

$$\bar{y} = \frac{\Sigma Ay}{\Sigma A} = \frac{205.5}{15.81} = 13.00 \text{ in.}$$

The moment of inertia from the lower-bound moment of inertia tables is $I_{LB} = 925$ in.⁴, verifying the calculated result.

ANSWER $I_{LB} = 925$ in.⁴   ■

■ **EXAMPLE 9.11**

Rework Example 9.8 with the aid of the tables in Part 5 of the *Manual*.

SOLUTION   **a. Beam design.** From Example 9.8, M_u = 153 ft-kips (excluding the beam weight). Assuming that a = 2 in., we get

$$Y2 = t - \frac{a}{2} = 4.75 - \frac{2}{2} = 3.75 \text{ in.}$$

From the Composite Beam Selection Table, any combination of steel shape, ΣQ_n, and $Y2$ that furnishes a design strength of more than 153 ft-kips will be an acceptable trial beam. Two possibilities are summarized in Table 9.10.

The W14 × 22 is the lighter shape, but because ΣQ_n is larger, it will require more shear connectors — more than twice as many. For this reason, **try the W16 × 26.** Refine the value for $Y2$:

$$a = \frac{C}{0.85 f'_c b} = \frac{\Sigma Q_n}{0.85 f'_c b} = \frac{69.1}{0.85(3)(90)} = 0.3011 \text{ in.}$$

$$Y2 = t - \frac{a}{2} = 4.75 - \frac{0.3011}{2} = 4.60 \text{ in.}$$

(where b = 90 in. is from Example 9.8).

$$\phi M_n = 164.4 \text{ ft-kips}$$

From Example 9.8, M_u = 156 ft-kips with the beam weight included. This is less than the design strength of 164.4 ft-kips, so this selection is acceptable. Also from Example 9.8, both the flexural strength during the construction phase and the shear strength are adequate for a W16 × 26.

ANSWER   **a.** Use a W16 × 26.

b. Shear connectors. First, **try ¾-in. × 3-in. studs.** The number of studs required is

$$N_1 = \frac{\Sigma Q_n}{Q_n} = \frac{69.1}{21.04} = 3.3 \qquad \text{use 4 for half the beam, or 8 total}$$

■ **TABLE 9.10**

Shape	PNA Location	ΣQ_n	ϕM_n (by interpolation)
W16 × 26	7	69.1 kips	160 ft-kips
W14 × 22	3	159 kips	159 ft-kips

Eight studs correspond to a spacing of

$$\frac{30(12)}{8} = 45 \text{ in.}$$

This spacing is greater than the maximum permissible spacing of 36 inches, so more studs must be used. If a stud is placed in every sixth rib, the spacing will equal 36 inches, and the total number of studs will be

$$\frac{30(12)}{36} = 10, \qquad N_1 = 5$$

The corresponding shear force to be transferred is

$$\Sigma Q_n = 5(21.04) = 105.2 \text{ kips}$$

For convenience in using the tables, we conservatively approximate this force as $\Sigma Q_n = 104$ kips. Then,

$$a = \frac{104}{0.85(3)(90)} = 0.4532 \text{ in.}$$

$$Y2 = 4.75 - \frac{0.4532}{2} = 4.523 \text{ in.}$$

From the Composite Beam Selection Table, the design strength is

$$\phi M_n = 182 \text{ ft-kips} > 156 \text{ ft-kips} \qquad \text{(OK)}$$

ANSWER   **b.** Use ten ¾-inch × 3-inch studs, equally spaced. To resist uplift, use spot welds at 18 inches, midway between the studs.

c. Deflections. From Example 9.8, the deflection of the steel beam before composite behavior has been achieved is

$$\Delta_1 = 1.098 \text{ in.} \qquad \text{(excluding construction loads)}$$

For deflections that occur after the concrete has cured, the lower-bound moment of inertia from the tables can be used. For a W16 × 26 with $\Sigma Q_n = 104$ kips (PNA location 6) and $Y2 = 4.523$ inches,

$$I_{LB} = 623 \text{ in.}^4$$

No separate moment of inertia is available for calculation of the additional deflection caused by creep. The lower-bound moment of inertia, however, is smaller than the actual transformed section moment of inertia, and the overall effect is to overestimate deflections. If the long-term dead loads are small, the lower-bound moment of inertia should be acceptable.

From Example 9.8, the load applied after the concrete has cured is

$$w = w_D + w_L = 0.100 + 0.400 = 0.500 \text{ kips/ft}$$

and the corresponding deflection is

$$\Delta_2 = \frac{5wL^4}{384EI_{LB}} = \frac{5(0.500/12)(30 \times 12)^4}{384(29,000)(623)} = 0.5044 \text{ in.}$$

The total deflection is

$$\Delta_1 + \Delta_2 = 1.098 + 0.5044 = 1.602 \text{ in.}$$

The maximum permissible deflection is

$$\frac{L}{240} = \frac{30(12)}{240} = 1.500 \text{ in.} < 1.602 \text{ in.} \qquad \text{(N.G.)}$$

At this point in the design, we have two options: (1) compute the deflection more accurately by using the transformed section; or (2) choose a combination of steel shape and shear connectors with a larger lower-bound moment of inertia. Since the purpose of this example is to illustrate the use of the tables, we will choose the second option.

Determine the required lower-bound moment of inertia. The deflection caused by the slab and beam weight will not change, so the maximum permissible deflection caused by loads applied after the concrete has cured is

$$\text{Maximum } \Delta_2 = 1.500 - \Delta_1 = 1.500 - 1.098 = 0.4020 \text{ in.}$$

From

$$\Delta_2 = \frac{5wL^4}{384EI_{LB}}$$

the required I_{LB} is

$$I_{LB} \geq \frac{5wL^4}{384E\Delta_2} = \frac{5(0.500/12)(30 \times 12)^4}{384(29,000)(0.4020)} = 782 \text{ in.}^4$$

For a W16 × 26 with PNA 3 and $Y2 = 4.5$ in., the lower-bound moment of inertia is $I_{LB} = 804$ in.^{4} From the Composite Design Selection Table, for PNA 3 the horizontal shear is

$$\Sigma Q_n = 208 \text{ kips}$$

To obtain the correct value of $Y2$ and I_{LB}, first determine the location of the compressive force in the concrete.

$$a = \frac{\Sigma Q_n}{0.85f'_c b} = \frac{208}{0.85(3)(90)} = 0.9063 \text{ in.}$$

$$Y2 = t - \frac{a}{2} = 4.75 - \frac{0.9063}{2} = 4.30 \text{ in.}$$

From the lower-bound moment of inertia tables, by interpolation,

$$I_{LB} = 788 \text{ in.}^4 > 782 \text{ in.}^4 \qquad \text{(OK)}$$

The number of shear connectors required is

$$N_1 = \frac{\sum Q_n}{Q_n} = \frac{208}{21.04} = 9.9 \qquad \text{use 10 for half the beam, or 20 total}$$

Because of the spacing of the deck ribs, one stud in every third rib, for a spacing of 18 inches, will provide a total of 20 studs.

ANSWER   **c.** To satisfy deflection requirements, increase the number of studs from 10 to 20, placed one in every third deck rib. ■

Example 9.11 illustrates the usefulness of the tables. In particular, the Composite Beam Selection Table greatly simplifies the design of a partially composite beam, in which the PNA lies within the steel shape.

9.9   CONTINUOUS BEAMS

In a simply supported beam, the point of zero moment is at the support. The number of connectors required between each support and the point of maximum positive moment will be half the total number required. In a continuous beam, the points of inflection are also points of zero moment and, in general, $2N_1$ connectors will be required for each span. Figure 9.28a shows a typical continuous beam and the regions in which shear connectors would be required. In the negative moment zones, the concrete slab will be in tension and therefore ineffective. In these regions there will be no composite behavior in the sense that we have considered thus far. The only type of composite behavior possible is that between the structural steel beam and the longitudinal reinforcing steel in the slab. The corresponding composite cross section is shown in Figure 9.28b. If this concept is used, a sufficient number of shear connectors must be provided to achieve a degree of continuity between the steel shape and the reinforcing.

The AISC Specification in Section I3.2 offers two alternatives for negative moment.

1. Rely on the strength of the steel shape only.

2. Include the reinforcing steel in the composite section, subject to the following conditions:

   a. The steel shape must be compact with adequate lateral support.

   b. Shear connectors must be present in the negative moment region (between the point of zero moment and the point of maximum negative moment).

   c. The reinforcement within the effective width must be adequately developed (anchored).

The strength of the composite section should be based on a plastic stress distribution with $\phi_b = 0.85$.

■ **FIGURE 9.28**

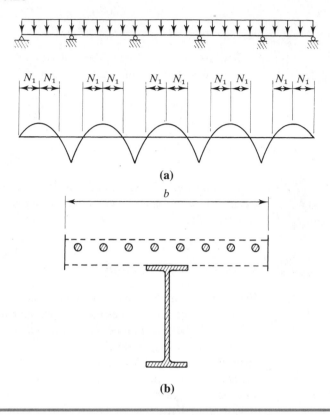

(a)

(b)

If composite behavior is accounted for, AISC I5.2 requires that the horizontal shear force to be transferred between the point of maximum negative moment and the point of zero moment be taken as the smaller of $A_r F_{yr}$ and ΣQ_n, where

A_r = area of reinforcing steel within the effective width of the slab

F_{yr} = yield stress of the reinforcing steel

The additional strength gained from including the reinforcing steel is relatively small, however, and cover plates are sometimes used in the negative moment regions.

9.10 COMPOSITE COLUMNS

Composite columns can take one of two forms: a pipe or tube filled with plain concrete or a rolled steel shape encased in concrete with vertical reinforcement and lateral ties, as in a reinforced concrete column. Figure 9.29 illustrates these two types.

The analysis of composite columns is done in the same way as for ordinary structural steel compression members, using the same equations from AISC Chapter E, but with values of F_y, E, and r that have been modified to give results that match experi-

■ **FIGURE 9.29**

mental and theoretical results. Before considering the AISC equations for these values, we examine the basis of the equations. If stability is ensured, the strength of a composite compression member can be expressed as the sum of the axial strengths of the steel shape, the reinforcing bars, and the concrete. This is called the "squash load" and is given by

$$P_n = A_s F_y + A_r F_{yr} + 0.85 f_c' A_c \qquad (9.5)$$

where

A_s = cross-sectional area of the rolled steel shape

A_r = total cross-sectional area of the vertical reinforcing steel

F_{yr} = yield stress of the reinforcing steel

A_c = cross-sectional area of the concrete

Modern reinforcing steel bars are *deformed* — that is, the surface of the bars has protrusions that help the steel grip or bond with the concrete. The cross-sectional area, A_r, used in computations is a nominal area that equals the area of a plain bar that has the same weight per foot of length as the deformed bar. Table 9.11 gives nominal diameters and areas for standard bar sizes as defined by ASTM (1996) and ACI (1995).

■ **TABLE 9.11**

Bar Number	Diameter (in.)	Area (in.²)
3	0.375	0.11
4	0.500	0.20
5	0.625	0.31
6	0.750	0.44
7	0.875	0.60
8	1.000	0.79
9	1.128	1.00
10	1.270	1.27
11	1.410	1.56
14	1.693	2.25
18	2.257	4.00

To obtain an overall stress, divide the load from Equation 9.5 by the area of the structural steel shape:

$$\frac{P_n}{A_s} = F_{my} = F_y + A_r \frac{F_{yr}}{A_s} + 0.85 f_c' \frac{A_c}{A_s} \tag{9.6}$$

The value of F_{my} obtained from Equation 9.6, when used in place of F_y in the compression member equations, gives good results for concrete-filled pipes and tubes, where the concrete is contained within the steel shape. (Longitudinal reinforcing bars are not normally used with pipes and tubes; therefore A_r would be zero for this type of composite column.)

For encased shapes, the benefit of confinement is absent, and the Structural Stability Research Council (SSRC, 1979) recommends that the ACI code (ACI, 1995) capacity reduction factor of 0.7 be applied to the reinforcing steel and concrete terms of Equation 9.6 as follows:

$$
\begin{aligned}
F_{my} &= F_y + 0.7A_r \frac{F_{yr}}{A_s} + 0.7(0.85) f_c' \frac{A_c}{A_s} \\
&= F_y + 0.7A_r \frac{F_{yr}}{A_s} + 0.595 f_c' \frac{A_c}{A_s}
\end{aligned}
\tag{9.7}
$$

To account for slenderness effects, the member flexural stiffness, which is proportional to the quantity EI/L, must be adjusted. This adjustment is made by modifying the value of E as follows:

$$E_m = E + \text{constant} \times E_c \frac{A_c}{A_s} \tag{9.8}$$

where

E = modulus of elasticity of the structural steel shape

E_c = modulus of elasticity of concrete

Although the stiffness is proportional to the moment of inertia, a ratio of areas gives better results for a composite column than a ratio of moments of inertia (SSRC, 1979). The constant in Equation 9.8 is 0.4 for a concrete-filled pipe or tube, representing an allowance of 40% of the concrete stiffness, and 0.2 for an encased shape.

The radius of gyration of a composite shape is larger than that of both the steel shape and the concrete area. The conservative approach is to use the larger of the radius of gyration of the steel shape or the radius of gyration of the concrete outline, which can be taken as 0.3 times the dimension in the plane of buckling. Denoting the radius of gyration of the composite shape r_m gives

$$r_m = r \geq 0.3b \tag{9.9}$$

where

r = radius of gyration of the steel shape in the plane of buckling

b = dimension of the concrete outline in the plane of buckling

Specification Requirements

The AISC provisions for composite columns are essentially the same as those just out-lined. Equations E2-1 and E2-3 from Chapter E of the Specification are used to determine the design strength, but the values of F_y, E, and r are modified. The modified values are based on Equations 9.6 through 9.9 in the preceding discussion. From AISC Section I2.2, the modified value of F_y is

$$F_{my} = F_y + c_1 F_{yr}(A_r/A_s) + c_2 f_c'(A_c/A_s) \qquad \text{(AISC Equation I2-1)}$$

where the constants c_1 and c_2 account for the differences between encased sections and concrete-filled pipes and tubes:

$c_1 = 1.0$ and $c_2 = 0.85$ for pipes and tubes

$c_1 = 0.7$ and $c_2 = 0.6$ for encased shapes

The AISC modified value of E is the same as that given by Equation 9.8, or

$$E_m = E + c_3 E_c(A_c/A_s) \qquad \text{(AISC Equation I2-2)}$$

where

$c_3 = 0.4$ for pipes and tubes

$ = 0.2$ for encased shapes

AISC I2.2 specifies the value of r_m to be that given by Equation 9.9 — that is,

$$r_m = r \geq 0.3b$$

To qualify as a composite column, the following limitations, given in AISC I2.1, must be observed.

1. The structural steel must make up at least 4% of the total cross-sectional area, or the compression member will act more like a concrete column than a composite column.

2. Encased sections must conform to the following details.

   a. Both longitudinal and lateral reinforcement must be used. The spacing of the lateral ties must be no greater than two thirds of the least dimension of the concrete. The cross-sectional area of both longitudinal and lateral reinforcement must be at least 0.007 square inches per inch of bar spacing.

   b. There must be at least $1\frac{1}{2}$ inches of clear concrete cover for both the lateral and the transverse reinforcement.

   c. *Load-carrying* longitudinal reinforcement must be continuous at framed levels. Longitudinal bars whose purpose is to help restrain the concrete may be spliced at framed levels.

3. The concrete strength f_c' must be between 3 ksi and 8 ksi for normal-weight concrete (no test data are available for f_c' greater than 8 ksi) and at least 4 ksi for lightweight concrete.

4. The yield stress of both the structural steel and the longitudinal reinforcing bars may not exceed 55 ksi in the computations. This limit can be derived from a consideration of local stability. As long as the steel is confined by the concrete, there will be no local buckling. The concrete will be able to confine the steel adequately if it does not spall (a disintegration at the concrete surface). If spalling is assumed to occur at a concrete strain of 0.0018, the corresponding stress in the steel is

$$F_{max} = \varepsilon_{max} E = 0.0018(29,000) = 52.2 \text{ ksi}$$

which is rounded to the limiting value of 55 ksi.

5. To prevent local buckling in pipes or tubes filled with concrete, the wall thickness should be at least

$$t = b\sqrt{F_y/3E}$$

for each rectangular section face of width b or

$$t = D\sqrt{F_y/8E}$$

for circular sections of outer diameter D.

■ **EXAMPLE 9.12**

A composite compression member consists of a W12 × 136 encased in a 20-inch × 22-inch concrete column, as shown in Figure 9.30. Four #10 bars are used for longitudinal reinforcement, and #3 ties spaced 13 inches center-to-center provide the lateral reinforcement. The steel yield stress is $F_y = 50$ ksi, and Grade 60 reinforcing bars are used. The concrete strength is $f_c' = 5000$ psi. Compute the design strength for an effective length of 16 feet with respect to both axes.

SOLUTION The modified values F_{my} and E_m will be determined from AISC Equations I2-1 and I2-2. Values needed for these equations are

$F_{yr} = 55$ ksi, the limiting value imposed by AISC I2.1
$A_r = 4(1.27) = 5.08 \text{ in.}^2$

■ **FIGURE 9.30**

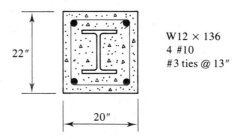

W12 × 136
4 #10
#3 ties @ 13"

22"

20"

$$A_c = \text{net area of concrete} = 20(22) - A_s - A_r = 440 - 39.9 - 5.08$$
$$= 395.0 \text{ in.}^2$$

For $f_c' = 5000$ psi,

$$E_c = w_c^{1.5} \sqrt{f_c'} = (145)^{1.5} \sqrt{5}$$
$$= 3904 \text{ ksi}$$

From AISC Equation I2-1, the modified yield stress is

$$F_{my} = F_y + c_1 F_{yr}\left(\frac{A_r}{A_s}\right) + c_2 f_c'\left(\frac{A_c}{A_s}\right)$$
$$= 50 + 0.7(55)\left(\frac{5.08}{39.9}\right) + 0.6(5)\left(\frac{395}{39.9}\right) = 84.60 \text{ ksi}$$

From AISC Equation I2-2, the modified modulus of elasticity is

$$E_m = E + c_3 E_c\left(\frac{A_c}{A_s}\right) = 29{,}000 + 0.2(3904)\left(\frac{395}{39.9}\right) = 36{,}730 \text{ ksi}$$

The radius of gyration r_m to be used in the compression member equations of AISC Chapter E will be either r for the steel shape or $0.3b$, whichever is larger. In this example, buckling will be about the y-axis of the member, so r for the shape $= r_y = 3.16$ inches. In the plane of buckling,

$$0.3b = 0.3(20) = 6 \text{ in.} \qquad \text{(controls)}$$

Therefore $r_m = 6$ in. The design strength can now be computed as for a noncomposite compression member by using the modified values F_{my}, E_m, and r_m in place of F_y, E, and r.

$$\lambda_c = \frac{KL}{r_m \pi}\sqrt{\frac{F_{my}}{E_m}} = \frac{16(12)}{6\pi}\sqrt{\frac{84.60}{36{,}730}} = 0.4888 < 1.5$$

$$F_{cr} = (0.658)^{\lambda_c^2} F_{my} = (0.658)^{(0.4888)^2} (84.60) = 76.55 \text{ ksi}$$

The nominal strength is

$$P_n = A_s F_{cr} = 39.9(76.55) = 3054 \text{ kips}$$

and the design strength is

$$\phi_c P_n = 0.85(3054) = 2600 \text{ kips}$$

ANSWER   The design compressive strength is 2600 kips.

Tables for Analysis and Design

Part 5 of the *Manual* contains tables that greatly simplify the analysis and design of composite columns. These tables are similar to the column strength tables in Part 3 of the *Manual*. Axial compressive design strengths are given as a function of effective length for concrete-filled pipes and tubes and for encased W-shapes. For the encased columns, longitudinal bars and transverse ties that satisfy AISC requirements are included. Values of r_{mx}/r_{my} are given for those cases where $K_x L \neq K_y L$.

■ **EXAMPLE 9.13**

An 18-foot-long compression member must support a total service load of 1000 kips, consisting of equal parts of dead load and live load. The member is pinned at both ends, with additional support at midheight in the weak direction. Use the tables in Part 5 of the *Manual* to select a square encased W-shape with the smallest area of concrete possible. Use A36 steel, Grade 60 reinforcing bars, and $f_c' = 3.5$ ksi.

SOLUTION The factored axial load is

$$P_u = 1.2(500) + 1.6(500) = 1400 \text{ kips}$$

An examination of the tables shows that for $f_c' = 3.5$ ksi, the value of r_{mx}/r_{my} varies from 1.0 to 1.22, with most values equaling 1.0. Since

$$\frac{K_x L}{K_y L} = 2 > 1.22$$

$K_x L$ will control. Assume that $r_{mx}/r_{my} = 1.0$ and enter the tables with

$$KL = \frac{K_x L}{r_{mx}/r_{my}} = \frac{18}{1.0} = 18 \text{ ft}$$

The following tabulation shows the possible choices.

Concrete Dimensions	Shape	r_{mx}/r_{my}	$\phi_c P_n$
18×18	W10 × 112	1.0	1450 kips
20×20	W12 × 87	1.0	1420 kips

Although the 20×20 column requires the lighter structural steel shape, the criterion is minimum concrete dimensions, so the 18×18 will be selected.

ANSWER Use an 18-inch × 18-inch column with a W10 × 112, four #8 bars, and #3 ties at 12 inches center-to-center. ■

■ PROBLEMS

NOTES   Unless otherwise indicated, the following conditions apply to all problems.

   1.   No shoring is used.

   2.   The slab formwork or deck provides continuous lateral support of the beam during the construction phase.

   3.   Normal-weight concrete is used.

Introduction. Analysis of Composite Beams

9.1-1   A W18 × 40 floor beam supports a 4-inch-thick reinforced concrete slab with an effective width b of 81 inches. Sufficient shear connectors are provided to make the beam fully composite. The 28-day compressive strength of the concrete is $f_c' = 4000$ psi.

   a.   Compute the moment of inertia of the transformed section.

   b.   For a positive service load moment of 375 ft-kips, compute the maximum tensile stress in the steel, the maximum compressive stress in the steel, and the maximum compressive stress in the concrete.

9.1-2   A W21 × 57 floor beam supports a 5-inch-thick reinforced concrete slab with an effective width b of 75 inches. Sufficient shear connectors are provided to make the beam fully composite. The 28-day compressive strength of the concrete is $f_c' = 4000$ psi.

   a.   Compute the moment of inertia of the transformed section.

   b.   For a positive service load moment of 300 ft-kips, compute the maximum tensile stress in the steel, the maximum compressive stress in the steel, and the maximum compressive stress in the concrete.

9.1-3   A W24 × 55 floor beam supports a 4½-inch-thick reinforced concrete slab with an effective width b of 78 inches. Sufficient shear connectors are provided to make the beam fully composite. The 28-day compressive strength of the concrete is $f_c' = 4000$ psi.

   a.   Compute the moment of inertia of the transformed section.

   b.   For a positive service load moment of 450 ft-kips, compute the maximum tensile stress in the steel, the maximum compressive stress in the steel, and the maximum compressive stress in the concrete.

9.1-4   Compute the flexural design strength of the composite beam in Problem 9.1-1 if A572 Grade 50 steel is used.

9.1-5   Compute the flexural design strength of the composite beam in Problem 9.1-2 if A36 steel is used.

9.1-6   Compute the flexural design strength of the composite beam in Problem 9.1-3 if A572 Grade 50 steel is used.

Strength of Unshored Composite Beams

9.2-1 A W14 × 22 acts compositely with a 4-inch-thick floor slab whose effective width b is 90 inches. The beams are spaced at 7 feet 6 inches, and the span length is 30 feet. No shoring is used, and the superimposed loads are as follows: construction load = 20 psf, partition load = 10 psf, weight of ceiling and light fixtures = 5 psf, and live load = 80 psf. A36 steel is used, and $f_c' = 3000$ psi. Determine whether the flexural strength is adequate.

9.2-2 A composite floor system consists of W18 × 97 beams and a 5-inch-thick reinforced concrete floor slab. The effective width of the slab is 84 inches. The span length is 30 feet, and the beams are spaced at 7 feet. In addition to the weight of the slab, there is a construction load of 20 psf and a uniform live load of 800 psf. No shoring is used. Determine whether the flexural strength is adequate. Assume that there is no lateral support during the construction phase. A572 Grade 50 steel is used, and $f_c' = 4000$ psi.

Effective Flange Width

9.3-1 A fully composite floor system consists of W12 × 16 floor beams supporting a 4-inch-thick reinforced concrete floor slab. The beams are spaced at 5 feet, and the span length is 25 feet. The superimposed loads consist of a construction load of 20 psf, a partition load of 15 psf, and a live load of 125 psf. A36 steel is used, and $f_c' = 3000$ psi. Determine whether this beam satisfies the provisions of the AISC Specification.

9.3-2 A fully composite floor system consists of W16 × 50 floor beams of A36 steel supporting a 4½-inch-thick reinforced concrete floor slab, with $f_c' = 3000$ psi. The beams are spaced at 8 feet, and the span length is 35 feet. The superimposed loads consist of a construction load of 20 psf and a live load of 160 psf. Determine whether this beam satisfies the provisions of the AISC Specification.

Shear Connectors

9.4-1 A floor system consists of W18 × 40 beams supporting a 4-inch-thick floor slab. The effective width b is 96 inches. The steel is A36, and the concrete strength is $f_c' = 4000$ psi. What is the total number of ¾-inch × 3-inch shear studs required for each beam for full composite behavior? The beam is simply supported and uniformly loaded.

9.4-2 How many ¾-inch × 3-inch shear studs are required for the beam in Problem 9.1-1 for full composite behavior? Use $F_y = 50$ ksi.

9.4-3 How many ⅞-inch × 3½-inch shear studs are required for the beam in Problem 9.1-3 for full composite behavior? Use $F_y = 50$ ksi.

9.4-4 Design shear connectors for the beam in Problem 9.3-1. Show the results on a sketch similar to Figure 9.13.

9.4-5   Design shear connectors for the beam in Problem 9.3-2. Show the results on a sketch similar to Figure 9.13.

Design

9.5-1   A fully composite floor system consists of 40-foot-long steel beams spaced at 9 feet center-to-center supporting a 6-inch-thick reinforced concrete floor slab. The steel is A572 Grade 50, and the concrete strength is $f_c' = 4000$ psi. There is a construction load of 20 psf and a live load of 250 psf.

    a. Select a W-shape.

    b. Select shear connectors and show the layout on a sketch similar to Figure 9.13.

9.5-2   The span length of a certain fully composite floor system is 30 feet, and the beam spacing is 9 feet. The floor slab is 4 inches thick. The superimposed live load is 100 psf, and there is a construction load of 20 psf. The concrete strength is $f_c' = 3000$ psi, and A36 steel is used.

    a. Select a W-shape.

    b. Select shear connectors and show the layout on a sketch similar to Figure 9.13.

9.5-3   A fully composite floor system consists of 27-foot-long steel beams spaced at 8 feet supporting a 4-inch-thick reinforced concrete floor slab. The steel is A572 Grade 50, and the concrete strength is $f_c' = 4000$ psi. There is a construction load of 20 psf, a partition load of 20 psf, and a live load of 120 psf.

    a. Select a W-shape.

    b. Select shear connectors and show the layout on a sketch similar to Figure 9.13.

Deflections

9.6-1   Compute the following deflections for the beam in Problem 9.3-1.

    a. Maximum deflection before the concrete has cured

    b. Maximum short-term deflection after composite behavior has been attained

    c. Maximum long-term deflection after composite behavior has been attained

9.6-2   Compute the following deflections for the beam in Problem 9.3-2.

    a. Maximum deflection before the concrete has cured

    b. Maximum deflection after composite behavior has been attained

9.6-3   For the beam designed in Problem 9.5-1,

    a. compute the deflections that occur before and after the concrete has cured.

    b. If the live load deflection exceeds $L/360$, select another steel shape.

9.6-4 For the beam designed in Problem 9.5-2,

    a. compute the deflections that occur before and after the concrete has cured.

    b. If the total deflection after the concrete has cured exceeds $L/240$, select another steel shape.

9.6-5 For the beam designed in Problem 9.5-3,

    a. compute the deflections that occur before and after the concrete has cured.

    b. If the live load deflection exceeds $L/360$, select another steel shape.

Composite Beams with Formed Steel Deck

9.7-1 A fully composite floor system consists of W18 × 35 beams spaced at 6 feet, spanning 30 feet, and supporting a formed steel deck with a concrete slab. The total depth of the slab is $4\frac{1}{2}$ inches, and the deck is 2 inches high. The steel and concrete strengths are $F_y = 50$ ksi and $f_c' = 4000$ psi.

    a. Compute the transformed moment of inertia and compute the deflection for a service load of 1 kip/ft.

    b. Compute the design strength of the composite section.

9.7-2 A fully composite floor system consists of W12 × 19 beams of A36 steel supporting a formed steel deck and concrete slab. The concrete is lightweight, with a unit weight of $w_c = 110$ pounds per cubic foot, and a 28-day compressive strength of $f_c' = 3000$ psi. The total thickness of the slab is 4 inches, and the deck rib height is $1\frac{1}{2}$ inches. The beams are spaced at 7 feet 6 inches, and the span length is 30 feet. Determine whether this system is adequate for a 20 psf construction load, a 10 psf partition load, ceiling and light fixtures weighing 5 psf, and a live load of 60 psf.

9.7-3 A fully composite floor system consists of W16 × 45 beams of A36 steel supporting a formed steel deck and concrete slab. The concrete has a 28-day compressive strength of $f_c' = 3000$ psi, and the total thickness of the slab is 4 inches. The deck rib height is 2 inches. The beams are spaced at 8 feet, and the span length is 36 feet. Determine whether this system is adequate for a 20 psf construction load, a 20 psf partition load, a ceiling load of 8 psf, and a live load of 100 psf.

9.7-4 A composite floor system is composed of steel beams spaced at 12 feet, spanning 35 feet, and supporting a formed steel deck and concrete slab. The 28-day compressive strength of the concrete is $f_c' = 4000$ psi. The total thickness of the slab is $5\frac{1}{2}$ inches, and the deck has the following properties: rib spacing = 6 inches, rib width = $2\frac{1}{4}$ inches, and rib height = 2 inches. There is a construction load of 20 psf, a partition load of 10 psf, and a live load of 100 psf. Assume that there is no limit on deflection.

    a. Select a W-shape of A572 Grade 50 steel.

    b. Select stud shear connectors and show their distribution on a sketch similar to Figure 9.22.

9.7-5 A composite floor system is composed of steel beams spaced at 8 feet, spanning 40 feet, and supporting a formed steel deck and concrete slab. Lightweight concrete with a unit weight of 110 pcf is used, and the 28-day compressive strength is $f_c' = 3000$ psi. The total thickness of the slab is 4 inches, and the deck has the following properties: rib spacing $= 6$ inches, rib width $= 2\frac{1}{4}$ inches, and rib height $= 1\frac{1}{2}$ inches. There is a construction load of 20 psf and a live load of 175 psf. The maximum permissible live load deflection is $L/360$.

    a. Select a W-shape of A36 steel.

    b. Select stud shear connectors and show their distribution on a sketch similar to Figure 9.22.

Tables for Composite Beam Analysis and Design

9.8-1 For the beam in Problem 9.1-2, use A36 steel and the tables in Part 5 of the *Manual* to obtain

    a. the design strength ϕM_n for full composite action.

    b. the design strength ϕM_n if the total number of $\frac{3}{4}$-inch-diameter shear studs is 24.

    c. the lower-bound moment of inertia I_{LB} if the total number of $\frac{3}{4}$-inch-diameter shear studs is 24.

Assume uniform loading and simple supports.

9.8-2 For the beam in Problem 9.1-3, use A572 Grade 50 steel and the tables in Part 5 of the *Manual* to obtain

    a. the design strength ϕM_n for full composite action.

    b. the design strength ϕM_n if the total number of $\frac{5}{8}$-inch-diameter shear studs is 34.

    c. the lower-bound moment of inertia I_{LB} if the total number of $\frac{5}{8}$-inch-diameter shear studs is 34.

Assume uniform loading and simple supports.

9.8-3 For the beam in Problem 9.7-1, use the tables in Part 5 of the *Manual* to determine the design strength if a total of twenty $\frac{3}{4}$-inch-diameter studs is used. Use a stud strength reduction factor of 1.0. Assume that the beam is simply supported and uniformly loaded.

9.8-4 If the maximum factored load moment acting on the composite beam in Problem 9.4-1 is 310 ft-kips, how many $\frac{3}{4}$-inch-diameter studs are required? Use partial composite action and try to stay as close to 310 ft-kips as possible. Use the tables in Part 5 of the *Manual*.

9.8-5 A composite floor system is composed of steel beams spaced at 8 feet, spanning 30 feet, and supporting a formed steel deck and concrete slab. Lightweight concrete with a unit weight of 110 pcf is used, and the 28-day compressive strength is $f_c' = 3000$ psi. The total thickness of the slab is 5 inches, and a cross section of the deck is shown in Figure P9.8-5. There is a construction load of 20 psf, a partition load of 20 psf, and a live load of 50 psf. There is no restriction on the deflection. Use A36 steel and select a W-shape with the aid of the tables in Part 5 of the *Manual*. Use partial composite action and the recommended minimum number of ¾-inch × 3½-inch studs.

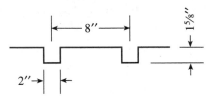

FIGURE P9.8-5

9.8-6 Use the tables in Part 5 of the *Manual* and select a W-shape and shear connectors for the following conditions:

Beam spacing = 5 ft - 6 in.

Span length = 30 ft

Total slab thickness = 4½ in.

Construction load = 20 psf

Partition load = 20 psf

Weight of ceiling = 5 psf

Live load = 150 psf

F_y = 50 ksi and f_c' = 4 ksi

Formed steel deck is used, a cross section of which is shown in Figure P9.8-6. The maximum live load deflection cannot exceed $L/360$ (a lower-bound moment of inertia may be used).

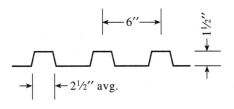

FIGURE P9.8-6

9.8-7 Use the tables in Part 5 of the *Manual* and select a W-shape and shear connectors for the following conditions:

Beam spacing = 9 ft - 6 in.

Span length = 35 ft

Total slab thickness = 5½ in.

Construction load = 20 psf

Partition load = 20 psf

Miscellaneous dead load = 10 psf

Live load = 50 psf

F_y = 50 ksi   and f_c' = 4 ksi

Formed steel deck is used, a cross section of which is shown in Figure P9.8-7. The maximum total deflection cannot exceed $L/360$ (a lower-bound moment of inertia may be used).

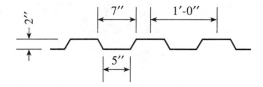

FIGURE P9.8-7

9.8-8   Use the tables in Part 5 of the *Manual* and select a W-shape and shear connectors for the following conditions:

Beam spacing = 12 ft

Span length = 40 ft

Total slab thickness = 5 in.

Construction load = 20 psf

Miscellaneous dead load = 10 psf

Live load = 80 psf

F_y = 50 ksi   and f_c' = 4 ksi   (lightweight concrete with a unit weight of w_c = 110 pcf is used)

The formed steel deck is the same as that in Problem 9.8-7. The maximum total deflection cannot exceed $L/240$ (a lower-bound moment of inertia may be used).

9.8-9   Use the tables in Part 5 of the *Manual* and select a W-shape and shear connectors for the following conditions:

Beam spacing = 12 ft

Span length = 25 ft

Total slab thickness = 6 in.   (the slab and deck combination weighs 43 psf)

Construction load $=$ 20 psf

Partition load $=$ 20 psf

Ceiling weight $=$ 5 psf

Flooring weight $=$ 2 psf

Live load $=$ 160 psf

$F_y = 50$ ksi   and $f_c' = 4$ ksi   (lightweight concrete with a unit weight of $w_c = 110$ pcf is used)

A cross section of the formed steel deck is shown in Figure P9.8-9. The maximum live load deflection cannot exceed $L/360$ (a lower-bound moment of inertia may be used).

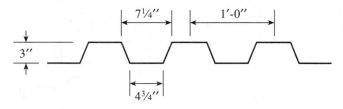

FIGURE P9.8-9

Composite Columns

9.10-1   A $7 \times 5 \times \frac{5}{16}$ structural tube filled with concrete is used as a composite column, as shown in Figure P9.10-1. The steel has a yield stress of $F_y = 46$ ksi, and the concrete has a compressive strength of $f_c' = 3500$ psi. Compute the design strength of the column and check your answer with the tables in Part 5 of the *Manual*.

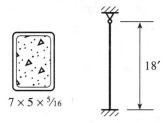

$7 \times 5 \times \frac{5}{16}$

FIGURE P9.10-1

9.10-2   The composite compression member shown in Figure P9.10-2 has lateral support at the one-third points in the weak direction only. Compute the design strength and check your answer with the tables in Part 5 of the *Manual*. The steel yield stress is 50 ksi, and $f_c' = 8000$ psi.

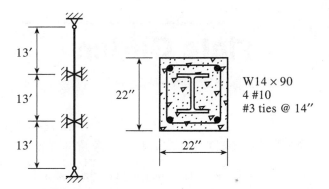

FIGURE P9.10-2

9.10-3 Use the tables in Part 5 of the *Manual* and select a concrete-filled pipe for the conditions shown in Figure P9.10-3. Select the smallest diameter pipe that will work. Use ASTM A501 steel and $f_c' = 3500$ psi.

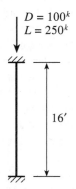

$D = 100^k$
$L = 250^k$

$16'$

FIGURE P9.10-3

9.10-4 Design a square composite column with an encased W-shape to resist a factored axial load of 1200 kips. The effective lengths with respect to the two principle axes are $K_x L = 30$ ft and $K_y L = 16$ ft. Select the smallest concrete dimensions possible. Use $F_y = 50$ ksi and $f_c' = 5000$ psi. The tables in Part 5 of the *Manual* may be used.

10 Plate Girders

10.1 INTRODUCTION

A plate girder is usually thought of as a flexural member whose cross section is composed of plate elements. (As discussed in Chapter 5 and to be discussed again presently, AISC distinguishes beams from plate girders on the basis of the web slenderness.) A plate girder is basically a large beam, both in span and cross section, the large cross section usually being a consequence of the long span. If the largest available hot-rolled steel shape is inadequate for a given span and loading, the first alternative would normally be a rolled shape with cover plates added to one or both flanges. If this configuration could not provide enough moment strength, a cross section with exactly the properties needed could be fabricated from plate elements. If the span is long enough, however, the depth and weight of a built-up girder may be excessive, and other alternatives, such as a truss, may be necessary.

A plate girder cross section can take several forms. Figure 10.1 shows some of the possibilities. The usual configuration is a single web with two equal flanges, with all parts connected by welding. The box section, which has two webs as well as two flanges, is a torsionally superior shape and can be used when large unbraced lengths are necessary. Hybrid girders, in which the steel in the flanges is of a higher strength than that in the web or webs, are sometimes used.

Before the widespread use of welding, connecting the components of the cross section was a major consideration in the design of plate girders. All of the connections were made by riveting, so there was no way to attach the flange directly to the web, and additional cross-sectional elements were introduced for the specific purpose of transmitting the load from one component to the other. The usual technique was to use a pair of angles, placed back-to-back, to attach the flange to the web; one pair of legs was attached to the web, and the other pair attached to the flange, as shown in Figure 10.1b. If web stiffeners were needed, pairs of angles were used for that purpose also. To avoid a conflict between the stiffener angles and the flange angles, filler plates were added to the web so that the stiffeners could clear the flange angles, as shown in Figure 10.1c. If a variable cross section was desired, one or more cover plates of different lengths were riveted to the flanges. (Although cover plates can also be used with welded plate girders, a simpler approach is to use different thicknesses of flange plate, welded end-to-end, at different locations along the length of the girder.) It should be evident that the welded plate girder is far superior to the riveted or bolted girder in terms of simplicity and efficiency. We consider only I-shaped welded plate girders in this chapter.

Before considering specific AISC Specification requirements for plate girders, we need to examine, in a very general way, the peculiarities of plate girders as opposed to

■ **FIGURE 10.1**

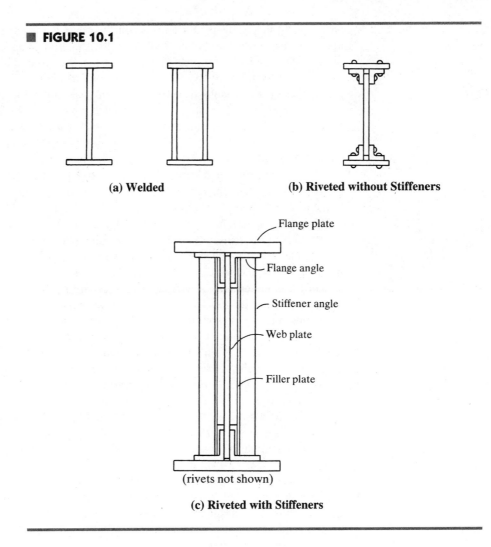

(a) Welded   (b) **Riveted without Stiffeners**

Flange plate

Flange angle

Stiffener angle

Web plate

Filler plate

(rivets not shown)

(c) **Riveted with Stiffeners**

ordinary rolled beams. Although we covered flexural members in Chapter 5, "Beams," plate girders have special flexural and shear strength requirements.

10.2 GENERAL CONSIDERATIONS

Structural steel design is largely a matter of providing for stability, either locally or in an overall sense. Many standard hot-rolled structural shapes are proportioned so that local stability problems have been eliminated or minimized. When a plate girder is used, however, the designer must account for factors that in many cases would not be a problem with a rolled shape. Deep, thin webs account for many of the special problems associated with plate girders, including local instability. A thorough understanding of the basis of the AISC provisions for plate girders requires a background in stability theory, particularly plate stability. Such a treatment is beyond the scope of this book, however, and the emphasis in this chapter is on the qualitative basis of the Specification requirements and their application. For those interested in delving deeper, the *Guide to Stability Criteria for Metal Structures* (Johnston, 1976) is a good starting point, and *Buckling Strength of Metal Structures* (Bleich, 1952) and *Theory of Elastic Stability* (Timoshenko and Gere, 1961) will provide the fundamentals of stability theory.

Plate girders rely on the strength available after the web has buckled, so most of the flexural strength will come from the flanges. The limit states considered are yielding of the tension flange and buckling of the compression flange. The compression flange buckling can take the form of vertical buckling into the web or flange local buckling (FLB), or it can be caused by lateral-torsional buckling (LTB).

At a location of high shear in a girder web, usually near the support and at or near the neutral axis, the principal planes will be inclined with respect to the longitudinal axis of the member, and the principal stresses will be diagonal tension and diagonal compression. The diagonal tension poses no particular problem, but the diagonal compression can cause the web to buckle. This problem can be addressed in one of three ways: (1) the depth-to-thickness ratio of the web can be made small enough that the problem is eliminated, (2) web stiffeners can be used to form panels with increased shear strength, or (3) web stiffeners can be used to form panels that resist the diagonal compression through *tension-field action*. Figure 10.2 illustrates the concept of tension-field action. At the point of impending buckling, the web loses its ability to support the diagonal compression, and this stress is shifted to the transverse stiffeners and the flanges. The stiffeners resist the vertical component of the diagonal compression, and the flanges resist the horizontal component. The web will need to resist only the diagonal tension, hence the term *tension-field action*. This behavior can be likened to that of a Pratt truss, in which the vertical web members carry compression and the diagonals carry tension, as shown in Figure 10.2b. Since the tension field does not actually exist until the web begins to buckle, its contribution to the web shear strength will not exist until the web buckles. The total strength will consist of the strength prior to buckling plus the post-buckling strength deriving from tension-field action.

If an unstiffened web is incapable of resisting the applied shear, appropriately spaced stiffeners are used to develop tension-field action. Cross-section requirements for these stiffeners, called *intermediate stiffeners,* are minimal because their primary purpose is to provide stiffness rather than resist directly applied loads.

Additional stiffeners may be required at points of concentrated loads for the purpose of protecting the web from the direct compressive load. These members are called *bearing stiffeners,* and they must be proportioned to resist the applied loads. They can also simultaneously serve as intermediate stiffeners. Figure 10.3 shows a bearing stiffener consisting of two rectangular plates, one on each side of the girder web. The plates are notched at the inside top and bottom corners so as to avoid the flange-to-web welds. If the stiffeners are conservatively assumed to resist the total applied load P (this as-

■ **FIGURE 10.2**

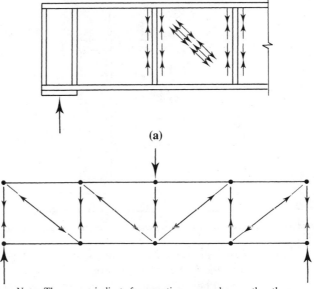

(a)

Note: The arrows indicate forces acting on members, rather than on joints

(b)

sumption neglects any contribution by the web), the bearing stress on the contact surfaces may be written as

$$f_p = \frac{P}{A_{pb}}$$

where

A_{pb} = projected bearing area

= $2at$ (see Figure 10.3)

or, expressing the bearing load in terms of the stress,

$$P = f_p A_{pb} \tag{10.1}$$

In addition, the pair of stiffeners, together with a short length of web, is treated as a column with an effective length less than the web depth and is investigated for compliance with the same Specification provisions as any other compression member. This cross section is illustrated in Figure 10.4. The compressive strength should always be

■ **FIGURE 10.3**

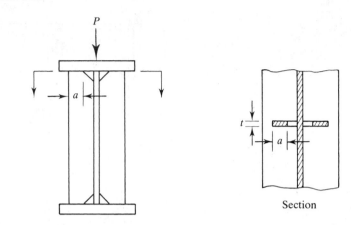

Section

based on the radius of gyration about an axis in the plane of the web, as instability about the other principal axis is prevented by the web itself.

Other limit states resulting from the application of concentrated loads to the top flange are web yielding, web crippling (buckling), and sidesway web buckling. Sidesway web buckling occurs when the compression in the web causes the *tension* flange to buckle laterally. This phenomenon can occur if the flanges are not adequately restrained against movement relative to one another by stiffeners or lateral bracing.

The welds for connecting the components of a plate girder are designed in much the same way as for other welded connections. The flange-to-web welds must resist the

■ **FIGURE 10.4**

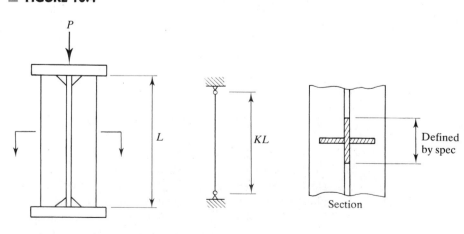

Section

horizontal shear at the interface between the two components. This applied shear, called the *shear flow,* is usually expressed as a force per unit length of girder to be resisted by the weld. From Chapter 5, the shear flow, based on elastic behavior, is given by

$$f = \frac{VQ}{I_x}$$

where Q is the moment, about the neutral axis, of the area between the horizontal shear plane and the outside face of the section. This expression is Equation 5.6 for shearing stress multiplied by the width of the shear plane. Since the applied shear force V will usually be variable, the spacing of intermittent welds, if used, can also vary.

10.3   AISC REQUIREMENTS

Requirements for plate girders are addressed in Chapter G of the AISC Specification. Chapter G does little more than refer you to Appendix G, so for all practical purposes, plate girders are covered in Appendix G.

Whether a flexural member is classified as a beam or a plate girder is a function of the web slenderness, h/t_w, where h is the depth of the web from inside face of flange to inside face of flange and t_w is the web thickness. If h/t_w is less than $970/\sqrt{F_{yf}}$, the member is classified as a beam, and the provisions of AISC Chapter F apply, regardless of whether the member is built up from plates or is a hot-rolled shape. If h/t_w is greater than $970/\sqrt{F_{yf}}$, the member is treated as a plate girder, and the provisions of Chapter G apply. Thus flexural members with slender webs, where slenderness is defined in AISC Chapter B, are treated as plate girders. Note that when hybrid girders are used, the limiting value of h/t_w is based on F_{yf}, the yield stress of the flange. The reason is that the stability of the web against flexural buckling is dependent on the strain in the flange (Zahn, 1987).

To prevent vertical buckling of the flange into the web, AISC Appendix G2 imposes an upper limit on the web width–thickness ratio, h/t_w. The limiting value is a function of the *aspect ratio, a/h,* of the girder panels, which is the ratio of intermediate stiffener spacing to web depth (see Figure 10.5):

For $a/h \leq 1.5$,

$$\frac{h}{t_w} \leq \frac{2000}{\sqrt{F_{yf}}} \qquad\qquad \text{(AISC Equation A-G1-1)}$$

For $a/h > 1.5$,

$$\frac{h}{t_w} \leq \frac{14{,}000}{\sqrt{F_{yf}(F_{yf} + 16.5)}} \qquad\qquad \text{(AISC Equation A-G1-2)}$$

where a is the clear distance between stiffeners.

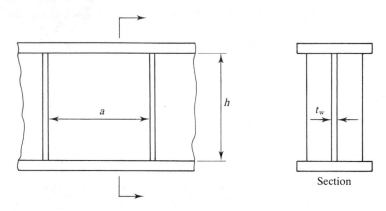

FLEXURAL STRENGTH

10.4

The design flexural strength of plate girders is $\phi_b M_n$, where $\phi_b = 0.90$. The nominal flexural strength, M_n, is based on either tension flange yielding or compression flange buckling. The compression flange buckling strength will be determined by either flange local buckling (FLB) or lateral-torsional buckling (LTB). Vertical buckling of the compression flange into the web has been eliminated from consideration by the limitations imposed by AISC Equations A-G1-1 and A-G1-2 (Cooper, Galambos, and Ravindra, 1978).

Tension Flange Yielding

From Chapter 5, the maximum bending stress in a flexural member bent about its strong axis is

$$f_b = \frac{M}{S_x}$$

where S_x is the elastic section modulus about the strong axis. Expressing the bending moment as a function of the section modulus and stress gives

$$M = S_x f_b \tag{10.2}$$

AISC Appendix G2 gives the nominal flexural strength based on tension flange yielding as

$$M_n = S_{xt} R_e F_{yt} \qquad \text{(AISC Equation A-G2-1)}$$

where

S_{xt} = elastic section modulus referred to the tension side

R_e = hybrid girder factor

F_{yt} = yield stress of tension flange

The hybrid girder factor R_e equals 1.0 for nonhybrid girders. For hybrid girders,

$$R_e = \frac{12 + (A_w/A_f)(3m - m^3)}{12 + 2(A_w/A_f)} \leq 1.0$$

where

A_w = area of the web $\leq 10A_f$

A_f = area of the compression flange

m = F_{yw}/F_{yf} (this definition is for the limit state of tension flange yielding; for compression flange buckling, the hybrid girder factor will also apply, but m will be defined differently.)

Compression Flange Buckling

The nominal flexural strength corresponding to compression flange buckling is also based on Equation 10.2. From AISC Appendix G2, this strength is expressed as

$$M_n = S_{xc}R_{PG}R_eF_{cr} \qquad \text{(AISC Equation A-G2-2)}$$

where

S_{xc} = elastic section modulus referred to the compression side

R_{PG} = strength reduction factor to account for elastic web buckling

F_{cr} = critical compression flange stress, based on either LTB or FLB

R_e = hybrid girder factor (computed by the same equation as for tension flange yielding, but with a value of $m = F_{yw}/F_{cr}$).

The hybrid girder factor again equals 1.0 for nonhybrid girders. The plate girder strength reduction factor R_{PG} is given by

$$R_{PG} = 1 - \frac{a_r}{1200 + 300a_r}\left(\frac{h_c}{t_w} - \frac{970}{\sqrt{F_{cr}}}\right) \leq 1.(\qquad \text{(AISC Equation A-G2-3)}$$

where

a_r = $A_w/A_f \leq 10$

h_c = twice the distance from the centroid to the inside face of the compression flange ($h_c = h$ for girders with equal flanges)

The critical stress F_{cr} is based on either lateral-torsional buckling or flange local buckling. The AISC Specification uses the generic notation λ, λ_p, and λ_r to deal with the slenderness parameters for both of these limit states, and a common set of equations is used for F_{cr}. The equations are presented here in a somewhat expanded form in an attempt to clarify the provisions. For lateral torsional buckling, the slenderness of a portion of the compression region of the girder is used, so

$$\lambda = \frac{L_b}{r_T}$$

$$\lambda_p = \frac{300}{\sqrt{F_{yf}}}$$

$$\lambda_r = \frac{756}{\sqrt{F_{yf}}}$$

(AISC Equations A-G2-7, 8 and 9)

where L_b is the unbraced length and r_T is the radius of gyration about the weak axis for a portion of the cross section consisting of the compression flange and one third of the compressed part of the web. For a doubly symmetrical girder, this dimension will be one sixth of the web depth. (See Figure 10.6.) Then:

If $\lambda \leq \lambda_p$, failure will be by yielding, and

$$F_{cr} = F_{yf}$$

(AISC Equation A-G2-4)

If $\lambda_p < \lambda \leq \lambda_r$, failure will be by inelastic LTB, and

$$F_{cr} = C_b F_{yf} \left[1 - \frac{1}{2} \left(\frac{\lambda - \lambda_p}{\lambda_r - \lambda_p} \right) \right] \leq F_{yf}$$

(AISC Equation A-G2-5)

where C_b is given by AISC Equation F1-3.

■ **FIGURE 10.6**

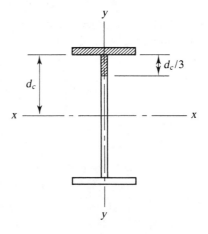

If $\lambda > \lambda_r$, failure will be by elastic LTB, and

$$F_{cr} = \frac{C_{PG}}{\lambda^2} \qquad \text{(AISC Equation A-G2-6)}$$

where

$$C_{PG} = 286{,}000C_b \qquad \text{(AISC Equation A-G2-10)}$$

For compression flange buckling based on flange local buckling, the relevant width–thickness ratio and its limits are

$$\lambda = \frac{b_f}{2t_f}$$

$$\lambda_p = \frac{65}{\sqrt{F_{yf}}} \qquad \text{(AISC Equations A-G2-11, 12, and 13)}$$

$$\lambda_r = \frac{230}{\sqrt{F_{yf}/k_c}}$$

where

$$k_c = 4/\sqrt{h/t_w}, \qquad 0.35 \leq k_c \leq 0.763$$

Then:

If $\lambda \leq \lambda_p$, failure will be by yielding, and

$$F_{cr} = F_{yf} \qquad \text{(AISC Equation A-G2-4)}$$

If $\lambda_p < \lambda \leq \lambda_r$, failure will be by inelastic FLB, and

$$F_{cr} = C_b F_{yf}\left[1 - \frac{1}{2}\left(\frac{\lambda - \lambda_p}{\lambda_r - \lambda_p}\right)\right] \leq F_{yf} \qquad \text{(AISC Equation A-G2-5)}$$

where $C_b = 1.0$.
    If $\lambda > \lambda_r$, failure will be by elastic FLB, and

$$F_{cr} = \frac{C_{PG}}{\lambda^2} \qquad \text{(AISC Equation A-G2-6)}$$

where

$$C_{PG} = 26{,}200k_c \qquad \text{(AISC Equation A-G2-14)}$$

The computation of flexural strength is illustrated in Example 10.1, part (a).

10.5

SHEAR STRENGTH

The design shear strength of a plate girder is $\phi_v V_n$, where $\phi_v = 0.9$. The shear strength is a function of the depth-to-thickness ratio of the web and the spacing of any intermediate stiffeners that may be present. The shear capacity has two components: the strength before buckling and the post-buckling strength. The post-buckling strength relies on tension field action, which is made possible by the presence of intermediate stiffeners. If stiffeners are not present or are spaced too far apart, tension-field action will not be possible, and the shear capacity will consist only of the strength before buckling. The nominal shear strength is given in AISC Appendix G3 as follows:

For $\dfrac{h}{t_w} \leq 187\sqrt{\dfrac{k_v}{F_{yw}}}$,

$$V_n = 0.6A_w F_{yw} \qquad\qquad \text{(AISC Equation A-G3-1)}$$

For $\dfrac{h}{t_w} > 187\sqrt{\dfrac{k_v}{F_{yw}}}$,

$$V_n = 0.6A_w F_{yw}\left(C_v + \frac{1 - C_v}{1.15\sqrt{1 + (a/h)^2}} \right) \qquad \text{(AISC Equation A-G3-2)}$$

where

$$k_v = 5 + \frac{5}{(a/h)^2}$$

$$= 5 \text{ if } \frac{a}{h} > 3 \qquad\qquad \text{(AISC Equation A-G3-4)}$$

$$= 5 \text{ if } \frac{a}{h} > \left[\frac{260}{(h/t_w)} \right]^2$$

The first of the above equations, AISC Equation A-G3-1, gives the shear strength when tension-field action is not present and web failure is by yielding. The second equation, AISC Equation A-G3-2, accounts for tension-field action. Equation A-G3-2 can also be written as

$$V_n = 0.6A_w F_{yw} C_v + 0.6A_w F_{yw}\frac{1 - C_v}{1.15\sqrt{1 + (a/h)^2}}$$

The first term in this equation gives the web shear buckling strength and the second term gives the post-buckling strength. The factor C_v is the ratio of the critical web buckling stress to the web shear yield stress and is defined as follows:

For $187\sqrt{\dfrac{k_v}{F_{yw}}} \leq \dfrac{h}{t_w} \leq 234\sqrt{\dfrac{k_v}{F_{yw}}},$

$$C_v = \frac{187\sqrt{k_v/F_{yw}}}{h/t_w} \qquad \text{(AISC Equation A-G3-5)}$$

For $\dfrac{h}{t_w} > 234\sqrt{\dfrac{k_v}{F_{yw}}},$

$$C_v = \frac{44,000k_v}{(h/t_w)^2 F_{yw}} \qquad \text{(AISC Equation A-G3-6)}$$

Solution of AISC Equations A-G3-1 and A-G3-2 is facilitated by tables given in the Numerical Values section of the Specification. Tables 9-36 and 10-36 relate the parameters of these two equations for A36 steel, and Tables 9-50 and 10-50 do the same for steels with a yield stress of 50 kips per square inch. We discuss details of the use of these tables in a later example.

A tension field cannot ordinarily be fully developed in an end panel. This can be understood by considering the horizontal components of the tension fields shown in Figure 10.7. (The vertical components are resisted by the stiffeners.) The tension field in panel CD is balanced on the left side in part by the tension field in panel BC. Thus interior panels are anchored by adjacent panels. Panel AB, however, has no such anchorage on the left side. Although anchorage could be provided by an end stiffener specially designed to resist the bending induced by a tension field, that usually is not done. (Since the tension field does not cover the full depth of the web, the interior stiffeners are also subjected to a certain amount of bending caused by the offset of the tension fields in the adjacent panels, but this bending is not significant.) Hence the anchorage for panel BC must be provided on the left side by a *beam-shear panel* rather than the tension-field panel shown. Thus the nominal shear strength is the same as for a flexural member without tension-field action — that is,

$$V_n = 0.6A_w F_{yw} C_v \qquad \text{(AISC Equation A-G3-3)}$$

■ **FIGURE 10.7**

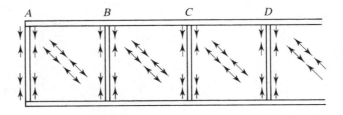

Note that this equation is the first term of AISC Equation A-G3-2. Tension-field action is also not permitted for hybrid girders or when $a/h > 3$ or when $a/h > [260/(h/t_w)]^2$. (Each of these last two cases corresponds to $k_v = 5$.) Therefore AISC Equation A-G3-3 applies in these situations.

AISC G4 states that intermediate stiffeners are not required when $h/t_w \leq 418/\sqrt{F_{yw}}$. This can be shown as follows. When there are no intermediate stiffeners, a/h is greater than 3, and $k_v = 5$, resulting in

$$C_v = \frac{187\sqrt{5/\sqrt{F_{yw}}}}{h/t_w} = \frac{418/\sqrt{F_{yw}}}{h/t_w}$$

and a nominal shear strength of

$$V_n = \frac{0.6A_w F_{yw}\left[418/\sqrt{F_{yw}}\right]}{h/t_w}$$

Thus when $h/t_w \leq 418/\sqrt{F_{yw}}$,

$$V_n \geq 0.6A_w F_{yw}$$

But this expression corresponds to the upper limit (shear yielding), so the greater-than symbol is dropped, and

$$V_n = 0.6A_w F_{yw} \qquad \text{(AISC Equation A-G3-1)}$$

which is the nominal shear strength when $h/t_w \leq 187\sqrt{k_v/F_{yw}}$.

If tension-field action is not used, the shear strength is obtained from AISC Appendix F2 as follows:

For $\dfrac{h}{t_w} \leq 187\sqrt{\dfrac{k_v}{F_{yw}}}$,

$$V_n = 0.6F_{yw}A_w \qquad \text{(AISC Equation A-F2-1)}$$

For $187\sqrt{\dfrac{k_v}{F_{yw}}} < \dfrac{h}{t_w} \leq 234\sqrt{\dfrac{k_v}{F_{yw}}}$,

$$V_n = 0.6F_{yw}A_w \frac{187\sqrt{k_v/F_{yw}}}{h/t_w} \qquad \text{(AISC Equation A-F2-2)}$$

For $\dfrac{h}{t_w} > 234\sqrt{\dfrac{k_v}{F_{yw}}}$,

$$V_n = A_w \frac{26,400k_v}{(h/t_w)^2} \qquad \text{(AISC Equation A-F2-3)}$$

To summarize, the nominal shear strength is determined as follows:

1. Compute the aspect ratio a/h.
2. Compute k_v and C_v.
3. For all panels of hybrid girders and for end panels of nonhybrid girders:

$$\text{If } \frac{h}{t_w} \le 187\sqrt{\frac{k_v}{F_{yw}}},$$

$$V_n = 0.6A_w F_{yw}$$

$$\text{If } \frac{h}{t_w} > 187\sqrt{\frac{k_v}{F_{yw}}},$$

$$V_n = 0.6A_w F_{yw} C_v$$

For other panels of nonhybrid girders with tension-field action:

$$\text{If } \frac{h}{t_w} \le 187\sqrt{\frac{k_v}{F_{yw}}},$$

$$V_n = 0.6A_w F_{yw}$$

$$\text{If } \frac{h}{t_w} > 187\sqrt{\frac{k_v}{F_{yw}}},$$

$$V_n = 0.6A_w F_{yw}\left(C_v + \frac{1 - C_v}{1.15\sqrt{1 + (a/h)^2}}\right)$$

If tension-field action is not used, the provisions of Appendix F2 will apply:

$$\text{If } \frac{h}{t_w} \le 187\sqrt{\frac{k_v}{F_{yw}}},$$

$$V_n = 0.6F_{yw}A_w$$

$$\text{If } 187\sqrt{\frac{k_v}{F_{yw}}} < \frac{h}{t_w} \le 234\sqrt{\frac{k_v}{F_{yw}}},$$

$$V_n = 0.6F_{yw}A_w\frac{187\sqrt{k_v/F_{yw}}}{h/t_w}$$

$$\text{If } \frac{h}{t_w} > 234\sqrt{\frac{k_v}{F_{yw}}},$$

$$V_n = A_w\frac{26{,}400k_v}{(h/t_w)^2}$$

Shear strength computation is illustrated in Example 10.1, part (b).

Intermediate Stiffeners

If intermediate stiffeners are necessary to obtain enough shear strength when tension-field action is being used, the minimum cross-sectional area of either a single stiffener or a pair is given by AISC Appendix G4 as

$$A_{st} = \frac{F_{yw}}{F_{yst}}\left[0.15Dht_w(1 - C_v)\frac{V_u}{\phi_v V_n} - 18t_w^2\right] \geq \iota \qquad \text{(AISC Equation A-G4-1)}$$

where

A_{st} = total cross-sectional area of the stiffener required when tension-field action is used

F_{yst} = yield stress of the stiffener

D = a function of the configuration of the stiffener

= 1.0 for stiffeners in pairs (angles or plates)

= 1.8 for single-angle stiffeners

= 2.4 for single-plate stiffeners

The area specified by AISC Equation A-G4-1 is needed to resist the vertical component of diagonal compression in the panel. Tables 10-36 and 10-50 in the Numerical Values section of the Specification, which give the shear strength based on tension-field action, also give required stiffener areas expressed as a percentage of the web area for various values of a/h and h/t_w.

The minimum moment of inertia of the stiffener, taken about an axis in the plane of the web (or for a single stiffener, about the face of the stiffener in contact with the web), is given in AISC Appendix F2.3 as

$$I_{st} = at_w^3 j$$

where

$$j = \frac{2.5}{(a/h)^2} - 2 \geq 0.5 \qquad \text{(AISC Equation A-F2-4)}$$

Although intermediate stiffeners are not designed as compression members, a width–thickness ratio limitation for avoiding local buckling can be used as a guide in proportioning the stiffener cross section. Table B5.1 in AISC B5 does not contain a recommendation for plate girder stiffeners. The width–thickness ratio limit for outstanding legs of pairs of angles in continuous contact will be used here — that is,

$$\frac{b}{t} = \frac{95}{\sqrt{F_y}}$$

Unless they also serve as bearing stiffeners, intermediate stiffeners are not required to bear against the tension flange, so their length can be somewhat less than the web depth h, and fabrication problems associated with close fit can be avoided. According

■ **FIGURE 10.8**

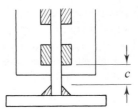

to Appendix F2.3 of the Specification, the length should be within limits established by the distance between the weld connecting the stiffener to the web and the weld connecting the web to the tension flange. This distance, labeled c in Figure 10.8, should be between four and six times the web thickness.

Proportioning the intermediate stiffeners by the AISC rules does not require the computation of any forces, but a force must be transmitted from the stiffener to the web, and the connection should be designed for this force. Basler (1961) recommends the use of a shear flow of

$$f = 0.045h\sqrt{\frac{F_y^3}{E}} \text{ kips/in.} \tag{10.3}$$

The minimum intermittent fillet weld will likely be adequate (Salmon and Johnson, 1996). The only guidance from AISC is in Appendix F2.3, which requires the clear distance between intermittent fillet welds be no more than $16t_w$ or 10 inches.

10.6   INTERACTION OF FLEXURE AND SHEAR

In general, the stress at a point in the web of a plate girder will be a combination of shear and bending, with a major principal stress larger than either component. This combined stress usually will not be a problem if there is no tension-field action; the interaction of shear and bending may be ignored in that case. If there is a tension field, the diagonal tensile stress will reach much higher levels, and the combination of stresses must be considered. AISC requires that the interaction be accounted for when there is a tension field with shear and bending moment levels of

$$0.6\phi V_n \leq V_u \leq \phi V_n \quad \text{and} \quad 0.75\phi M_n \leq M_u \leq \phi M_n$$

where $\phi = 0.90$. The following interaction equation must then be satisfied:

$$\frac{M_u}{\phi M_n} + 0.625\frac{V_u}{\phi V_n} \leq 1.375 \qquad \text{(AISC Equation A-G5-1)}$$

The check for interaction of flexure and shear is illustrated in Example 10.1, part (c).

10.7   BEARING STIFFENERS

Bearing stiffeners are required when the web has insufficient strength for any of the limit states of web yielding, web crippling, or sidesway web buckling. These limit states are covered in Chapter K of the Specification ("Concentrated Forces, Ponding, and Fatigue"). For web yielding, the design strength of the web is ϕR_n, where $\phi = 1.0$, and when the load is at least a distance equal to the girder depth from the end,

$$R_n = (5k + N)F_{yw}t_w \qquad \text{(AISC Equation K1-2)}$$

When the load is less than this distance from the end,

$$R_n = (2.5k + N)F_{yw}t_w \qquad \text{(AISC Equation K1-3)}$$

where

k = distance from the outer face of the flange to the toe of the fillet in the web (for rolled beams) or to the toe of the weld (for welded girders)

N = length of bearing of the concentrated load, measured in the direction of the girder longitudinal axis (not less than k for an end reaction)

(We covered this limit state in Chapter 5.)

For web crippling, the resistance factor $\phi = 0.75$, and when the load is at least *half* the girder depth from the end,

$$R_n = 135t_w^2\left[1 + 3\left(\frac{N}{d}\right)\left(\frac{t_w}{t_f}\right)^{1.5}\right]\sqrt{\frac{F_{yw}t_f}{t_w}} \qquad \text{(AISC Equation K1-4)}$$

When the load is less than this distance from the end of the girder,

$$R_n = 68t_w^2\left[1 + 3\left(\frac{N}{d}\right)\left(\frac{t_w}{t_f}\right)^{1.5}\right]\sqrt{\frac{F_{yw}t_f}{t_w}}, \qquad \text{for } \frac{N}{d} \le 0. \qquad \text{(AISC Equation K1-5a)}$$

and

$$R_n = 68t_w^2\left[1 + \left(4\frac{N}{d} - 0.2\right)\left(\frac{t_w}{t_f}\right)^{1.5}\right]\sqrt{\frac{F_{yw}t_f}{t_w}}, \qquad \frac{N}{d} > 0. \quad \text{(AISC Equation K1-5b)}$$

where

d = overall depth of girder

t_f = thickness of girder flange

(We also covered this limit state in Chapter 5.)

Bearing stiffeners are required to prevent sidesway web buckling only under a limited number of circumstances. Sidesway web buckling should be checked when the compression flange is not restrained against movement relative to the tension flange. The design strength is ϕR_n, where $\phi = 0.85$.

If the flange is restrained against rotation,

$$R_n = \frac{C_r t_w^3 t_f}{h^2}\left[1 + 0.4\left(\frac{h/t_w}{\ell/b_f}\right)^3\right]$$ (AISC Equation K1-6)

[This equation need not be checked if $(h/t_w)/(\ell/b_f) > 2.3$.]
If the flange is not restrained against rotation,

$$R_n = \frac{C_r t_w^3 t_f}{h^2}\left[0.4\left(\frac{h/t_w}{\ell/b_f}\right)^3\right]$$ (AISC Equation K1-7)

[This equation need not be checked if $(h/t_w)/(\ell/b_f) > 1.7$.]

where

C_r = 960,000 when $M_u < M_y$ at the location of the force

    = 480,000 otherwise

ℓ = largest unbraced length of the flange

Although the web can be proportioned to resist directly any applied concentrated loads, bearing stiffeners are usually provided. If stiffeners are used at each concentrated load, the limit states of web yielding, web crippling, and sidesway web buckling do not need to be checked.

The bearing strength of a stiffener is given in AISC J8 as ϕR_n, where $\phi = 0.75$ and

$$R_n = 1.8F_y A_{pb}$$ (AISC Equation J8-1)

This equation is the same as Equation 10.1 with a bearing stress of $f_p = 1.8F_y$.

AISC K1.9 requires that full-depth bearing stiffeners be used in pairs and analyzed as axially loaded columns subject to the following guidelines.

1. The cross section of the axially loaded member consists of the stiffener plates and a symmetrically placed length of the web (see Figure 10.4). This length can be no greater than 12 times the web thickness for an end stiffener or 25 times the web thickness for an interior stiffener.

2. The effective length should be taken as 0.75 times the actual length — that is, $KL = 0.75h$.

The provisions of AISC Chapter E apply.

AISC K1.9 also gives the following additional criteria for bearing stiffeners.

- The width–thickness ratio should satisfy the following requirement:

$$\frac{b}{t} \le \frac{95}{\sqrt{F_y}}$$

- The weld connecting the stiffener to the web should have the capacity to transfer the unbalanced shear force. Conservatively, the weld can be designed to carry the entire concentrated load.

Bearing stiffener analysis is illustrated in Example 10.1, part (d).

■ EXAMPLE 10.1

The plate girder shown in Figure 10.9 must be investigated for compliance with the AISC Specification. The loads are service loads with a live-load to dead-load ratio of 3.0. The uniform load of 4 kips/ft includes the weight of the girder. The compression flange has lateral support at the ends and at the points of application of the concentrated loads. The compression flange is restrained against rotation at these same points. Bearing stiffeners are provided as shown at the ends and at the concentrated loads. They are clipped one inch at the inside edge, both top and bottom, to clear the flange-to-web welds. There are no intermediate stiffeners, and A36 steel is used throughout. Assume that all welds are adequate and check

    a. flexural strength.

    b. shear strength.

    c. flexure-shear interaction.

    d. bearing stiffeners.

SOLUTION   Figure 10.10 shows the loading, shear, and bending moment diagrams based on *factored* loads. Verification of the values shown is left as an exercise for the reader.

■ FIGURE 10.9

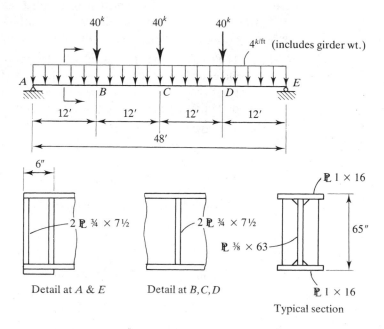

Detail at *A & E*   Detail at *B, C, D*   Typical section

■ **FIGURE 10.10**

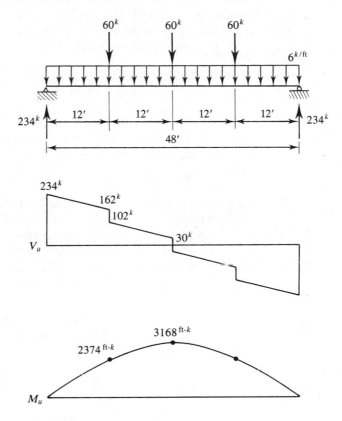

The first step in the analysis is to determine whether this member satisfies the AISC definition of a plate girder:

$$\frac{h}{t_w} = \frac{63}{3/8} = 168$$

$$\frac{970}{\sqrt{F_{yf}}} = \frac{970}{\sqrt{36}} = 161.7$$

Since $h/t_w > 970/\sqrt{F_{yf}}$, this flexural member is a plate girder, and the provisions of AISC Appendix G apply.

The web must satisfy the slenderness limitation of AISC G1. The limiting value of h/t_w will depend on the aspect ratio, a/h. For this plate girder, the bearing stiffeners will serve as intermediate stiffeners, and

$$\frac{a}{h} \approx \frac{12(12)}{63} = 2.286$$

This ratio is approximate because a is not exactly 12 feet. In the interior panels, 12 feet is the center-to-center stiffener spacing rather than the clear spacing. In the end panels, a is less than 12 feet because of the double stiffeners at the supports.

Because a/h is greater than 1.5, AISC Equation A-G1-2 governs:

$$\frac{14,000}{\sqrt{F_{yf}(F_{yf} + 16.5)}} = \frac{14,000}{\sqrt{36(36 + 16.5)}} = 322 > \frac{h}{t_w} = 168 \quad \text{(OK)}$$

a. **Flexural strength.** The flexural strength will be limited by the strength of either the tension flange or the compression flange. In either case, the elastic section modulus will be needed. Because of symmetry,

$$S_{xt} = S_{xc} = S_x$$

The computations for I_x, the moment of inertia about the strong axis, are summarized in Table 10.1. The moment of inertia of each flange about its centroidal axis was neglected because it is small relative to the other terms. The elastic section modulus is

$$S_x = \frac{I_x}{c} = \frac{40,570}{32.5} = 1248 \text{ in.}^3$$

The hybrid girder factor R_e will be needed for both the tension flange strength and the compression flange strength. This girder is nonhybrid, so

$$R_e = 1.0$$

From AISC Equation A-G2-1, the tension flange strength, based on yielding, is

$$M_n = S_{xt}R_eF_{yt} = 1248(1.0)(36) = 44,930 \text{ in.-kips}$$
$$= 3744 \text{ ft-kips}$$

The compression flange buckling strength is given by AISC Equation A-G2-2:

$$M_n = S_{xc}R_{PG}R_eF_{cr}$$

TABLE 10.1

Component	A	$\bar{I}$	d	$\bar{I} + Ad^2$
Web	—	7814	—	7,814
Flange	16	—	32	16,380
Flange	16	—	32	16,380
				40,574

where the critical buckling stress F_{cr} is based on either lateral-torsional buckling or flange local buckling. To check lateral-torsional buckling, we need the radius of gyration, r_T. From Figure 10.11,

$$I_y = \frac{1}{12}(1)(16)^3 + \frac{1}{12}(10.5)(3/8)^3 = 341.4 \text{ in.}^4$$

$$A = 16(1.0) + 10.5(3/8) = 19.94 \text{ in.}^2$$

$$r_T = \sqrt{\frac{I_y}{A}} = \sqrt{\frac{341.4}{19.94}} = 4.138 \text{ in.}$$

The unbraced length of the compression flange is 12 feet, and the slenderness parameters for lateral-torsional buckling are

$$\lambda = \frac{L_b}{r_T} = \frac{12(12)}{4.138} = 34.80$$

$$\lambda_p = \frac{300}{\sqrt{F_{yf}}} = \frac{300}{\sqrt{36}} = 50$$

Since $\lambda < \lambda_p$, $F_{cr} = F_{yf} = 36$ ksi.

The critical buckling stress F_{cr} will now be computed for flange local buckling. The relevant slenderness parameters are

$$\lambda = \frac{b_f}{2t_f} = \frac{16}{2(1.0)} = 8$$

$$\lambda_p = \frac{65}{\sqrt{F_{yf}}} = \frac{65}{\sqrt{36}} = 10.83$$

Again, $\lambda < \lambda_p$ and $F_{cr} = F_{yf} = 36$ ksi.

■ **FIGURE 10.11**

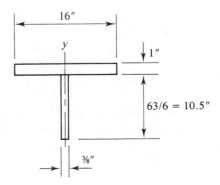

To compute the plate girder strength reduction factor R_{PG}, the value of a_r will be needed:

$$a_r = \frac{A_w}{A_f} = \frac{63(3/8)}{16(1.0)} = \frac{23.62}{16} = 1.477 < 10$$

From AISC Equation A-G2-3,

$$R_{PG} = 1 - \frac{a_r}{1200 + 300a_r}\left(\frac{h_c}{t_w} - \frac{970}{\sqrt{F_{cr}}}\right) \le 1.0$$

$$= 1 - \frac{1.477}{1200 + 300(1.477)}\left(168 - \frac{970}{\sqrt{36}}\right) = 0.9943$$

This value is almost 1.0 because this flexural member barely qualifies as a plate girder and is very close to being classified as a beam. From AISC Equation A-G2-2, the nominal flexural strength for the compression flange is

$$M_n = S_{xc}R_{PG}R_eF_{cr} = 1248(0.9943)(1.0)(36)$$
$$= 44{,}670 \text{ in.-kips} = 3723 \text{ ft-kips}$$

This result is slightly smaller than the nominal strength corresponding to the tension flange and therefore controls. The design strength is

$$\phi_b M_n = 0.90(3723) = 3350 \text{ ft-kips}$$

From Figure 10.10, the maximum factored load moment is

$$M_u = 3168 \text{ ft-kips} < 3350 \text{ ft-kips} \qquad \text{(OK)}$$

ANSWER
a. The flexural strength is adequate.

b. **Shear strength.** The shear strength is a function of the web slenderness ratio h/t_w and the aspect ratio a/h. We will first determine whether tension-field action can be used in regions other than the end panels. It can be used when a/h is less than 3.0 and less than

$$\left[\frac{260}{(h/t_w)}\right]^2 = \left[\frac{260}{(168)}\right]^2 = 2.395$$

The approximate value of a/h is 2.286; therefore tension-field action can be used. Determine k_v and C_v. From AISC Equation A-G3-4,

$$k_v = 5 + \frac{5}{(a/h)^2} = 5 + \frac{5}{(2.286)^2} = 5.957$$

Establish the range of h/t_w:

$$187\sqrt{\frac{k_v}{F_{yw}}} = 187\sqrt{\frac{5.957}{36}} = 76.07$$

$$234\sqrt{\frac{k_v}{F_{yw}}} = 234\sqrt{\frac{5.957}{36}} = 95.19$$

Since $h/t_w = 168 > 234\sqrt{k_v/F_{yw}}$, C_v is found from AISC Equation A-G3-6:

$$C_v = \frac{44{,}000k_v}{(h/t_w)^2 F_{yw}} = \frac{44{,}000(5.957)}{(168)^2(36)} = 0.2580$$

Because $h/t_w > 187\sqrt{k_v/F_{yw}}$, AISC Equation A-G3-2, which accounts for tension-field action, will be used to determine the nominal shear strength (except for the end panels):

$$V_n = 0.6A_w F_{yw}\left(C_v + \frac{1 - C_v}{1.15\sqrt{1 + (a/h)^2}} \right)$$

$$= 0.6(23.62)(36)\left[0.2580 + \frac{1 - 0.2580}{1.15\sqrt{1 + (2.286)^2}} \right] = 263.6 \text{ kips}$$

The design shear strength is

$$\phi_v V_n = 0.90(263.6) = 237 \text{ kips}$$

From Figure 10.10, the maximum factored load shear in the middle half of the girder is 102 kips, so the shear strength is adequate where tension-field action is permitted.

For the end panels, tension-field action is not permitted, and the shear strength must be computed from AISC Equation A-G3-3:

$$V_n = 0.6A_w F_{yw} C_v = 0.6(23.62)(36)(0.2580) = 131.6 \text{ kips}$$

The design strength is

$$\phi_v V_n = 0.90(131.6) = 118 \text{ kips}$$

The maximum factored load shear in the end panel is

$$V_u = 234 \text{ kips} > 118 \text{ kips} \qquad \text{(N.G.)}$$

Two options are available for increasing the shear strength: Either decrease the web slenderness (probably by increasing its thickness) or decrease the aspect ratio of each end panel by adding an intermediate stiffener. Stiffeners are added in this example. (Decreasing the web slenderness enough to gain sufficient shear strength would bring this flexural member into the beam category, changing the analysis completely.)

The location of the first intermediate stiffener will be determined by the following strategy: First, equate the shear strength from AISC Equation A-G3-3 to the required shear strength and solve for the required value of C_v. Next, solve for k_v from Equation A-G3-6, then solve for a/h from Equation A-G3-4. Doing so, we have

$$\phi_v V_n = \phi_v(0.6A_w F_{yw} C_v) \qquad \text{(AISC Equation A-G3-3)}$$

$$C_v = \frac{\phi_v V_n}{\phi_v(0.6A_w F_{yw})} = \frac{234}{0.90(0.6)(23.62)(36)} = 0.5096$$

$$C_v = \frac{44,000 k_v}{(h/t_w)^2 F_{yw}} \qquad \text{(AISC Equation A-G3-6)}$$

$$k_v = \frac{C_v(h/t_w)^2 F_{yw}}{44,000} = \frac{0.5096(168)^2(36)}{44,000} = 11.77$$

$$k_v = 5 + \frac{5}{(a/h)^2} \qquad \text{(AISC Equation A-G3-4)}$$

$$\frac{a}{h} = \sqrt{\frac{5}{k_v - 5}} = \sqrt{\frac{5}{11.77 - 5}} = 0.8594$$

The required stiffener spacing is

$$a = 0.8594h = 0.8594(63) = 54.1 \text{ in.}$$

Although a is defined as the clear spacing, we will treat it conservatively as a center-to-center spacing and place the first intermediate stiffener at 54 inches from the end of the girder. This placement will give a design strength that approximately equals the maximum factored load shear of 234 kips. No additional stiffeners will be needed, since the factored load shear outside of the end panels is less than the design strength of 237 kips.

The determination of the stiffener spacing can be facilitated by the use of the tables in the Numerical Values section of the Specification; we illustrate this technique in Example 10.2.

ANSWER **b.** The shear strength is inadequate. Add one intermediate stiffener 54 inches from each end of the girder.

c. Flexure–shear interaction. Flexure–shear interaction must be checked when there is a tension field (that is, outside of the end panels) and when both of the following conditions are satisfied.

1. The factored load shear is in the range

$$0.6\phi V_n \leq V_u \leq \phi V_n$$

For this girder, this range of shear outside of the end panels would be

$$0.6(237) \leq V_u \leq 237$$
$$142 \leq V_u \leq 237$$

2. The factored load bending moment is in the range

$$0.75\phi M_n \leq M_u \leq \phi M_n$$

For this girder, the range would be

$$0.75(3350) \leq M_u \leq 3350$$

$$2510 \leq M_u \leq 3350$$

An examination of Figure 10.10 shows that there is no combination of shear and bending moment that satisfies these conditions.

ANSWER   **c.** Interaction of flexure and shear does not need to be checked.

       d. **Bearing stiffeners.** Bearing stiffeners are provided at each concentrated load, so there is no need to check the provisions of AISC Chapter K for web yielding, web crippling, or sidesway web buckling. For both the interior and support bearing stiffeners,

$$\frac{b}{t} = \frac{7.5}{0.75} = 10.0$$

$$\frac{95}{\sqrt{F_y}} = \frac{95}{\sqrt{36}} = 15.8 > 10.0 \qquad \text{(OK)}$$

For the interior bearing stiffeners, first compute the bearing strength. From Figure 10.12,

$$A_{pb} = 2at = 2(7.5 - 1)(0.75) = 9.750 \text{ in.}^2$$

■ **FIGURE 10.12**

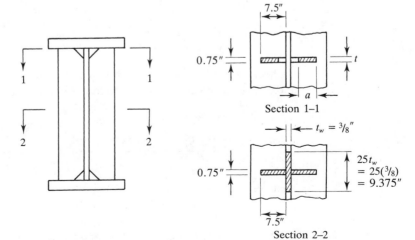

From AISC Equation J8-1,

$$R_n = 1.8F_yA_{pb} = 1.8(36)(9.750) = 631.8 \text{ kips}$$
$$\phi R_n = 0.75(631.8) = 474 \text{ kips} > 60 \text{ kips} \qquad (\text{OK})$$

Check the strength of the stiffener as a compression member. Referring to Figure 10.12, we can use a web length of 9.375 inches, resulting in a cross-sectional area for the "column" of

$$A = 2(0.75)(7.5) + (3/8)(9.375) = 14.77 \text{ in.}^2$$

The moment of inertia of this area about an axis in the web is

$$I = \Sigma(\bar{I} + Ad^2)$$
$$= \frac{9.375(3/8)^3}{12} + 2\left[\frac{0.75(7.5)^3}{12} + 7.5(0.75)\left(\frac{7.5}{2} + \frac{3/8}{2}\right)^2\right] = 227.2 \text{ in.}^4$$

and the radius of gyration is

$$r = \sqrt{\frac{I}{A}} = \sqrt{\frac{227.2}{14.77}} = 3.922 \text{ in.}$$

The slenderness ratio is

$$\frac{KL}{r} = \frac{Kh}{r} = \frac{0.75(63)}{3.922} = 12.05$$

From AISC Table 3-36 in the Numerical Values section of the Specification,

$$\phi_c F_{cr} = 30.37 \text{ ksi}$$

and

$$\phi_c P_n = \phi_c F_{cr}A = 30.37(14.77) = 449 \text{ kips} > 60 \text{ kips} \qquad (\text{OK})$$

For the bearing stiffeners at the supports, from Figure 10.13, the design bearing strength is

$$\phi R_n = \phi(1.8F_yA_{pb}) = 0.75(1.8)(36)[4(6.5)(0.75)]$$
$$= 948 \text{ kips} > 234 \text{ kips} \qquad (\text{OK})$$

Check the stiffener–web assembly as a compression member. Referring to Figure 10.13, the moment of inertia about an axis in the plane of the web is

$$I = \Sigma(\bar{I} + Ad^2)$$
$$= \frac{4.5(3/8)^3}{12} + 4\left[\frac{0.75(7.5)^3}{12} + 7.5(0.75)\left(\frac{7.5}{2} + \frac{3/8}{2}\right)^2\right] = 454.3 \text{ in.}^4$$

and the area and radius of gyration are

$$A = 4.5(3/8) + 4(0.75)(7.5) = 24.19 \text{ in.}^2$$

■ **FIGURE 10.13**

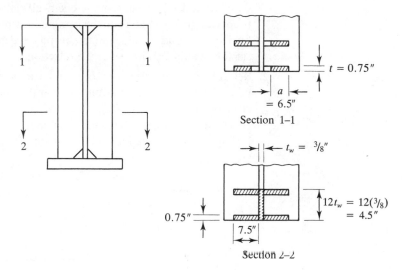

Section 1–1

$t_w = {}^3/_8''$

$12t_w = 12(^3/_8)$
$= 4.5''$

0.75''

7.5''

Section 2–2

and

$$r = \sqrt{\frac{I}{A}} = \sqrt{\frac{454.3}{24.19}} = 4.334 \text{ in.}$$

The slenderness ratio is

$$\frac{Kh}{r} = \frac{0.75(63)}{4.334} = 10.90$$

From AISC Table 3-36, $\phi_c F_{cr} = 30.41$ kips. The design strength is

$$\phi_c P_n = \phi_c F_{cr} A = 30.41(24.19) = 736 \text{ kips} > 234 \text{ kips} \qquad \text{(OK)}$$

ANSWER **d.** The bearing stiffeners are adequate. ■

| 10.8 | **DESIGN** |

The primary task in plate girder design is to determine the size of the web and the flanges. If a variable moment of inertia is desired, decisions must be made regarding the method of varying the flange size — that is, whether to use cover plates or different thicknesses of flange plate at different points along the length of the girder. A decision about whether to use intermediate stiffeners must be made early in the process because it will affect the web thickness. If bearing stiffeners are needed, they must be designed. Finally, the various components must be connected by properly designed welds. The following step-by-step procedure is recommended.

1. **Select the overall depth.** As a rule of thumb, a well-proportioned girder will have a depth of one tenth to one twelfth the span length. As with any beam design, constraints on the maximum depth could establish the depth by default. Building code limitations on the depth-to-span ratio or the deflection would also influence the selection.

2. **Select a trial web size.** The web depth can be estimated by subtracting twice the flange thickness from the overall depth selected. Of course, at this stage of the design, the flange thickness must also be estimated, but the consequences of a poor estimate are minor. The web thickness t_w can then be found by using the following limitations as a guide:

   For $a/h \leq 1.5$,

   $$\frac{h}{t_w} \leq \frac{2000}{\sqrt{F_{yf}}} \qquad \text{(AISC Equation A-G1-1)}$$

   For $a/h > 1.5$,

   $$\frac{h}{t_w} \leq \frac{14,000}{\sqrt{F_{yf}(F_{yf} + 16.5)}} \qquad \text{(AISC Equation A-G1-2)}$$

3. **Estimate the flange size.** The required flange area can be estimated from a simple formula derived as follows. Let

   $$I_x = I_{web} + I_{flanges}$$
   $$\approx \frac{1}{12} t_w h^3 + 2A_f y^2 \approx \frac{1}{12} t_w h^3 + 2A_f (h/2)^2 \qquad (10.4)$$

   where

   A_f = cross-sectional area of one flange

   y = distance from the elastic neutral axis to the centroid of the flange

   The contribution of the moment of inertia of each flange about its own centroidal axis was neglected in Equation 10.4. The section modulus can be estimated as

   $$S_x = \frac{I_x}{c} \approx \frac{t_w h^3/12}{h/2} + \frac{2A_f (h/2)^2}{h/2} = \frac{t_w h^2}{6} + A_f h$$

   If we assume that compression flange buckling will control the design, we can find the required section modulus from AISC Equation A-G2-2:

   $$M_n = S_{xc} R_{PG} R_e F_{cr}$$
   $$S_{xc} = \frac{M_n}{R_{PG} R_e F_{cr}} = \frac{M_u/\phi_b}{R_{PG} R_e F_{cr}}$$

where M_u is the maximum factored load bending moment. Equating the required section modulus to the approximate value, we have

$$\frac{M_u/\phi_b}{R_{PG}R_eF_{cr}} = \frac{t_wh^2}{6} + A_f l$$

and

$$A_f = \frac{M_u}{\phi_b h R_{PG}R_eF_{cr}} - \frac{t_w l}{6}$$

If we assume that $R_{PG} = 1.0$, $R_e = 1.0$, and $F_{cr} = F_y$, the required area of one flange is

$$A_f = \frac{M_u}{0.90 h F_y} - \frac{A_w}{6} \tag{10.5}$$

where A_w is the web area. Once the required flange area has been determined, select the width and thickness. If the thickness used in the estimate of the web depth is retained, no adjustment in the web depth will be needed. At this point, an estimated girder weight can be computed, and M_u and A_f should be recomputed.

4. **Check the bending strength of the trial section.**
5. **Check shear.** If an end panel is being considered, or if intermediate stiffeners are not used, the shear strength can be found from AISC Equation A-G3-3, which gives the strength in the absence of a tension field. Table 9-36 or 9-50 in the Numerical Values section of the Specification can be used for this purpose. If the tables are not used, the intermediate stiffener spacing can be determined as follows.

    a. Equate the required shear strength to the shear strength given by AISC Equation A-G3-3 and solve for the required value of C_v.

    b. Solve for k_v from AISC Equation A-G3-5 or A-G3-6.

    c. Solve for the required value of a/h from AISC Equation A-G3-4.

    If tension-field action is used, either a trial-and-error approach or AISC Table 10-36 or 10-50 can be used to obtain the required a/h. (The required cross-sectional area of the stiffeners, expressed as a percentage of the web area, is also given in these tables for certain values of h/t_w and a/h.) Establish a trial stiffener size that will satisfy the area requirement and check the moment of inertia requirement of AISC Appendix F2.3.

6. **Check for interaction of shear and bending.**
7. **Check the web resistance to any applied concentrated loads** (web yielding, web crippling, and web sidesway buckling). If bearing stiffeners are required, the following design procedure is recommended.

    a. Try a width that brings the edge of the stiffener near the edge of the flange and a thickness that satisfies the width–thickness requirement

$$\frac{b}{t} \le \frac{95}{\sqrt{F_y}}$$

    b. Compute the cross-sectional area needed for bearing strength. Compare this area with the trial area and revise if necessary.

    c. Check the stiffener–web assembly as a compression member.

  8. **Design the flange-to-web welds, stiffener-to-web welds, and any other connections** (flange segments, web splices, etc.).

■ EXAMPLE 10.2

Design a simply supported plate girder to span 60 feet and support the service loads shown in Figure 10.14a. The maximum permissible depth is 65 inches. Use A36 steel and E70XX electrodes, and assume that the girder has continuous lateral support. The ends have bearing-type supports and are not framed.

■ FIGURE 10.14

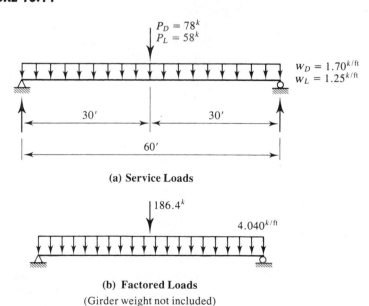

(a) Service Loads

(b) Factored Loads
(Girder weight not included)

SOLUTION The factored loads, excluding the girder weight, are shown in Figure 10.14b. Determine the overall depth:

$$\frac{\text{Span length}}{10} = \frac{60(12)}{10} = 72 \text{ in.}$$

$$\frac{\text{Span length}}{12} = \frac{60(12)}{12} = 60 \text{ in.}$$

Use the maximum permissible depth of 65 inches.

Try a flange thickness of $t_f = 1.5$ inches and a web depth of

$$h = 65 - 2(1.5) = 62 \text{ in.}$$

To determine the web thickness, first examine the limiting values of h/t_w. For this flexural member to qualify as a plate girder,

$$\frac{h}{t_w} \geq \frac{970}{\sqrt{F_{yf}}} = \frac{970}{\sqrt{36}} = 161.7$$

$$t_w < \frac{h}{161.7} = \frac{62}{161.7} = 0.383 \text{ in.}$$

From AISC Equations A-G1-1 and A-G1-2:

For $a/h \leq 1.5$,

$$\frac{h}{t_w} \leq \frac{2000}{\sqrt{F_{yf}}} = \frac{2000}{\sqrt{36}} = 333.3$$

$$t_w \geq \frac{62}{333.3} = 0.186 \text{ in.}$$

For $a/h > 1.5$,

$$\frac{h}{t_w} \leq \frac{14,000}{\sqrt{F_{yf}(F_{yf} + 16.5)}} = \frac{14,000}{\sqrt{36(36 + 16.5)}} = 322.0$$

$$t_w \geq \frac{62}{322.0} = 0.192 \text{ in.}$$

Try a ¼ × 62 web plate. Determine the required flange size. From Figure 10.14b, the maximum factored load bending moment is

$$M_u = \frac{186.4(60)}{4} + \frac{4.040(60)^2}{8} = 4614 \text{ ft-kips}$$

From Equation 10.5, the required flange area is

$$A_f = \frac{M_u}{0.90 h F_y} - \frac{A_w}{6}$$

$$= \frac{4614(12)}{0.90(62)(36)} - \frac{62(1/4)}{6} = 24.98 \text{ in.}^2$$

The girder weight can now be estimated.

Web area: $62(1/4) = 15.5$ in.^{2}
Flange area: $2(24.98) = \underline{49.96\ \text{in.}^2}$
Total: 65.46 in.^{2}

Weight: $\dfrac{65.46}{144}(490) = 222.7$ lb/ft   say 250 lb/ft

The adjusted bending moment is

$$M_u = 4614 + \frac{(1.2 \times 0.250)(60)^2}{8} = 4749 \text{ ft-kips}$$

and the required flange area is

$$A_f = \frac{4749(12)}{0.90(62)(36)} - \frac{62(1/4)}{6} = 25.79 \text{ in.}^2$$

If the flange thickness estimate of 1.5 inches is retained, the required width will be

$$b_f = \frac{A_f}{t_f} = \frac{25.79}{1.5} = 17.2 \text{ in.}$$

Try a 1½ × 18 flange plate. Figure 10.15 shows the trial section, and Figure 10.16 shows the shear and bending moment diagrams for the factored loads, which include an approximate girder weight of 250 lb/ft.

Check the flexural strength of the trial section. From Figure 10-15, the moment of inertia about the axis of bending is

$$I_x = \frac{(1/4)(62)^3}{12} + 2(1.5)(18)(31.75)^2 = 59,400 \text{ in.}^4$$

and the elastic section modulus is

$$S_x = \frac{I_x}{c} = \frac{59,400}{32.5} = 1828 \text{ in.}^3$$

An examination of AISC Equations A-G2-1 and A-G2-2 shows that for a nonhybrid girder with a symmetrical cross section, the flexural strength will never be controlled by tension flange yielding; therefore only compression flange buckling will be investigated. Furthermore, as this girder has continuous lateral support, lateral-torsional buckling need not be considered. For the limit state of flange local buckling,

$$\lambda = \frac{b_f}{2t_f} = \frac{18}{2(1.5)} = 6$$

$$\lambda_p = \frac{65}{\sqrt{F_{yf}}} = \frac{65}{\sqrt{36}} = 10.83$$

■ **FIGURE 10.15**

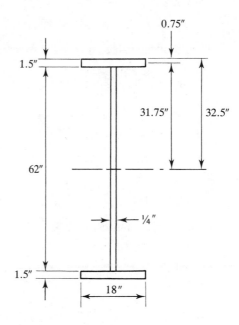

■ **FIGURE 10.16**

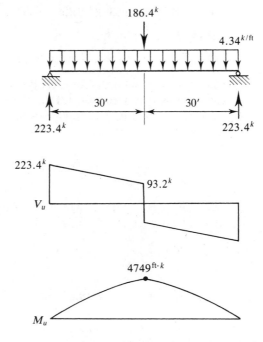

Since $\lambda < \lambda_p$,

$$F_{cr} = F_{yf} = 36 \text{ ksi}$$

The following values are needed for computing the strength reduction factor R_{PG}:

$$A_w = 62\left(\frac{1}{4}\right) = 15.5 \text{ in.}^2$$

$$A_f = 18(1.5) = 27 \text{ in.}^2$$

$$a_r = \frac{A_w}{A_f} = \frac{15.5}{27} = 0.5741 < 10$$

$$\frac{h}{t_w} = \frac{62}{1/4} = 248$$

From AISC Equation A-G2-3,

$$R_{PG} = 1 - \frac{a_r}{1200 + 300a_r}\left(\frac{h}{t_w} - \frac{970}{\sqrt{F_{cr}}}\right)$$

$$= 1 - \frac{0.5741}{1200 + 300(0.5741)}\left(248 - \frac{970}{\sqrt{36}}\right) = 0.9639$$

From AISC Equation A-G2-2, the nominal flexural strength is

$$M_n = S_{xc}R_{PG}R_eF_{cr}$$
$$= 1828(0.9639)(1.0)(36) = 63{,}430 \text{ in.-kips} = 5286 \text{ ft-kips}$$

and the design strength is

$$\phi_b M_n = 0.90(5286) = 4757 \text{ ft-kips} > 4749 \text{ ft-kips} \quad \text{(OK)}$$

Check the shear strength. The shear is maximum at the support, but tension-field action cannot be used in an end panel. Table 9-36 in the Numerical Values section of the Specification will be used to obtain the required size of the end panel. The table will be entered with $h/t_w = 248$ and

$$\frac{\phi_v V_n}{A_w} = \frac{223.4}{15.5} = 14.41 \text{ ksi}$$

This value would require an a/h of less than 0.5 and is beyond the range of the table, indicating that a thicker web is needed. **Try a $\frac{5}{16} \times 62$ web:**

$$\frac{h}{t_w} = \frac{62}{5/16} = 198.4$$

$$A_w = 62\left(\frac{5}{16}\right) = 19.38 \text{ in.}^2$$

$$\frac{\phi_v V_n}{A_w} = \frac{223.4}{19.38} = 11.5 \text{ ksi}$$

From AISC Table 9-36, for $h/t_w = 198$ and $a/h = 0.6$, the interpolated value of $\phi_v V_n/A_w$ is 11.5 ksi. Therefore, use $a/h = 0.6$ and

$$a = 0.6h = 0.6(62) = 37.2 \text{ in.}$$

Although the required distance a is a clear spacing, the use of center-to-center distances is somewhat simpler and will be slightly conservative. Use a distance of 36 inches from the center of the end bearing stiffener to the center of the first intermediate stiffener.

Before proceeding with the shear strength analysis, determine the effect of the change in web thickness.

First, find the girder weight.

Web area: $62(5/16) = 19.38 \text{ in.}^2$

Flange area: $2(1.5)(18) = \underline{54.00 \text{ in.}^2}$

Total: 73.38 in.^2

Weight: $\dfrac{73.38}{144} (490) = 250 \text{ lb/ft}$ (same as the previous assumption)

Then

$$I_x = \frac{(5/16)(62)^3}{12} + 2(1.5)(18)(31.75)^2 = 60,640 \text{ in.}^4$$

$$S_x = \frac{I_x}{c} = \frac{60,640}{32.5} = 1866 \text{ in.}^3$$

$$a_r = \frac{A_w}{A_f} = \frac{19.38}{27} = 0.7178 < 10$$

From AISC Equation A-G2-3,

$$R_{PG} = 1 - \frac{a_r}{1200 + 300a_r} \left(\frac{h_c}{t_w} - \frac{970}{\sqrt{F_{cr}}} \right)$$

$$= 1 - \frac{0.7178}{1200 + 300(0.7178)} \left(198.4 - \frac{970}{\sqrt{36}} \right) = 0.9814$$

The nominal flexural strength is

$$M_n = S_{xc} R_{PG} R_e F_{cr}$$
$$= (1866)(0.9814)(1.0)(36)$$
$$= 65,930 \text{ in.-kips} = 5494 \text{ ft-kips}$$

The design strength is

$$\phi_b M_n = 0.90(5494) = 4944 \text{ ft-kips}$$

Although this capacity is somewhat more than needed, it will compensate for the weight of the stiffeners and other incidentals that we have not accounted for.

ANSWER   Use a $\frac{5}{16} \times 62$ web and $1\frac{1}{2} \times 18$ flanges, as illustrated in Figure 10.17.

■ **FIGURE 10.17**

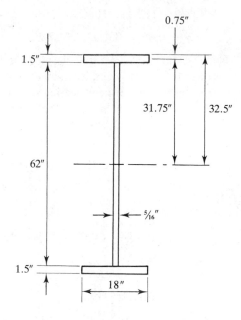

Determine the intermediate stiffener spacings needed for shear strength outside the end panels. At a distance of 36 inches from the left end, the shear is

$$V_u = 223.4 - 4.34\left(\frac{36}{12}\right) = 210.4 \text{ kips}$$

$$\frac{\phi_v V_n}{A_w} = \frac{210.4}{19.38} = 10.86 \text{ ksi}$$

Tension-field action can be used outside the end panels, so AISC Table 10-36 will be used. For $h/t_w = 200$ and $a/h = 1.6$,

$$\frac{\phi_v V_n}{A_w} = 11.2 \text{ ksi} > 10.86 \text{ ksi} \qquad \text{(OK)}$$

Use $a/h = 1.6$ and

$$a = 1.6h = 1.6(62) = 99.2 \text{ in.}$$

Note that no values are given in the table when $h/t_w = 200$ and $a/h > 1.6$. The reason is that tension-field action is not permitted when

$$\frac{a}{h} > \left[\frac{260}{(h/t_w)}\right]^2 = \left(\frac{260}{200}\right)^2 = 1.69$$

For this reason, a maximum stiffener spacing of 99.2 inches will apply for the remainder of the girder, but to achieve uniform spacing between the end panels, a center-to-center spacing of 81 inches, as shown in Figure 10.18, will be used. With this reduced spacing,

$$\frac{a}{h} = \frac{81}{62} = 1.306$$

To compute the corresponding shear strength, enter Table 10-36 with $a/h = 1.3$ and $h/t_w = 200$. By interpolation,

$$\frac{\phi_v V_n}{A_w} = 12.6$$

and

$$\phi_v V_n = 12.6A_w = 12.6(19.38) = 244.2 \text{ kips}$$

The cross section of the intermediate stiffeners is based on three criteria: (1) a minimum area, (2) a minimum moment of inertia, and (3) a maximum width–thickness ratio. The required area of a pair of stiffeners is a function of h/t_w and a/h, and it can be found from AISC Table 10-36 by interpolation.

For $a/h = 1.3$ and $h/t_w = 200$,

$$A_{st} = 2.3\% \text{ of the web area}$$
$$= 0.023(19.38) = 0.446 \text{ in.}^2$$

From AISC Equation A-F2-4,

$$j = \frac{2.5}{(a/h)^2} - 2 \geq 0.5$$

$$= \frac{2.5}{(1.306)^2} - 2 = -0.534$$

This is less than 0.5, so use $j = 0.5$.

■ **FIGURE 10.18**

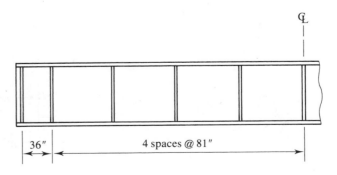

36"   4 spaces @ 81"

The required moment of inertia is

$$I_{st} = at_w^3 j = 81(5/16)^3(0.5) = 1.24 \text{ in.}^4$$

Use a maximum value of b/t of

$$\frac{95}{\sqrt{F_y}} = \frac{95}{\sqrt{36}} = 15.8$$

Try two ¼ × 4 plates:

$$\frac{b}{t} = \frac{4}{1/4} = 16 \approx 15.8 \qquad \text{(say OK)}$$

$$A_{st} \text{ provided} = 2(4)\left(\frac{1}{4}\right) = 2.0 \text{ in.}^2 > 0.446 \text{ in.}^2 \qquad \text{(OK)}$$

From Figure 10.19 and the parallel axis theorem,

$$I_{st} = \Sigma(\bar{I} \times Ad^2)$$

$$= \left[\frac{0.25(4)^3}{12} + 0.25(4)(2 + 5/32)^2\right] \times 2 \text{ stiffeners}$$

$$= 11.97 \text{ in.}^4 > 1.24 \text{ in.}^4 \qquad \text{(OK)}$$

■ **FIGURE 10.19**

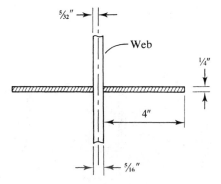

To determine the length of the stiffeners, first compute the distance between the stiffener-to-web weld and the web-to-flange weld (see Figure 10.8):

$$\text{Minimum distance} = 4t_w = 4\left(\frac{5}{16}\right) = 1.25 \text{ in.}$$

$$\text{Maximum distance} = 6t_w = 6\left(\frac{5}{16}\right) = 1.875 \text{ in.}$$

If we assume a flange-to-web weld size of ⁵⁄₁₆ inch and 1.25 inches between welds, the approximate length of the stiffener is

$$h - \text{weld size} - 1.25 = 62 - 0.3125 - 1.25$$
$$= 60.44 \text{ in.} \qquad \text{say } 60 \text{ in.}$$

ANSWER   Use two plates ¼ inch × 4 inches × 5 ft 0 inch for the intermediate stiffeners.

Check for interaction of flexure and shear, which needs to be done only where there is a tension field. The range of shear to be investigated is

$$0.6\phi V_n \leq V_u \leq \phi V_n \quad \text{or} \quad 0.6(244.2) \leq V_u \leq 244.2$$
$$146.5 \leq V_u \leq 244.2$$

The range of bending moment to be checked is

$$0.75\phi M_n \leq M_u \leq \phi M_n \quad \text{or} \quad 0.75(4944) \leq M_u \leq 4944$$
$$3708 \leq M_u \leq 4944$$

From Figure 10.16, $V_u = 146.5$ kips when

$$223.4 - 4.34x = 146.5 \text{ kips}$$

where

$$x = \text{distance from left end of girder} = 17.72 \text{ ft}$$

At this same location, the bending moment is

$$M_u = 223.4(17.72) - \frac{4.34(17.72)^2}{2} = 3277 \text{ ft-kips}$$

In the region where the shear is greater than 146.5 kips, the moment is less than 3708 ft-kips; therefore interaction of flexure and shear need not be considered.

    Bearing stiffeners will be provided at the supports and at midspan. Since there will be a stiffener at each concentrated load, there is no need to investigate the resistance of the web to these loads. If the stiffeners were not provided, the web would need to be protected from yielding and crippling. To do so, enough bearing length, N, as required by AISC Equations K1-2 through K1-5, must be provided. Sidesway web buckling would not be an applicable limit state because the girder has continuous lateral support (making the unbraced length $\ell = 0$ and $(h/t_w)/(\ell/b_f) > 2.3$).

Try a stiffener width, b, of 8 inches. The total combined width will be $2(8) + \frac{5}{16}$ (the web thickness) $= 16.31$ inches, or slightly less than the flange width of 18 inches. From AISC K1.9,

$$\frac{b}{t} \le \frac{95}{\sqrt{F_y}} \quad \text{or} \quad t \ge \frac{b\sqrt{F_y}}{95} = \frac{8\sqrt{36}}{95} = 0.505 \text{ in.}$$

Try two $\frac{3}{4} \times 8$ stiffeners. Assume a $\frac{5}{16}$-inch web-to-flange weld and a $\frac{1}{2}$-inch cutout in the stiffener. Check the stiffener at the support. The bearing strength is

$$
\begin{aligned}
\phi R_n &= 0.75(1.8 F_y A_{pb}) \\
&= 0.75(1.8)(36)(0.75)(8 - 0.5) \times 2 \\
&= 547 \text{ kips} > 223.4 \text{ kips} \quad \text{(OK)}
\end{aligned}
$$

Check the stiffener as a column. The length of web acting with the stiffener plates to form a compression member is 12 times the web thickness for an end stiffener (AISC K1.9). As shown in Figure 10.20, this length is $12(\frac{5}{16}) = 3.75$ in. Because the stiffener should be centrally located within this length, the point of support (location of the girder reaction) must be approximately $\frac{3.75}{2} = 1.875$ inches from the end of the girder. Use 3 inches, as shown in Figure 10.21, but base the computations on a total length of web of 3.75 inches, which gives

$$A = 2(8)\left(\frac{3}{4}\right) + \left(\frac{5}{16}\right)(3.75) = 13.17 \text{ in.}^2$$

$$I = \frac{3.75(5/16)^3}{12} + 2\left[\frac{0.75(8)^3}{12} + 8\left(\frac{3}{4}\right)\left(4 + \frac{5}{32}\right)^2\right] = 271.3 \text{ in.}^4$$

$$r = \sqrt{\frac{I}{A}} = \sqrt{\frac{271.3}{13.17}} = 4.539 \text{ in.}$$

$$\frac{KL}{r} = \frac{Kh}{r} = \frac{0.75(62)}{4.539} = 10.24$$

From AISC Table 3-36, $\phi_c F_{cr} = 30.43$ ksi. The design strength is

$$\phi_c P_n = \phi_c F_{cr} A = 30.43(13.17) = 401 \text{ kips} > 223.4 \text{ kips} \quad \text{(OK)}$$

■ **FIGURE 10.20**

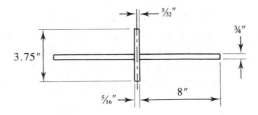

■ FIGURE 10.21

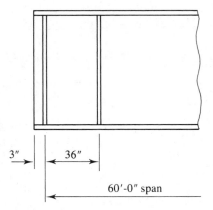

3″   36″

60′-0″ span

Since the load at midspan is smaller than the reaction, use the same stiffener at midspan.

ANSWER Use two plates $\frac{3}{4} \times 8$ for the bearing stiffeners.

At this point, all components of the girder have been sized. The connections of these elements will now be designed. E70 electrodes, with a design strength of $\phi F_w = 31.5$ ksi, will be used for all welds.

For the flange-to-web welds, compute the horizontal shear flow at the flange-to-web junction:

Maximum $V_u = 223.4$ kips

Q = flange area $\times 31.75$ (see Figure 10.17)

    = $1.5(18)(31.75) = 857.2$ in.^{3}

$I_x = 60,640$ in.^{4}

Maximum $\dfrac{V_u Q}{I_x} = \dfrac{223.4(857.2)}{60,640} = 3.158$ kips/in.

For the plate thicknesses being welded, the minimum weld size, w, is $\frac{5}{16}$ inch. If intermittent welds are used, their minimum length is

$L_{\min} = 4 \times w \geq 1.5$ in.

    = $4\left(\dfrac{5}{16}\right) = 1.25$ in.   ∴ use 1.5 in.

Try $\frac{5}{16}$-in. $\times$ $1\frac{1}{2}$-in. fillet welds:

Capacity per inch = $0.707 \times w \times \phi F_W \times 2$ welds

    = $0.707(5/16)(31.5)(2) = 13.92$ kips/in.

The shear capacity of the base metal is

$$t(\phi F_{BM}) = t[0.90(0.60F_y)] = t(0.54F_y) = \left(\frac{5}{16}\right)(0.54)(36)$$
$$= 6.075 \text{ kips/in.} < 13.92 \text{ kips/in.}$$

Use a total weld capacity of 6.075 kips/in. The capacity of a 1.5-inch length of a pair of welds is

$$6.075(1.5) = 9.112 \text{ kips}$$

To determine the spacing, let

$$\frac{9.112}{s} = \frac{V_u Q}{I_x}$$

where s is the center-to-center spacing of the welds in inches and

$$s = \frac{9.112}{V_u Q / I_x} = \frac{9.112}{3.158} = 2.89 \text{ in.}$$

Using a center-to-center spacing of 2.75 inches will give a clear spacing of 2.75 − 1.5 = 1.25 inches. The AISC Specification gives a maximum permissible spacing of intermittent fillet welds for this application in Section B10, "Proportions of Beams and Girders." The provisions for built-up compression members (AISC E4) and built-up tension members (AISC D2) are to be used for the compression flange and tension flange connections. For compression,

$$d \leq \frac{127t}{\sqrt{F_y}}, \qquad \text{but no greater than 12 in.}$$

For tension,

$$d \leq 24t, \qquad \text{but no greater than 12 in.}$$

where

$$d = \text{clear spacing in inches}$$
$$t = \text{thickness of the thinner outside connected plate in a built-up shape}$$

Adapting these limits to the present case yields

$$\frac{127t}{\sqrt{F_y}} = \frac{127(1.5)}{\sqrt{36}} = 31.8 \text{ in.} > 12 \text{ in.}$$
$$24t = 24(1.5) = 36 \text{ in.} > 12 \text{ in.}$$

The maximum permissible clear spacing is therefore 12 inches, and the required clear spacing of 1.25 inches is satisfactory.

Although the 2.75-inch center-to-center spacing can be used for the entire length of the girder, an increased spacing can be used where the shear is less than the maximum of 223.4 kips. Three different spacings will be investigated:

1. The closest required spacing of 2.75 inches.
2. The maximum permissible center-to-center spacing of $12 + 1.5 = 13.5$ inches.
3. An intermediate spacing of 5 inches.

The 8-inch spacing can be used when

$$\frac{V_u Q}{I_x} = \frac{9.112}{s} \text{ or } V_u = \frac{9.112 I_x}{Qs} = \frac{9.112(60,640)}{857.2(5)} = 128.9 \text{ kips}$$

Refer to Figure 10.16 and let x be the distance from the left support, giving

$$V_u = 223.4 - 4.34x = 128.9 \text{ kips}$$
$$x = 21.77 \text{ ft}$$

The 13.5-inch spacing can be used when

$$V_u = \frac{9.112 I_x}{Qs} = \frac{9.112(60,640)}{857.2(13.5)} = 47.75 \text{ kips}$$

Figure 10.16 shows that the shear never gets this small, so the maximum spacing never controls.

ANSWER Use $\tfrac{5}{16}$-inch $\times$ $1\tfrac{1}{2}$-inch fillet welds for the flange-to-web welds, spaced as shown in Figure 10.22.

■ **FIGURE 10.22**

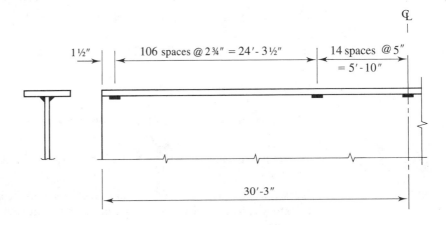

For the intermediate stiffener welds:

Minimum weld size $= \dfrac{3}{16}$ in. (based on the thickness $t_w = \tfrac{5}{16}$ in. and $t = \tfrac{1}{4}$ in.

Minimum length $= 4\left(\dfrac{3}{16}\right) = 0.75$ in. < 1.5 in.　　use 1.5 in.

The capacity per inch for 4 welds (2 per stiffener plate) is

$$0.707\left(\frac{3}{16}\right)(31.5)(4) = 16.70 \text{ kips/in.}$$

The shear capacity of the base metal is 6.075 kips/in. (See the previous computation for $t = \tfrac{5}{16}$ in.) Use a strength of 6.075 kips/in.

From Equation 10.3, the shear to be transferred is

$$f = 0.045h\sqrt{\frac{F_y^3}{E}} = 0.045(62)\sqrt{\frac{(36)^3}{29,000}} = 3.539 \text{ kips/in.}$$

Use intermittent welds. The capacity of a 1.5-inch length of the 4 welds is

$$1.5(6.075) = 9.112 \text{ kips}$$

Equating the shear strength per inch and the required strength gives

$$\frac{9.112}{s} = 3.539 \text{ kips/in. or } s = 2.57 \text{ in.}$$

From AISC Appendix F2.3, the maximum clear spacing is 16 times the web thickness but no greater than 10 inches, or

$$16t_w = 16\left(\frac{5}{16}\right) = 5 \text{ in.}$$

Use a center-to-center spacing of 2.5 inches, resulting in a clear spacing of

$$2.5 - 1.5 = 1 \text{ in.} < 5 \text{ in.}　　(\text{OK})$$

ANSWER　　Use $\tfrac{3}{16}$-inch $\times$ $1\tfrac{1}{2}$-inch fillet welds for intermediate stiffeners, spaced as shown in Figure 10.23.

For the bearing stiffener welds:

Minimum size $= \dfrac{5}{16}$ in. (based on the thicknesses $t_w = \tfrac{5}{16}$ in. and $t = \tfrac{3}{4}$ in.)

Minimum length $= 4\left(\dfrac{5}{16}\right) = 1.25$ in. < 1.5 in.　　use 1.5 in.

Use two welds per stiffener for a total of four. As with the intermediate stiffeners, the strength will be controlled by the base metal shear strength of 6.075 kips/in., or 9.112 kips for a 1.5-inch length.

■ **FIGURE 10.23**

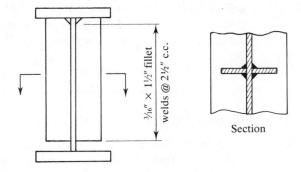

For the end bearing stiffener, the applied load per inch is

$$\frac{\text{Reaction}}{\text{Length available for weld}} = \frac{223.4}{62 - 2(0.5)} = 3.662 \text{ kips/in.}$$

From $\dfrac{9.112}{s} = 3.662$, $s = 2.49$ in.

ANSWER Use $\frac{3}{16}$-inch $\times$ $1\frac{1}{2}$-inch fillet welds for all bearing stiffeners, spaced as shown in Figure 10.24.

■ **FIGURE 10.24**

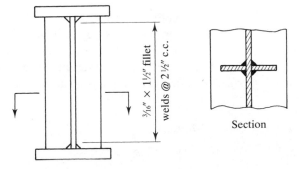

The girder designed in this example is not necessarily the most economical one. Other possibilities include a girder with a thinner web and more intermediate stiffeners, and a girder with a thicker web and no intermediate stiffeners. Variables that affect economy include weight (volume of steel required) and fabricating costs. Although girders with intermediate stiffeners will usually require less steel, the savings can be offset by the additional fabrication cost. Variable flange thicknesses can also be considered. This alternative will definitely save weight, but here also, fabrication costs must be considered. A practical approach to achieving an economical design is to prepare several alternatives and compare their costs, using estimates of material and fabricating costs. *Design of Welded Structures* (Blodgett, 1966) contains many useful suggestions for the design of economical welded plate girders.

■ PROBLEMS

Flexural Strength

10.4-1   Compute the design flexural strength of a plate girder comprising a $\frac{5}{16}$-inch × 70-inch web and $\frac{3}{4}$-inch × 18-inch flanges. A572 Grade 50 steel is used, and the compression flange has continuous lateral support.

10.4-2   A plate girder of A572 Grade 50 steel has a $\frac{1}{2}$-inch × 70-inch web and 3-inch × 22-inch flanges.

  a. Compute the design flexural strength if there is continuous lateral support of the compression flange.

  b. Compute the design flexural strength if the unbraced length is 40 feet. Use $C_b = 1.30$.

10.4-3   A plate girder of A36 steel has a $\frac{5}{16}$-inch × 58-inch web and $\frac{5}{8}$-inch × 16-inch flanges.

  a. Compute the design flexural strength if there is continuous lateral support of the compression flange.

  b. Compute the design flexural strength if the unbraced length is 10 feet. Use $C_b = 1.0$.

10.4-4   A plate girder of A36 steel has a $\frac{7}{16}$-inch × 80-inch web and 2-inch × 23-inch flanges. The span length is 70 feet. Lateral bracing is provided at the ends and at the one-third points. The loading and the shear and bending moment diagrams are shown in Figure P10.4-4. The loads are factored and the girder weight is included. Determine whether the design flexural strength is adequate (the shear strength will be investigated in a subsequent problem).

Shear Strength

10.5-1   For the girder in Problem 10.4-2,

  a. compute the design shear strength of the end panel if the first intermediate stiffener is placed 66 inches from the support.

  b. compute the design shear strength of an interior panel with a stiffener spacing of 180 inches.

  c. What is the design shear strength if no intermediate stiffeners are used?

10.5-2   For the girder in Problem 10.4-3,

  a. compute the design shear strength of the end panel if the first intermediate stiffener is placed 60 inches from the support.

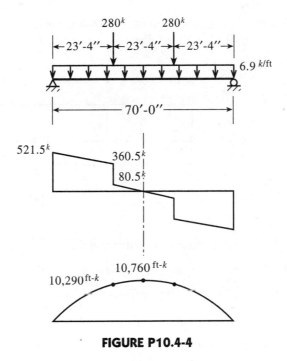

FIGURE P10.4-4

b. compute the design shear strength of an interior panel with a stiffener spacing of 108 inches.

c. What is the design shear strength if no intermediate stiffeners are used?

10.5-3 A plate girder of A572 Grade 50 steel consists of a ⅜-inch × 84-inch web and ⅞-inch × 20-inch flanges. The span length is 46 feet, and stiffeners are placed at 3 feet, 9 feet, and 15 feet from each end, as shown in Figure P10.5-3. Compute the design shear strength for each panel defined by the stiffeners.

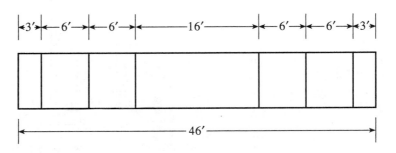

FIGURE P10.5-3

10.5-4 The girder in Problem 10.4-4 is provided with stiffeners spaced as shown in Figure P10.5-4. Determine whether the shear strength is adequate.

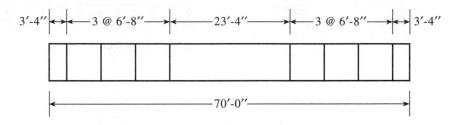

FIGURE P10.5-4

Interaction of Flexure and Shear

10.6-1 Does the girder in Problem 10.4-4 with the stiffeners provided in Problem 10.5-4 satisfy the AISC provisions for the interaction of flexure and shear?

Bearing Stiffeners

10.7-1 The details of an end bearing stiffener are shown in Figure P10.7-1. The stiffener plates are $9/16$-inch thick, and the web is $3/16$-inch thick. The stiffeners are clipped $\frac{1}{2}$ inch to provide clearance for the flange-to-web welds. If A572 Grade 50 steel is used, determine the maximum reaction that can be supported.

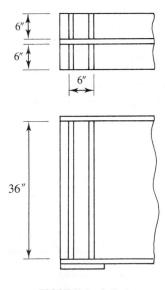

FIGURE P10.7-1

10.7-2 Bearing stiffeners are provided for the girder in Problem 10.4-4 at the supports and at the points of concentrated load application. At each location, the stiffener is a pair of $\frac{3}{4}$-inch × 10-inch plates, clipped $\frac{1}{2}$ inch to clear the flange-to-web welds. A36 steel is used. Determine whether the bearing stiffeners are adequate.

Design

10.8-1 The preliminary design of a plate girder has resulted in a $\frac{3}{8}$-inch × 68-inch web. The maximum factored load bending moment is 5900 ft-kips. Use this information to select a preliminary flange size for A572 Grade 50 steel. Assume continuous lateral support of the compression flange.

10.8-2 A plate girder must be designed to resist a factored load bending moment of 23,800 ft-kips. The total depth of the girder will be 101 inches, and the unbraced length is 40 feet. Use A36 steel and select a trial cross section. Assume that the moment includes an accurate estimate of the girder weight. Use $C_b = 1.0$.

10.8-3 A plate girder must be designed to resist a factored load bending moment of 19,850 ft-kips. Assume that the moment includes an accurate estimate of the girder weight. The total depth of the girder will be 78 inches, and the compression flange has continuous lateral support. Use A572 Grade 50 steel.

    a. Select a trial cross section based on the bending moment.

    b. If the factored load shear is 794 kips at the support, where must the first intermediate stiffener be located? Assume that the shear includes an accurate estimate of the girder weight.

10.8-4 A plate girder consists of a $\frac{1}{2}$-inch × 90-inch web and $2\frac{1}{2}$-inch × 28-inch flanges. The factored load reaction is 369 kips. A36 steel is used.

    a. Determine the distance from the end of the girder to the first intermediate stiffener.

    b. Design a bearing stiffener to be located at the support.

10.8-5   **a.** Each intermediate stiffener for the girder in Problem 10.4-4 is a pair of bars, 4 inches × $\frac{1}{2}$-inch. Design welds for the stiffener-to-web connection.

    b. Each bearing stiffener for the girder in Problem 10.4-4 is a pair of plates, $\frac{3}{4}$-inch × 10 inches. Design welds for the stiffener-to-web connections for both the end and interior stiffeners.

10.8-6 Design flange-to-web welds for the girder in Problem 10.4-4.

10.8-7 A plate girder must be designed for the following conditions: the span length is 55 feet, the service live load is 2.1 kips/ft (there is no dead load other than the weight of the girder), and the steel is A572 Grade 50. Lateral support is provided only at the ends.

    a. Design the cross section. Use a total depth of 69 inches.

    b. Determine the location and size of intermediate stiffeners.

    c. Determine the required size of bearing stiffeners at the supports.

    d. Design welds for all components.

10.8-8 A 48-foot-long plate girder must support a uniformly distributed load and a concentrated load at midspan. The uniform load consists of a 1.0-kip/ft dead load and a 2.0-kip/ft live load. The concentrated load consists of a 50-kip dead load and a 100-kip live load. There is lateral support at the ends and at midspan. Use A572 Grade 50 steel for all components.

    a. Design the cross section.

    b. Determine the location and size of intermediate stiffeners.

    c. Determine the required size of bearing stiffeners at the supports and at the point of concentrated load application.

    d. Design welds for all components.

10.8-9 A 66-foot-long plate girder must support a uniformly distributed load and concentrated loads at the one-third points. The uniform load consists of a 1.3-kip/ft dead load and a 2.3-kip/ft live load. Each concentrated load consists of a 28-kip dead load and a 49-kip live load. There is lateral support at the ends and at the points of concentrated load application. Use A572 Grade 50 steel for all components.

    a. Design the cross section.

    b. Determine the location and size of intermediate stiffeners.

    c. Determine the required size of bearing stiffeners at the supports and at the points of concentrated load application.

    d. Design welds for all components.

10.8-10 A plate girder *ABCDE* will be used in a building to provide a large column-free area, as shown in Figure P10.8-10. The uniform load on the girder consists of 1.9 kips/ft of

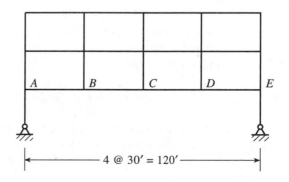

FIGURE P10.8-10

dead load (not including the weight of the girder) and 2.8 kips/ft of live load. In addition, the girder must support column loads at *B, C,* and *D* consisting of 112 kips dead load and 168 kips live load at each location. Assume that the girder is simply supported and that the column loads act as concentrated loads; that is, there is no continuity with the girder. Assume lateral support of the compression flange at 10-foot intervals. Use steel with F_y = 50 ksi for all components.

a. Design the girder cross section. Use a total depth of 10 feet.

b. Determine the location and size of intermediate stiffeners.

c. Determine the required size of bearing stiffeners at *A, B, C, D,* and *E.*

d. Design the welds for all components.

Plastic Analysis and Design

INTRODUCTION

We introduced the concept of plastic collapse in Section 5.2, "Bending Stress and the Plastic Moment." Failure of a structure will take place at a load that forms enough plastic hinges to create a mechanism that will undergo uncontained displacement without any increase in the load. In a statically determinate beam, only one plastic hinge is required. As shown in Figure A.1, the hinge will form where the moment is maximum — in this case, at midspan. When the bending moment is large enough to cause the entire cross section to yield, any further increase in moment cannot be countered, and the plastic hinge has formed. This hinge is similar to an ordinary hinge except that the plastic hinge will have some moment resistance, much like a "rusty" hinge.

The plastic moment capacity, denoted M_p, is the bending moment at which a plastic hinge forms. It equals and is opposite to the internal resisting moment corresponding to the stress distribution shown in Figure A.1c. The plastic moment can be computed for a given yield stress and cross-sectional shape, as indicated in Figure A.2. If the stress distribution in the fully plastic condition is replaced by two equal and opposite statically equivalent concentrated forces, a couple is formed. The magnitude of each of these forces equals the yield stress times one half of the total cross-sectional area. The moment produced by this internal couple is

$$M_p = F_y \frac{A}{2} a = F_y Z_x$$

where A is the total cross-sectional area, a is the distance between centroids of the two half areas, and Z_x is the plastic section modulus. The factor of safety between first yielding and the fully plastic state can be expressed in terms of the section moduli. From Figure A.1b, the moment causing first yield can be written as

$$M_y = F_y S_x \quad \text{and} \quad \frac{M_p}{M_y} = \frac{F_y Z_x}{F_y S_x} = \frac{Z_x}{S_x}$$

This ratio is a constant for a given cross-sectional shape and is called the *shape factor*. For a beam designed by allowable stress theory, it is a measure of the reserve capacity and has an average value of 1.12 for W-shapes.

■ **FIGURE A.1**

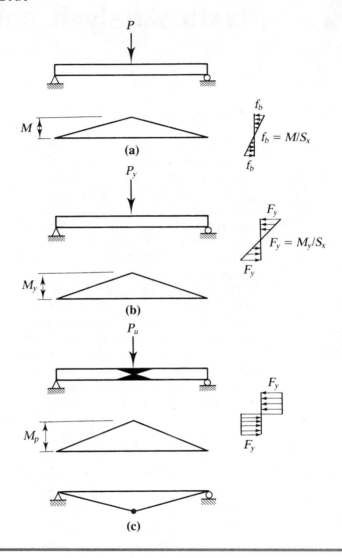

In a statically indeterminate beam or frame, more than one plastic hinge will be required for the formation of a collapse mechanism. These hinges will form sequentially, although it is not always necessary to know the sequence. The analysis of statically indeterminate structures will be considered after a discussion of specification requirements.

■ **FIGURE A.2**

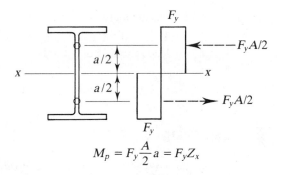

$$M_p = F_y \frac{A}{2} a = F_y Z_x$$

AISC REQUIREMENTS

The AISC Specification permits plastic analysis and design to be used when the structure can remain stable, both locally and overall, to the point of plastic collapse. Because the beam or frame could be required to undergo very large deformations as plastic hinges are being formed, lateral bracing requirements are especially severe.

To prevent local buckling, AISC B5.2 requires that the member have a compact cross-sectional shape — that is, $\lambda \leq \lambda_p$ for both the web and flanges. For I-shaped members such as W- and S-shapes, the limiting width–thickness ratios from Table B5.1 are

$$\frac{b_f}{2t_f} \leq \frac{65}{\sqrt{F_y}} \text{ and } \frac{h}{t_w} \leq \frac{640}{\sqrt{F_y}}$$

To prevent lateral buckling, AISC F1.2d limits the maximum unbraced length L_b at plastic hinge locations to L_{pd}, where, for I-shaped members,

$$L_{pd} = \frac{3600 + 2200(M_1/M_2)}{F_y} r_y \qquad \text{(AISC Equation F1-17)}$$

In this equation, M_1 is the smaller moment at the end of the unbraced length and M_2 is the larger. The ratio M_1/M_2 is positive when M_1 and M_2 bend the segment in reverse curvature and negative when they cause single curvature bending.

For compact shapes with adequate lateral bracing, M_n can be taken as M_p for use in plastic analysis. However, AISC F1.2d specifies that in the region of the last plastic hinge to form and in regions not adjacent to plastic hinges, the usual methods must be used to compute M_n.

Other AISC Specification provisions related to plastic analysis and design are as follows.

A5.1 Plastic analysis is permitted only for $F_y \leq 65$ ksi.

C2.2 The axial force in columns caused by factored gravity and horizontal loads should not exceed $0.75\phi_c A_g F_y$.

E1.2 For columns, the slenderness parameter λ_c must not exceed $1.5K$, where K is the effective length factor.

A.3   ANALYSIS

If more than one collapse mechanism is possible, as with the continuous beam illustrated in Figure A.3, the correct one can be found and analyzed with the aid of three basic theorems of plastic analysis, given here without proof.

1. **Lower-bound theorem (static theorem):** If a safe distribution of moment (one in which the moment is less than or equal to M_p everywhere) can be found, and it is statically admissible with the load (equilibrium is satisfied), then the corresponding load is less than or equal to the collapse load.

2. **Upper-bound theorem (kinematic theorem):** The load that corresponds to an assumed mechanism must be greater than or equal to the collapse load. As a consequence, if all possible mechanisms are investigated, the one requiring the smallest load is the correct one.

■ **FIGURE A.3**

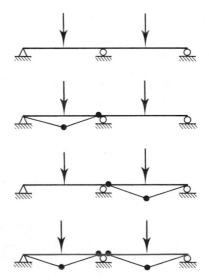

3. **Uniqueness theorem:** If there is a safe and statically admissible distribution of moment in which enough plastic hinges form to produce a collapse mechanism, the corresponding load is the collapse load; that is, if a mechanism satisfies both the upper- and lower-bound theorems, it is the correct one.

Analysis based on the lower-bound theorem is called the *equilibrium method* and is illustrated in Example A.1.

■ EXAMPLE A.1

Find the ultimate load for the beam shown in Figure A.4a by the equilibrium method of plastic analysis. Assume continuous lateral support and use A36 steel.

SOLUTION

A W30 × 99 of A36 steel is compact and, with continuous lateral support, the lateral bracing requirement is satisfied; therefore, plastic analysis is acceptable.

The loading history of the beam, from working load to collapse load, is traced in Figure A.4a–d. At working loads, before yielding begins anywhere, the distribution of bending moments will be as shown in Figure A.4a, with the maximum moment occurring at the fixed ends. As the load is gradually increased, yielding begins at the supports when the bending moment reaches $M_y = F_y S_x$. Further increase in the load will cause the simultaneous formation of plastic hinges at each end at a moment of $M_p = F_y Z_x$. At this level of loading, the structure is still stable, the beam having been rendered statically determinate by the formation of the two plastic hinges. Only when a third hinge forms will a mechanism be created. This happens when the maximum positive moment attains a value of M_p. By virtue of the uniqueness theorem, the corre-

■ FIGURE A.4

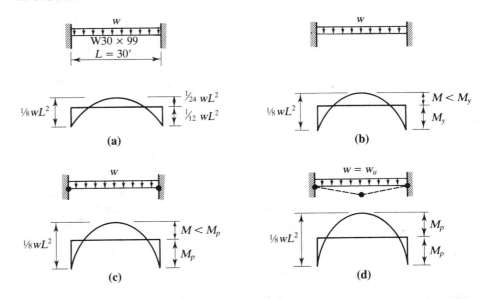

sponding load is the collapse load because the distribution of moment is safe and statically admissible.

At all stages of loading, the sum of the absolute values of the maximum negative and positive moments is $wL^2/8$. At collapse, this sum becomes

$$M_p + M_p = \frac{1}{8} w_u L^2 \text{ or } w_u = \frac{16M_p}{L^2}$$

Factored loads must be compared with factored strengths, so $\phi_b M_p$ rather than M_p should have been used in the preceding equations. To keep notation simple, however, we use M_p in all examples until the final step, when we substitute $\phi_b M_p$ for it. The correct result for this example is

$$w_u = \frac{16\phi_b M_p}{L^2}$$

For a W30 × 99,

$$M_p = F_y Z_x = \frac{36(312)}{12} = 936 \text{ ft-kips}$$

and

$$\phi_b M_p = 0.9(936) = 842.4 \text{ ft-kips}$$

The value of $\phi_b M_p$ can also be obtained directly from the Load Factor Design Selection Table in Part 4 of the *Manual*.

ANSWER

$$w_u = \frac{16(842.4)}{(30)^2} = 15.0 \text{ kips/ft.}$$

■

■ EXAMPLE A.2

If the beam in Example A.1 does not have continuous lateral support, determine where it must be braced.

SOLUTION

The plastic hinges at the ends form simultaneously and before the hinge forms at midspan. Hence the maximum unbraced length should be checked with reference to the ends (i.e., the last hinge to form is not subject to the bracing requirements for plastic analysis).

With respect to a hinge at the left end, assume that the brace point is at midspan. In this case, $M_1 = M_2 = M_p$, and the beam is bent in reverse curvature (the two moments are of opposite sign), so $M_1/M_2 = +1$. From AISC Equation F1-17, the maximum unbraced length is

$$L_{pd} = \frac{3600 + 2200(M_1/M_2)}{F_y} r_y = \frac{3600 + 2200(1)}{36} \quad (2.10)$$

$$= 338.3 \text{ in.} = 28.2 \text{ ft.}$$

(Note that the beam is almost satisfactory with no lateral bracing.)

With one lateral brace at midspan,

$$L_b = 15 \text{ ft.} < 28.2 \text{ ft.} \qquad \text{(OK)}$$

The unbraced length under consideration encompasses the hinge at midspan. There are no regions not adjacent to a plastic hinge, so no other design strength computations are necessary.

ANSWER Use one lateral brace at midspan. ■

The *mechanism method* is based on the upper-bound theorem and requires that all possible collapse mechanisms be investigated. The one requiring the smallest load will control, and the corresponding load is the collapse load. The analysis of each mechanism is accomplished by application of the principle of virtual work. An assumed mechanism is subjected to virtual displacements consistent with the possible mechanism motion, and the external work is equated to the internal work. A relationship can then be found between the load and the plastic moment capacity, M_p. This technique will be illustrated in Examples A.3 and A.4.

■ EXAMPLE A.3

The continuous beam shown in Figure A.5 has a compact cross section with a design strength $\phi_b M_p$ of 1040 ft-kips. Use the mechanism method to find the collapse load P_u. Assume continuous lateral support.

SOLUTION In keeping with the notation adopted in Example A.2, we will use M_p in the solution and substitute $\phi_b M_p$ in the final step.

There are two possible failure mechanisms for this beam. As indicated in Figure A.5, they are similar, with each segment undergoing a rigid-body motion. To investigate the mechanism in span *AB*, impose a virtual rotation θ at *A*. The corresponding rotations at the plastic hinges will be as shown in Figure A.5b, and the vertical displacement of the load will be 10θ. From the principle of virtual work,

External work = internal work

$$P_u(10\theta) = M_p(2\theta) + M_p\theta$$

(No internal work is done at *A* because there is no plastic hinge.) Solving for the collapse load gives

$$P_u = \frac{3M_p}{10}$$

The mechanism for span *BC* is slightly different: All three hinges are plastic hinges. The external and internal virtual work in this case are

$$2P_u(15\theta) = M_p\theta + M_p(2\theta) + M_p\theta \text{ and } P_u = \frac{2}{15}M_p$$

■ **FIGURE A.5**

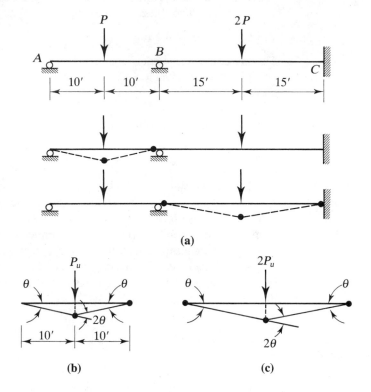

(a)

(b)   (c)

This second possibility requires the smaller load and is therefore the correct mechanism. The collapse load will be obtained using $\phi_b M_p$ in place of M_p.

ANSWER $P_u = \dfrac{2}{15} \phi_b M_p = \dfrac{2}{15} (1040) = 139$ kips.

■

■ **EXAMPLE A.4**

Determine the collapse load P_u for the rigid frame shown in Figure A.6. Each member of the frame is a W21 × 147 with $F_y = 50$ ksi. Assume continuous lateral support.

SOLUTION   A W21 × 147 is compact for $F_y = 50$ ksi, and with continuous lateral support provided, the conditions for the use of plastic analysis are satisfied.

As indicated in Figure A.6, there are three possible failure modes for this frame: a beam mechanism in member *BC*, a sway mechanism, and one that is a combination of the first two. We begin the analysis of each mechanism by imposing a virtual rota-

■ **FIGURE A.6**

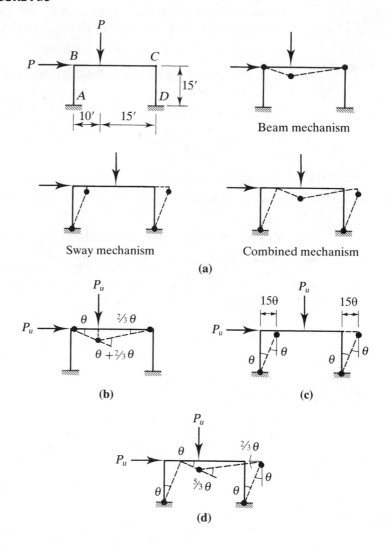

Beam mechanism

Sway mechanism

Combined mechanism

(a)

(b)

(c)

(d)

tion of θ at one of the hinges and expressing the remaining notations and displacements as a function of this angle.

The virtual displacement of the beam mechanism is shown in Figure A.6b. From the equivalence of external and internal work,

$$P_u(10\theta) = M_p\theta + M_p\left(\frac{5}{3}\theta\right) + M_p\left(\frac{2}{3}\theta\right)$$

where M_p has been used in place of $\phi_b M_p$. Solving for P_u gives

$$P_u = 0.3333M_p$$

If axial strains in member BC are neglected, the sway mechanism will displace in the manner shown in Figure A.6c, with the same horizontal displacement at B and C. As a consequence, the rotations of all the plastic hinges are the same:

$$P_u(15\theta) = M_p(4\theta) \quad \text{or} \quad P_u = 0.2667M_p$$

From Figure A.6d, the principle of virtual work for the combined mechanism gives

$$P_u(15\theta) + P_u(10\theta) = M_p\theta + M_p\left(\frac{5}{3}\theta\right) + M_p\left(\frac{2}{3}\theta + \theta\right) + M_p\theta$$

$$P_u = 0.2133M_p \quad \text{(controls)}$$

ANSWER   The collapse load for the frame is $P_u = 0.2133\phi_b M_p = 0.2133(1400) = 299$ kips.   ■

Note that there is a certain similarity between the two methods of analysis. Although the equilibrium method does not require a consideration of all possible mechanisms, it does require that you recognize a mechanism when the assumed distribution of moment is consistent with one. Both methods require the assumption of failure mechanisms, but in the equilibrium method each assumption is checked for a safe and statically admissible distribution of moment, and it may not be necessary to investigate all possible mechanisms.

A.4   DESIGN

The design process is similar to analysis, except that the unknown being sought is the required plastic moment capacity, M_p. The collapse load is known in advance, having been obtained by multiplying the service loads by the load factors.

■ EXAMPLE A.5

The three-span continuous beam shown in Figure A.7 must support the gravity service loads given. Each load consists of 25% dead load and 75% live load. Cover plates will be used in spans BC and CD to give the relative moment strengths indicated. Assume continuous lateral support and select a shape in A36 steel.

SOLUTION   The collapse loads are obtained by multiplying the service loads by the appropriate load factors. For the 45-kip service load,

$$P_u = 1.2(0.25 \times 45) + 1.6(0.75 \times 45) = 67.5 \text{ kips}$$

For the 57-kip service load,

$$P_u = 1.2(0.25 \times 57) + 1.6(0.75 \times 57) = 85.5 \text{ kips}$$

■ **FIGURE A.7**

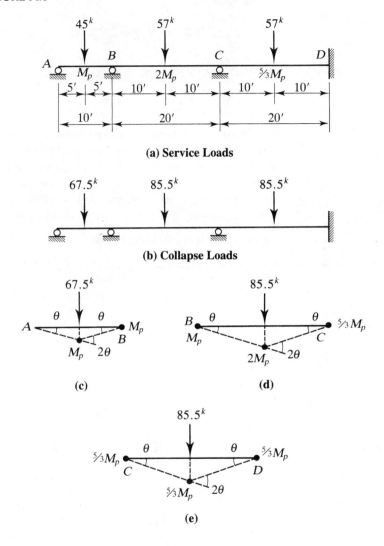

(a) Service Loads

(b) Collapse Loads

(c)

(d)

(e)

Three mechanisms must be investigated, one in each span. Figure A.7c–e shows each mechanism after being subjected to a virtual displacement. When a plastic hinge forms at a support where members of unequal strength meet, it will form when the bending moment equals the plastic moment capacity of the weaker member.

For span AB,

External work = internal work

$67.5(5\theta) = M_p(2\theta + \theta)$ or $M_p = 112.5$ ft-kips

For span BC,

$$85.5(10\theta) = M_p\theta + 2M_p(2\theta) + \frac{5}{3}M_p\theta \text{ or } M_p = 128.2 \text{ ft-kips}$$

For span CD,

$$85.5(10\theta) = \frac{5}{3}M_p(\theta + 2\theta + \theta) \text{ or } M_p = 128.2 \text{ ft-kips}$$

The upper-bound theorem may be interpreted as follows: The value of the plastic moment corresponding to an assumed mechanism is less than or equal to the plastic moment for the collapse load. Thus the mechanism requiring the largest moment capacity is the correct one. Both of the last two mechanisms evaluated in this design problem require the same controlling value of M_p and will therefore occur simultaneously. The required strength is actually the required *design* strength, so

Required $\phi_b M_p = 128.2$ ft-kips

From the Load Factor Design Selection Table, the lightest shape is a W16 × 31 with a design strength of $\phi_b M_p = 146$ ft-kips

Try a W16 × 31 and check the shear (refer to Figure A.8).

■ **FIGURE A.8**

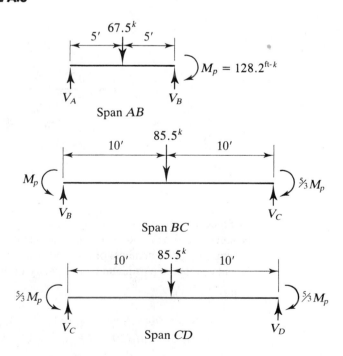

Span AB

Span BC

Span CD

For span *AB*,

$$\Sigma\, M_B = V_A(10) - 67.5(5) + 128.2 = 0$$
$$V_A = 20.93 \text{ kips}$$
$$V_B = 20.93 - 67.5 = -46.57 \text{ kips}$$

For span *BC*,

$$\Sigma\, M_B = -M_p + 85.5(10) + \left(\frac{5}{3}\right)M_p - V_C(20) = 0$$

$$V_C = \frac{85.5(10) + (2/3)M_p}{20} = \frac{855 + (2/3)(128.2)}{20} = 47.02 \text{ kips}$$

$$V_B = 85.5 - 47.02 = 38.48 \text{ kips}$$

For span *CD*,

$$\Sigma\, M_C = -\frac{5}{3}M_p + \frac{5}{3}M_p + 85.5(10) - V_D(20) = 0$$

$$V_D = 42.75 \text{ kips} = V_C$$

The maximum shear is therefore V_C from span *BC*, or 47.02 kips.

From the factored uniform load tables in Part 4 of the *Manual,* the shear design strength of a W16 × 31 is

$$\phi_v V_n = 84.9 \text{ kips} > 47.02 \text{ kips} \qquad \text{(OK)}$$

ANSWER   Use a W16 × 31.   ■

A.5   CONCLUDING REMARKS

Analysis of a mechanism subjected to distributed loads presents an additional complication that has not been covered here. Any realistic problem in plastic analysis or design will certainly include this type of loading. In addition, members subjected to both axial compression and bending, as in the rigid frame in Example A.4, should be investigated for interaction of these effects.

With regard to analysis methods in general, a more formal and systematic approach to the equilibrium method exists and is discussed in considerable detail in *The Plastic Methods of Structural Analysis* (Neal, 1977). A more rigorous formulation of the mechanism method is also possible. With this approach, known as the *method of inequalities,* the correct mechanism can be determined directly by linear programming techniques. For most of the structures for which plastic design would ordinarily be used, however, the mechanism method as presented in this appendix usually will be adequate.

B

Structural Steel Design Based on Allowable Stress

INTRODUCTION

The chief difference between allowable stress design and load and resistance factor design is in the factors of safety.* In LRFD, load factors are applied to the loads, and a resistance factor is applied to the strength; furthermore, the values of the load factors depend on the type of load and the load combination. In allowable stress design (ASD), only one factor of safety is used, and it is applied to the stress that exists at the limit state. Limit states for allowable stress design are the same as for LRFD: yielding, fracture, and buckling. The principle of allowable stress analysis and design is as follows: The stress at the limit state is divided by a factor of safety to obtain an allowable stress, and the maximum stress caused by service loads must not exceed this allowable stress. For example, for axial tension,

$$f_t = \frac{P}{A} \leq F_t \tag{B.1}$$

where

f_t = computed tensile stress

P = *service* axial tensile load

F_t = allowable tensile stress

The allowable tensile stress will be either the yield stress divided by a factor of safety or the ultimate tensile stress divided by a different factor of safety. We cover tension members in detail in Section B.2.

Allowable stress design for structural steel was the norm before the introduction of the LRFD Specification in 1989. The last edition of the *ASD Specification* (AISC, 1989b) and the corresponding *Manual of Steel Construction* (AISC, 1989a) were published in 1989. Both documents are organized in the same way as their LRFD counterparts: the Specification is divided into chapters (for example, "Chapter D, Tension Members") and the *Manual* is divided into parts (such as "Part 2, Beam and Girder Design"). The Specification is accompanied by a Commentary.

---

*We assume that you are familiar with the AISC LRFD Specification and *Manual*.

The notation of Equation B.1 is the same as that used in the Specification. A lowercase f is used for computed (actual) stress, and an uppercase F is used for allowable stress. The subscript indicates the type of stress.

Because this appendix is an introduction only, we do not use AISC Specification section numbers or equation numbers. Although the material is taken from the Specification, the equation numbers are ours. Furthermore, when the terms *Specification* or *Manual* are used in this appendix, the allowable stress versions are implied, unless otherwise stated.

Many elements of steel design are the same for both ASD and LRFD. For example, the net area provisions for tension members are the same, including the $s^2/4g$ technique for staggered holes and the U factor for shear lag (although the ASD Specification uses the average U values and places the equation for U in the Commentary — the reverse of the way it is done in the LRFD Specification). The definitions of compact, noncompact, and slender members are virtually the same, but the latest LRFD Specification rules are more up-to-date. Generally, when there is a conflict between an ASD and LRFD provision, it should be resolved in favor of the LRFD Specification, which is more current.

Although there are no load factors in allowable stress design, the relative importance of various loads can still be taken into account in combinations. For example, the following load combinations for roof structures are frequently used: $D + S$, $D + W$, $D + (S/2) + W$, and $D + S + (W/3)$. In addition, the Specification permits allowable stresses to be increased by one third when wind or seismic loads are included. Many building codes also contain this provision.

The *ASD Manual* contains many tables and charts similar to the ones in the *LRFD Manual* that simplify design. Only the more important ones are discussed in this brief introduction.

B.2 TENSION MEMBERS

From Equation B.1, the computed axial tensile stress is $f_t = P/A$. The allowable stress is based on the more critical of the limit states of yielding and fracture. For yielding of the gross section, the applied stress is

$$f_t = \frac{P}{A_g} \tag{B.2}$$

where A_g is the gross cross-sectional area. The factor of safety for this limit state is 5/3, and the allowable stress is

$$F_t = \frac{F_y}{F.S.} = \frac{F_y}{5/3} = 0.60F_y \tag{B.3}$$

For fracture of the net section,

$$f_t = \frac{P}{A_e} \tag{B.4}$$

where A_e is the effective net area. The factor of safety is 2.0, resulting in an allowable stress of

$$F_t = \frac{F_u}{F.S.} = \frac{F_u}{2} = 0.50F_u \tag{B.5}$$

■ EXAMPLE B.1

Check the stresses in the tension member shown in Figure B.1 caused by the 50-kip service load. A36 steel and ⅞-inch diameter bolts are used.

SOLUTION   From Equations B.2 and B.3, the applied stress on the gross area is

$$f_t = \frac{P}{A_g} = \frac{50}{2.48} = 20.2 \text{ ksi}$$

and the allowable stress is

$$F_t = 0.60F_y = 0.60(36) = 21.6 \text{ ksi} > 20.2 \text{ ksi} \qquad \text{(OK)}$$

The stress on the net area is

$$A_n = A_g - (\text{thickness} \times \text{hole diameter})$$

$$= 2.48 - \frac{3}{8}\left(\frac{7}{8} + \frac{1}{8}\right) = 2.105 \text{ in.}^2$$

If we use the average value of U, the effective net area is

$$A_e = UA_n = 0.85A_n = 0.85(2.105) = 1.789 \text{ in.}^2$$

■ FIGURE B.1

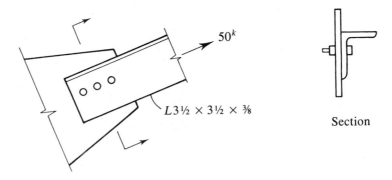

50^k

$L3\tfrac{1}{2} \times 3\tfrac{1}{2} \times \tfrac{3}{8}$

Section

From Equations B.4 and B.5,

$$f_t = \frac{P}{A_e} = \frac{50}{1.789} = 27.9 \text{ ksi}$$

$$F_t = 0.50F_u = 0.50(58) = 29 \text{ ksi} > 27.9 \text{ ksi} \qquad \text{(OK)}$$

ANSWER   The member is satisfactory.   ■

B.3   ## COMPRESSION MEMBERS

The stress in an axially loaded compression member is

$$f_a = \frac{P}{A_g}$$

The allowable stress, denoted F_a, is obtained by dividing the critical buckling load by a factor of safety. The factor of safety for elastic columns (slender columns) is a constant, and the factor for inelastic columns is variable. In ASD, the compressive strength is expressed as a function of the slenderness ratio, KL/r, whereas in LRFD, the strength is a function of $\lambda_c = (KL/r\pi)\sqrt{F_y/E}$. In the elastic range, the critical stress is the Euler buckling load divided by the area, or

$$F_{cr} = \frac{P_{cr}}{A_g} = \frac{\pi^2 EA_g}{(KL/r)^2} \div A_g = \frac{\pi^2 E}{(KL/r)^2} \tag{B.6}$$

For the inelastic range, the proportional limit is assumed to be at $F_y/2$, and the following empirical equation is used in lieu of the tangent modulus formula:

$$F_{cr} = F_y \left[1 - \frac{(KL/r)^2}{2C_c^2} \right] \tag{B.7}$$

where C_c is the value of KL/r corresponding to a stress of $F_y/2$. Equation B.7 represents a parabola that is tangent to the Euler curve at $KL/r = C_c$ and tangent to a horizontal line at $KL/r = 0$. We can find an expression for C_c by setting the right-hand side of Equation B.6 to $F_y/2$:

$$\frac{F_y}{2} = \frac{\pi^2 E}{(KL/r)^2} = \frac{\pi^2 E}{C_c^2}$$

which results in

$$C_c = \sqrt{\frac{2\pi^2 E}{F_y}} \tag{B.8}$$

The column strength over the full range of slenderness, before the application of safety factors, is represented graphically in Figure B.2.

To obtain allowable compressive stresses, we divide Equations B.6 and B.7 by factors of safety. For elastic columns, the factor of safety is 23/12. For inelastic columns, the following variable factor is used:

$$F.S. = \frac{5}{3} + \frac{3(KL/r)}{8C_c} - \frac{(KL/r)^3}{8C_c^3}$$

This expression has a value of 5/3 at $KL/r = 0$ (the same as for yielding of a tension member) and a value of 23/12 at $KL/r = C_c$ (approximately 15% more than 5/3). Dividing the strength equations by the appropriate factors of safety, we obtain the following allowable stresses:

For $KL/r < C_c$,

$$F_a = \frac{F_y\left[1 - \dfrac{(KL/r)^2}{2C_c^2}\right]}{\dfrac{5}{3} + \dfrac{3(KL/r)}{8C_c} - \dfrac{(KL/r)^3}{8C_c^3}} \tag{B.9}$$

For $KL/r > C_c$,

$$F_a = \frac{\pi^2 E}{(KL/r)^2} \div \frac{23}{12} = \frac{12\pi^2 E}{23(KL/r)^2} \tag{B.10}$$

■ **FIGURE B.2**

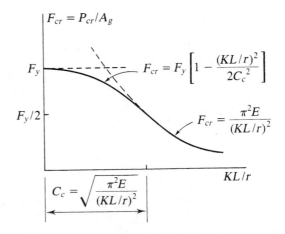

For members with slender cross-sectional elements, a reduction must be made in the allowable stress to account for the possibility of local buckling. This is accomplished with reduction factors from an appendix to the Specification.

■ **EXAMPLE B.2**

Determine the allowable service load P for the compression member shown in Figure B.3.

SOLUTION

Check for slender cross-sectional elements. The width–thickness ratio limits for compression members given in the ASD Specification are the same as those in the LRFD Specification:

$$\frac{b_f}{2t_f} = 6.4 \qquad \text{(from the properties tables in the \textit{Manual})}$$

$$\frac{95}{\sqrt{F_y}} = \frac{95}{\sqrt{36}} = 15.8 > 6.4 \qquad \text{(OK)}$$

$$\frac{h}{t_w} = 25.3$$

$$\frac{253}{\sqrt{F_y}} = \frac{253}{\sqrt{36}} = 42.2 > 25.3 \qquad \text{(OK)}$$

■ **FIGURE B.3**

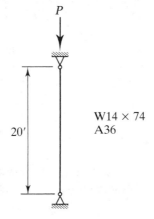

P

W14 × 74
A36

20'

The stress is $f_a = P/A_g$, so the corresponding load is $P = f_a A_g$, and the allowable compressive load is $F_a A_g$. From Equation B.8,

$$C_c = \sqrt{\frac{2\pi^2 E}{F_y}} = \sqrt{\frac{2\pi^2 (29,000)}{36}} = 126.1$$

The maximum slenderness ratio is

$$\frac{KL}{r} = \frac{KL}{r_{min}} = \frac{KL}{r_y} = \frac{1.0(20)(12)}{2.48} = 96.77$$

The result is less than C_c, so F_a is found from Equation B.9:

$$F_a = \frac{F_y\left[1 - \dfrac{(KL/r)^2}{2C_c^2}\right]}{\dfrac{5}{3} + \dfrac{3(KL/r)}{8C_c} - \dfrac{(KL/r)^3}{8C_c^3}} = \frac{36\left[1 - \dfrac{(96.77)^2}{2(126.1)^2}\right]}{\dfrac{5}{3} + \dfrac{3(96.77)}{8(126.1)} - \dfrac{(96.77)^3}{8(126.1)^3}}$$

$$= 13.38 \text{ ksi}$$

ANSWER $P = F_a A_g = 13.38(21.8) = 292 \text{ kips.}$ ■

Design Aids

The *ASD Manual* contains column design aids similar in form to those found in the *LRFD Manual*. Foremost among these aids are the tables for allowable axial loads. When these tables are entered with the effective length KL and a required service load capacity, a shape with sufficient capacity can be found quickly. As with the LRFD column load tables, the effective length with respect to the minimum radius of gyration r_y should be used; that is, $K_y L$ should be used. Alternatively, the tables can be entered with $K_x L/(r_x/r_y)$.

When the effective length factor K is found from the Jackson–Mooreland alignment charts, a stiffness reduction factor can be applied if the column is inelastic at failure ($KL/r < C_c$). A table for this purpose is provided in the *Manual*.

B.4   BEAMS

The maximum bending stress in a homogeneous beam that has not been stressed beyond the proportional limit is given by the *flexure formula:*

$$f_b = \frac{Mc}{I} = \frac{M}{I/c} = \frac{M}{S}$$

(B.11)

where

M = maximum bending moment in the beam

c = distance from the neutral axis to the extreme fiber

I = moment of inertia about the axis of bending

S = elastic section modulus

The coverage in this section is limited to hot-rolled I- and H-shaped cross sections bent about an axis perpendicular to the web (the x-axis).

The allowable bending stress is denoted F_b and is based on one of the following limit states: yielding, local buckling, or lateral-torsional buckling. In allowable stress design, it is convenient to divide beams into two categories: laterally supported and laterally unsupported. If a beam has adequate lateral support, the allowable stress will be based on yielding if the shape is compact and will be based on local buckling if the shape is noncompact. The allowable bending stress for laterally unsupported beams will be based on lateral-torsional buckling.

Lateral Support

A beam with an unbraced length of L_b is considered to have enough lateral support to prevent lateral-torsional buckling when $L_b \leq L_c$, where L_c is the smaller of

$$\frac{76b_f}{\sqrt{F_y}} \quad \text{or} \quad \frac{20,000}{(d/A_f)F_y}$$

That is,

$$L_c = \frac{76b_f}{\sqrt{F_y}} \leq \frac{20,000}{(d/A_f)F_y} \tag{B.12}$$

We use this criterion to classify beams as *laterally supported* or *laterally unsupported.*

Laterally Supported Beams

If a laterally supported beam can be stressed to the yield point without local buckling, the factor of safety is 5/3, and the allowable stress is

$$F_b = \frac{F_y}{F.S.} = \frac{F_y}{5/3} = 0.60F_y$$

This corresponds to a shape whose flange width–thickness ratio is at the upper limit for noncompactness; that is, $b_f/2t_f = 95/\sqrt{F_y}$. (This limit is different from the LRFD limit, but it will be used here since it is incorporated in an AISC equation for allowable stress.) If the shape is compact, the fully plastic condition can be reached without local buckling, and an additional 10% is permitted. The allowable stress for this case is

$$F_b = 1.10(0.60F_y) = 0.66F_y$$

For noncompact shapes, AISC uses a linear transition between $0.60F_y$ and $0.66F_y$, based on the value of $b_f/2t_f$. (All hot-rolled I- and H-shapes in the *Manual* have compact webs.) This results in the following equation for the allowable stress:

$$F_b = F_y\left(0.79 - 0.002\,\frac{b_f}{2t_f}\,\sqrt{F_y}\right)$$

Figure B.4 shows the relationship between flange width–thickness ratio and allowable stress for laterally supported beams. Slender shapes are dealt with in an appendix to the Specification, but none of the hot-rolled I- and H-shapes in the *Manual* are slender.

In summary, the allowable stresses for laterally supported beams are as follows:

If the shape is compact,

$$F_b = 0.66F_y \qquad\qquad (B.13)$$

If the shape is noncompact,

$$F_b = F_y\left(0.79 - 0.002\,\frac{b_f}{2t_f}\,\sqrt{F_y}\right) \qquad\qquad (B.14)$$

Laterally Unsupported Beams

The strength of laterally unsupported beams is based on the limit state of lateral-torsional buckling. In ASD, this state has two components: uniform warping and nonuniform warping. Uniform warping is elastic and the limit stress is

$$f_u = \frac{0.65E}{L_b d/A_f} \qquad\qquad (B.15)$$

■ **FIGURE B.4**

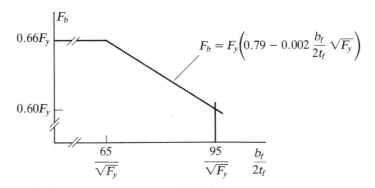

where

d = overall depth of beam

A_f = area of compression flange

Nonuniform warping can be either inelastic or elastic. For elastic warping, the failure stress is

$$f_{nu} = \frac{\pi^2 E}{(L_b/r_y)^2} \tag{B.16}$$

For inelastic warping, the following empirical equation, similar to the one for compression members, is used:

$$f_{nu} = \frac{10}{9} F_y \left[1 - \frac{(L_b/r_y)^2}{2C^2} \right] \tag{B.17}$$

where

C = maximum value of L_b for which the nonuniform warping is inelastic (if $L_b > C$, the warping is elastic)

$$= 3\pi \sqrt{\frac{E}{5F_y}}$$

The buckling stresses given by equations B.15–B.17 are subject to an upper limit of F_y. Figure B.5 shows the uniform warping stress as a function of L_b, and Figure B.6 shows the nonuniform warping stress.

■ **FIGURE B.5**

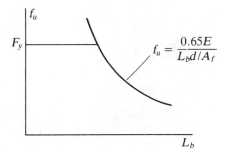

■ **FIGURE B.6**

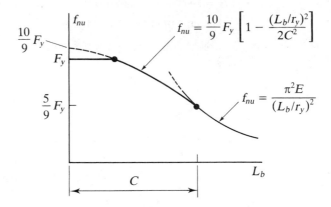

To obtain the AISC equations for allowable bending stress based on lateral-torsional buckling, the following adjustments are made to the foregoing equations.

1. All failure stresses are divided by a factor of safety of 5/3.

2. The radius of gyration r_y is replaced with r_T, the radius of gyration about the weak axis for a portion of the cross section consisting of the compression flange and one third of the compressed part of the web. This value does not differ appreciably from r_y and is tabulated in the *ASD Manual*.

3. All equations are expressed in terms of the ratio L_b/r_T.

4. The factor C_b is incorporated to account for variation of bending moment over the unbraced length (the warping equations are based on uniform moment).

5. Although the lateral-torsional buckling strength derives from both the uniform and nonuniform warping components, AISC conservatively uses only one component — the larger one.

The AISC equations for allowable bending stress for laterally unsupported beams can be summarized as:

For $\dfrac{L_b}{r_T} < \sqrt{\dfrac{102{,}000C_b}{F_y}}$,

$F_b = 0.60F_y$

For $\sqrt{\dfrac{102{,}000C_b}{F_y}} \leq \dfrac{L_b}{r_T} \leq \sqrt{\dfrac{510{,}000C_b}{F_y}}$, use the larger of

$$F_b = \left[\frac{2}{3} - \frac{F_y(L_b/r_T)^2}{1{,}530{,}000C_b}\right]F_y \leq 0.60F_y \qquad \text{(inelastic nonuniform warping)} \quad \text{(B.18)}$$

and

$$F_b = \frac{12,000C_b}{L_b d / A_f} \le 0.60F_y \qquad \text{(uniform warping)} \qquad \text{(B.19)}$$

For $\dfrac{L_b}{r_T} > \sqrt{\dfrac{510,000C_b}{F_y}}$, use the larger of

$$F_b = \frac{170,000C_b}{(L_b / r_T)^2} \le 0.60F_y \qquad \text{(elastic nonuniform warping)} \qquad \text{(B.20)}$$

and

$$F_b = \frac{12,000C_b}{L_b d / A_f} \le 0.60F_y \qquad \text{(uniform warping)} \qquad \text{(B.19)}$$

The ASD Specification gives an equation for the modifier C_b that is different from the one in the LRFD Specification, but either equation can be used. Note that, although the flexural strength according to LRFD is directly proportional to C_b, that is not the case for the allowable stresses given by equations B.18–B.20. This somewhat complicates allowable stress design of beams.

Shear

The shearing stress is computed as the maximum service load shear force divided by the web area, or

$$f_v = \frac{V}{A_w} \approx \frac{V}{t_w d}$$

The allowable shearing stress is based on shear yielding and is taken as two thirds of the allowable tensile stress on the gross section, or

$$F_v = \frac{2}{3}F_t = \frac{2}{3}(0.60F_y) = 0.40F_y \qquad \text{(B.21)}$$

■ **EXAMPLE B.3**

A W16 × 100 is used for a simply supported, uniformly loaded beam that is laterally braced only at its ends. If A36 steel is used, determine the maximum service load bending moment that can be resisted for span lengths of (a) 10 ft, (b) 15 ft, and (c) 40 ft.

SOLUTION   First, determine L_c. From Equation B.12,

$$\frac{76b_f}{\sqrt{F_y}} = \frac{76(10.42)}{\sqrt{36}} = 132.0 \text{ in.} = 11.0 \text{ ft}$$

$$\frac{20,000}{(d/A_f)F_y} = \frac{20,000}{\dfrac{16.97}{10.42(0.985)}(36)} = 336.0 \text{ in.} = 28.0 \text{ ft}$$

The smaller value controls; therefore $L_c = 11.0$ ft.

a. 10-ft span,

$L_b = 10$ ft $< L_c$ $\quad \therefore \quad$ the beam is laterally supported.

Since a W16 × 100 is a compact shape for A36 steel, the allowable stress from Equation B.13 is

$$F_b = 0.66F_y = 0.66(36) = 23.76 \text{ ksi}$$

The maximum bending stress for a moment M is given by Equation B.11 as $f_b = M/S$, so the maximum moment occurs when the stress f_b equals the allowable stress F_b:

$$M = F_b S = 23.76(175) = 4158 \text{ in.-kips} = 346 \text{ ft-kips}$$

ANSWER   **a.** Maximum moment = 346 ft-kips.

b. For the 15-ft span,

$L_b = 15$ ft $> L_c = 11.0$ ft $\quad \therefore \quad$ the beam is laterally unsupported.

$r_T = 2.81$ in.   (This value is given in the properties tables in the *ASD Manual*.)

$$\frac{L_b}{r_T} = \frac{15(12)}{2.81} = 64.06$$

For a simply supported, uniformly loaded beam with lateral bracing at the ends, $C_b = 1.14$ (calculated with the LRFD Specification equation, but the ASD equation could be used). Determine the limits for L_b/r_T:

$$\sqrt{\frac{102,000C_b}{F_y}} = \sqrt{\frac{102,000(1.14)}{36}} = 56.8$$

$$\sqrt{\frac{510,000C_b}{F_y}} = \sqrt{\frac{510,000(1.14)}{36}} = 127$$

Because $56.8 < L_b/r_T < 127$, use equations B.18 and B.19:

$$F_b = \left[\frac{2}{3} - \frac{F_y(L_b/r_T)^2}{1,530,000C_b} \right]F_y \le 0.60F_y$$

$$= \left[\frac{2}{3} - \frac{36(64.06)^2}{1,530,000(1.14)} \right]36 = 20.95 \text{ ksi}$$

or

$$F_b = \frac{12,000C_b}{L_b d/A_f} \le 0.60F_y$$

$$= \frac{12,000(1.14)}{(15 \times 12)(16.97)/(10.42 \times 0.985)} = 45.97 \text{ ksi}$$

The larger value is greater than $0.60F_y = 0.60(36) = 21.6$ ksi, so use

$$F_b = 0.60F_y = 21.6 \text{ ksi}$$

The maximum bending moment is

$$M = F_b S = 21.6(175) = 3780 \text{ in.-kips} = 315 \text{ ft-kips}$$

ANSWER   **b.** Maximum moment = 315 ft-kips

c. For the 40-ft span,

$$\frac{L_b}{r_T} = \frac{40(12)}{2.81} = 170.8 > \sqrt{\frac{510,000C_b}{F_y}} = 127$$

Use equations B.20 and B.19:

$$F_b = \frac{170,000C_b}{(L_b/r_T)^2} = \frac{170,000(1.14)}{(170.8)^2} = 6.643 \text{ ksi} < 0.60F_y$$

or

$$F_b = \frac{12,000C_b}{L_b d/A_f} = \frac{12,000(1.14)}{(40 \times 12)(16.97)/(10.42 \times 0.985)} = 17.24 \text{ ksi} < 0.60F_y$$

Use $F_b = 17.24$ ksi. The maximum moment is

$$M = F_b S = 17.24(175) = 3017 \text{ in.-kips} = 251 \text{ ft-kips}$$

ANSWER   **c.** Maximum moment = 251 ft-kips. ■

Design Aids

Most of the design aids for beams in the *LRFD Manual* have their counterparts in the *ASD Manual*. These include design charts that give allowable bending moment as a function of unbraced length for shapes normally used for beams. These curves are based on $C_b = 1.0$, but they cannot be used directly for other values of C_b, since the allowable bending stress F_b is not directly proportional to C_b.

B.5   ## BEAM-COLUMNS

Structural members subjected to both bending and axial stress are analyzed with interaction equations incorporating ratios of actual stress to allowable stress. The ASD Specification equations are variations of

$$\frac{f_a}{F_a} + \frac{f_{bx}}{F_{bx}} + \frac{f_{by}}{F_{by}} \leq 1.0$$

where x and y denote the axes of bending. Two equations are used in the Specification: one that is evaluated with a bending stress based on the absolute maximum moment in the member, and one with a bending stress based on the maximum end moment. A single amplification factor is used; there are no separate factors to account for the sway and nonsway components. This amplification factor has the form

$$\frac{C_m}{1 - (f_a/F'_e)}$$

where C_m is defined as:

For members subject to sidesway,

$$C_m = 0.85$$

For members not subject to sidesway, if there are no transverse loads on the member,

$$C_m = 0.6 - 0.4(M_1/M_2) \tag{B.22}$$

where M_1 and M_2 are the bending moments at the ends of the member, with M_1 the smaller in absolute value. The ratio M_1/M_2 is positive if the member is bent in reverse curvature and is negative for single curvature bending.

For members restrained against sidesway and subjected to transverse loads,

$$C_m = 0.85 \text{ if the ends are restrained against rotation}$$
$$= 1.0 \text{ if the ends are not restrained}$$

The factor F'_e is the Euler buckling stress divided by a factor of safety of 23/12:

$$F'_e = \frac{12\pi^2 E}{23(KL_b/r_b)^2} \tag{B.23}$$

The subscript b refers to the axis of bending. If bending about the x-axis is being considered, $F'_e = F'_{ex}$ and $KL_b/r_b = K_x L/r_x$. Similarly, for F'_{ey}, use $K_y L/r_y$.

The following interaction equations must be checked:

If $f_a/F_a \leq 0.15$, moment amplification is not required, and

$$\frac{f_a}{F_a} + \frac{f_{bx}}{F_{bx}} + \frac{f_{by}}{F_{by}} \leq 1.0 \tag{B.24}$$

If $f_a/F_a > 0.15$, *both* of the following equations must be checked:

$$\frac{f_a}{F_a} + \frac{C_{mx} f_{bx}}{\left(1 - \dfrac{f_a}{F'_{ex}}\right) F_{bx}} + \frac{C_{my} f_{by}}{\left(1 - \dfrac{f_a}{F'_{ey}}\right) F_{by}} \leq 1. \tag{B.25}$$

and

$$\frac{f_a}{0.60F_y} + \frac{f_{bx}}{F_{bx}} + \frac{f_{by}}{F_{by}} \leq 1.0 \tag{B.26}$$

Equation B.25 is a stability check, and the maximum bending moments, regardless of where they occur, must be used for computing f_{bx} and f_{by}. Equation B.26, which does not use amplification factors, is a stress check, and the maximum *end* moments should be used to compute f_{bx} and f_{by}. Note that $0.60F_y$ is used in Equation B.26 in lieu of F_a because yielding, rather than buckling, is the limit state. For the same reason, the member can be considered as laterally supported for computing F_{bx} for Equation B.26, but the actual lateral bracing conditions must be accounted for when Equation B.25 is being evaluated.

The purpose of the factor C_{mx} in Equation B.25 is to account for the x-axis moment gradient in the member. In a laterally unsupported member, the factor C_b, used in computing F_{bx}, also accounts for this moment gradient, so the Specification requires that C_b be taken as unity when F_{bx} is being computed for use in Equation B.25 *for members braced against joint translation.*

■ EXAMPLE B.4

The beam-column shown in Figure B.7 is part of a braced frame. Bending is about the x-axis, and lateral bracing is provided at the ends. Assume that $K_x = K_y = 1.0$ and review this member for compliance with the AISC Specification.

SOLUTION   Determine which interaction equation(s) to use. The axial compressive stress is

$$f_a = \frac{P}{A_g} = \frac{100}{14.4} = 6.944 \text{ ksi}$$

■ FIGURE B.7

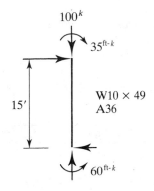

Compute the allowable compressive stress. The maximum slenderness ratio is

$$\frac{K_y L}{r_y} = \frac{1.0(15)(12)}{2.54} = 70.87$$

From Equation B.8,

$$C_c = \sqrt{\frac{2\pi^2 E}{F_y}} = \sqrt{\frac{2\pi^2 (29,000)}{36}} = 126.1$$

Since $KL/r < C_c$, compute the allowable compressive stress with Equation B.9:

$$F_a = \frac{F_y\left[1 - \dfrac{(KL/r)^2}{2C_c^2}\right]}{\dfrac{5}{3} + \dfrac{3(KL/r)}{8C_c} - \dfrac{(KL/r)^3}{8C_c^3}} = \frac{36\left[1 - \dfrac{(70.87)^2}{2(126.1)^2}\right]}{\dfrac{5}{3} + \dfrac{3(70.87)}{8(126.1)} - \dfrac{(70.87)^3}{8(126.1)^3}} = 16.34 \text{ ksi}$$

$$\frac{f_a}{F_a} = \frac{6.944}{16.34} = 0.4250 > 0.15 \qquad \therefore \text{ Check Equations B.25 and B.26.}$$

$$f_{bx} = \frac{M_x}{S_x} = \frac{60(12)}{54.6} = 13.19 \text{ ksi}$$

$$f_{bv} = 0$$

Compute the allowable bending stress. From Equation B.12,

$$\frac{76 b_f}{\sqrt{F_y}} = \frac{76(10.00)}{\sqrt{36}} = 126.7 \text{ in.} = 10.6 \text{ ft}$$

$$\frac{20,000}{(d/A_f)F_y} = \frac{20,000}{\dfrac{9.98}{0.560(10.00)}(36)} = 311.7 \text{ in.} = 26.0 \text{ ft}$$

The smaller value controls, so $L_c = 10.6$ ft. This length is less than the unbraced length of $L_b = 15$ ft, so the member will be treated as laterally unsupported. Because this member is braced against sidesway, use $C_b = 1.0$.

$$\sqrt{\frac{102,000 C_b}{F_y}} = \sqrt{\frac{102,000(1.0)}{36}} = 53.2$$

$$\sqrt{\frac{510,000 C_b}{F_y}} = \sqrt{\frac{510,000(1.0)}{36}} = 119$$

$$\frac{L_b}{r_T} = \frac{15(12)}{2.74} = 65.69 \qquad (r_T \text{ is tabulated in the } Manual)$$

Because $53.2 < L_b/r_T < 119$, use the larger of the values computed with equations B.18 and B.19, subject to an upper limit of

$$0.60F_y = 0.60(36) = 21.6 \text{ ksi}$$

From Equation B.18,

$$\left[\frac{2}{3} - \frac{F_y(L_b/r_T)^2}{1,530,000C_b} \right] F_y = \left[\frac{2}{3} - \frac{36(65.69)^2}{1,530,000(1.0)} \right] 36 = 20.34 \text{ ksi}$$

From Equation B.19,

$$\frac{12,000C_b}{L_b d/A_f} = \frac{12,000(1.0)}{(15 \times 12)(9.98)/(0.560 \times 10.00)} = 37.4 \text{ ksi}$$

The larger value is greater than $0.60F_y$, so

$$F_{bx} = 0.60F_y = 21.6 \text{ ksi}$$

Check Equation B.26 first. In this equation, which is a check on the stress condition at the support, the allowable bending stress may be computed as if the member had full lateral support of its compression flange. A W10 × 49 is compact for A36 steel, so the allowable stress can be taken as $0.66F_y$. Bending is about the x-axis, so the terms involving y-axis bending will be omitted. Then,

$$\frac{f_a}{0.60F_y} + \frac{f_{bx}}{F_{bx}} = \frac{6.944}{0.60(36)} + \frac{13.19}{0.66(36)} = 0.877 < 1.0 \qquad \text{(OK)}$$

Check Equation B.25. From Equation B.22,

$$C_m = 0.6 - 0.4 \frac{M_1}{M_2} = 0.6 - 0.4\left(-\frac{35}{60}\right) = 0.8333$$

The slenderness ratio for use in computing F'_{ex} is

$$\frac{KL_b}{r_b} = \frac{K_x L}{r_x} = \frac{1.0(15)(12)}{4.35} = 41.38$$

and

$$F'_{ex} = \frac{12\pi^2 E}{23(K_x L/r_x)^2} = \frac{12\pi^2 (29,000)}{23(41.38)^2} = 87.21 \text{ ksi}$$

$$\frac{f_a}{F_a} + \frac{C_{mx}f_{bx}}{\left(1 - \dfrac{f_a}{F'_{ex}}\right)F_{bx}} = 0.4250 + \frac{0.8333(13.19)}{\left(1 - \dfrac{6.944}{87.21}\right)21.6} = 0.978 < 1.0 \qquad \text{(OK)}$$

ANSWER   The W10 × 49 is satisfactory.   ■

Design Aids

Other than the tables and charts for column and beam design, the principal *Manual* design aid for beam-columns is a table of constants for use in making a preliminary member selection (Burgett, 1973). These constants enable the designer to transform the bending moment into an equivalent axial load that can be combined with the actual load to obtain a total effective axial load. The column allowable load tables can then be entered with this effective axial load and a trial shape can be selected for evaluation.

B.6 CONCLUDING REMARKS

Although allowable stress design is rapidly being supplanted by load and resistance factor design, it is still authorized by AISC and will be in use for some time to come. Those interested in pursuing ASD beyond the brief introduction in this appendix are referred to *Design of Steel Structures* (Gaylord, Gaylord, and Stallmeyer, 1992), which incorporates the latest AISC Specification provisions.

References

Abbreviations

AASHTO	American Association of State Highway and Transportation Officials
ACI	American Concrete Institute
AISC	American Institute of Steel Construction
AISI	American Iron and Steel Institute
AREA	American Railway Engineering Association
ASCE	American Society of Civil Engineers
ASTM	American Society for Testing and Materials
AWS	American Welding Society
BOCA	Building Officials and Code Administrators International
FEMA	Federal Emergency Management Agency
ICBO	International Conference of Building Officials
RCSC	Research Council on Structural Connections
SBCC	Southern Building Code Congress International
SJI	Steel Joist Institute
SSRC	Structural Stability Research Council

Ad Hoc Committee on Serviceability. 1986. "Structural Serviceability: A Critical Appraisal and Research Needs." *Journal of Structural Engineering,* American Society of Civil Engineers 112 (no. 12):2646–2664.

American Association of State Highway and Transportation Officials. 1992. *Standard Specifications for Highway Bridges,* 15th ed. Washington, D.C.

American Association of State Highway and Transportation Officials. 1994. *AASHTO LRFD Bridge Design Specifications.* Washington, D.C.

American Concrete Institute. 1995. *Building Code Requirements for Structural Concrete* (ACI 318–95). Detroit.

American Institute of Steel Construction. 1983. *Detailing for Steel Construction.* Chicago.

American Institute of Steel Construction. 1989a. *Manual of Steel Construction: Allowable Stress Design,* 9th ed. Chicago.

American Institute of Steel Construction. 1989b. *Specification for Structural Steel Buildings: Allowable Stress Design and Plastic Design*. Chicago.

American Institute of Steel Construction. 1993. *Load and Resistance Factor Design Specification for Structural Steel Buildings*. Chicago.

American Institute of Steel Construction. 1994. *Manual of Steel Construction: Load and Resistance Factor Design,* 2d ed. Chicago.

American Institute of Steel Construction. 1997. *Torsional Analysis of Structural Steel Members*. Chicago.

American Iron and Steel Institute. 1996. *Specification and Commentary for the Design of Cold-Formed Steel Structural Members*. Washington, D.C.

American Railway Engineering Association. 1992. *Specifications for Steel Railway Bridges*. Chicago.

American Society for Testing and Materials. 1996a. *Annual Book of ASTM Standards*. Philadelphia.

American Society for Testing and Materials. 1996b. "ASTM A6, Standard Specification for General Requirements for Rolled Steel Plates, Shapes, Sheet Piling, and Bars for Structural Use." *Annual Book of ASTM Standards*. Vol. 1.04. Philadelphia.

American Society of Civil Engineers. 1996. *Minimum Design Loads for Buildings and Other Structures*. ASCE 7-95. New York.

American Welding Society. 1996. *Structural Welding Code* (ANSI/AWS D1.1-96). Miami.

Basler, K. 1961. "Strength of Plate Girders in Shear." *Journal of the Structural Division,* ASCE 87 (no. ST7):151–197.

Bethlehem Steel. 1968. *Cable Roof Structures*.

Bethlehem Steel. 1969. *High-strength Bolting for Structural Joints*.

Bickford, John H. 1981. *An Introduction to The Design and Behavior of Bolted Joints*. New York: Marcel Dekker.

Bjorhovde, R., Galambos, T. V., and Ravindra, M. K. 1978. "LRFD Criteria for Steel Beam-Columns." *Journal of the Structural Division,* ASCE 104 (no. ST9):1371–87.

Bleich, Friedrich. 1952. *Buckling Strength of Metal Structures*. New York: McGraw-Hill.

Blodgett, O. W. 1966. *Design of Welded Structures*. Cleveland: The James F. Lincoln Arc Welding Foundation.

Building Officials and Code Administrators International. 1996. *The BOCA National Building Code,* 13th ed. Chicago.

Burgett, Lewis B. 1973. "Selection of a 'Trial' Column Section." *Engineering Journal,* AISC 10 (no. 2):54–59.

Butler, L. J., Pal, S., and Kulak, G. L. 1972. "Eccentrically loaded Welded Connections." *Journal of the Structural Division,* ASCE 98 (no. ST5):989–1005.

Cochrane, V. H. 1922. "Rules for Riveted Hole Deduction in Tension Members." *Engineering News Record* (November).

Cooper, P. B., Galambos, T. V., and Ravindra, M. K. 1978. "LRFD Criteria for Plate Girders." *Journal of the Structural Division,* ASCE 104 (no. ST9):1389–1407.

Crawford, S. H., and Kulak, G. L. 1971. "Eccentrically Loaded Bolted Connections." *Journal of the Structural Division,* ASCE 97 (no. ST3):765–83.

Darwin, D. 1990. *Design of Steel and Composite Beams with Web Openings.* Chicago: American Institute of Steel Construction.

Disque, Robert O. 1973. "Inelastic K-factor for Column Design." *Engineering Journal,* AISC 10 (no. 2):33–35.

Easterling, W. S., and Giroux, L. G. 1993. "Shear Lag Effects in Steel Tension Members." *Engineering Journal,* AISC 30 (no. 3):77–89.

Federal Emergency Management Agency. 1995. *Interim Guidelines: Evaluation, Repair, Modification and Design of Steel Moment Frames.* FEMA-267, Report No. SAC-95-02. Sacramento.

Fisher, J. W., Galambos, T. V., Kulak, G. L., and Ravindra, M. K. 1978. "Load and Resistance Factor Design Criteria for Connectors." *Journal of the Structural Division,* ASCE 104 (no. ST9):1427–41.

Galambos, T. V., ed. 1988. *Guide to Stability Design Criteria for Metal Structures,* 4th ed. Structural Stability Research Council. New York: Wiley-Interscience.

Galambos, T. V., and Ravindra, M. K. 1978. "Properties of Steel for Use in LRFD." *Journal of the Structural Division,* ASCE 104 (no. ST9):1459–68.

Gaylord, Edwin H., Gaylord, Charles N., and Stallmeyer, James E., 1992. *Design of Steel Structures,* 3d ed. New York: McGraw-Hill.

Hansell, W. C., Galambos, T. V., Ravindra, M. K., and Viest, I. M. 1978. "Composite Beam Criteria in LRFD." *Journal of the Structural Division,* ASCE 104 (no. ST9):1409–26.

Hendrick, A., and Murray, T. M. 1984. "Column Web Compression Strength at End-Plate Connections." *Engineering Journal,* AISC 21 (no. 3):161–9.

Higdon, A., Ohlsen, E. H., and Stiles, W. B. 1960. *Mechanics of Materials.* New York: John Wiley & Sons.

International Conference of Building Officials. 1997. *Uniform Building Code.* Whittier, Calif.

Johnston, Bruce G., ed. 1976. *Guide to Stability Design Criteria for Metal Structures,* 3d ed. Structural Stability Research Council. New York: Wiley-Interscience.

Krishnamurthy, N. 1978. "A Fresh Look at Bolted End-Plate Behavior and Design." *Engineering Journal,* AISC 15 (no. 2):39–49.

Kulak, G. L., and Timler, P. A. 1984. "Tests on Eccentrically Loaded Fillet Welds." Department of Civil Engineering, University of Alberta, Edmonton (December).

Kulak, G. L., Fisher, J. W., and Struik, J. H. A. 1987. *Guide to Design Criteria for Bolted and Riveted Joints,* 2d ed. New York: John Wiley & Sons.

Lothars, J. E. 1972. *Design in Structural Steel,* 3d ed. Englewood Cliffs, N.J.: Prentice-Hall.

McGuire, W. 1968. *Steel Structures.* Englewood Cliffs, N.J.: Prentice-Hall.

Munse, W. H., and Chesson, E., Jr. 1963. "Riveted and Bolted Joints: Net Section Design." *Journal of the Structural Division,* ASCE 89 (no. ST1):107–126.

Murphy, G. 1957. *Properties of Engineering Materials,* 3d ed. Scranton, Pa.: International Textbook Co.

Murray, Thomas M. 1983. "Design of Lightly Loaded Steel Column Base Plates." *Engineering Journal,* AISC 20 (no. 4):143–52.

Neal, B. G. 1977. *The Plastic Methods of Structural Analysis,* 3d ed. London: Chapman and Hall.

Ollgaard, J. G., Slutter, R. G., and Fisher, J. W. 1971. "Shear Strength of Stud Connectors in Lightweight and Normal-Weight Concrete." *Engineering Journal,* AISC 8 (no. 2):55–64.

Ravindra, M. K., and Galambos, T. V. 1978. "Load and Resistance Factor Design for Steel." *Journal of the Structural Division.* ASCE 104 (no. ST9):1337–53.

Ravindra, M. K., Cornell, C. A., and Galambos, T. V. 1978. "Wind and Snow Load Factors for Use in LRFD." *Journal of the Structural Division,* ASCE 104 (no. ST9):1443–57.

Research Council on Structural Connections. 1994. *Load and Resistance Factor Design Specification for Structural Joints Using ASTM A325 or A490 Bolts.* Chicago.

Salmon, C. G., and Johnson, J. E. 1996. *Steel Structures, Design and Behavior,* 4th ed. New York: HarperCollins.

Shanley, F. R. 1947. "Inelastic Column Theory." *Journal of Aeronautical Sciences* 14 (no. 5):261.

Sherman, D. R. 1997. "Designing with Structural Tubing." *Modern Steel Construction.* AISC 37 (no. 2):36–45.

Southern Building Code Congress International. 1994. *Standard Building Code*. Birmingham.

Steel Joist Institute. 1994. *Standard Specifications, Load Tables, and Weight Tables for Steel Joists and Joist Girders*. Myrtle Beach, S.C.

Structural Stability Research Council, Task Group 20. 1979. "A Specification for the Design of Steel–Concrete Composite Columns." *Engineering Journal*, AISC 16 (no. 4):101–15.

Tall, L., ed. 1964. *Structural Steel Design*. New York: Ronald Press.

Thornton, W. A. 1990a. "Design of Small Base Plates for Wide Flange Columns." *Engineering Journal*, AISC 27 (no. 3):108–10.

Thornton, W. A. 1990b. "Design of Base Plates for Wide Flange Columns—A Concatenation of Methods." *Engineering Journal*, AISC 27 (no. 4):173–4.

Timoshenko, Stephen P. 1953. *History of Strength of Materials*. New York: McGraw-Hill.

Timoshenko, Stephen P., and Gere, James M. 1961. *Theory of Elastic Stability*, 2d ed. New York: McGraw-Hill.

Yura, Joseph A. 1971. "The Effective Length of Columns in Unbraced Frames." *Engineering Journal*, AISC 8 (no. 2):37–42.

Yura, Joseph A. 1988. *Elements for Teaching Load and Resistance Factor Design*. Chicago: American Institute of Steel Construction.

Yura, J. A., Galambos, T. V., and Ravindra, M. K. 1978. "The Bending Resistance of Steel Beams." *Journal of the Structural Division*, ASCE 104 (no. ST9):1355–70.

Zahn, C. J. 1987. "Plate Girder Design Using LRFD." *Engineering Journal*, AISC 24 (no. 1):11–20.

Answers to Selected Problems

NOTES **1.** Answers are given for all odd-numbered problems except

    a. design problems in which there is an element of trial and error in the solution procedure or for which there is more than one acceptable answer; and

    b. problems for which knowledge of the answer leads directly to the solution.

2. All answers are rounded to three significant figures.

Chapter 1: Introduction

1.5-1	a. 120 ksi b. 13.3% c. 38.9%
1.5-3	c. Approximately 30,100 ksi
1.5-5	c. Approximately 30,000,000 psi d. Approximately 44,000 psi

Chapter 2: Concepts in Structural Steel Design

2.3-1	a. 28.4 kips b. 28.4 kips c. 33.4 kips
2.3-3	a. 155 ft-kips b. 172 ft-kips
2.3-5	46.8 psf

Chapter 3: Tension Members

3.2-1	85.0 kips
3.2-3	136 kips
3.3-1	a. 4.13 in.^{2} b. 2.34 in.^{2} c. 3.12 in.^{2} d. 2.50 in.^{2} e. 2.50 in.^{2}
3.3-3	a. 80.4 kips b. 80.4 kips
3.3-5	Not adequate: 220 kips > 172 kips
3.3-7	304 kips
3.4-1	162 kips
3.4-3	97.2 kips
3.4-5	299 kips
3.5-1	67.1 kips
3.5-3	120 kips
3.7-1	Required $d = 1.57$ in.

3.7-3	Required $d = 0.661$ in.
3.7-5	Required $d = 0.143$ in.
3.8-3	6.95 kips
3.8-5	Required $d = 0.181$ in.

Chapter 4: Compression Members

4.3-1	a. 693 kips
4.3-3	112 kips
4.3-5	151 kips
4.3-7	252 kips (calculated value)
4.5-1	250 kips
4.5-5	a. 2.1 b. 220 kips c. 2.0 (case f)
4.5-7	a. 1.44 b. 512 kips c. 1.2 (case c)
4.5-9	a. 1.73 b. 168 kips
4.6-1	423 kips
4.7-1	$r_x = 1.58$ in., $r_y = 1.49$ in.
4.7-3	$r_x = 4.63$ in., $r_y = 1.12$ in.
4.7-5	1810 kips
4.7-7	428 kips
4.7-9	131 kips

Chapter 5: Beams

5.2-1	a. $M_p = 386$ ft-kips, $Z = 92.7$ in.^{3} b. $S = 80.9$ in.^{3}, $M_y = 337$ ft-kips
5.2-3	101 in.^{3}
5.4-3	160 ksi
5.5-1	Adequate: 460 ft-kips < 464 ft-kips
5.5-3	0.583 kips/ft (including beam weight)
5.5-5	20.1 kips, 2.01 kips/ft (in addition to the beam weight)
5.5-7	1.67
5.5-9	a. 313 ft-kips b. 190 ft-kips
5.5-11	0.429 kips/ft (including the beam weight)
5.5-13	Not adequate: 132 ft-kips > 64.8 ft-kips
5.6-1	28.8 ft-kips
5.6-3	29.8 kips
5.8-1	50.9 kips

5.8-3	Shear strength not adequate: 120 kips > 86.6 kips
5.8-5	34.3 kips
5.8-7	63.7 kips
5.11-1	a. 482 in.^{3} b. 563 in.^{3}
5.11-3	a. S_x (top) = 46.7 in.^{3}, S_x (bot) = 48.3 in.^{3} b. 53.4 in.^{3}
5.14-1	Not adequate: Result of interaction equation = 1.73.
5.14-5	Adequate: Result of interaction equation = 0.744.

Chapter 6: Beam-Columns

6.2-1	Satisfactory: Result of interaction equation = 0.924.
6.3-1	Satisfactory: Result of interaction equation = 0.943.
6.4-1	Yes.
6.6-1	Satisfactory: Result of interaction equation = 0.973.
6.6-3	Satisfactory: Result of interaction equation = 0.916.
6.6-5	Unsatisfactory: Result of interaction equation = 1.04.
6.6-7	Unsatisfactory: Result of interaction equation = 1.06.
6.6-9	Unsatisfactory: Result of interaction equation = 1.03.
6.6-11	Unsatisfactory: Result of interaction equation = 1.21.
6.7-1	Satisfactory: Result of interaction equation = 0.497.

Chapter 7: Simple Connections

7.3-1	a. Satisfactory: s = 3 in. > 2.33 in.; L_e = 1.5 in. = min. L_e. b. 61.0 kips
7.4-1	23.9 kips
7.4-3	7.6 bolts (use 8)
7.6-1	63.6 kips
7.6-3	44.2 kips
7.7-1	83.5 kips
7.7-3	a. 2.3 bolts each side (use 4 for symmetry, 8 total)
7.8-1	Bolts are adequate: B_c = 20.4 kips < 29.8 kips. Tee is not adequate: Required t_f = 0.750 in. > 0.680 in.
7.9-1	Adequate: Result of interaction equation = 0.933.
7.9-3	3.6 bolts (use 4)
7.11-1	64.8 kips
7.11-3	33.4 kips
7.11-5	Required L = 12.8 in.; use 13 in.

Chapter 8: Eccentric Connections

8.2-1   35.0 kips

8.2-3   Shear: Required $d = 0.951$ in. Slip: Required $T_m = 68.6$ kips. Use $d = 1\frac{1}{4}$ in.

8.2-5   Slip: $T_m \geq 62.0$ kips. Use $d = 1\frac{1}{4}$ in.

8.2-7   19.9 kips

8.2-9 $T_m \geq 80.0$ kips. Use $d = 1\frac{1}{8}$ in.

8.2-11   a. 29 kips   b. 34.1 kips (18%)

8.2-13 $T_m \geq 51.3$ kips. Use $d = 1\frac{1}{8}$ in.

8.3-1   Not adequate: Result of interaction equation = 1.47.

8.3-3   116 kips

8.3-5   Adequate: Result of interaction equation = 0.967.

8.3-7   Required $d = 0.866$ in. Use $\frac{7}{8}$ in.

8.4-1   8.89 kips/in.

8.4-3 $0.616P_u$

8.4-5   Not adequate: 4.50 kips/in. > 4.18 kips/in.

8.4-7 $0.00528P_u$

8.4-9   5.57 kips/in.

8.4-11 $0.281P_u$

8.4-13   a. $0.0407P_u$ b. $0.0177P_u$

8.5-1   0.213 in. Use $\frac{1}{4}$ in.

8.5-3   Not adequate: 7.18 kips/in. > 5.57 kips/in.

8.5-5   a. 0.409 in.

8.6-1   58.5 ft-kips

8.7-1   Stiffeners not required.

8.8-1   Adequate: $\phi R_{str} = 14.5$ kips > 11.2 kips; result of interaction equation = 0.792.

Chapter 9: Composite Construction

9.1-1   a. 1760 in.^{4} b. f_s (tension) = 44.5 ksi (at bottom of steel), f_s (compression) = 1.21 ksi (at top of steel), f_c = 1.43 ksi (compression)

9.1-3   a. 3750 in.^{4} b. f_s (tension) = 31.7 ksi (at bottom of steel), f_s (compression) = 2.21 ksi (at top of steel), f_c = 1.08 ksi (compression)

9.1-5   611 ft-kips

9.2-1   Not adequate: Before concrete has cured, $\phi_b M_n = 89.6$ ft-kips > 80.6 ft-kips (OK); after concrete has cured, $\phi_b M_n = 171$ ft-kips < 177 ft-kips (N.G.)

9.3-1   Satisfactory: Before concrete has cured, $\phi_b M_n = 54.3$ ft-kips > 37.4 ft-kips; after concrete has cured, $\phi_b M_n = 113$ ft-kips > 110 ft-kips; $\phi_v V_n = 51.3$ kips > 17.6 kips.

9.4-1	34
9.4-3	46
9.6-1	a. 1.08 in. b. 1.33 in. c. 1.34 in.
9.7-1	a. 1560 in.^{4}, 0.404 in. b. 449 ft-kips
9.7-3	Adequate: Before concrete has cured, $\phi_b M_n = 222$ ft-kips > 128 ft-kips; after concrete has cured, $\phi_b M_n = 376$ ft-kips > 337 ft-kips; $\phi_v V_n = 108$ kips > 37.5 kips.
9.8-1	a. 611 ft-kips b. 540 ft-kips c. 2440 in.^{4}
9.8-3	393 ft-kips
9.10-1	196 kips

Chapter 10: Plate Girders

10.4-1	3570 ft-kips
10.4-3	a. 1830 ft-kips b. 1830 ft-kips
10.5-1	a. 451 kips b. 465 kips c. 212 kips
10.5-3	End panel: $\phi_v V_n = 481$ kips; second and third panels: $\phi_v V_n = 621$ kips; middle panel: $\phi_v V_n = 74.6$ kips.

Index